CAMBRIDGE LIBRARY COLLECTION

Books of enduring scholarly value

Travel and Exploration

The history of travel writing dates back to the Bible, Caesar, the Vikings and the Crusaders, and its many themes include war, trade, science and recreation. Explorers from Columbus to Cook charted lands not previously visited by Western travellers, and were followed by merchants, missionaries, and colonists, who wrote accounts of their experiences. The development of steam power in the nineteenth century provided opportunities for increasing numbers of 'ordinary' people to travel further, more economically, and more safely, and resulted in great enthusiasm for travel writing among the reading public. Works included in this series range from first-hand descriptions of previously unrecorded places, to literary accounts of the strange habits of foreigners, to examples of the burgeoning numbers of guidebooks produced to satisfy the needs of a new kind of traveller - the tourist.

Catalogue of the Special Loan Collection of Scientific Apparatus at the South Kensington Museum 1876

In 1876 the South Kensington Museum held a major international exhibition of scientific instruments and equipment, both historical and contemporary. Many of the items were retained and eventually formed the basis of important collections now held at the Science Museum, London. This is the 1877 third edition of the exhibition catalogue, which was expanded to include a 'large number of objects' received since the publication of the second edition, and which also included corrections in order to 'afford a complete record of the collection for future reference'. In two volumes and twenty sections comprising over 4,500 entries, the catalogue lists a huge variety of items, ranging from slide rules and telescopes to lighthouse parts and medical equipment. It gives detailed explanations of how they were used, and notes of their ownership and provenance, while the opening pages comprehensively record the contributing individuals and institutions in Britain, Europe and America.

Cambridge University Press has long been a pioneer in the reissuing of out-of-print titles from its own backlist, producing digital reprints of books that are still sought after by scholars and students but could not be reprinted economically using traditional technology. The Cambridge Library Collection extends this activity to a wider range of books which are still of importance to researchers and professionals, either for the source material they contain, or as landmarks in the history of their academic discipline.

Drawing from the world-renowned collections in the Cambridge University Library, and guided by the advice of experts in each subject area, Cambridge University Press is using state-of-the-art scanning machines in its own Printing House to capture the content of each book selected for inclusion. The files are processed to give a consistently clear, crisp image, and the books finished to the high quality standard for which the Press is recognised around the world. The latest print-on-demand technology ensures that the books will remain available indefinitely, and that orders for single or multiple copies can quickly be supplied.

The Cambridge Library Collection brings back to life books of enduring scholarly value (including out-of-copyright works originally issued by other publishers) across a wide range of disciplines in the humanities and social sciences and in science and technology.

Catalogue of the Special Loan Collection of Scientific Apparatus at the South Kensington Museum 1876

VOLUME 2

ANON

CAMBRIDGE
UNIVERSITY PRESS

CAMBRIDGE UNIVERSITY PRESS

Cambridge, New York, Melbourne, Madrid, Cape Town,
Singapore, São Paolo, Delhi, Tokyo, Mexico City

Published in the United States of America by Cambridge University Press, New York

www.cambridge.org
Information on this title: www.cambridge.org/9781108042420

© in this compilation Cambridge University Press 2012

This edition first published 1877
This digitally printed version 2012

ISBN 978-1-108-04242-0 Paperback

SECTION 13.—CHEMISTRY.

WEST GALLERY, UPPER FLOOR, ROOM P.

I. HISTORICAL.

2401. Balance used in his experiments by Dr. Joseph Black, Professor of Chemistry in the University of Edinburgh, from 1766 to 1799. Dr. Black was the discoverer of fixed air (carbonic acid), and author of the theory of latent heat.

Edinburgh Museum of Science and Art.

2398. Balance used by **Cavendish.**

The Royal Institution of Great Britain.

This balance, of rude exterior but singular perfection, was made by Harrison according to the plan and by order of Henry Cavendish, Esq., and passed at his death to his cousin and heir, Lord George Cavendish. By him it was presented to Sir Humphry Davy, together with the greater part of Mr. Cavendish's philosophical apparatus. Presented to the Royal Institution of Great Britain by Mr. Felix R. Garden.

2401a. Balance used by, or formerly belonging to, Dr. Priestley. *William Sykes Ward.*

This is remarkable as a good specimen of early work, having friction pulleys for suspending the beam ; also an interchangeable piece with steel planes, gearing for raising the beam concealed in foot, extra pans and glass bucket for taking specific gravities. The successive owners of the balance are believed to have been : Dr. Priestley (Leeds), Abraham Sharpe (Bradford), Joshua Muff (Leeds), and the Exhibitor.

2400. Balance used by **Sir Humphry Davy.** Presented to Professor Roscoe by Mrs. F. Crace-Calvert.

Prof. Roscoe, F.R.S.

2399. Balance used by **Dr. Thomas Young** and **Sir Humphry Davy.** *The Royal Institution of Great Britain.*

A balance made by Fidler for the Royal Institution, nearly resembling those of Ramsden and Troughton. " Lectures on Natural Philosophy," by Thomas Young, M.D., 1807. " Works of Sir Humphry Davy," vol. 5, page 17.

2537. Pneumatic Trough used in his experiments by Dr. Joseph Black, Professor of Chemistry in the University of Edinburgh, from 1766 to 1799. Dr. Black was the discoverer of fixed air (carbonic acid), and the author of the theory of latent heat.

Edinburgh Museum of Science and Art.

2538. Glass Chemical Vessels (retort, bottle, and flask or receiver) used in the chemical laboratory of the University of Edinburgh during the latter half of last century. Exhibited to show the contrast between them and vessels used for similar purposes at the present day.

Edinburgh Museum of Science and Art.

2396. Apparatus employed by **John Dalton** in his **Researches.**

The Council of the Literary and Philosophical Society of Manchester.

The apparatus employed by John Dalton in his classical researches, whether physical or chemical, was of the simplest and even of the rudest character. Most of it was made with his own hands, and that which is exhibited has been chosen as illustrating this fact, and as indicating the genius which with so insignificant and incomplete an experimental equipment was able to produce such great results. The Society has in its possession a large quantity of apparatus used by Dalton, most of which however consists of electrical apparatus, models of mechanical powers, models of steam engines, air pumps, a Gregorian telescope, and other apparatus of a similar kind, which was either bought or presented to him. It has not been thought necessary to exhibit these, but rather to show the home-made apparatus with which Dalton obtained his most remarkable results.

I. *Meteorological and Physical Apparatus made and used by Dr. Dalton.*

Throughout his life Dalton devoted much time and attention to the study of meteorology; indeed his first work, published in 1793, was entitled " Me-" teorological Observations and Essays," and his last paper, printed in 1842,* (Mem. Lit. and Phil. Soc. VI. 617) consists of auroral observations. Hence the first of Dalton's apparatus which claim attention are the meteorological instruments.

No. 1 is Dalton's mountain barometer, with accompanying thermometer, made for him by the late Mr. Lawrence Buchan of Manchester. The barometer is enclosed in a wooden case, which Dalton was accustomed to carry in his hand.

Several home-made barometers used by Dalton in his observations are in possession of the Society. They are all of them filled, and the scales prepared, by Dalton himself, and are simple siphon tubes with a bulb blown on at the bottom to serve as a mercury reservoir. These are attached to plain pieces of deal, upon the upper part of which the paper scale is pasted. One of these, which has probably also served for tension experiments (No. 2), has been placed in the collection.

Many of the thermometers appear also to have been home-made.

No. 3 is a mercurial thermometer, evidently made and graduated by Dr. Dalton, and marked with his initials, J. D. The freezing point of this thermometer was tested recently by Mr. Baxendell, who found that it had not altered since the instrument was graduated.

Another (No. 4) is of the same kind, and bears the date 1823.

---

* *Vide* Life of Dalton by Dr. Henry, published by the Cavendish Society; Memoir of Dr. Dalton and the History of the Atomic Theory, published in the Memoirs of the Literary and Philosophical Society of Manchester, 2nd Series, Vol. I.; Dr. Lonsdale's Life of Dalton, Longmans, 1874.

No. 5 is a third mercurial thermometer with long stem and wooden scale.

No. 6 is an alcohol thermometer with wooden scale.

No. 7 a registering maximum and minimum thermometer employed by Dalton; maker's name J. Renchetti, 29, Balloon Street, Manchester.

II. *Apparatus constructed and used by Dalton in his Researches.*

(1) " On the constitution of mixed gases," (2) " On the force of steam or " vapour from water or other liquids at different temperatures both in a Torri- " cellian vacuum and in air," (3) on evaporation, and (4) " On the expansion " of gases by heat."*

No. 8 is an apparatus used for the determination of the tension of volatile liquids at low temperatures; it consists of a siphon tube, at the upper end of which is a scale in inches in Dalton's handwriting. He describes it thus :

" I took a barometer tube 45 inches in length, and having sealed it her- metically at one end, bent it into a siphon shape, making the legs parallel, the end that was closed being 9 inches long, the other 36 inches. I then con- veyed two or three drops of ether to the end of the closed leg, and filled the rest of the tube with mercury, except about 10 inches at the open-end. This done, I immersed the whole of the short leg containing the ether into a tall glass containing hot water."

No. 9 is a smaller tube containing another liquid, also having a graduated scale written on paper and attached to the tube.

Nos. 10, 11, 12, 13, 14, are tubes used by Dalton for measuring the tension of vapour from water and other liquids at higher temperatures than their boiling points, both in a vacuum and air.

No. 15 is a tube used by Dalton for measuring the tension of the vapour of bisulphide of carbon, labelled " Sulphuret carb.," with a paper scale in Dalton's handwriting, and a cork showing that the upper portion of the tube containing the bisulphide of carbon could be heated in a water bath to various temperatures.

No. 16 is a manometer tube, fixed into a board, divided and numbered by Dalton.

No. 17 is an apparatus used by Dalton for the determination of the tension of the vapour of ether, and is interesting as being the instrument by means of which Dalton arrived at one of his most important experimental laws. It is described as follows (p. 564) : —

" The ether I used boiled in the open air at 102°. I filled a barometer tube with mercury moistened by agitation in ether; after a few minutes a portion of the ether rose to the top of the mercurial column, and the height of the column became stationary. When the whole had acquired the tempera- ture of the room (62°) the mercury stood at 17·00 inches, the barometer being at the same time 29·75 inches. Hence the force of the vapour from ether at 62° is equal to 12·74 of aqueous vapour at 172° temperature, which are 40° from the respective boiling points of the liquids."

This is generally known as Dalton's law of tensions, since shown by Reg- nault not to be rigorously true.

No. 18 is a wet and dry bulb mercurial thermometer made by H. H. Watson, of Bolton.

III. *Apparatus for Measuring Gases and for determining the Solubility of Gases in Water.*

No. 19 is an apparatus with a graduated tube, probably used by Dalton for the determination of the laws regulating " the absorption of gases by water

---

* Experimental essays on the above subjects, by John Dalton, read October 2nd, 16th, and 30th, 1801, and published in the 1st series, vol. 5, part 2, of the Memoirs of the Literary and Philosophical Society of Manchester.

" and other liquids," read October 21st, 1803. " Manchester Memoirs." 2nd Series. Vol. 1.

No. 20 is a graduated glass tube attached to a bottle of india-rubber, also probably used in his researches on the absorption of gases by water.

No. 21, No. 22, are divided eudiometer tubes, employed by Dalton for measuring the volumes of gases.

No. 23 is a spark eudiometer.

Nos. 24, 25, 26 are glass tubes, pipettes, and funnels graduated by Dr. Dalton and used by him for measuring gases.

No. 27 is a graduated glass bell-jar, used for measuring gases.

No. 28 is a phial, with graduated tube attached by cement, for collecting and measuring gases.

Nos. 29, 30 are stoppered phials, with the bottoms cut off, used as gas jars for collecting and measuring gases.

No. 31 is a thousand grains specific gravity bottle, with its counterpoise of lead stamped " 175 " by Dalton, and paper labelled in his handwriting " bottle balance."

No. 32 is a pipette.

No. 33, square bottle of thin glass, fitted with brass caps, and probably used for the determination of the specific gravities of gases.

No. 34 is an earthenware cup, used by Dalton as a mercury-trough, and containing a small phial with mercury.

Nos. 35, 36 are bulb tubes, with graduated scales, serving for the determination of the coefficients of expansion of gases.

No. 37 is a Florence flask with cork and valve for determining the specific gravity of gases.

No. 38 is a glass alembic.

IV. *Weights, Balances, Apparatus, Reagents, and Specimens used by Dalton.*

No. 39, eleven phials, containing creosote, iodine, amalgam of bismuth and mercury, quercitron bark, grana sylvestra, cochineal, and other substances, labelled in Dalton's handwriting.

No. 40, three divided blocks, used by Dalton for the illustration of his lectures; these are not, however, the balls an inch in diameter (referred to in his latest memoir on the " Analysis of Sugar ") which he employed occasionally in his lectures, as illustrating his newly-discovered laws of combination and the atomic theory; these appear, unfortunately, to be no longer in existence.

No. 41 is a common pair of scales used by Dalton.

No. 42, a pair of apothecary's scales and weights employed by Dalton, with a paper of weights made of wire, labelled in his handwriting, " 100th grains."

No. 43 is a box of weights used by Dalton, and containing a pill box labelled " Platina," another pill box labelled " Hund," and containing 100th of grains, and another wooden box containing brass gramme weights, labelled " Weights, French;" the other ordinary weights are of lead.

No. 44 is Dalton's pocket balance, consisting of a small pair of apothecaries' scales, with beam about 4 inches long, and having the pans attached by common string; it is contained in a tin case for the pocket.

No. 45 is a penholder used by Dalton.

No. 46, leaden grain weights made by Dalton from sheet lead, and stamped in numbers by him.

No. 47, iron punches used by Dalton for this purpose.

No. 48, a glass lens, wrapped in a piece of paper, labelled in Dalton's writing, " Sun's focus 4·2 inches."

No. 49 is a paper containing " 10th of grains," made by Dr. Dalton of iron wire. The paper in which these are wrapped is part of a note from one of

Dr. Dalton's pupils (as is well known, he lived by teaching mathematics at half-a-crown per lesson), in which the writer presents his "complements to "Mr. Dalton, and is sorry that he will not be able to wait upon him to-day, "as he is going to Liverpool with a few friends who are trying the railway "for the first time. Mr. D. may fully expect him on Monday at the usual "time."

No. 50 are bottles of tin, earthenware, and silver, some of them being common penny pot ink bottles. Each has a thermometer tube cemented into the neck of the bottle, and these tubes are provided with paper scales. These were used by Dalton probably for experiments on radiant heat.

No. 51 is a manometer tube used by Dalton; it consists of a tin vessel attached on either side to leaden tubing, and having a thermometer-tube closed at the upper end, and provided with a divided scale, fixed into the upper portion of the tin vessel.

No. 52, Dalton's Balance, made by Accum, and capable of arrangement as hydrostatic-balance with weights and counterpoises.

2397. Frame containing proofs in **John Dalton's** handwriting of part of the "New System of Chemistry." Vol. II., Part I., pages 347, 349, and 352. *Prof. Roscoe, F.R.S.*

2397a. Receipt by **John Dalton** for **instruction in Chemistry.** *C. Law.*

2418. Press used by Dr. W. H. Wollaston, to compress Platina, obtained in a fine state of division from precipitated chloride. *G. H. Wollaston.*

By means of this press the force obtained when the handle is at an angle of 5° is very nearly 60 × by power, and at an angle of ·1° very nearly 300 × by power.

2418a. Rhodium extracted from Platinum Ore by Dr. W. H. Wollaston. Probably also contains Iridium and Ruthenium. *G. H. Wollaston.*

2418b. Platinum Crucible and Platinum Trough, used by Dr. W. H. Wollaston. *G. H. Wollaston.*

1738. Portion of the **Battery used by Sir Humphry Davy** in decomposing the alkalies. *The Royal Institution of Great Britain.*

Phil. Trans. 1808.

687b. Sir Humphry Davy's Laboratory Note Book. *The Royal Institution of Great Britain.*

Davy's record of his decomposition of potash by the voltaic battery, October 19, 1807.

2405. Faraday's Apparatus for the **Condensation** and **Liquefaction** of **Gases.** *The Royal Institution of Great Britain.*

Apparatus used by Faraday for the condensation and liquefaction of gases, consisting of condensing pump and connexions, conducting and other tubes, gauges, sealed tubes for containing the liquefied gases, &c. &c.—Phil. Trans. 1845.

2406. Original Tubes containing Gases liquefied by Faraday. *The Royal Institution of Great Britain.*

1.⎫
2.⎪
3.⎬ Muriatic Acid.
4.⎪
5.⎭
6. Carbonic Acid.
7. Nitrous Oxide.
8. Sulphurous Acid.
9.⎫
10.⎬ Cyanogen.
11. Ammonia.
(Phil. Trans. 1845.)

12. Arseniuretted Hydrogen.
13. Ammonia.
14.⎫
15.⎬ Hydrobromic Acid.
16. Chlorine.
17.⎫
18.⎬ Chlorine and Sulph. Acid.
19. Sulphuretted Hydrogen.
20. Hydriodic Acid.
21. Arseniuretted Hydrogen.

687c. Faraday's Laboratory Note Book.
The Royal Institution of Great Britain.

Faraday's record of his condensation and liquefaction of gases, March 19, 1823.

Apparatus used by Brande for the continuous preparation of Ether.
The Royal Institution of Great Britain.

2409a. Original Apparatus, by M. Dumas, for ascertaining the density of gases. *M. J. Dumas, Paris.*

2409b. Apparatus, by M. Dumas, for ascertaining the density of vapours. *M. J. Dumas, Paris.*

APPARATUS EMPLOYED BY THE LATE THOMAS GRAHAM, F.R.S., MASTER OF THE MINT, IN HIS PRINCIPAL RESEARCHES BETWEEN THE YEARS 1834 AND 1866. The series is interesting as showing the simplicity of the appliances with which Graham worked, and by the aid of which he discovered facts and established laws which have since proved to be of so much importance.
W. Chandler Roberts, F.R.S.

2539. Tubes with discs of graphite and hydrophane employed by Graham in experiments on " diffusion " of gases.

These experiments were commenced in 1834, when the discs were formed of plaster of Paris. The instrument consists of a graduated glass tube, open at one end and closed at the other by the porous substance. When such a diffusion tube is filled with a gas over mercury an interchange of the gas and the air takes place through the porous septum. By experiments such as these Graham determined the diffusion rates of different gases, and developed the law that their diffusibilities vary in the inverse ratio of the square roots of their densities.

2540. The Apparatus employed for ascertaining the diffusion rates of liquids (Bakerian Lecture, 1849).

The saline solution to be diffused was placed in the inner vessel, which communicated freely with distilled water in the outer vessel. It was shown by this means that, when two liquids of different density and capable of mixing are placed in contact, diffusion takes place between them, much in the same manner as between gases, except that the rate of diffusion, which varies with the nature of the liquids, the temperature, and the degree of concentration, is slower. The phenomena are governed by several well ascertained laws, for a brief account of which *see* Watts's Dictionary of Chemistry, Vol. III., p. 705.

2541. Osmometers, or apparatus employed in Graham's researches on **" Osmotic Force."**

When a solution of a salt in a liquid is separated by a membrane or porous septum from a mass of water, a flow of liquid takes place from one side of the septum to the other. In some cases this flow is sufficiently powerful to sustain a column of water several inches in height. This diffusion is termed " Osmose." The experiments of Graham led to the conclusion that osmose depends essentially on the chemical action of the liquid on the septum. The base of the bell glass is covered either with thin unglazed earthenware or animal membrane, and is of such a size that the area of the opening is 100 times that of the tube, in order to facilitate the observations.

2542. Apparatus employed in experiments on **Liquid Diffusion** applied to analysis and on dialysis.

It consists simply of a gutta-percha hoop, about 9 inches in diameter, on which is stretched a sheet of parchment paper. The solution to be dialysed is placed in the hoop, and the whole floated in a considerable volume of water. Animal membrane, or a mere film of a colloidal septum, permits crystalloids to diffuse freely through it, but is entirely impervious to any colloids, for example gelatine, which may be in solution. The septum does not act in any way as a filter, but permits only the permeation of molecules and not masses. This mode of diffusion Graham termed Dialysis. By its aid he obtained hydrated silicic acid, oxides of iron and alumina, &c., all soluble in water, and he succeeded in separating crystalloid poisons from organic matter in toxicological investigations.

2543. Apparatus by which Graham studied **Capillary Liquid Transpiration** in relation to chemical composition.

The globe was filled with the liquid under examination, and the force employed to impel it through the capillary was the weight of 1 atmosphere, and was obtained from compressed air contained in the large reservoir, which was provided with a mercurial gauge.

2544. Tube Atmolyser, or instrument for the **separation of Gases by diffusion.**

As gases differ in the rate at which they transpire through a porous septum, it follows that, when a mixture of gases is in contact with a septum, while the other side is vacuous, the per-centage composition will be changed. The atmolyser consists of a bundle of tobacco-pipe stems, enclosed in a glass tube, which can be rendered vacuous. The mixed gases · are passed through the tube. For example, an explosive mixture of 66 per cent. by volume of hydrogen and 33 per cent. of oxygen was passed through the tube. The resulting mixture contained only 9·3 per cent.. of hydrogen, and a taper burned in it without explosion.

2545. Barometrical Diffusiometer, used for the investigation of the **Molecular Mobility of Gases.**

The gas under examination was allowed to enter either a perfect or partial vacuum through a thin porous septum. The results proved that the diffusion rates of the gases closely correspond with the law which has already been given, and quite excluded the idea of capillary transpiration, which would give for the same gases an entirely different result.

2546. Apparatus employed in experiments on **Absorption** and **dialytic separation of Gases by Colloid Septa.**

(1.) Waterproof silk bag for the dialysis of air. When the bag is exhausted by means of a Sprengel pump the air which penetrates the walls of the bag is found to contain 42 per cent. of oxygen, that is, the per-centage of oxygen present is doubled.

(2.) The penetration of metals by gases was studied by the aid of metallic tubes, of which the palladium tube shown is one. It was enclosed in a tube of glass or porcelain, and rendered vacuous by the aid of a Sprengel exhauster; a stream of gas was then passed through the annular space between the tubes. In the cold no hydrogen will pass through palladium, but at a red heat the gas passes at a rate of nearly 4,000 cubic cent. per square metre of surface in a minute.

2603. E. Mitscherlich's Vapour Density Apparatus.
Prof. A. Mitscherlich, Münden.

This apparatus is described in E. Mitscherlich's text-book of chemistry, p. 308, and was employed by him in determining the specific gravity of vapours.

II. MODELS, DIAGRAMS, APPARATUS, AND CHEMICALS EMPLOYED IN OR RESULTING FROM CHEMICAL RESEARCH.

2407. Apparatus by which Dr. Andrews and Professor Tait proved that ozone is a condensed form of oxygen.
Dr. Andrews, F.R.S.

Dry oxygen gas is exposed to the action of the silent electrical discharge in a glass tube terminating in a U tube containing sulphuric acid, which indicates by its change of position the change of volume in the gas.

2409. Apparatus used for ascertaining the **Density of Ozone.**
J. Louis Soret, Geneva.

This process, based upon the velocity of diffusion, has been described in detail in the "Recherches sur le Densité de l'Ozone" (Archives des Sciences physiques et naturelles, 1867, Vol. 30, p. 328. Annales de Chimie et de physique, 1868, Vol. 13, p. 264). Constructed by the Geneva Association for the Construction of Scientific Instruments.

2547. Apparatus for Gas Experiment, as used for the determination of the Composition of Ozone.
Sir B. C. Brodie.

This apparatus will be described in the form in which it was employed for the determination of the composition of ozone, the object being to determine the changes which electrized oxygen underwent when subjected to the action of heat or of chemical reagents.

A current of pure and dry oxygen was passed through an induction tube, fundamentally of the kind devised by Siemens, but the inner tube was nearly filled with water, in which was placed one of the terminal wires of Ruhmkorff's coil, the tube itself being immersed in a vessel of water connected with the other terminal wire of the coil. The tube is delineated in Fig. 1. The gas enters the apparatus at h, and, passing over anhydrous-phosphoric acid contained in the three bulbs i, traverses the narrow space k between the two tubes, and is there submitted to the electric action, after which the electrized gas is again passed over anhydrous phosphoric acid contained in the three bulbs l, and is delivered at m.

Fig. 1.

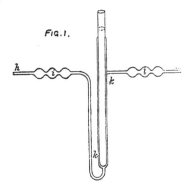

The electrized gas is collected and preserved for the purpose of experiment in a gas-holder, delineated in Fig. 2. On this side of the induction tube connexions of caoutchouc can no longer be employed, this substance being instantaneously corroded by even the minutest trace of ozone, and the friction between the gas-holder and the induction-tube is effected by means of what may be termed a paraffin joint. Over the tubes to be connected, which are placed close together, is slipped a piece of glass tube into which they exactly fit, and from which they are separated by a capillary space; a fragment of pure paraffin is placed at the external junction of the tubes. The union of the tubes is effected by gently melting the paraffin; the liquid paraffin is extremely limpid, and runs into and fills up the narrow space between the tubes. When the paraffin is solidified, the tubes are united by a joint, which is perfectly air-tight, which will resist very considerable pressure, and which is quite unaffected by the passage of the ozone. This simple joint is an essential feature of this arrangement, and will doubtless be of great service in many forms of gas apparatus.

The gas-holder consists of a glass bell p, contained in a glass cylinder q, in which it is suspended, being supported by a knob of glass passing through a cap fitted to the top of the jar. This cap is made in two pieces, which are subsequently united so as to be readily placed in a proper position as a support to the glass bell.

At a superior level is placed a glass jar r, containing pure and concentrated sulphuric acid; this jar is connected by a syphon tube s (in which is placed a glass stop cock) with the lower cylinder q. This upper jar, which I shall term the reservoir, is closed by a cap t, through which the siphon tube passes, and in which is also fitted a second glass tube u. The gas from the induction tube is delivered at n, whence it passes into the gas-holder by an arrangement of tubes, which is best understood from the drawing. The passage of the gas is regulated by means of an air pump, connected with the glass tube u by a tube of caoutchouc.

The volume of gas submitted to experiment was measured in a glass pipette, of which a drawing is given in Fig. 3. The capacity of the pipette (290·8 sub-centimes) between the two marks b and c was estimated by calibration with mercury. It was then welded to a glass tube of the form given in the figure; e is a reservoir of sulphuric acid with a syphon tube attached; d a cylinder containing water, in which the pipette a is immersed, and in

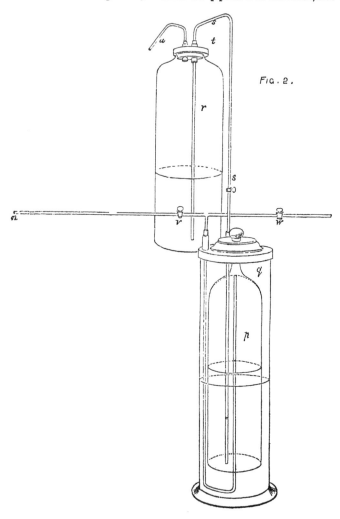

FIG. 2.

which a thermometer is placed. The gas pipette is connected with the gas holder by a paraffin joint at f. The arrangements for working this pipette are similar to those in the case of the gas-holder.

In order to estimate the changes in bulk which the electrized gas underwent in the various experiments hereafter described, the measuring apparatus was employed, of which a drawing is given, Fig. 4. In this apparatus, which I shall term the aspirator, the volumes of gas at 0° C. and 760 millims. pressure is ascertained by determining the pressure which it is necessary to put upon the gas in order to cause it to occupy a known space at a known temperature. This is the principle of Regnault's apparatus for gas analysis, and also of Frankland's apparatus.

The pipette was, placed on a table, being separated from the aspirator by an interval of about 8 or 10 inches. In this interval the experiment to which the electrized gas was submitted was made.

The aspirator consists of a cylinder of strong glass a, connected by an iron tube with an iron reservoir, b, containing an amount of mercury rather more than sufficient to fill the cylinder a. In the iron tube connecting the cylinder and the reservoir is intercalated a stopcock c, by which the connexion

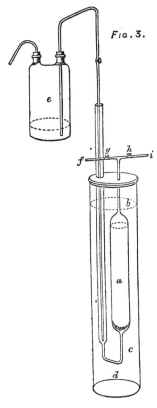

Fig. 3.

between the cylinder and reservoir may at pleasure be made or cut off. In the reservoir b a small iron tube d is inserted, connected by a tube of caoutchouc with a forcing pump firmly fixed to the table on which the apparatus is placed. By means of this forcing pump the air contained in the upper part of the reservoir may be compressed, and any required pressure put upon the mercury contained in it. The cylinder a is cemented by means of a resinous cement into two steel caps, e and f; the lower cap f is screwed firmly upon the support of the apparatus, which is made of iron. The cylinder is connected by means of a channel cut in the lower part of the steel cap f and continued through the iron frame, with a glass tube g, which is

Fig. 4.

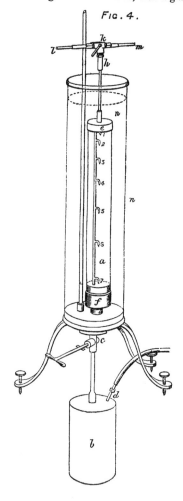

about half an inch in diameter and graduated in millimetres. This tube I shall speak of as the pressure tube. The cylinder *a* and the pressure tube *g* are thus in permanent connexion, and constitute one vessel, which is broken into parts solely for facility of construction. This apparatus is supported upon three screws, as shown in the figure, by the adjustment of which the pressure tube *g* is placed in a perpendicular position before the commencement of the experiment. The steel stopcock *k* is a three-way stopcock, in which the channels are so cut that a communication may be made between the tube *l* and the cylinder *a*, or between the tube *m* and the cylinder *a*, or between the tubes *l* and *m* (all other communications being shut off) at pleasure; or the communications may be entirely closed.

In the cylinder *a* is placed a thin piece of glass rod, to which seven points are attached, also of glass, as shown in the figure. The capacity of the cylinder *a* between each point is ascertained by calibration with mercury.

The cylinder, and pressure tube are enclosed in a second glass cylinder *n* filled with water, in which a thermometer is placed.

For further detail as to the construction, calibration, and mode of operating with the apparatus, *see* Phil. Trans., 1872, p. 3.

In Fig. 5 a drawing is given of the whole apparatus as arranged for experiment.

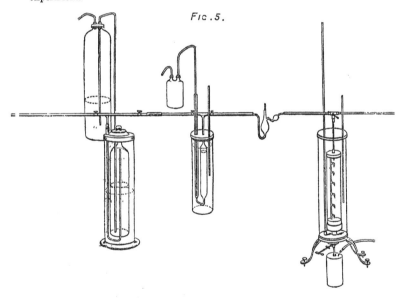

FIG.5.

2409a. Apparatus by which the body in the atmosphere which acts upon Schönbein's test-papers was first certainly proved to be identical with ozone. *Dr. Andrews, F.R.S.*

Atmospheric air, in fine weather and acting freely upon the test-papers, was drawn steadily through a glass globe, capable of being heated to 300° C., when all chemical reactions disappeared, which could only occur in the case of ozone.

2562. Hofmann's Vapour Density Apparatus.
Julius Schober, Berlin.

2411. Apparatus for rapidly saturating a small quantity of a liquid with a gas. Rough model.   *George Gore, F.R.S.*

2412. Apparatus for rapidly saturating a small quantity of a liquid with a gas.   *George Gore, F.R.S.*

2413. Apparatus for preparing small quantities of liquid carbonic acid gas, or liquid hydrochloric acid gas, and examining its action upon various substances.   *George Gore, F.R.S.*

(Philosophical Transactions of the Royal Society, 1861, and Proceedings of the Royal Society, May 1865.)

2414. Apparatus for preparing small quantities of liquefied cyanogen or liquefied ammonia gases, and examining their action upon various substances.   *George Gore, F.R.S.*

(Proceedings of the Royal Society, Vol. 20, p. 67 ; also No. 141, 1873.)

2415. Rod of Selenium, eight feet high, used in some experiments upon light.   *George Gore, F.R.S.*

2419. One hundred and eleven vessels and **articles** of **Platinum** (weight 117¼ ounces), used in Researches on Anhydrous Hydrofluoric acid and the Fluorides.

George Gore, F.R.S.

2419a. Pure Fluorides in bottles of platinum, &c., used in researches on the Fluorides.   *George Gore, F.R.S.*

Four platinum bottles containing pure fluorides of sodium, cerium, manganese, and bismuth.
Four gutta-percha bottles containing fluorides of glucinum, thallium, and cadmium, and the double fluoride of hydrogen and potassium.
Two glass bottles containing fluorides of lithium and rubidium.
Used in researches on the fluorides. (See Philosophical Transactions of the Royal Society, 1869.)

2419b. Specimen of Alibert's Native Graphite, from Mount Batougal, Irkoutsk, Siberia.   *George Gore, F.R.S.*

2419c. Bottles (53) containing Fluorides used in researches on the Fluorides.   *George Gore, F.R.S.*

37 gutta-percha bottles, 12 glass ditto, and 4 platinum ditto, containing various pure fluorides used in researches on anhydrous hydrofluoric acid and the fluorides. (See Philosophical Transactions of the Royal Society, 1869.)

2419d. Vessels and Articles (35) of baked and fused fluoride, cryolite, fluor-spar, native graphite, gas carbon, and wood charcoal, for researches on Fluorides.   *George Gore, F.R.S.*

13 boats formed of baked and fused fluorides.
6 „ „ „ fluoride of calcium.
5 „ „ fluor-spar.
2 „ „ cryolite.
1 „ „ native graphite.
1 „ „ wood charcoal.
1 tube „ „
1 crucible „ gas carbon.
1 cup „ „
4 plates „ polished fluor-spar.

All these articles for use in researches on the fluorides. (See Philosophical Transactions of the Royal Society, 1869.)

2419e. Three Specimens of Melted Silicon in bottle.
George Gore, F.R.S.

2419f. Vessels and Articles of wood-charcoal, native graphite, and gas carbon, used in researches on the Fluorides.
George Gore, F.R.S.

11 crucibles of wood charcoal, in bottle.
9 boats of native graphite, in box.
2 rods of lignum vitæ charcoal.
1 boat of gas carbon.

Used in researches on the fluorides. (See Philosophical Transactions of the Royal Society, 1869, &c.)

2419g. Glass Vessel for **collecting** and **Measuring Gases.**
George Gore, F.R.S.

2419h. Glass Beehive Shelf used in **collecting Gases.**
George Gore, F.R.S.

2419i. Specimens of Native Rock Salt from Jhelum, in Cabul, Affghanistan; Wiclicska, in Poland; and Erfurt, in Prussia. *George Gore, F.R.S.*

2419k. Specimens of Native Iron. *George Gore, F.R.S.*

2419l. Specimens of Native Antimony.
George Gore, F.R.S.

2421. Pseudo-crystal of **Native Ceylon Graphite. Bar** of extremely pure **metallic Tellurium,** weighing 2,000 grains.
George Gore, F.R.S.

2431. Collection of Vanadium Compounds.
Prof. H. E. Roscoe, F.R.S.

1. Vanadium ore, containing 2 per cent. vanadium.
2. Roasted vanadium ore.
3. Ammonium meta-vanadate (3rd crystallization).
4. Pure vanadium pentoxide, V_2O_5.
5. Vanadium tetroxide, V_2O_4.
6. Vanadium trioxide, V_2O_3.
7. Vanadium pentoxide, containing phosphorus.

40075. O o

8. Vanadium mononitride, VN.
9. Vanadium platinum alloy.
10. Vanadium silicon alloy.
11. Ammonium meta-vanadate (pure), NH_4VO_3.
12. Lead meta-vanadate, $Pb2VO_3$.
13. Di-vanadyl monochloride, V_2O_2Cl.
14. Vanadium tetrachloride (VCl_4) decomposed into vanadium trichloride (VCl_3) and Chlorine (Cl).
15. Vanadous sulphate. }
16. Copper sulphate. }
17. Potassium anhydro-chromate. }
18. Potassium anhydro-vanadate. }
19. Sodium ortho-vanadate, $Na_3VO_4 + 16H_2O$.
20. Sodium ortho-vanadate (fused mass).
21. Artificial vanadinite 3 $(Pb_3V_2O_8) + PbCl_2$.
22. Fused mass of artificial vanadinite.
23. Silver ortho-vanadate, Ag_3VO_4.
24. Sodium pyro-vanadate, $Na_4V_2O_7 + 18H_2O$.
25. Sodium pyro-vanadate, crystallized from alcohol.
26. Barium pyro-vanadate, $Ba_2V_2O_7$.
27. Basic pyro-vanadate of lead $2(Pb_2V_2O_7) + PbO$.
28. Silver pyro-vanadate, $Ag_4V_2O_7$.
29. Meta-vanadic acid, HVO_3.
30. Ammonium meta-vanadate NH_4VO_3 (crystallized).
31. Sodium anhydro-vanadate $2(NaVO_3) + V_2O_5$ (v. Hauer).
32. Calcium di-vanadate, $Ca2VO_3 + V_2O_5 + 18H_2O$ (v. Hauer).
33. Vanadium metal, $V = 51 \cdot 3$.
34. Pure vanadium pentoxide.   From $VOCl_3$.
35. Vanadium nitride, VN.   From metal and VCl_4.
36. Vanadium nitride, VN.   From pure VCl_2.
37. Vanadium nitride, VN.   From V_2O_3.
38. Vanadium nitride, VN.   From NH_4VO_3.
39. Vanadyl dichloride, $VOCl_2$.
40. Vanadium sesquioxide, undergoing oxidation.
41. Ammonium magnesium phosphate.   From Berzelius's vanadium (1831).
42. Thallium tetra-vanadate, $Tl_{12}V_4O_{16}$ (or $4Tl_3VO_4$).
43. Thallium hexa-vanadate, $Tl_{12}V_6O_{21}$.
44. Thallium octa-vanadate, $Tl_{12}V_8O_{26}$.
45. Silver octa-vanadate, $Ag_{12}V_8O_{26}$.
46. Sodium octa-vanadate, $Na_{12}V_8O_{26}$.
47. Thallium deca-vanadate, $Tl_{12}V_{10}O_{31}$.
48. Thallium dodeca-vanadate, $Tl_{12}V_{12}O_{36}$.
49. Thallium tetra-kai-deca-vanadate, $Tl_{12}V_{14}O_{41}$.
50. Vanadium metal, $V = 51 \cdot 3$.
51. Vanadium oxidibromide, $VOBr_2$.

52. VCl_4 decomposed into $VCl_3 + Cl$.
53. Vanadium oxitrichloride, $VOCl_3$.
54. VCl_4 undergoing decomposition.
55. Vanadium dichloride, VCl_2.
56. Vanadium trichloride, VCl_3.
57. Vanadium pentoxide, V_2O_5.
58. V_2O_5.
59. V_2O_4 by reduction with SO_2.
60. V_2O_3 „ „ Mg.
61. V_2O_2 „ „ Na or Zn.

⎱ Solutions of the oxides in sulphuric acid.

62. $VOCl_3$
63. VCl_4 ⎱ Obtained by throwing these
64. VCl_3 ⎰ compounds into water.
65. VCl_2

2423. Mercury Distilling Apparatus, constructed according to Dr. A. Weinhold's system.
J. Bosscha, Professor, Royal Polytechnic School, Delft.

(Carl' Repertorium der Physik, vol. ix., p. 69), with slight modifications of the mercury reservoirs, giving per hour about 500 grammes of distilled mercury, constructed by M. M. P. J. Kipp and Sons, Delft.

2425. The Original Air Battery.
J. H. Gladstone and Alfred Tribe.

When a plate of copper is connected with metallic silver, and immersed in a solution of copper nitrate, the oxygen of the air dissolved in the liquid takes part in the re-action, the copper is eaten away, and cuprous oxide is deposited upon the silver. This gives rise to the ordinary voltaic phenomena.

2426. The Original Copper-Zinc Couple. A.D. 1871.
J. H. Gladstone and Alfred Tribe.

This consists of coils of sheet-zinc on which copper has been deposited in a finely divided condition, by pouring copper sulphate upon them, and washing away the salt produced. In this case the coils are separated by layers of muslin. The couple having been immersed in water, has decomposed it, and the zinc is now thoroughly oxidized.

2427. The Original Copper-zinc Couple of the present arrangement. *J. H. Gladstone and Alfred Tribe.*

In this case the zinc foil is crumpled, and covered with the deposited copper. The zinc is now completely covered with oxide through the decomposition of water.

2428. Tubes showing that the hydrogen prepared from water by the copper-zinc couple does not contain arsenic, though the zinc is so greatly contaminated with arsenic as to give much arseniuretted hydrogen when dissolved in acid.
J. H. Gladstone and Alfred Tribe.

2429. Original Specimens of Ethylo-bromide of Zinc, and **Ethylo-chloride of Zinc,** substances prepared by means of the copper-zinc couple. *J. H. Gladstone and Alfred Tribe.*

2430. Essential Oils and their derivatives. *J. H. Gladstone.*

1. Menthene. An oil of the $C_{10} H_{16}$ group.
2. Cedrene. An oil of the $C_{15} H_{24}$ group.
3. Hydrosulphate of menthole.
4. Hydrate of oil of turpentine. $C_{10} H_{16}, 3H_2O$.
5. The same recrystallized from water.
6. Cœruleine, a deep blue colouring matter found in the essential oils of Matricaria Chamomilla, Artemisia Absinthium, and several others of the $C_{15} H_{24}$ group.

2430a. Specimens of **Chemicals illustrating Researches** on :—

1. The transformation of cyanogen (CN) into oxatyl (COHo).
2. Polymerisation of ethylic cyanide.
3. The isolation of the organic radicals.
4. Organo-metallic bodies.
5. The substitution of $'N_2''$ for C^{iv} in organic compounds.
6. Organo-boron compounds.
7. The synthesis of ethers, acids of the acetic series, and ketones.
8. The synthesis of acids of the lactic series.
9. The synthesis of acids of the acrylic series.

E. Frankland, F.R.S.

I.—Transformation of Cyanogen (CN) into Oxatyl (COHo) in Organic Compounds.— (Kolbe & Frankland.)

1. Acetic acid $\left(\left\{ \begin{array}{l} \textbf{CH}_3 \\ \textbf{COHo} \end{array} \right. \right)$ made from β methylic cyanide $\left(\left\{ \begin{array}{l} \textbf{CH}_3 \\ \textbf{CN} \end{array} \right. \right)$.

2. Argentic acetate $\left(\left\{ \begin{array}{l} \textbf{CH}_3 \\ \textbf{COAgo} \end{array} \right. \right)$ made from the above acid.

3. Propionic acid $\left(\left\{ \begin{array}{l} \textbf{CMeH}_2 \\ \textbf{COHo} \end{array} \right. \right)$ made from β ethylic cyanide $\left(\left\{ \begin{array}{l} \textbf{CMeH}_2 \\ \textbf{CN} \end{array} \right. \right)$.

The following salts made from this acid :—

4. Argentic propionate $\left(\left\{ \begin{array}{l} \textbf{CMeH}_2 \\ \textbf{COAgo} \end{array} \right. \right)$.

5. Baric propionate $\left\{ \begin{array}{l} \textbf{CMeH}_2 \\ \textbf{CO} \\ \textbf{CO} \\ \textbf{CMeH}_2 \end{array} \right.$ Bao''.

6. Plumbic propionate $\left\{ \begin{array}{l} \textbf{CMeH}_2 \\ \textbf{CO} \\ \textbf{CO} \\ \textbf{CMeH}_2 \end{array} \right.$ Pbo''.

7. Caproic acid $\left(\left\{ \begin{array}{l} \textbf{CBuH}_2 \\ \textbf{COHo} \end{array} \right. \right)$ made from β amylic cyanide $\left\{ \begin{array}{l} \textbf{CBuH}_2. \\ \textbf{CN} \end{array} \right.$

The following salts made from this acid :—

8. Argentic caproate $\left\{ \begin{array}{l} \textbf{CBuH}_2 \\ \textbf{COAgo} \end{array} \right.$.

9. Baric caproate $\left\{ \begin{array}{l} \textbf{CBuH}_2 \\ \textbf{CO} \\ \textbf{CO} \\ \textbf{CBuH}_2 \end{array} \right.$ Bao''.

II.—POLYMERISATION OF ETHYLIC CYANIDE.—(Kolbe & Frankland.)

   1. Cyanethine $^{vi}(N_3)^{ix}(CEt)'''_3$.

   2. Cyanethine nitrate $NO_2(N_3C_9H_{16}O)$.

   3. Cyanethine oxalate $\begin{cases} CO(N_3C_9H_{16}O) \\ CO(N_3C_9H_{16}O) \end{cases}$.

   4. Cyanethine platinic chloride $2N_3C_9H_{16}Cl,\ PtCl_4$.

III.—ISOLATION OF THE ORGANIC RADICALS.—(Frankland.)

   1. Methyl $\begin{cases} CH_3 \\ CH_3 \end{cases}$ or Me_2.

   2. Ethyl $\begin{cases} CMeH_2 \\ CMeH_2 \end{cases}$ or Et_2.

   3. Ethylic hydride $CMeH_3$ or EtH.

   4. Amyl $\begin{cases} CBuH_2 \\ CBuH_2 \end{cases}$ or Ay_2.

   5. Amylic hydride $CBuH_3$ or AyH.

   6. Eudiometer in which ethyl was first analysed.

IV.—ORGANO-METALLIC BODIES.—(Frankland).

   1. Zincic methide $ZnMe_2$.
   2. Zincic ethide $ZnEt_2$.
   3. Zincic ethylate $ZnEto_2$.
   4. Zincic amylide $ZnAy_2$.
   5. Stannous ethide $SnEt_2$.
   6. Stannic ethide $SnEt_4$.
   7. Stannic ethylo-dimethide $SnEt_2Me_2$.
   8. Stannic dichlorethide $SnEt_2Cl_2$.
   9. Stannic iodo-diethide $SnEt_2I_2$.
  10. Stannic bromo-triethide $SnEt_3Br$.
  11. Distannic iodo-tetrethide $Sn_2Et_4I_2$.
  12. Mercuric methide $HgMe_2$.
  13. Mercuric iodomethide $HgMeI$.
  14. Mercuric chlormethide $HgMeCl$.
  15. Mercuric ethide $HgEt_2$.
  16. Mercuric chlorethide $HgEtCl$.
  17. Mercuric amylide $HgAy_2$.
  18. Mercuric chloramylide $HgAyCl$.
  19. Mercuric iodoamylide $HgAyI$.

V.—THE SUBSTITUTION OF $'N_2''$ FOR C^{iv} IN ORGANIC COMPOUNDS.—
(Frankland.)

1. Ethylozincic dinitroethylate $N_2EtO(OZnEt)$.

2. Zincic dinitroethylate $\begin{matrix} N_2EtO \\ N_2EtO \end{matrix} Zno''$ analogous to zincic propionate

$\begin{matrix} CEtO \\ CEtO \end{matrix} Zno''$.

3. Calcic dinitroethylate $\begin{matrix} N_2EtO \\ N_2EtO \end{matrix} Cao''$ analogous to calcic propionate

$\begin{matrix} CEtO \\ CEtO \end{matrix} Cao''$.

4. Baric dinitroethylate $\frac{N_2EtO}{N_2EtO}$Bao″.

5. Magnesic dinitroethylate $\frac{N_2EtO}{N_2EtO}$Mgo″.

6. Cupric dinitroethylate $\frac{N_2EtO}{N_2EtO}$Cuo″.

7. Sodic dinitroethylate N_2EtONao analogous to sodic propionate CEtONao.

VI.—ORGANO-BORON COMPOUNDS.—(Frankland.)

1. Boric methide BMe_3.
2. Ammonia-boric methide NH_3, BMe_3.
3. Boric ethide BEt_3.
4. Ammonia boric ethide NN_3, BEt_3.
5. Boric etho-diethylate BEtEto$_2$.
6. Boric etho-dihydrate BEtHo$_2$.

VII.—SYNTHESIS OF ETHERS, ACIDS OF THE ACETIC SERIES AND KETONES.—(Frankland & Duppa.)

Ethers.

1. Ethylic methaceto-acetate	$C_7H_{12}O_3$	or $\begin{cases} \text{COMe} \\ \text{CMeH.} \\ \text{COEto} \end{cases}$
2. Ethylic dimethaceto-acetate	$C_8H_{14}O_3$	or $\begin{cases} \text{COMe} \\ \text{CMe}_2. \\ \text{COEto} \end{cases}$
3. Ethylic ethaceto-acetate	$C_8H_{14}O_3$	or $\begin{cases} \text{COMe} \\ \text{CEtH.} \\ \text{COEto} \end{cases}$
4. Ethylic diethaceto-acetate	$C_{10}H_{18}O_3$	or $\begin{cases} \text{COMe} \\ \text{CEt}_2. \\ \text{COEto} \end{cases}$
5. Ethylic isopropaceto-acetate	$C_9H_{16}O_3$	or $\begin{cases} \text{COMe} \\ \text{C}\beta\text{PrH.} \\ \text{COEto} \end{cases}$
6. Ethylic dimethacetate (isobutyrate)	$C_6H_{12}O_2$	or $\begin{cases} \text{CMe}_2\text{H} \\ \text{COEto} \end{cases}$.
7. Ethylic ethacetate (butyrate)	$C_6H_{12}O_2$	or $\begin{cases} \text{CEtH}_2 \\ \text{COEto} \end{cases}$.
8. Ethylic diethacetate (isocaproate)	$C_8H_{16}O_2$	or $\begin{cases} \text{CEt}_2\text{H} \\ \text{COEto} \end{cases}$.
9. Ethylic isopropacetate (isova.erate)	$C_7H_{14}O_2$	or $\begin{cases} \text{C}\beta\text{PrH}_2 \\ \text{COEto} \end{cases}$.
10. Ethylic amylacetate (œnanthylate)	$C_9H_{18}O_2$	or $\begin{cases} \text{CAyH}_2 \\ \text{COEto} \end{cases}$.

Acids and their Salts.

1. Dimethacetic acid (isobutyric acid)	$C_4H_8O_2$	or $\begin{cases} \text{CMe}_2\text{H} \\ \text{COHo} \end{cases}$.
2. Argentic dimethacetate	$C_4H_7AgO_2$	or $\begin{cases} \text{CMe}_2\text{H} \\ \text{COAgo} \end{cases}$.

3. Ethacetic acid (butyric acid) $C_4H_8O_2$ or $\begin{cases} CEtH_2 \\ COHo \end{cases}$.

4. Argentic ethacetate $C_4H_7AgO_2$ or $\begin{cases} CEtH_2 \\ COAgo \end{cases}$.

5. Diethacetic acid (isocaproic acid) $C_6H_{12}O_2$ or $\begin{cases} CEt_2H \\ COHo \end{cases}$.

6. Argentic diethacetate $C_6H_{11}AgO_2$ or $\begin{cases} CEt_2H \\ COAge \end{cases}$.

7. Baric diethacetate $C_{12}H_{22}BaO_4$ or $\begin{cases} CEt_2H \\ CO \\ CO \\ CEt_2H \end{cases}$Bao″.

8. Isopropacetic acid (isovaleric acid) $C_5H_{10}O_2$ or $\begin{cases} C\beta PrH_2 \\ COHo \end{cases}$.

9. Argentic isopropacetate $C_5H_9AgO_2$ or $\begin{cases} C\beta PrH_2 \\ COAgo \end{cases}$

10. Amylacetic acid (œnanthylic acid) $C_7H_{14}O_2$ or $\begin{cases} CAyH_2 \\ COHo \end{cases}$.

11. Argentic amylacetate $C_7H_{13}AgO_2$ or $\begin{cases} CAyH_2 \\ COAgo \end{cases}$.

12. Baric amylacetate $C_{14}H_{26}BaO_4$ or $\begin{cases} CAyH_2 \\ CO \\ CO \\ CAyH_2 \end{cases}$Bao″.

Ketones.

1. Methylated acetone C_4H_8O or $\begin{cases} COMe \\ CMeH_2 \end{cases}$.

2. Dimethylated acetone $C_5H_{10}O$ or $\begin{cases} COMe \\ CMe_2H \end{cases}$.

3. Ethylated acetone $C_5H_{10}O$ or $\begin{cases} COMe \\ CEtH_2 \end{cases}$.

4. Diethylated acetone $C_7H_{14}O$ or $\begin{cases} COMe \\ CEt_2H \end{cases}$.

5. Isopropylated acetone $C_6H_{12}O$ or $\begin{cases} COMe \\ C\beta PrH_2 \end{cases}$.

VIII.—SYNTHESIS OF ACIDS OF THE LACTIC SERIES.—(Frankland & Duppa.)

1. Dimethoxalic acid $C_4H_8O_3$ or $\begin{cases} CMe_2Ho \\ COHo \end{cases}$.

2. Argentic dimethoxalate $C_4H_7AgO_3$ or $\begin{cases} CMe_2Ho \\ COAgo \end{cases}$.

3. Baric dimethoxalate $C_8H_{14}BaO_6$ or $\begin{cases} CMe_2Ho \\ CO \\ CO \\ CMe_2Ho \end{cases}$Bao″.

4. Ethomethoxalic acid $C_5H_{10}O_3$ or $\begin{cases} CEtMeHo \\ COHo \end{cases}$.

5. Argentic ethomethoxalate $\quad C_5H_9AgO_3 \quad$ or $\left\{\begin{array}{l}\mathbf{C}\mathrm{EtMeHo}\\\mathbf{C}\mathrm{OAgo}\end{array}\right.$.

6. Ethylic ethomethoxalate $\quad C_7H_{14}O_3 \quad$ or $\left\{\begin{array}{l}\mathbf{C}\mathrm{EtMeHo}\\\mathbf{C}\mathrm{OEto}\end{array}\right.$.

7. Baric ethomethoxalate $\quad C_{10}H_{18}BaO_6 \quad$ or $\left\{\begin{array}{l}\mathbf{C}\mathrm{EtMeHo}\\\mathbf{C}\mathrm{O}\\\mathbf{C}\mathrm{O}\end{array}\mathrm{Bao''}\right.$
$\left.\right\}\mathbf{C}\mathrm{EtMeHo}$

8. Diethoxalic acid $\quad C_6H_{12}O_3 \quad$ or $\left\{\begin{array}{l}\mathbf{C}\mathrm{Et_2Ho}\\\mathbf{C}\mathrm{OHo}\end{array}\right.$.

9. Argentic diethoxalate $\quad C_6H_{11}AgO_3 \quad$ or $\left\{\begin{array}{l}\mathbf{C}\mathrm{Et_2Ho}\\\mathbf{C}\mathrm{OAgo}\end{array}, \tfrac{1}{2}\mathbf{O}\mathrm{H_2}.\right.$

10. Methylic diethoxalate $\quad C_7H_{14}O_3 \quad$ or $\left\{\begin{array}{l}\mathbf{C}\mathrm{Et_2Ho}\\\mathbf{C}\mathrm{OMeo}\end{array}\right.$.

11. Ethylic diethoxalate $\quad C_8H_{16}O_3 \quad$ or $\left\{\begin{array}{l}\mathbf{C}\mathrm{Et_2Ho}\\\mathbf{C}\mathrm{OEto}\end{array}\right.$.

12. Amylic diethoxalate $\quad C_{11}H_{22}O_3 \quad$ or $\left\{\begin{array}{l}\mathbf{C}\mathrm{Et_2Ho}\\\mathbf{C}\mathrm{OAyo}\end{array}\right.$.

13. Baric diethoxalate $\quad C_{12}H_{22}BaO_6 \quad$ or $\left\{\begin{array}{l}\mathbf{C}\mathrm{Et_2Ho}\\\mathbf{C}\mathrm{O}\\\mathbf{C}\mathrm{O}\end{array}\mathrm{Bao''}.\right.$
$\left.\right\}\mathbf{C}\mathrm{Et_2Ho}$

14. Cupric diethoxalate $\quad C_{12}H_{22}CuO_6 \quad$ or $\left\{\begin{array}{l}\mathbf{C}\mathrm{Et_2Ho}\\\mathbf{C}\mathrm{O}\\\mathbf{C}\mathrm{O}\end{array}\mathrm{Cuo''}.\right.$
$\left.\right\}\mathbf{C}\mathrm{Et_2Ho}$

15. Zincic diethoxalate $\quad C_{12}H_{22}ZnO_6 \quad$ or $\left\{\begin{array}{l}\mathbf{C}\mathrm{Et_2Ho}\\\mathbf{C}\mathrm{O}\\\mathbf{C}\mathrm{O}\end{array}\mathrm{Zno''}.\right.$
$\left.\right\}\mathbf{C}\mathrm{Et_2Ho}$

16. Ethylic zincmonethyl diethoxalate $\quad C_{10}H_{20}ZnO_3 \quad$ or $\left\{\begin{array}{l}\mathbf{C}\mathrm{Et_2(OZnEt)}\\\mathbf{C}\mathrm{OEto}\end{array}\right.$.

17. Amylhydroxalic acid $\quad C_7H_{14}O_3 \quad$ or $\left\{\begin{array}{l}\mathbf{C}\mathrm{AyHHo}\\\mathbf{C}\mathrm{OHo}\end{array}\right.$.

18. Ethylic amylhydroxalate $\quad C_9H_{18}O_3 \quad$ or $\left\{\begin{array}{l}\mathbf{C}\mathrm{AyHHo}\\\mathbf{C}\mathrm{OEto}\end{array}\right.$.

19. Calcic amylhydroxalate $\quad C_{14}H_{26}CaO_6 \quad$ or $\left\{\begin{array}{l}\mathbf{C}\mathrm{AyHHo}\\\mathbf{C}\mathrm{O}\\\mathbf{C}\mathrm{O}\end{array}\mathrm{Cao''}.\right.$
$\left.\right\}\mathbf{C}\mathrm{AyHHo}$

20. Baric amylhydroxalate $\quad C_{14}H_{26}BaO_6 \quad$ or $\left\{\begin{array}{l}\mathbf{C}\mathrm{AyHHo}\\\mathbf{C}\mathrm{O}\\\mathbf{C}\mathrm{O}\end{array}\mathrm{Bao''}.\right.$
$\left.\right\}\mathbf{C}\mathrm{AyHHo}$

21. Cupric amylhydroxalate $\quad C_{14}H_{26}CuO_6 \quad$ or $\left\{\begin{array}{l}\mathbf{C}\mathrm{AyHHo}\\\mathbf{C}\mathrm{O}\\\mathbf{C}\mathrm{O}\end{array}\mathrm{Cuo''}.\right.$
$\left.\right\}\mathbf{C}\mathrm{AyHHo}$

22. Amylethoxalic acid $\quad C_9H_{18}O_3 \quad$ or $\left\{\begin{array}{l}\mathbf{C}\mathrm{AyEtHo}\\\mathbf{C}\mathrm{OHo}\end{array}\right.$.

23. Argentic amylethoxalate $C_9H_{17}AgO_3$ or $\begin{cases} \mathbf{C}AyEtHo \\ \mathbf{C}OAgo \end{cases}$.

24. Ethylic amylethoxalate $C_{11}H_{22}O_3$ or $\begin{cases} \mathbf{C}AyEtHo \\ \mathbf{C}OEto \end{cases}$.

25. Baric amylethoxalate $C_{18}H_{34}BaO_6$ or $\begin{cases} \mathbf{C}AyEtHo \\ \mathbf{C}O \\ \mathbf{C}O\,Bao' \\ \mathbf{C}AyEtHo \end{cases}$.

26. Diamyloxalic acid $C_{12}H_{24}O_3$ or $\begin{cases} \mathbf{C}Ay_2Ho \\ \mathbf{C}OHo \end{cases}$.

27. Ethylic diamyloxalate $C_{14}H_{28}O$ or $\begin{cases} \mathbf{C}Ay_2Ho \\ \mathbf{C}OEto \end{cases}$.

28. Amylic diamyloxalate $C_{17}H_{34}O_3$ or $\begin{cases} \mathbf{C}Ay_2Ho \\ \mathbf{C}OAyo \end{cases}$.

29. Baric diamyloxalate $C_{24}H_{46}BaO_3$ or $\begin{cases} \mathbf{C}Ay_2Ho \\ \mathbf{C}O \\ \mathbf{C}O\,Bao'' \\ \mathbf{C}Ay_2Ho \end{cases}$.

IX.—Synthesis of Acids of the Acrylic Series.—(Frankland & Duppa.)

1. Methacrylic acid $C_4H_6O_2$ or $\begin{cases} \mathbf{C}Me''Me \\ \mathbf{C}OHo \end{cases}$.

2. Argentic methacrylate $C_4H_5AgO_2$ or $\begin{cases} \mathbf{C}Me''Me \\ \mathbf{C}OAgo \end{cases}$.

3. Ethylic methacrylate $C_6H_{10}O_2$ or $\begin{cases} \mathbf{C}Me''Me \\ \mathbf{C}OEto \end{cases}$.

4. Baric methacrylate $C_8H_{10}BaO_4$ or $\begin{cases} \mathbf{C}Me''Me \\ \mathbf{C}O \\ \mathbf{C}O\,Bao' \\ \mathbf{C}Me''Me \end{cases}$.

5. Cupric methacrylate $C_8H_{10}CuO_4$ or $\begin{cases} \mathbf{C}Me''Me \\ \mathbf{C}O \\ \mathbf{C}O\,Cuo'' \\ \mathbf{C}Me''Me \end{cases}$.

6. Methyl-crotonic acid $C_5H_8O_2$ or $\begin{cases} \mathbf{C}Et''Me \\ \mathbf{C}OHo \end{cases}$.

7. Argentic methyl-crotonate $C_5H_7AgO_2$ or $\begin{cases} \mathbf{C}Et''Me \\ \mathbf{C}OAgo \end{cases}$.

8. Ethylic methyl-crotonate $C_7H_{12}O_2$ or $\begin{cases} \mathbf{C}Et''Me \\ \mathbf{C}OEto \end{cases}$.

9. Baric methyl-crotonate $C_{10}H_{14}BaO_4$ or $\begin{cases} \mathbf{C}Et'Me \\ \mathbf{C}O \\ \mathbf{C}O\,Bao'' \\ \mathbf{C}Et''Me \end{cases}$.

10. Ethyl-crotonic acid $C_6H_{10}O_2$ or $\begin{cases} \mathbf{C}Et''Et \\ \mathbf{C}OHo \end{cases}$.

11. Potassic ethyl-crotonate $C_6H_9KO_2$ or $\begin{cases} \mathbf{C}Et''Et \\ \mathbf{C}OKo \end{cases}$.

12. Argentic ethyl-crotonate $C_6H_9AgO_2$ or $\begin{cases} CEt''Et \\ COAgo \end{cases}$.

13. Ethylic ethyl-crotonate $C_8H_{14}O_2$ or $\begin{cases} CEt''Et \\ COEto \end{cases}$.

14. Cupric ethyl-crotonate $C_{12}H_{18}CuO_4$ or $\begin{cases} CEt''Et \\ CO \\ CO \\ CEt'Et \end{cases}Cuo''$.

15. Plumbic ethyl-crotonate $C_{12}H_{18}PbO_4$ or $\begin{cases} CEt''Et \\ CO \\ CO \\ CEt''Et \end{cases}Pbo''$.

2536. Specimens of **Chemicals,** illustrating some of W. H. Perkin's researches in organic chemistry.

W. H. Perkin, F.R.S.

I. On the action of chloride of cyanogen on naphthylamine.

II. On the action of bromine on acetic acid and artificial formation of glycocol.   (Perkin and Duppa.)

III. On the action of bromine on bromacetic acid.

(Perkin and Duppa.)

IV. On dibromosuccinic acid, and the artificial production of tartaric acid.   (Perkin and Duppa.)

V. On azodinaphthyldiamin.   (Perkin and Church.)

VI. On mauve or aniline purple.

VII. On the basicity of tartaric acid.

VIII. On some derivatives of the hydride of salicyl.

IX. On the action of acetic anhydride on the hydride of salicyl.

X. On the artificial production of coumarin and its homologues.

XI. On some new derivatives of coumarin.

XII. On chlorinated chloride of methyl.

XIII. On some derivatives of anthracene.

XIV. On artificial alizarin.

XV. On bromo-alizarin.

XVI. On nitro-alizarin.

XVII. On anthrapurpurin.

XVIII. On anthraflavic acid.

XIX. On the formation of anthrapurpurin.

XX. On glyoxylic acid.

I.

Menaphthylamine (carbodinaphthyltriamine) $C_{21}H_{17}N_3$, product of the action of chloride of cyanogen on naphthylamine $C_{10}H_9N$.

‹II.

Bromacetic acid $C_2H_3BrO_2 =$ $\left|\begin{array}{l}CO(OH)\\CH_2Br\end{array}\right.$

„ ether $C_4H_7BrO_2 =$ $\left|\begin{array}{l}CO(OEt)\\CH_2Br.\end{array}\right.$

(Perkin and Duppa.)

III.

Dibromacetic acid $C_2H_2Br_2O_2 -$ $\left|\begin{array}{l}CO(OH)\\CHBr_2\end{array}\right.$

Dibromacetamide $C_2H_3Br_2NO =$ $\left|\begin{array}{l}CO-NH_2\\CHBr_2.\end{array}\right.$

(Perkin and Duppa.)

IV.

Dibromosuccinic acid $C_4H_4Br_2O_4 = \left|\begin{array}{l}CO(OH)\\C_2H_2Br_2\\CO(OH)\end{array}\right.$ prepared from the chloride

of dibromosuccinyl.

Tartaric acid (inactive) $C_4H_6O_6, H_2O = \left|\begin{array}{l}CO(OH)\\C_2H_2(OH)_2,\\CO(OH)\end{array}\right.$ obtained by boiling

argentic dibromosuccinate with water.   (Perkin and Duppa.)

V.

Azodinaphthyldiamin $C_{20}H_{15}N_3 = \left\{\begin{array}{l}(C_{10}H_7)_2)\\N'''\\H\end{array}\right\} N_2.$

(Perkin and Church.)

VI.

Mauveine $C_{27}H_{24}N_4.$
„ hydrochlorate $C_{27}H_{24}N_4-Cl.$
„ sulphate $(C_{27}H_{24}N_4)_2H_2SO_4.$
„ nitrate $C_{27}H_{24}N_4HNO_3.$
„ hydriodate $C_{27}H_{24}N_4HI.$
„ carbonate $C_{27}H_{24}N_4H_2CO_3$?

VII.

Diacetotartaric ether $C_{12}H_{18}O_8 = \left|\begin{array}{l}CO(OEt)\\C_2H_2\left\{\begin{array}{l}(OAc)\\(OAc)\end{array}\right.\\CO(OEt).\end{array}\right.$

Diacetoparatartaric ether   do.   do.

Diacetotartaric anhydride $C_8H_8O_8 = C_2H_2\left\{\begin{matrix}(OAc)\\(OAc)\end{matrix}\right.$

$$\begin{matrix}CO\\|\\C_2H_2\left\{\begin{matrix}(OAc)\\(OAc)\end{matrix}\right.O\\|\\CO\end{matrix}\quad O$$

Benzotartaric ether $C_{15}H_{18}O_7 = C_2H_2\left\{\begin{matrix}(OBz)\\(OH)\end{matrix}\right.$

$$\begin{matrix}CO(OEt)\\|\\C_2H_2\left\{\begin{matrix}(OBz)\\(OH)\end{matrix}\right.\\|\\CO(OEt)\end{matrix}$$

Acetobenzotartaric ether $C_{17}H_{20}O_8 = C_2H_2\left\{\begin{matrix}(OBz)\\(OAc)\end{matrix}\right.$

$$\begin{matrix}CO(OEt)\\|\\C_2H_2\left\{\begin{matrix}(OBz)\\(OAc)\end{matrix}\right.\\|\\CO(OEt).\end{matrix}$$

Succinotartaric ether $C_{20}H_{30}O_{14} = C_2H_2\left\{\begin{matrix}O-Sü-O\\(OH)(HO)\end{matrix}\right\}C_2H_2$

$$\begin{matrix}CO(OEt)\\|\\C_2H_2\left\{\begin{matrix}O-Sü-O\\(OH)(HO)\end{matrix}\right\}C_2H_2\\|\\CO(OEt)\end{matrix}\quad\begin{matrix}CO(OEt)\\|\\C_2H_2\\|\\CO(OEt).\end{matrix}$$

VIII.

Hydride of methylsalicyl $C_8H_8O_2 = $
$$\begin{matrix}COH\\|\\C_6H_4(O\ Me)\end{matrix}$$

Hydride of ethylsalicyl $C_9H_{10}O_2 = $
$$\begin{matrix}COH\\|\\C_6H_4\ (OEt)\end{matrix}$$

IX.

Hydride of salicyl and acetic anhydride $C_{11}H_{12}O_5 = $
$$\begin{matrix}CH\left\{\begin{matrix}(OAc)\\(OAc)\end{matrix}\right.\\|\\C_6H_4(OH).\end{matrix}$$

X.

Coumarin $C_9H_6O_2$, artificially prepared from hydride of sodium-salicyl, and acetic anhydride. It is probably ortho-oxyphenyl acrylic anhydride $C_6H_4\left\{\begin{matrix}O\\C_2H_2\end{matrix}\right\}CO$.

Propionic coumarin, ortho-oxyphenylcrotonic anhydride $C_{10}H_8O_2 = C_6H_4\left\{\begin{matrix}O\\C_3H_4\end{matrix}\right\}CO$, produced as above, but employing propionic anhydride.

Butyric coumarin, ortho-oxyphenylangelic anhydride $C_{11}H_{10}O_2 = C_6H_4\left\{\begin{matrix}O\\C_4H_6\end{matrix}\right\}CO$, prepared as above, but employing butyric anhydride.

Valeric coumarin, ortho-oxyphenylpyroterebic anhydride (?) $C_{12}H_{12}O_2 = C_6H_4\left\{\begin{matrix}O\\C_5H_8\end{matrix}\right\}CO$, prepared as above, but employing valeric anhydride.

XI.

α Bromocoumarin $C_9H_5BrO_2 = C_6H_4 \left\{ \begin{array}{c} O \\ C_2HBr \end{array} \right\} CO.$

β Bromocoumarin $C_9H_5BrO_2 = C_6H_3Br \left\{ \begin{array}{c} O \\ C_2H_2 \end{array} \right\} CO.$

α Dibromocoumarin $C_9H_4Br_2O_2 = C_6H_3Br \left\{ \begin{array}{c} O \\ C_2HBr \end{array} \right\} CO.$

β Dibromocoumarin $C_9H_4Br_2O_2 = C_6H_2Br_2 \left\{ \begin{array}{c} O \\ C_2H_2 \end{array} \right\} CO.$

α Chlorocoumarin $C_9H_5ClO_2 = C_6H_4 \left\{ \begin{array}{c} O \\ C_2HCl \end{array} \right\} CO.$

Tetrachlorocoumarin $C_9H_2Cl_4O_2.$

Coumarilic acid $C_9H_6O_3 = C_6H_4 \left\{ \begin{array}{c} O \\ C_2H(OH) \end{array} \right\} CO.$

Potassic coumarilate $C_9H_5KO_3 = C_6H_4 \left\{ \begin{array}{c} O \\ C_2H(OK) \end{array} \right\} CO.$

Sulphocoumarilic acid $C_9H_6O_2SO_3 2H_2O$

Baric sulphocoumarilate $C_{18}H_{10}BaO_4S_2O_6 = 5H_2O.$

Baric disulphocoumarilate $C_9H_4BaO_2S_2O_6H_2O.$

Bromopropionic coumarin $C_{10}H_7BrO_2.$

XII.

Chlorinated chloride of methyl, dichloride of methylene $CH_2Cl_2.$

XIII.

Anthracene $C_{14}H_{10}.$

Dichloranthracene $C_{14}H_8Cl_2.$

Dibromanthracene $C_{14}H_8Br_2.$

Dichloranthracene combined with picric acid $C_{14}H_8Cl_2C_6H_3(NO_2)_3O.$

Sodic disulphodichloranthracenate $C_{14}H_6Cl_2 \left\{ \begin{array}{c} NaSO_3 \\ NaSO_3 \end{array} \right.$

Baric disulphodichloranthracenate $C_{14}H_6Cl_2Ba2SO_3.$

Strontic disulphodichloranthracenate $C_{14}H_6Cl_2Sr2SO_3.$

Baric disulphodibromanthracenate $C_{14}H_6Br_2Ba2SO_3.$

Anthraquinone $C_{14}H_8O_2.$

Baric disulphanthraquinonate $C_{14}H_6O_2Ba2SO_3.$

XIV.

Artificial alizarin $C_{14}H_8O_4 = C_6H_4 \left\{ \begin{array}{c} CO \\ CO \end{array} \right\} C_6H_2(OH)_2.$

Acetyl alizarin $= C_{16}H_{10}O_5 = C_6H_4 \left\{ \begin{array}{c} CO \\ CO \end{array} \right\} C_6H_2 \left\{ \begin{array}{c} (OH) \\ (OAc) \end{array} \right. .$

Diacetyl alizarin $= C_{18}H_{12}O_6 = C_6H_4 \left\{ \begin{array}{c} CO \\ CO \end{array} \right\} C_6H_2(OAc)_2.$

XV.

Bromalizarin $C_{14}H_7BrO_4 = C_6H_4 \left\{ \begin{array}{c} CO \\ CO \end{array} \right\} C_6HBr(OH)_2.$

Diacetylbromalizarin $C_{18}H_{11}BrO_6 = C_6H_4 \left\{ \begin{array}{c} CO \\ CO \end{array} \right\} C_6HBr(OAc_2).$

XVI.

Nitroalizarin $C_{14}H_7(NO_2)O_4 = C_6H_4 \left\{ {CO \atop CO} \right\} C_6H(NO_2)(OH)_2$.

Amidoalizarin $C_{14}H_7(NH_2)O_4 = C_6H_4 \left\{ {CO \atop CO} \right\} C_6H(NH_2)(OH)_2$.

XVII.

Anthrapurpurin $C_{14}H_8O_5 = C_6H_3(OH) \left\{ {CO \atop CO} \right\} C_6H_2(OH)_2$.

Triacetylanthrapurpurin $C_{20}H_{14}O_8 = C_6H_3(OAc) \left\{ {CO \atop CO} \right\} C_6H_2(OAc)_2$.

XVIII.

Anthraflavic acid $= C_{14}H_8O_4 = C_6H_3(OH) \left\{ {CO \atop CO} \right\} C_6H_3(OH)$.

Baric anthraflavate $2C_{14}H_6BaO_4, 3H_2O$.

Diacetoanthraflavic acid $= C_{18}H_{12}O_6 = C_6H_3(OAc) \left\{ {CO \atop CO} \right\} C_6H_3(OAc)$.

XIX.

New isomer of alizarin yielding anthrapurpurin on fusion with potassic hydrate $C_{14}H_8O_4$.

XX.

Crystallized glyoxylic acid obtained from dibromacetic acid $C_2H_4O_4 = O(OH)$

$CH \left\{ {OH \atop OH} \right.$.

Calcic glyoxylate $C_4H_6Ca''O_4$.

2431a. 1. **Calorimeter,** by M. Berthelot, for thermo-chemical investigations, with electric mover of the agitator.

2. Air Thermometer, by M. Berthelot, glass, showing the temperature up to 550 degrees.

3. Air Thermometer, by M. Berthelot, silver, showing the temperature up to 900 degrees.

4. Apparatus for proving the disengagement of heat in the reaction of two gases.

5. Apparatus, by M. Berthelot, for the endothermic decomposition of formic acid by heat.

6. Two sealed globes, for the synthesis of formic acid by direct union of the oxide of carbon with potash, and sample of formate of lead, obtained by synthesis.

7. Apparatus, by M. Berthelot, for the synthesis of acetylene.

8. Bell glass, curved, in which was effected the synthesis of benzine, by the condensing of acetylene, by M. Berthelot; also another bell glass, empty.

9. Alcohol, obtained synthetically, by the union of olefiant gas with water.

10. Sealed tubes, containing organic matter, hydrogenized by iodhydric acid.   *Laboratory of the College of France.*

2432. Series of Cryohydrates. *Frederick Guthrie.*

These are solutions of various salts in water, of such strengths that, when reduced below 0° C. to the temperature marked on each flask, the salt and the water solidify together at that temperature. The same bodies in all cases result from the mixture of the respective salt with ice as a freezing mixture.

2433. Graphic Diagram, showing (1) the solubility of ice in the cryohydrates of various salts, (2) the composition of the cryohydrates and their melting points, (3) the solubility of the anhydrous salts in the cryohydrates and in water, (4) the temperature which the salt produces when used in a freezing mixture with ice or as an ice cryogen. *Frederick Guthrie.*

2435. Nitroxide of Amylen. *Frederick Guthrie.*

Discovered by the exhibitor. Of historical interest as being the first instance in which peroxide of nitrogen NO_2 was shown to behave as a halogen in uniting directly with an olefine to form a body homologous with "Dutch liquid." The composition of the body is $C_5H_{10}2NO_2$.

2436. Sulphide of Œnanthyl. *Frederick Guthrie.*

Discovered by the exhibitor, and of historical interest as being the first instance in which a term of a higher alcohol series was made from terms of lower alcohols. Formed by the action of zinc ethyl on sulpho-chloride of amylen.

2437. Nitrite of Amyl. *Frederick Guthrie.*

Discovered by M. Balard. Its therapeutic action discovered, and its introduction into the pharmacopœia recommended, by the exhibitor. Now coming into use in tetanic and other nervous affections.

2438. Collection of Preparations relating to chemical history of podocarpic acid (Annalen der Chemie und Pharmacie, vol. 170, p. 213). *Prof. A. C. Oudemans, Delft.*

2439. Hydrocarbons and Derivatives from Pennsylvania Petroleum. 23 specimens. *C. Schorlemmer, F.R.S.*

Group I.—Normal Paraffins.
Pentane. Pentene. Pentyl chlorides.
Hexane. Hexene. Hexyl chlorides. Hexyl acetates. Hexyl alcohols.
Heptane. Heptene. Heptyl chlorides. Heptyl alcohols. Octane.
Group II.—Isoparaffins.
Isopentane. Isohexane. Isoheptane.
Group III.—Aromatic Nitro-compounds.
Dinitrobenzene. Dinitrotoluene,
Nitro-compounds from fraction
boiling at 110°—120°
120°—130°
130°—140° solid and liquid.
140°—150°.

2440. Ethyl-Compounds, derived from dimethyl, which was obtained by the electrolysis of potassium acetate. Four specimens. *W. H. Darling.*

Ethyl chloride. Ethyl alcohol. Dichlorethane. Sodium Acetate.

2441. Aurin and Derivatives. 14 specimens.

C. Schorlemmer, F.R.S., and R. S. Dale, B.A.

Two specimens of commercial aurin.

Three specimens of pure aurin crystallised from alcohol.

Four specimens of pure aurin crystallised from acetic acid.

One specimen of the compound of aurin and sulphur dioxide.

One specimen of aurin-potassium sulphite.

One specimen of leucaurin.

One specimen of triacetyl-leucaurin.

One specimen of red aurin or peonin.

2485. Organic Compounds relating to the dry distillation of the " Galipot " resin, and to the constitution of pimaric acid (personal researches of the author). *Gustave Bruylants, Louvain.*

2486. Organic Compounds relating specially, 1° to the question of isomerism in glycerin derivatives; 2° to indicate the trialcoholic character of glycerin; 3° to show the reciprocal relations of allium compounds with glycerin compounds; 4° to propargylic compounds, and to the relations of allium compounds with propargylic compounds; 5° to binary compounds of allium; 6° to the binary propargylic compounds, and their relations with binary compounds of allium; 7° to the structure of the addition-products of hypochlorous acid.

Certain groups of these products are intended show: 8° the alcoholic character of alcohol acids, of basic ethers, and polyatomic alcohols, &c.; 9° the stability of the groupings Oxy-alcoholic C_n, H O in presence of PCl_5, PBr_5, &c., in opposition to the energetic reaction to which the corresponding hydro-oxygen compounds (O H) are subjected on the part of these re-agents, &c., and as a consequence the synthesis of oxalic acid, &c.

Louis Henry, Louvain.

2487. Mono-benzyl Urea, obtained by the action of ammonia on isocyanate of benzyl. Crystallized from water. (*See* Journal of Chemical Society, June 1872.)

$$Formula \quad CO \begin{cases} NHC_7H_7 \\ NH_2 \end{cases}$$

Dr. E. A. Letts and J. Fulton.

2488. Isocyanurate of Benzyl, obtained by the action of chloride of benzyl on cyanate of silver. Crystallized from alcohol. (*See* Journal of Chemical Society, June 1872.)

$$Formula \quad (C_7H_7-NCO)_3$$

Dr. E. A. Letts and J. Fulton.

2489. Benzyl-phenyl Urea, obtained by the action of aniline on isocyanate of benzyl. The compound is unstable. (*See* Journal of Chemical Society, June 1872.)

$$Formula \quad CO \begin{cases} NHC_7H_7 \\ NHC_6H_5 \end{cases}$$

Dr. E. A. Letts and J. Fulton.

2490. Dibenzyl Urea, obtained by the action of water on isocyanate of benzyl. The specimen was crystallized from alcohol. (*See* Journal of Chemical Society, June 1872.)

$$Formula \quad CO \begin{cases} NHC_7H_7 \\ NHC_7H_7 \end{cases}$$

Dr. E. A. Letts and J. Fulton.

2491. Isocyanate of Benzyl mixed with chloride of benzyl, obtained by heating chloride of benzyl with cyanate of silver. The isocyanate of benzyl cannot be obtained in the pure condition, as it rapidly becomes polymerized into isocyanurate of benzyl. (*See* Journal of Chemical Society, June 1872.)

Formula of the isocyanate $\quad C_7H_7 - N = C = O$

Dr. E. A. Letts and J. Fulton.

2492. Hydrobromate of Dimethyl Thetine, obtained by the action of sulphide of methyl on bromacetic acid. The specimen was crystallized from alcohol. (*See* Proceedings of Royal Society of Edinburgh, vol. viii., Nos. 87 and 89.)

$$Formula \quad Br—\overset{\displaystyle CH_3}{\underset{\displaystyle CH_3}{S}}—CH_2—COOH$$

Prof. A. Crum Brown and Dr. E. A. Letts.

2493. Hydrochlorate of Dimethyl Thetine, obtained by the action of chloride of barium on sulphate of dimethyl thetine. The specimen was separated from an aqueous solution by a mixture of alcohol and ether.

$$Formula \quad Cl - \overset{\displaystyle CH_3}{\underset{\displaystyle CH_3}{S}} - CH_2 - COOH$$

(*See* Proceedings of the Royal Society of Edinburgh, vol. viii., Nos. 87 and 89.)

Prof. A. Crum Brown and Dr. E. A. Letts.

2494. Sulphate of Dimethyl Thetine, obtained by the action of sulphate of silver on hydrobromate of dimethyl thetine. The specimen was crystallized from water.

$$Formula \quad HOOC—CH_2—\overset{\displaystyle CH_3}{\underset{\displaystyle CH_3}{S}}—SO_4—\overset{\displaystyle CH_3}{\underset{\displaystyle CH_3}{S}}—CH_2—COOH$$

(*See* Proceedings of the Royal Society of Edinburgh, vol. viii., Nos. 87 and 89.)

Prof. A Crum Brown and Dr. E. A. Letts.

2495. Poly-iodide of Dimethyl Thetine, obtained by the slow action of hydriodic acid on the base dimethyl thetine.
Not yet analysed.

(*See* Proceedings of the Royal Society of Edinburgh, vol. viii., Nos. 87 and 89.)

Prof. A. Crum Brown and Dr. E. A. Letts.

2496. Nitrate of Dimethyl Thetine, obtained by the action of nitrate of silver on hydrobromate of dimethyl thetine. The specimen was crystallized from alcohol.

$$\text{Formula} \quad NO_3 - \overset{\displaystyle CH_3}{\underset{\displaystyle CH_3}{S}} - CH_2 - COOH$$

(*See* Proceedings of the Royal Society of Edinburgh, vol. viii., Nos. 87 and 89.)

Prof. A. Crum Brown and Dr. E. A. Letts.

2497. Dimethyl Thetine, obtained by the action of water and oxide of silver on the hydrobromate of dimethyl thetine, or by the action of carbonate of barium on the sulphate. The base loses a molecule of water at 100° C.

$$\text{Formula} \quad HO - \overset{\displaystyle CH_3}{\underset{\displaystyle CH_3}{S}} - CH_2 - COOH$$

(*See* Proceedings of the Royal Society of Edinburgh, vol. viii., Nos. 87 and 89.)

Prof. A. Crum Brown and Dr. E. A. Letts.

2498. Lead Salt of Hydrobromate of Dimethyl Thetine, obtained by the action of the latter on carbonate of lead. The specimen was crystallized from hot water.

$$\text{Formula} \quad Br - \overset{\displaystyle CH_3}{\underset{\displaystyle CH_3}{S}} - CH_2 - COOPbBr, \ 2PbBr_2$$

(*See* Proceedings of the Royal Society of Edinburgh, vol. viii., Nos. 87 and 89.)

Prof. A. Crum Brown and Dr. E. A. Letts.

2499. Glass Digester in which to heat substances under pressure.

The apparatus consists of a well-annealed glass cylinder, closed at one end and drawn out at the other to a tube. Its capacity is about 600 cubic centimetres; the thickness of its walls about $\frac{1}{2}$ an inch throughout. The tube forming the neck is about $\frac{1}{6}$ of an inch in bore, and is closed by an accurately ground glass plate. The glass cylinder is mounted in a framework of brass, provided with a screw, which when turned presses the glass plate tightly against the tube, and thus hermetically closes the digester. The apparatus has been employed in the preparation of bromacetic acid, and large quantities of that substance have been prepared by its aid. The digester is intended as a substitute for sealed tubes, and has the great advantage over these that large quantities of substance can be heated at one operation, and that the danger of unsealing is avoided.

The apparatus is heated in an oil bath. *Dr. E. A. Letts.*

2500. Methyl-Sulphate of Calcium.
Dr. E. A. Letts and C. Abraham.

2501. Methyl-sulphate of Zinc.
Dr. E. A. Letts and C. Abraham.

2502. Methyl-sulphate of Ammonium.
Dr. E. A. Letts and C. Abraham.

2503. Specimens of Acetamide, Butyramide, Isobutyramide, and Valeramide; also of Valero-nitrile and Benzo-nitrile. These have been obtained by the action of the corresponding acids on sulphocyanate of potash, according to the equation

$$R-COOH+KCNS=HCNS+R-COOK.$$
$$R-COOH+HCNS=COS\ +R-CONH_2$$
$$R-COOH+HCNS=H_2S+CO_2+R-CN.$$

(In the case of the fatty acids the amide is the principal product, whilst the nitrile alone appears to be produced from the acid of the aromatic series. The process is easily and rapidly performed, and the yield good in all cases.

(*See* Proceedings of the Royal Society, No. 140, 1873.)
Dr. E. A. Letts and R. S. Marsden.

2504. Hyposulphite of Copper, Ammonium and Sodium, obtained by mixing a strong solution of hyposulphite of soda with ammonio-sulphate of copper. The salt separates from the solution spontaneously. Its formula is doubtful.
Dr. E. A. Letts and W. J. Nicol.

2505. Hyposulphite of Magnesium, obtained from hyposulphite of strontium by double (decomposition with sulphate of magnesia, and subsequent evaporation of the solution on a water bath.

Formula $Mg\ S_2\ O_3,\ 6\ H_2\ O.$

Dr. E. A. Letts and W. J. Nicol.

2506. Hyposulphite of Strontium, obtained by mixing hot concentrated solutions of hyposulphite of soda and chloride of strontium. The salt separates spontaneously from the solution.

Formula $Sr\ S_2\ O_3,\ 5\ H_2\ O.$

Dr. E. A. Letts and W. J. Nicol.

2507. Hyposulphite of Barium, obtained by mixing solutions of hyposulphite of soda and chloride of barium.

Formula $Ba\ S_2\ O_3,\ H_2\ O.$

Dr. E. A. Letts and W. J. Nicol.

2508. Hyposulphite of Nickel, obtained from hyposulphite of strontium by double decomposition with sulphate of nickel, and subsequent evaporation of the solution in vacuo. The salt readily decomposes.

Formula $Ni\ S_2\ O_3,\ 6\ H_2\ O.$

Dr. E. A. Letts and W. J. Nicol.

2509. Hyposulphite of Cobalt, obtained from hyposulphite of strontium by double decomposition with sulphate of cobalt, and subsequent evaporation of the solution in vacuo. The salt readily decomposes.

Formula $Co\ S_2\ O_3,\ 6H_2\ O.$

Dr. E. A. Letts and W. J. Nicol.

2510. Liquid Hydrochlorate of Turpentine, distilled over with water vapour, and dried with chloride of calcium.

Formula $C_{10}\ H_{16}\ HCl.$

Dr. E. A. Letts.

2511. Solid Hydrochlorate of Turpentine, crystallized from alcohol.

Formula $C_{10}\ H_{16}\ HCl.$

Dr. E. A. Letts.

2512. Hydrate of Turpentine (Terpene), crystallized from water.

Formula $C_{10}\ H_{16}\ 2H_2\ O.$

Dr. E. A. Letts and James Davidson.

2513. Hydrate of Hydrocarbon contained in Oil of Lavender (exotic); obtained by allowing a mixture of the oil with dilute alcohol and nitric acid to remain for some weeks at rest. The specimen was crystallized from water. Not yet analysed.   *Dr. E. A. Letts and James Davidson.*

2514. Hydrochlorate of the Hydrocarbon contained in Oil of Lavender (exotic); obtained by saturating oil of lavender with hydrochloric acid, and distilling with water vapour. No solid modification seems to exist.

Formula $C_{10} H_{16}$ 2 HCl.

Dr. E. A. Letts and James Davidson.

2515. Sulphide of Potassium, obtained by allowing a very concentrated aqueous solution (prepared in the usual way) to remain at rest for some time. Sulphide of potassium is usually stated to be red and uncrystallizable from water.

Dr. E. A. Letts.

2516. Mono-sodium Glycerine, obtained by mixing glycerine with an alcoholic solution of ethylate of sodium. It contains a molecule of alcohol (acting as water of crystallization), which it loses at 100° C.

(*See* Berichte der deutschen chemischen Gesellschaft, 1872.)

Formula $C_3 H_5 (OH)_2 (ONa) + C_2 H_6 O.$

Dr. E. A. Letts.

2517. Crystallized Pyro-sulphuric Acid, obtained by adding the equivalent quantity of sulphuric anhydride to oil of vitriol.

Formula $H_2 S_2 O_7$

Dr. E. A. Letts.

2518. Pure Oxalate of Methyl, used for the preparation of pure methyl alcohol.

Formula
$$\begin{array}{l} COOCH_3 \\ | \\ COOCH_3 \end{array}$$

Dr. E. A. Letts and R. M. Morrison.

2519. Pure Methyl Alcohol, obtained by the action of boiling water on the oxalate of methyl, and subsequent dehydration with lime and sodium.

Formula $H—CH_2 OH$

Dr. E. A. Letts and R. M. Morrison.

COLLECTION OF PREPARATIONS RESULTING FROM TECHNICAL AND SCIENTIFIC RESEARCH. CONTRIBUTED BY MEMBERS OF THE GERMAN CHEMICAL SOCIETY. (The names of the members are given below their contributions).

2677.

1. Anthracene.
2. Anthraquinone.
3. Dinitrobenzene.
4. Eosine potassium.
5. Methyl Violet.
6. Rubin (Rosanilinchlorhydrate).
7. Benzylchloride.
8. Aniline.
9. Hydrochlorate of Safranine.
10. Hofmann's Violet (Iodine Violet).
11. Toluene.
12. Coralline (Ammonium-rosolate).
13. Nitrobenzene.

14. Aurine (Rosolic acid).
15. Phenylene Brown.
16. Azobenzol.
17. Phosphine (Nitrate of Chrysaniline).
18. Nitronaphthaline.
19. Picric acid.
20. Diphenylamine.
21. Rosanilinbase.
22. Methyl Green, crystallized.
23. Nitrotoluene, solid.
24. Phtalic acid.
25. Toluidine, liquid.
26. Aniline Blue, soluble (Sodium-triphenylrosaniline-sulfate).
27. Toluidine, crystallized.
28. Phenol, crystallized.

29. Dimethylaniline.
30. Aniline Blue (Triphenylrosaniline-chlorhydrate).
31. Martius Yellow (Dinitronaphthol-calcium).
32. Dinitrocresol potassium.
33. Benzene, crystallized.
34. Naphtol.
35. Naphtylamine.
36. Methyldiphenylamine.
37. Naphtalene.
38. Aurantia.
39. Nigrosin, soluble in spirits.
40. Nigrosin, soluble in water.
41. Magdala-Red.
42. Mauve.
43. Cresol.

The compounds are intended for illustrating the present state of the coaltar colour manufacture, and the above collection embraces all the preparations of this class manufactured at present.

Actien-Gesellschaft für Anilinfabrikation, Rummelsburg, near Berlin.

Monochlorcresol $C^6H^3\begin{cases}CH^3\\Cl\\OH\end{cases}$ (Fusing point 56°.)
Mesitol $C^6H^2(CH^3)^3OH$. (Fus. p. 68°—69°.)
Nitronaphtol-Potassium $C^{10}H^6(NO^2)OK$.
Nitronaphtol-Sodium $C^{10}H^6(NO^2)ONa + 2H^2O$.
Nitronaphtol-Barium $[C^{10}H^6(NO^2)O]^2Ba + H^2O$.
Nitronaphtol-Lead $[C^{10}H^6(NO^2)O]^2Pb$.
Bibromnaphtol $C^{10}H^5Br^2(OH)$. (F. p. 111°.)
Mononitrophenylendiamine $C^6H^3(NO^2)(NH^2)^2$.

Dr. Rud. Biedermann, Berlin.

Magnesium-platino-cyanide
　　　　　　Mg Pt (CN)$_4$
Magnesium carbonate (crystallised).
Benzomonochloride $C_6H_5CH_2Cl$.
Benzodichloride $C_6H_5CHCl_2$.
Benzotrichloride $C_6H_5CCl_3$.
Monochlortoluene $C_6H_4ClCH_3$.
Monochlorbenzylchloride
　　　　　　$C_6H_4ClCH_2Cl$.
Benzylsulphhydrate $C_6H_5CH_2SH$.
Benzylsulphide $(C_6H_5CH_2)_2S$.
Mercury-Benzylsulphide
　　　　　　$(C_6H_5CH_2S)_2Hg$.
Benzylsulphoxyde $(C_6H_5CH_2)_2SO$.
Benzylsulphoxyde and very remarkable from their different power of crystallization.
Disulphobenzyle $(C_6H_5CH_2)_2S_2$.
Benzylxylene $C_{15}H_{16}$.

Benzylxylolketone
　　　　　　$C_6H_5COC_6H_3(CH_3)_2$.
Benzoylisophtalic acid
　　　　　　$C_6H_5COC_6H_3(COOH)_2$.
Benzoylisophtalate of barium
　　　　　　$C_6H_5COC_6H_3BaC_2O_4$.
Benzoylisophtalate of Ethyl ⎱
Benzoylisophtalate of Methyl ⎰
　　$C_6H_5COC_6H_3\begin{cases}COOC_2H_5.\\COOC_2H_5.\end{cases}$
Isomeric anhydride of benzoylisophtalic acid $C_{15}H_{10}O_4$.
Benzylisophtalic acid
　　　　　　$C_6H_5CH_2C_6H_3(COOH)_2$.
Benzylisophtalate of Barium, Calcium.
Barium-Ethylsulphate, Ba$\begin{cases}C_2H_5SO_4.\\C_2H_5SO_4.\end{cases}$
(Crystals of rare beauty.)

Anthracene.
Red Chromate of lead.
Green Chromate of lead.

Yellow Chromate of lead (chemically pure).
Turkish red.

The lead preparations are manufactured by means of chemically pure Ceruse.

Dr. Alex. Aug. Blatzbecker, Cologne.

Elaidic Acid. Fus. p. 45. Obtained from the oleine of stearin, manufactured by pressing it once in cold and once in heat.

Franz Bornemann, Verden.

Angelic acid; well shaped crystals, 1·2 cm. long, 3 mm. thick $C_5H_8O_2$.
Sodium-Ethylhyposulphite
$$SO_2 \begin{cases} C_2H_5 \\ ONa \end{cases}$$
Taurine Crystals, 5 cm. long $C_2H_7NSO_3$.
Cubebene-Stearoptene $C_{15}H_{26}O$.
Hydrochlorate of Cubebene $C_{15}H_{24}HCl$.
Hydrochlorate of Turpentine $C_{10}H_{16}HCl$.
Terpin $C_{10}H_{20}O + H_2O$.
Chelidonic acid $C_7H_4O_6 + H_2O$.
Bromalhydrate crystals, long 4 cm.,high 2 cm., thick 1 cm. $C_2HCl_3OH_2O$.
Arbutine $C_{25}H_{34}O_{14}$.

Iodine cyanide CNI.
Barium-Amylsulphate
$$(SO_4C_5H_{11})_2Ba.$$
Potassium-dichloracetate
$$C_2HCl_2KO_2.$$
Potassium-Stannocyanide
$$Sn (CN)_2 2KCN$$
Ethyl-dichloracetate $C_2HCl_2C_2H_5O_2$.
Octyl-Butyrate, Ethereal oil of Pastia.
Octyl-Butyrate, Ethereal oil of Hera cleum Spondyl.
Octylalcohol norm. $C_8H_{18}O$.
Octylacetate norm. $C_8H_{17}OC_2H_3O$.
Iodine-trichloride-potassium-chloride
$$KClCl_3.$$

Breslau, Pharmaceutical Institute of the University; Prof. Dr. Poleck.

Tetramethylammoniumtriiodide
$$(CH_3)_4NI_3.$$
Tetramethylammoniumpentaiodide
$$(CH_3)_4NI_5.$$
Trimethylethylammoniumtriiodide
$$(CH_3)_3(C_2H_5)NI_5.$$
Trimethylethylammoniumpentaiodide $(CH_3)_3(C_2H_5)NI_5$.
Methyltriethylammoniumtriiodide
$$(CH_3)(C_2H_5)_3N_5I_3.$$
Tetrethylammoniumpentaiodide
$$(C_2H_5)_4NI_5.$$
Tetraethylammoniumtribromide
$$(C_2H_5)_4NBr_3.$$
Molybdenumdichloride Mo_2Cl_4.
Molybdenumdichloride $Mo_2Cl_4 3H_2O$.

Molybdenumtrichloride Mo_2Cl_6.
Molybdenumtetrachloride $MoCl_4$.
Molybdenumpentachloride Mo_2Cl_5.
Molybdenum-oxychloride MoO_2Cl_2.
Molybdenum-oxytetrachloride
$$MoOCl_4.$$
The Polyiodides of the ammonium-bases are resulting from the researches of Weltzien.
The Molybdenum-chlorides and oxychlorides are prepared by Liechti and Kempe.
The large tube of Molybdenum-pentachloride has been prepared recently by Aronheim and Bornemann.

Carlsruhe, Chemical Laboratory of the Polytechnic Academy; Prof. L. Meyer.

Phosphenylchloride $C_6H_5PCl_2$.
Phosphenyltetrachloride $C_6H_5PCl_4$.
Phosphenylchlorobromide
$$C_6H_5PCl_2Br_2.$$
Phosphenylchlorotetrabromide
$$C_6H_5PCl_2Br_4.$$
Phosphenyloxychloride $C_6H_5PCl_2O$.

Phosphenylic acid $C_6H_5PO(OH)_2$.
Calcium-phosphenylate
$$C_6H_5PO(O_2Ca).$$
Phosphenylous acid
$$C_6H_5PO(OH)H.$$
Trichlortolylphosphinic acid
$$C_6H_4CCl_3PO(OH)_2.$$

Diethylphenylphosphine
$$C_6H_5(C_2H_5)_2P.$$
Monophenylphosphoric chloride
$$POCl_2(OC_6H_5).$$
Diphenylphosphoric chloride
$$POCl(OC_6H_5)_2.$$
Monophenylphosphoric acid
$$PO(OC_6H_5)(OH)_2.$$

Phosphorchlorotetrabromide
$$PCl_3Br_4.$$
Amidophosphenylic acid
$$C_6H_4(NH_2)PO(OH)_2.$$
Nitrophosphenylic acid
$$C_6H_4(NO_2)PO(OH)_2.$$

Carlsruhe, Chemical Laboratory of the Polytechnic Academy ; Prof. A. Michaelis.

1. Chlorbromacetic acid.
2. Chloral-cyanide-cyanate.
3. Chloral-anilide.

Dr. C. V. Cech, Berlin.

1. Triphenylbenzene ⎫
2. Triphenylbenzene ⎬ different crystallisations.
3. Triphenylbenzene ⎭
4. Sulphide of benzophenone.
5. Triphenylic ether.
6. Crystallized Pinakon of the Acetophenone.

Prof. Engler, Halle.

Benzoic acid sublimed from urine. Fine crystals with scarcely any smell ; possessing a pure white colour.

Furtenbach and Oelhafen, Reichelsdorf near Nuremberg.

Ammonium bicarbonate $CO\left\{\begin{array}{l} OH \\ ONH^4 \end{array}\right.$ This compound, hitherto found in the working apparatus of gas manufactories where the apparatus has the temperature of the surrounding air, is interesting on account of its having been observed in the gas-discharge pipes of hydraulic machines. The pipes have a temperature of 50° C. and upwards, at which temperature carbonate of ammonia for the most part dissociates into its components. No iodine was found in the compound ; the dark colour was owing to coal-dust.

Oxalic acid from lighting gas, $C^2O^4H^2,2H^2O$, obtained by treating larger quantities of coal-gas with fuming nitric acid, after having withdrawn from the gas the vapours of hydrocarbons by means of intense cold.

Oils from lighting gas from Silesian coals obtained in very small quantity by exposure of large quantities of lighting gas to the intense cold of $-21°$ R.

Gaswerke Städtische, Berlin (Dr. Tieftrunk).

1. Stannous chloride, Tin salt, $SnCl^2 + 2H^2O$ chemically pure, contains 52 per cent. of Sn.
2. Stannic chloride, $SnCl_4 + 5H_2O$, contains 43 per cent. of Sn.
3. Sodium-stannate contains 42–44 per cent. of Stannic oxide in soluble form, $Na_2SnO_3 + 2H_2O.$

Th. Goldschmidt, Berlin.

Sample of the first Boron ever made, Bo.
   „   Aluminium, Al.
   „   Urea, $CO(NH_2)_2.$
Wolfram phosphide.

Silicon-calcium.
Amorphous Silicon, from Silicon-hydride.
Titaniumcyanochloride $TiCl_4C\,NCl.$
Chromic oxide $Cr_2O_3.$

Cyanotitaniumnitride Ti_5CN_4.

Crystal of Arsenous anhydride
$$As_2O_3.$$

Silicon-Nitride.

Phosphoruscyanide $P(CN)_3$.

* Bariumorthonitrobenzoate
$$(C_6H_4(NO_2)(NO_2)^oCOO)_2Ba + 3H_2O.$$

* Ethylorthonitrobenzoate
$$C_6H_4(NO_2)oCOOC_2H_5.$$

Sodiummetanitrobenzoate
$$C_6H_4(NO_2)^mCOONa + 3H_2O.$$

* Ethylmetanitrobenzoate
$$C_6H_4(NO_2)^mCOOC_2H_5.$$

* Bariummetanitrobenzoate
$$(C_6H_4(NO_2)^mCOO)_2Ba + 4H_2O.$$

* Bariumparanitrobenzoate
$$(C_6H_4NO_2)^pCOO)_2Ba + 5H_2O.$$

*Ethylparanitrobenzoate
$$C_6H_4(NO_2)^pCOOC_2H_5.$$

Sodiumsulphbenzoate
$$C_6H_4COOHSO_2ONa.$$

Orthonitrobenzanilide from Benzani-
lide $C_6H_4\begin{Bmatrix} NO_2^o \\ NH \end{Bmatrix} (COC_6H_5)$.

Metanitrobenzanilide, from Benzani-
lide $C_6H_4\begin{Bmatrix} NO_2^m \\ NH \end{Bmatrix} (COC_6H_5)$.

Paranitrobenzanilide, from Benzani-
lide $C_6H_4\begin{Bmatrix} NO_2^p \\ NH \end{Bmatrix} (COC_6H_5)$.

Parabromtoluene $C_6H_4CH_3Br^p$.

Those bodies marked with * are prepared for the crystallographic investigation of isomeric compounds.

Göttingen ; Universitäts Laboratorium (Prof. Wöhler and Prof. Hübner).

1. Triacetonamine.
2. Triacetonamine nitrate.
3. Triacetonamine sulfate.
4. Triacetonamine oxalate.
5. Triacetonamine bioxalate.
6. Chlorhydrate of Triacetonammonium-platinochloride.
7. Chlorhydrate of Triacetonammonium-platinchloride.
8. Diacetonamine nitrate.
9. Diacetonamine binoxalate.
10. Chlorhydrate of Tri-isotriacetonammoniumplatinochloride.
11. Diacetonammoniumplatinochloride.
12. Diacetonic alcohol.

Prof. Dr. W. Heintz, Halle a. S.

Metallic modification of phosphorus; the crystalline leaves are on lead and isolated from the lead by means of diluted nitric acid.

Prof. W. Hittorf, Münster, Westfalia.

1. Dimethylammoniumchloride.
2. Trimethylammoniumchloride.
3. Tetramethylammoniumiodide.
4. Diethylamine.
5. Triethylamine.
6. Tetraethylammoniumiodide.
7. Ethylendiaminchlorhydrate.
8. Monophenyloxamide.
9. Phenylated diacetamide.
10. Formamide.
11. Phosphoniumiodide.
12. Ethylphosphiniodide.
13. Dibenzylphosphine.
14. Ethenyldiphenyldiamine.
15. Melaniline.
16. Melanilinnitrate.
17. Melanilinsulphate.
18. Diphenylparabanic acid.
19. Ethylic sulfocyanate (Ethylmustardoil).
20. Methylic sulfocyanate (Methylmustard oil).
21. Phenylic sulfocyanate (Phenylmustardoil).
22. Allylacetate.
23. Allylalcohol.
24. Allylamine.
25. Allylsulphide.
26. Carbanilide.
27. Sulphocarbanilide.

28. Cyananiline.
29. Monobromaniline.
30. Toluylendiamine.
31. Dinitrobenzene.
32. Base from the residues of the Aniline manufacture, $C_{19}H_{26}N_2$.
33. Toluidinehydroxalate.
34. Toluidinechlorhydrate.
35. Toluidinesulphate.
36. Melanilinehydroxalate.

37. Toluylic acid from Lepidium sativum.
38. Nitrophenol.
39. Rosanilinchlorhydrate.
40. Iodine-combination of Methylgreen.
41. Potassium salt of Eosine.
42. Xylidine pure.
43. Tetraphenylmelamine.

Prof. A. W. Hofmann, Berlin.

Collection of Substances discovered or investigated by Liebig.

Bariumperoxyd.
Phosphorus nitride.
Nitrogen chlorophosphide.
Acetal.
Acetone.
Aldehydeammonia.
Metaldehyde.
Formic acid.
Asparagin.
Asparic acid.
Aethylenechloride.
Butyric acid.
Calciumchloride, Acetic ether.
Carbothialdine.
Comenic acid.
Chloral.
Chloralhydrate.
Citraconic acid.
Acetate of Ethyle.
Lactose.
Sodium ethylate.
Barium methionate.
Manganeseoxalate.
Oxalate of Ethyle.
Ferrous oxalate.
Oxamate of Ethyle.
Oxamide.
Mercaptide of mercury.
Cane sugar.
Stearic acid.
Mucic acid.
Thialdine.
Vinylphosphate of barium.
Vinylsulphate of barium.
Vinylsulphate of calcium.
Tartaric acid.
Potassium-cyanate.
Cyanuric acid.
Fulminate of ammonium.
Fulminate of barium.
Persulphocyanic acid.
Sulphocyanate of silver.

Sulphocyanate of potassium.
Sulphocyanate of ammonium.
Urea.
Mercuric oxyde-urea.
Uric acid.
Alloxane.
Alloxanate of barium.
Alloxantine.
Allantoine.
Ammeline-nitrate.
Ammelide.
Dialurate of ammonium.
Melam.
Melamine.
Mellonate of potassium.
Mesoxalate of barium.
Murexide, pure.
    „   commercial.
Oxalurate of ammonium.
Parabanic acid.
Uramil.
Kreatine.
Sarkosinechlorhydrate.
Thionurate of ammonium.
Xanthogenate of potassium.
Benzoic aldehyde.
Benzoylchloride.
Benzoylbromide.
Benzoylcyanide.
Benzoyliodide.
Benzil.
Benzilic acid.
Benzoin.
Benzoate of Ethyle.
Mandelic acid.
Picric acid.
Atropine.
Brucine.
Caffeine.
Cinchonine.
Morphine.
Meconic acid.

Narcotine.
Nicotine.
Strychnine.
Opianic acid.
Piperine.

Calcined soda, 98 per cent. Na^2CO^3.
Calcined soda, 100 per cent. Na^2CO^3.
Sodium bicarbonate, $HNaCO^3$.

Metagallic acid.
Mycomelic acid.
Valeric acid.
Fulminate of potassium.

Prof. A. W. Hofmann, Berlin

The preparations have been obtained by the exhibitor by an ammonia-soda process carried out 'and developed by himself. This method has been introduced in the exhibitor's factory at Grevenberg, near Aix-la-Chapelle; at the Nuremberg Soda Manufactory, near Nuremberg; at the Soda Manufactory of the Rothenfeld Saline and Salt Bath Company at Rothenfelde, near Osnabrück, and at the Aalborger chemiske Fabrikker og Gjödningsfabrik at Aalborg in Denmark. *Moritz Honigmann, Aachen.*

Dextrin.
Monobasic sugar-lime.
Tribasic   „

Double phosphate of manganese and ammonium.
Artificial wax.
Paraffin.

Dr. Franz Hulwa, Breslau.

Barium-platino cyanide, $Pt(CN)_2Ba(CN)_24H_2O$.
Barium-chlorate, $BaCl_2O6$.
Cuprous oxide, Cu_2O.
Stannous oxide, SnO.

J. Hutstein, Manufacturer, Breslau.

1. Quinine puriss crystall.
2. Cotoin, the crystalline principle of the Coto bark.
3. Coto Bark.
4. Quinine salicylic. cryst.
5. Quinine phenylo-muriat.
6. Quinine phenylosulfuric.
7. Sulphophenylate of Cinchonidine.

8. Echicerin.
9. Echitin.
10. Echitëin, crystalline principle of the Dita bark.
11. Dita bark.
12. Santoninic Acid.
13. Quinamine.
14. Laudamine.

The salts of Quinine exhibited were at first prepared in the Stuttgart manufactory, and the above-mentioned new preparations—Cotoin, Echicerin, Echitëin—have been made out by the investigations of MM. Jobst and Hesse, Stuttgart.

Fried. Jobst, Stuttgart.

1. Methylic alcohol pure.
2. Formic acid cryst.
3. Methylsulphocarbylamin,
   $CH_3-N=CS$.
4. Sulphocarbamid.
5. Guanidine, CH_5N_3.
6. Iodoform.
7. Nitroethane, $C_2H_5-N=O_2$.
8. Ethylsulphocarbylamin,
   $C_2H_5-N=CS$.
9. Acetaldehyde.
10. Paraldehyde.
11. Aldehyde.
12. Isopropylic alcohol.
13. Isobutylic alcohol.
14. Allylic alcohol.
15. Propionic acid.

16. Allylamine.
17. Methylethylketon,
    $CH_3-CO--C_2H_5$.
18. Isobutylic alcohol.
19. Trimethylcarbinol.
20. Monochlorcrotonic acid.
21. Amylic ether.
22. Furfurol.
23. Capronitrile.
24. Capronic acid from the nitrile.
25. Hexyliodide from mannite.
26. Orcine.
27. Pyrocatechin.
28. Resorcin.
29. Hydroquinone.
30. Quinone.

C. A. F. Kahlbaum, Berlin.

Acidum Benzoicum ex urina, C_7H_5O, $\left.\right\} O$.

Kaufmann's manufactory of Benzoic acid is the largest on the Continent.

Karl Joseph Kaufmann, Königsberg, in Prussia.

Sinistrin.
Albumin, soluble from barley.
Albumin, coagulated from barley.
Oil fatty, from barley.

Sugar from barley in form of syrup.
Sugar from barley crystallised.
Cerealic acid.

Dr. G. Kühnemann, Dresden (later Görlitz).

1. Didymiumsulfate.
2. Iridium-sodium-sesquichloride.
3. Rhodium-sodium-sesquichloride.
Used for measuring the crystals.

Prof. von Lasaulx, Breslau.

One specimen of bromine.
One specimen of bitter salt.
Two specimens of magnesiumchloride.
One specimen of potassiumchloride, 98 per cent.
One specimen of potassiumchloride, 80 per cent.

Leopoldshaller Vereinigte chemische Fabriken, Actien-Gesellschaft, Leopoldshall near Stassfurt.

1. Cörulignone $C_{16}H_{16}O_6$.
2. Hydrocorulignone $C_{16}H_{18}O_6$.
   White in small crystals.
3. Hydrocorulignone $C_{16}H_{18}O_6$.
   Brownish in large crystals.
4. Bibromanthracenetetrabromide
          $C_{14}H_8Br_6$.
5. Bibromanthracene $C_{14}H_8Br_2$.
6. Tetrabromanthracene $C_{14}H_6Br_4$.
7. Anthraquinone $C_{14}H_8O_2$.
8. Anthracene $C_{14}H_{10}$. Blue fluorescence.
9. Chrysene (yellowish) $C_{18}H_{12}$.
10. Chrysene (white) $C_{18}H_{12}$.
11. Chrysoquinone $C_{18}H_{10}O_2$.
12. Xylindein
13. Monooxyanthraquinone
          $C_{14}H_7(OH)O_2$.
14. Dichloranthracene $C_{14}H_8Cl_2$.
15. Binitroanthraquinone
          $C_{14}H_6(NO_2)_2O_2$.
16. Biamidoanthraquinone
   Sublimed. $C_{14}H_6(NH_2)_2O_2$.
17. Hydrochrysamide
          $C_{14}H_2(NH_2)_4(OH)_2O_2$.

Prof. C. Liebermann, Berlin.

1. *Hyoscyamine crystals.*—Crystallised alkaloid from Hyoscyamus niger and alb., produced for the first time in larger quantities and brought into the market; crystallised in snow-white needles. According to the investigations of Dr. Harnack at Strassburg, and Dr. Fronmüller in Fürth (which, however, have not yet been brought to a conclusion), this preparation is exceedingly effective, and is in this respect almost equal to atropine.

The smallest dose which paralyses the termination of the vagus of the heart of a frog, consists of 1/200 milligramme (0·000005). (Raising of the suspended beating of the heart by Muscarin) the dose which still produces genlarement of the pupil of a rabbit, is 1/250 milligramme (0·000004) [Harnack].)

2. *Amorphous Hyoscyamine.*—Amorphous alkaloid from Hyoscyamus niger and alb. The smallest dose which paralyses the terminations of the vagus of the heart of a frog is in this instance 1/100 milligramme (0·000001, Harnack).

3. *Ditaine crystal.*—Crystallised bodies from the cortex dita of Echites scholaris, only recently introduced into medicine.

4. *Ditamine.*—Amorphous alkaloid from the same bark. Both bodies quite new in trade. They represent the effective substances of this drug.

5. *Kamalin crystal.*—Crystallised body from Kamala, the fibres and glands of the fruit of Rottlera tinctoria. Effective substance of this drug. New in trade.

6. *Muscarin nitricum.*—Nitrate of the alkaloid contained in the toadstool (Agaricus muscarinus). Muscarin is likewise new in commerce.

7. *Pilocarpin.*—Alkaloid from the leaves of Policarpus pinnatifolius, *Saborandi*, since a short time only employed in medicine. New in commerce.

8. *Veratric acid,* from the seeds of Sabadilla.

9. *Kosin crystallisatum.*—The acting principle of the Koso plant crystallises partly in needles, partly in short thick prisms belonging to the rhombic system. Almost insoluble in water, it is easily dissolved in ether, benzol, sulphide of carbon, and chloroform. Less easily in glacial acetic acid and alcohol.

For further particulars *see* Schroff, "Ausstellungsbericht" (Vienna), page 32 ; Flückiger, "Archiv der Pharmacie," 1874, part 2, and "Pharmaceutical Journal, transactions, 1875, No. 238."

10. *Apomorphia hydrochlor. crystal.*—Pure and certain acting preparation, *see* Harnack, "Archiv für experimentale Pathologie und Pharmakologie." Dr. Juratz, "Zeitschrift für Medizin."

Emanuel Merck, Darmstadt.

1. Telluric acid $H_2TO4 + 4H_2O$.
2. Tellurous anhydride TO_2.
3. Iodbromide of mercury $HgIBr$.
4. The same, larger crystals.
5. Phosphide of cadmium.
6. Methyl-benz-acetol
$$CH_3 - C(C_7H_5O)_2 - CH_3.$$
Prepared from monochlorpropyleniodhydrate by means of silverbenzoate.
7. Menthol $C_{10}H_{19}OH$.
8. Menthyl-acetate $C_{10}H_{19}OC_2H_3O$.
9. Menthene $C_{10}H_{18}$.
10. Borneol $C_{10}H_{17}OH$ (natural).
11. Benzoic aldehyde C_7H_6O.
Prepared synthetically from $C_6H_5CHCl_2$ by means of sulphuric acid.
12. Phthalic anhydride $C_8H_4O_3$.
Large crystals formed by slow decomposition of the chloride in moist air.
13. Ethyl aceto-acetate
$$CH_3COCH_2COOC_2H_5$$
Prepared synthetically from acetic ether.

14. Dehydracetic acid $C_8H_8O_4$.
15. Oxyuvitic acid $C_9H_8O_5$.
Prepared synthetically from acetic ether.
16. Metacresol C_7H_7OH.
Prepared synthetically from acetic ether.
17. Metacresol-ethylic ether
$$C_7H_7OC_2H_5.$$
Prepared synthetically from acetic ether.
18. Oxybenzoic acid $C_6H_5OHCOOH$.
Prepared synthetically from. acetic ether.
19. Methyl-oxybenzoic acid
$$C_6H_5OCH_3COOH.$$
20. Methyloxybenzoate of calcium
$$(C_6H_5OCH_3COO)_2Ca + 4H_2O.$$
21. Hydro-oxybenzoic acid
$$C_6H_6OHCOOH.$$
22. Cresotic acid $C_7H_6OHCOOH$.
23. Trinitrocresol $C_7H_4(NO_2)_3OH$.

Prof. A. Oppenheim, Berlin.

Butylchloral $C_4H_5Cl_3O$.
Butylchloral-hydrate $C_4H_7Cl_3O_2$.
Trichlorbutyric acid $C_4H_6Cl_3O_2$.
Chlorangelactic acid $C_5H_7ClO_3$.

Chloralcyan-hydrate $C_3H_2Cl_3NO$.
Trichlorolactate of ethyl $C_5H_7Cl_3O_3$.
Chloracrylate of ethyl $C_5H_7ClO_3$.
Bromalcyanhydrate $C_3H_2Br_3NO$.

Dr. A. Pinner, Berlin.

Chloralhydrate, large crystals ⎫
Chloralhydrate, small crystals ⎪
Chloralhydrate, in powder - ⎬ $C_2HCl_3O + H_2O.$
Chloralhydrate, in crusts   - ⎭
Chloroform from Chloral, $CHCl_3.$
All these preparations are perfectly pure, the chloralhydrate being a special product of the manufactory.

Saame and Co., Ludwigshafen on the Rhine.

Phenol, pure, is perfectly free from cresol, and therefore *perfectly* soluble in 20 times its quantity of water.

Phosphoric acid, chemically pure; the commercial so-called fused phosphoric acid contains up to 25°/₀ sodium-pyrophosphate.

Salicylic acid cryst., perfectly soluble in water, alcohol, and ether.

Salicylic acid subl.

Tannin leviss. contains only a very trifling quantity of glucose, is therefore, according to H. Schiff, nearly pure digallic acid.

Monobromcamphor.

Trichloride of carbon.

Salicylate of quinine.

Chloralhydrate in plates.

Chloralhydrate in crystals, prepared according to the special direc-

tions of O. Liebreich, perfectly free of other chlorinated compounds, and is unchangeable for years.

Chloralide.

Butylchloral.

Saccharate of iron, soluble, not containing any free sugar, but a pure chemical compound of ferric oxyde, soda, and sugar.

Bromide of potassium, entirely unaffected by barium salts.

Iodide of potassium, like the above, and free from iodic acid, therefore entirely unchangeable.

Potassium hydrate.

Permanganate of potassium.

Salicylate of sodium, easily soluble in water and alcohol.

Salicylate of zinc.

E. Schering, Berlin.

Hyposulphate of sodium
        $Na_2S_2O_6 + 2H_2O.$
Hyposulphate of barium
        $BaS_2O_6 + 4H_2O.$
Hyposulphate of strontium
        $SrS_2O_6 + H_2O.$
Hyposulphate of lead $PbS_2O_6 + 4H_2O.$
Selenium, precipitated.
Cube of selenium.
Tellurium in rods Te.
Tellurium in cubes.
Telluric acid $H_2TeO_4.$
Iodide of carbon $CI_4.$
Silicon-sulphide $SiS_2.$
Chlorate of sodium, three different crystalline forms, $NaClO_3.$
Periodate of sodium $NaIO_4 + 3H_2O.$
Arseniate of sodium
        $HNa_2AsO_4 + 12H_2O.$
Lithium in wire, Li.
Lithium in balls, Li.
Lithium in bars.
Beryllium metal Be.
Berylliumchloride, sublim, $BeCl_2.$
Sulphate of Beryllium $BeSO_4 + 2H_2O.$
Nitroprusside of sodium
        $Cy_3Fe_2NO_2, 2CyNa + 2H_2O.$
Zirconium metallic, Zr.

Calcium met.,obtained by electrolysis, Ca.
Strontium met., obtained by electrolysis, Sn.
Yttriumplatinocyanide
        $YCy_4Pt + 7H_2O.$
Didymiumplatinocyanide
        $Di_2Cy_{12}Pt_318H_2O.$
Sulphate of Ceroso-ceric oxide
        $Ce_5S_6O_2 + 21H_2O.$
Nitrate of Cerium-magnesia
        $MgN_2O_6 + CeN_2O_6 + 8H_2O.$
Nitrate of Cerium-nickel
        $NiN_2O_6 + CeN_2O_6 + 8H_2O.$
Nitrate of cerium and didymium
        $DiN_2O_6 + 4H_2O.$
Oxalate of yttrium $YC_2O_4 + 4H_2O.$
Oxalate of erbium $ErC_2O_4 + 3H_2O.$
Oxalate of cerium $CeC_2O_4.$
Oxalate of didymium $DiC_2O_4 + 4H_2O.$
Oxalate of thorium $ThC_4O_8 + 2H_2O.$
Sulphate of thorium $ThS_2O_8 + 9H_2O.$
Manganese metal Mn.
Nickelbromide subl. $NiBr_2.$
Sulphate of nickel $NiSO_4 + 7H_2O.$
Sulphate of nickel and potash
        $NiSO_4 + K_2SO_4 + 6H_2O.$

Selenate of copper $CuSeO_4 + 5H_2O$.
Sulphate of copper-oxyde-ammonia
$\qquad CuSO_4 + 4NH_3 + H_2O$.
Arsenious acid crystallised from hot
hydrochloric acid.
Iodide of arsenic, subl. As_2I_3.
Iodide of antimony, subl. SbI_3.
Fused vanadic acid V_2O_5.
Acid vanadate of sodium
$\qquad Na_2V_4O_{11} + 9H_2O$.
Acid vanadate of ammonium
$\qquad (NH_4)2V_4O_{11} + 4H_2O$.
Molybdic acid, subl. MoO_3.
Titaniumsesquichloride Ti_2Cl_6.
Acetate of calcium and copper
$\qquad \left. \begin{matrix} (C_2H_3O_2)_2Ca \\ (C_2H_3O_2)_2Cu \end{matrix} \right\} 5H_2O$
Sulphocarbonate of potassium
$\qquad CS < {SK \atop SK}$
Sulphocarbonate of sodium
$\qquad CS < {SNa \atop SNa}$
Xanthogenate of sodium
$\qquad CS < {OC_2H_5 \atop SNa}$

Xanthogenate of potassium
$\qquad CS < {OC_2H_5 \atop SK}$
Hippuric acid $CH_2 \begin{cases} NHC_7H_5O \\ CO_2H \end{cases}$
Carbonate of guanidine
$\qquad CH_5N_3H_2CO_3$.
Levulinate of calcium
$\qquad (C_5H_7O_3)2Ca + 2H_2O$.
Nitrate of cytisine.
Lithofellinic acid $C_{20}H_{36}O_4$.
Phthalic anhydride $C_6H_4 \begin{cases} CO \\ CO \end{cases} O$.
Stilbene, pure $\begin{cases} CH - C_6H_5. \\ CH - C_6H_5. \end{cases}$
Stilbene, common $\begin{cases} CH - C_6H_5. \\ CH - C_6H_5. \end{cases}$
Triphenylbenzene $C_6H_3(C_6H_5)_3$.
Styracene.
Nitrosodimethylaniline
$\qquad C_6H_4(NO)N(CH_3)_2$.
Nitrosodimethylaniline aniline
$\qquad [C_6H_4(NO)N(CH_3)_2]_2 + C_6H_5NH_2$,
Phenanthrene $C_{14}H_{10}$.
Acetophenon C_6H_5, CO, CH_3
Benzophenon C_6H_5, CO, C_6H_5.

Cases containing 64 preparations of the metals and other elements.

Dr. *Theodor Schuchardt, Görlitz, Silesia.*

Coniferin $C_{16}H_{22}O_8 + 2H_2O$.
Vanillin $C_8H_8O_3$, small crystals are
in this shape, the commercial product.
Vanillin $C_8H_8O_3$, slowly crystallised
from dilute solutions.
Vanillin $C_8H_8O_3$, from residues of
the artificial preparation of vanillin, and sublimed in the same
manner as the vanillin of the
Vannilla pods at the ordinary temperature.
Vanillic acid, slowly crystallised from
dilute solution.

Vanillic acid $C_8H_8O_4$, finely crystal-
lised as commercial commodity.
Sugar-vannillic acid $C_{14}H_{18}O_9$, pre-
pared by Reimer and Tiemann.
The compound is crystallised in the
bottle itself.
Acetoeugenol $C_{12}H_{14}O_3$, prepared by
Nagai and Tiemann.
Ferulic acid $C_{10}H_{10}O_4$, prepared by
Nagai and Tiemann. The above
acid is prepared synthetically from
Vanilla; the white opaque crystals
are re-crystallised from alcohol,
the yellowish brilliant from water.

Dr. F. Tiemann and Dr. W. Haarmann, Berlin and
Holzninden.

Allylic-Alcohol $\begin{cases} CH^2. \\ CH. \\ CH^2OH. \end{cases}$
Prepared by distilling glycerine with oxalic acid and traces of sal ammoniae.

Prof. B. Tollens, Göttingen.

Chromic acid.
Molybdic acid, fused and sublimed
specimens.
Molybdic acid, fused.

Molybdic acid, sublimed.
Selenious acid.
Tellurous acid.
Tungstic acid.

Vanadic acid.
Aluminiumsulphate.
Bariumbromide.
Bariumperoxyde cristall.
Berylliumsulphate.
Cadmiumbromide.
Cadmiumsulphide.
Caesium-alum.
Calciumchloride fus.
Ceriumprotosulphate.
Ceriumpersulphate.
Chromiumsulphate.
Cobalt sulphate.
Ammonium copper tartrate.
Potassium copper chromate.
Copper oxyde.
Ferric chloride, subl.
Didymsulphate.
Sulphide of iron.
Lanthansulphate.
Lead-dithionate.
Leadnitrite.
Lithiumcitrate.
Lithiumsulphate.
Metallic manganese.
Manganese chloride.
Metallic molybdenum.
Ammonmolybdate.
Nickelsulphate.
Potassiumhydrate in rods.
Potassiumhydrate in cakes.
Potassiumhydrate, pure.
Potassiumnitrite.
Rubidium alum.
Sodiumbromate.
Sodiumhydrate.
Sodiumhydrate, pure.
Sodiumdithionate.
Sodiumphosphotungstate.
Phosphotungstic acid.
Sodiumsilicate, crist.
Strontiumhydrate.
Tellurium.
Thalliumcarbonate.
Thalliumsulphate.
Thoriumsulphate.
Potassium titanfluoride.
Metallic Tungsten.
Ammoniumtungstate.
Ammonium Uranocarbonate.
Uraniumnitrate.
Yttr. Erb. oxid.
Yttr. Erb. sulfate.
Zincchloride fus.
Zircon. Potass. fluorid.
Ammoniummellilate.

Paramide.
Copper cyanide violet.
Copper cyanide green cryst.
Mercury-cyanide.
Barium platinocyanide.
Magnesium platinocyanide.
Potassium platinocyanide.
Potassiumcyanate.
Sulphocyanammonium.
Nitroprusside of sodium.
Cyanuric acid.
Ureanitrate.
Alloxan.
Ammonthionurate.
Carbontetrachloride.
Iodoform.
Potassium sulphomethylate.
Oxalate of methyle.
Copperformiate.
Strontiumformiate.
Butyric ether.
Capronic ether.
Ethyloxalate.
Sebacic ether.
Ethylenechloride.
Ethylenebromide.
Bariumsulphethylate.
Carbonhexachloride.
Glycol.
Aldehydeammonia.
Monochloracetic acid.
Acetamide.
Ammonium acetate.
Propylic alcohol norm.
Propylacetate.
Propionic acid from propylic alcohol.
Aceton. Natr. bisulfit.
Dichlorhydrin.
Lactic acid.
Zinc lactate.
Isobutylic alcohol.
Butyric acid from fermentation.
Amylchloride.
Bariumsulphamylate.
Valeral-sodium-bisulphite.
Capronic acid from fermentation.
Capronic acid from valerianic acid—residues.
Caprylic acid.
Caprinic acid.
Bariumcapronate.
Caprylic alcohol.
Methyloenanthol.
Capronic acid from caprylic alcohol.
Ethal.

Sebacic acid.
Malic acid.
Calciumbimalate.
Aconitic acid.
Anisic acid.
Asparaginic acid.
Camphoric acid.
Cinnamic acid.
Cuminic acid.
Fumaric acid.
Picramic acid.
Salicylous acid.
Sorbic acid.
Camphormonobromide.
Benzolbichloride.
Benzoin.
Styracin.
Styrol.
Benzoylalcohol from crude cinnamic
  alcohol.
Phenylpropylic alcohol.
Monochlortoluene.
Xylene.
Naphtalinbichloride.
Mesitylene.
Cymene.
Chrysophanic acid.
Emodine.
Aporetin.
Erythroretin.
Phaoretin.
Gentisine.
Haematoxyline.
Indigotin.
Taurin.
Glycyrrhizin.
Melampyrite (Dulcit).
Glycocoll.
Inosite.
Allantoin.
Taurin.
Hyoglycocholic acid.
Aconitine.
Aesculin.

Amygdalin.
Anemonin.
Anemonic acid.
Arbutin.
Methylhydroquinone from Ar butine
Green Hydroquinone from Ar butine
Asaron.
Atropine.
Tropinsulphate.
Atropa acid.
Tropa acid.
Berberinnitrate.
Cantharidin.
Caffeine, raw.
Caffeine, pure crist.
Colchicin.
Columbin.
Conine.
Corydalin.
Coumarin.
Cubebin.
Daturine.
Delphinine.
Elaterine cryst.
Filicin.
Filimelissic acid.
Ononine.
Helcnin.
Lactucerin.
Imperatorin.
Peucedanin.
Picrotoxin.
Piperine.
Potassiumpiperinate.
Quercitrin.
Quininvalerate.
Saponin.
Solanin.
Stramonine.
Syringin.
Theobromine.
Urson.
Veratrine.
Rubidiumbitartrate.

Dr. H. Trommsdorff, Erfurt.

Stafsfurt " Abraumsalz."
Potassium chloride, obtained by means
  of hot extraction.
Kieserite, obtained by cold lixiviation
  of the residues of the above process.
Picromerite, obtained by action of
  Kieserite on Potassium chloride in
  the heat.
Potassium sulphate, prepared by decomposition of the picromerite by
  means of potassium chloride.

40075.

Potash (crude fused), by fusion of
  the potassium sulphate with lime
  and charcoal (Leblanc's Process).
Potash 95°/₀, by evaporation and
  calcination of the leys from the
  above.
Potash hydrated, free from sulphuric
  acid; by refinement of the potash
  at 95°/₀.

Q q

These specimens show the various stages in the preparation of potash from the mineral salts.

Vorster and Grüneberg, Kalk b. Cöln.

Phenoquinone $C_{18}H_{14}O_4$.
Pyrogalloquinone $C_{18}H_{14}O_8$.
Purpurogallin $C_{18}H_{14}O_9$.
Triacetamide $C_6H_9O_3N$.
Bicyannaphtalene $C_{12}H_6N_2$.

Bioxynaphtalene $C_{10}H_8O_2$.
Picric acid Bromonaphtalene
$C_6H_2(NO_2)_3OHC_{10}H_7Br$.
Binitronaphtalene $C_{10}H_6N_2O_4$.

Prof. Wichelhaus, Berlin.

1. Bars of nickel; 2, bars of cobalt. These metal bars are about 170mm. long and 42mm. broad and brightly polished. They form the first cast pieces of cobalt and nickel of a large size which have ever been produced, and date from the year 1866. W. Hankel made use of the same in his investigations on the magnetic bearing of nickel and cobalt. (Berichte der Königl. Sächsischen Gesellschaft der Wissenschaften, mathematisch-physikalische Klasse. Meeting of the 21st July 1875.)

Prof. Winkler, Freiberg, Saxony.

2536a. Chemical Preparations (61) from the works of Engelhardt, Latschinow, Wolkow, Maïkupar. Produced in the Laboratory of the Agricultural Institute of St. Petersburg.
Chemical Society of Russia, University of St. Petersburg.

COLLECTION OF RUSSIAN CHEMICALS.

1. α Naphtol.
2. β Naphtol.
3. α Naphtylsulphuric chloride.
4. β Naphtylsulphuric chloride.
5. α Benzoylnaphtol.
6. β Benzoylnaphtol.
   Maikopar.
7. Mononitrochlorbenzol.
8. Binitrochlorbenzol.
9. β Cresotinic acid.
10. γ Cresotinic acid.
11. γ Ethyl-cresol.
12. α Benzoyl-cresol.
13. γ Benzoyl-cresol.
14. Potassium-α.cresylsulphate.
15. Potassium-γ cresylsulphate.
16. Barium-γ cresylsulphite.
17. Potassium-α cresylsulphite.
18. Barium-α cresylsulphite.
19. Barium-sulphocresylate.
20. Potassium-bisulphocresylate.
21. β Benzokresid, $C_7H_7O(C_7H_5O)$.
22. Potassium-α thymolsulphate.
23. Barium-α thymolsulphate.
24. Phosphate of thymol.
25. Potassium-β thymolsulphate.
26. Potassium-thiobenzoate.
27. Barium-thiobenzoate.

28. Thiobenzoic-ethylic-ether.
29. Thiobenzoic-amylic-ether.
30. Thiobenzoic anhydride.
31. Dithiobenzoate of silver.
32. Tetrachlorbenzol.
33. Benzoylbisulphide.
34. Binitrophenol.
35. Picrylchloride.
36. Picrotoluid.
37. Binitrophenyltoluid.
38. Propylenbromide.
   Engelhardt and Latschinoff.
39. Toluidinsulphuric acid, $C_7H_9NSO_3 + \frac{1}{2}H_2O$.
40. Toluidinsulphuric acid, $C_7H_9NSO_3 + H_2O$.
   Malyscheff.
41. $N \begin{cases} C_7H_7SO_2m. \\ C_7H_5O. \\ H \end{cases}$
42. $N \begin{cases} C_7H_7SO_2p. \\ C_7H_5O. \\ H \end{cases}$
43. α Cresol.
44. α Methyl-cresol.
46. β Cresol.
47. β Ethyle-cresol.
48. Cresol-phosphate.

49. *p* Cresylsulphuric chloride.
50. *p* Cresylsulphuric amide.
51. Nitrocresylsulphuric amide.
52. $C_7H_7(NHKSO_2)p$.
53. $C_7H_6(NH_2)m(SHO_3)p + H_2O$.
54. *m* Cresylsulphuric amide.
55. $C_7H_7(NHKSO_2)m$.
     Miss A. Wolkoff.

Rakowitsch's apparatus for determination of alcohol and sugar in liqueurs, as well as of the fat in butter, and for determination of water.
R. Nippe.

2536b. Preparations from Acids (8), by Sokolow and Latschinow (from acids).
Chemical Laboratory of the Agricultural Institute of St. Petersburg.

1. Hydrochlorate of diacetonhydramine.
2. Chloroplatinate of the above.
3. Neutral oxalate.
4. Bioxalate.
5. Bioxalate.
6. Picrate.
7. Chloroplatinate of the triacetonhydramin.
8. Chloroplatinate of triacetonamin.
   Sokolow and Latschinow.

2536c. Chemical Preparations (80) from the works of Butleroff, Menschutkin, Gustavson, Lwoff, Wazne, Wyschnegradsky, Jermalajew, Popoff, Prianichnikoff, Nakapetian, Gagarin. *Chemical Laboratory, University of St. Petersburg.*

1. Uramidobenzoic acid.
2. Uramidobenzamide.
   Both compounds illustrate the action of Potassiumcyanate on Amido-acids.
3. Mercury- and silver-compound of the succinamide.
4. Aethylsuccinamide and Methylsuccinamide.
5. Succinanil.
6. Succinamid-mercuric oxide, $C_4H_4O_2(NH_2)_2HgO$, $\frac{1}{2}$aq.
7. Phenylsuccinamide.
8. Succinanilide.
9. Salts of the succinamic acid.
10. Succinanilic acid.
11. Bariumsuccinanilate $[C_4H_4O_2(C_6H_5)HN]_2CaO_2 + 4H_2O$.
12. Calciumsuccinanilate.
13. Leadsuccinanilate (No. 5–13 comp. Liebig's Annalen, 162, 165).
14. Ammonium-parabanate.
15. Oxaluramid.
16. Potassium-parabanate.
17. Potassium-oxalurate.
18. Silver-parabanate.
19. Silver-parabanate + aq. (No. 14–19 comp. Liebig's Annalen, 172.)
20. Dimethylparabanic acid.
21. Ethyl-succidcyanate.
22. Methyl-succidcyanate.
23. Ethylsuccinuric acid together with its ammonium- and silversalts.
24. Ethylsuccinuramid (No. 20–24 comp. Liebig's Annalen, 174).
25. Dialurate of Ammonium and of Potassium.
26. Ammonium-dialurate.
27. Sodium-dialurate.
28. Several Dialurates.
29. Tartronamic acid.
30. Tartronamates of barium, potassium and silver.
31. Tartronamate of lead.
    W. Menschutkin.
32. Heyamethylenamine (Liebig's Ann. 144).
33. Oxymethylene (Liebig's Ann. 120).
36. Trimethylcarbinol (Liebig's Ann. 144, 162).
41. Pinakoline.
42. Cynonetol.
43. Isocrotylbromide.
44. Trimethylacetic acid.
45. Ethers of the above acid.
46. Potassium-trimethylacetate.

47. Silver-trimethylacetate.
48. Silver-trimethylacetate.
49. Ammonium-trimethylacetate.
50. Sodium-trimethylacetate.
51. Solution of zinc-trimethyl-acetate.
52. Zinc-salt and Cadmium-salt of the same acid.
53. Magnesium-salt.
54. Barium-salt.
55. Calcium-salt.
56. Mercury-salt.
57. Lead-salt.
58. Copper-salt.
59. Trimethylacetamide.
66. Trimethylacetic anhydride.
61. Pentamethylated ethylic alcohol.
62. Pentamethylaetholhydrat, $C_2(CH_3)_5OH + H_2O$.
63. Pentamethylated chloraethyl.
64. Pentamethylaetholhydrate.
68. Methylenchloride.
69. Di-isobutylene.

A. Butleroff.
34. Di-isopropoxalic acid.
Markovnikoff.
35. Tetraiodide of carbon.
39. Ethylideniodide.

40. Vinyliodide.
Gustavson.
37. Isohexylene.
Tschaikovsky.
38. Tetramethylformene.
Lwoff.
60. Dimethylethylacetate of silver.
71. Tertiary amylic alcohol.
72. Methylpropylcarbinol.
73. Tertiary amylic chloride.
74. Tertiary amylic iodide.
75. Ethylbutylpinakolin.
76. Ethylamylpinakolin.
80. Diamylene.
Wyschnegradsky.
64. Methylaethylcarbinol.
Wagner.
65. Isoamylen.
Jermalajew.
67. Triaethylcarbinol.
Nakapetian.
70. Dimethylaethylcarbinol, synthetic (Liebig's Ann. 145).
Popoff.
77. Dimethylisobutylcarbinol (Liebig's Ann. 162.)
Prianischnikoff.
78, 79. Aethylenjodbromide.
Prinz Gagarin.

2536d. Original Chemical Preparations (38), by C. Claus, obtained in his researches on Platina.

Chemical Laboratory, Imperial Berg-Institute, St. Petersburg.

(1.) $PtCy_2$.
(2.) $3KCl.OsO_3.SO_3$.
(3.) $2KCy.OsCy.3HO$.
(4.) $Ru_2Cl_3.3HCl$.
(5.) $2KCy.OsCy.3HO$.
(6.) $3NH_4Cl.Ir_2Cl_6$.
(7.) $3Hg_2Cl.Ir_2Cl_3$.
(8.) $3NH_4Cl.Ru_2Cl_6$.
(9.) Rh.
(10.) $3Hg_2Cl.Ir_2Cl_3$.
(11.) $3NH_4Cl.Ir_2Cl_3.3HO$.
(12.) $KCy.PtCy$.
(13.) $2KCy.RuCy.3HO$.
(14.) $KCy.PtCy.HO$.
(15.) $HgCy.PtCy.Hg_2ONO_5$.
(16.) $RuCy$.
(17.) $3KClRuCl_3$.
(18.) Silicate of Ruthenium and Potassium.
(19.) Ru_2S_3.
(20.) $KClRuCl_2$.
(21.) Ru_2O_33HO.

(22.) $RuCy$.
(23.) $2KCyRuCy$.
(24.) $3NH_4Cl.Ru_2Cl_3$.
(25.) Ru.
(26.) $2KCyRuCy3HO$.
(27.) $IrO_2.2HO$.
(28.) $Ru_2O_3.3LiO_3.KOSiO_3$.
(29.) $3NH_4O.SO_2.ZnO_2.SO_2.5HO$.
(30.) $IrCl_2.KCl$.
(31.) $2NH_4Cl.NH_4OSO_2$.

$$2IrO \left\{ \begin{matrix} 2SO_2 \\ Cl \end{matrix} \right\} 12HO.$$

(32.) $3KCl.Ir_2Cl_3.3HO$.
(33.) $RuCl_2.KCl$.
(34.) $IrCl_2.KCl$.
(35.) $IrCl_2.KCl$.
(36.) $2KCl.IrOSO_2$.
(37.) $IrCy.2KCy$.
(38.) $3KCl.Ir_2Cl_3.6HO$.
Chem. Laboratory of the Imp. Mining Institute in St. Petersburg.

2536e. Chemical Preparations (3), by L. Chishkoff, obtained in his researches on **Fulminating Mercury.**
Chemical Laboratory, Michailoff Artillery Academy.

1. Dibromnitroacetonitril.
2. Potassium-isocyanurate.
3. Silver-isocyanurate.

L. Schischkoff.

2536f. Chemical Preparations (41), by Fritzshe and Zinin.
Chemical Laboratory, Imperial Academy of Sciences, St. Petersburg.

1, 2, 3. Crystallized Zinc.
4. Anthracene.
5. Chrysogen + naphtalene.
6. Chrysogen + anthracene.
7. Phosen + oxybinitrophosen.
8. Phosen + oxybinitrophosen.
9. Harmin. $C_{13}H_{12}N_2O$.
Staatsrath Fritzsche.
10. Lepidene.
11. Isolepidene.
12. Oxylepidene.
13. The same, Cryst. in leaves.
14. The same, Cryst. in needles.
15. 16. Bioxylepidene.
17. Hydroxylepidene.
18. Bichlorlepidene.
19. Bibromlepidene.
20. Bichloroxylepiden, needles.
21. Bichloroxylepiden, octaedric.
22. Bibromoxylepiden, needles.
23. Bibromoxylepiden, octaedric.

24. Hydrobichloroxylepiden.
25. Oxylepidinic acid.
26. Bichloroxylepidinic acid.
27. Benzoin.
28. Hydrobenzoin.
29. Desoxybenzoin.
30. Benzil.
31. Binitrotetrachlorbenzil.
32. Tetrachlorbenzil.
33. Chlorbenzil.
34. Formobenzoic acid.
35. Formobenzoilamid + Benzaldehyd.
36. Bichlortolan, in needles.
37. Bichlortolan, in leaves.
38. Harmalin, hydrocyanate.
39. Amaric acid.
40. Amaric anhydride.
41. Benzamaron, $C_{70}H_{56}O_4$.
Professor Zinin.

2536g. Chemical Preparations (124), by Beilstein, Kuhlberg, Kurbatoff, Hemilian, Rudneff, and 81 preparations by Wroblesky.
Chemical Laboratory, Technological Institute, St. Petersburg.

The methylic group occupies always the place 1.
1. $C_7H_6BrNO_2$. Boil. p. 255°—256°. $= C_6H_3Br(NO_2).CH_3$.
$$\underset{4}{}\underset{3}{}\underset{1}{}$$
2. $C_7H_6BrNO_2$. Fus. p. 55°. $= C_6H_3Br(NO_2).CH_3$.
$$\underset{3}{}\underset{6}{}\underset{1}{}$$
3. C_7H_6BrI. Boil. p. 260°.
$$\underset{3\ 6}{}$$
4. $\alpha\ C_7H_6Cl.OC_2H_5$. Boil. p. 210°.
$$\underset{4}{}\underset{2}{}$$
5. $\beta\ C_7H_6Cl.OC_2H_5$. Boil. p. 210°.
6. $C_7H_3Cl(NO_2)_3.OK$.
$$\underset{3\ (2,5,6)}{}\underset{4}{}$$
7. $C_7H_3Cl(NO_2)_3OH$. Fus. p. 82·5°.
$$\underset{3\ (2,5,6)}{}\underset{4}{}$$
8. $C_6H_3(CH_3)_2OH$. Fus. p. 75°. Boil. p. 212·5°.
9. $C_6H_3(CH_3)_2OH$. Boil. p. 211·5°.

10. $C_6H_2BrI(NO_2).CH_3$.   Fus. p. 118°.
    $_{3\ 4}\quad _5$

11. $C_6H_2BrI(NO_2).CH_3$.
    $_{3\ 6}\quad _?$

12. $C_6H_3Cl(NO_2)CH_3$.   Boil. p. 249°.
    $_3\quad _?$

13. $C_6H_3BrI.CH_3$.   Boil. p. 265°.
    $_{3\ 4}$

14. $C_6H_3Br(NO_2)CH_3$.   Boil. p. 269°.
    $_3\quad _4$

15. $C_6H_3Br(NO_2)CH_3$.   Fus. p. 124°.
    $_3\quad _5$

16. $C_6H_2Br_2(NO_2).CH_3$.   Fus. p. 87·5°.
    $_{(3,5)}\quad _?$

17. $C_6H_2Br_2(NO_2)CH_3$.   Fus. p. 79°.
    $_{(2,6)}\quad _?$

18. $C_6H_2Br_2(NO_2)CH_3$.   Fus. p. 86°–87°.
    $_{(3,4)}\quad _?$

19. $C_6HBrBrBr(NO_2).CH_3$.   Fus. p. 215°.
    $_{2\ 6\ 4}\quad _3$

20. $C_6H_2BrBr(NO_2)CH_3$.   Fus. p. 46°–47°.
    $_{2\ 6}\quad _?$

21. $C_6HBrBrBr(NH_2)CH_3$.   Fus. p. 97°.
    $_{2\ 6\ 4}\quad _3$

22. $C_6H_2BrBr.NH_2.CH_3$.   Fus. p. 92·5°.
    $_{2\ 6}\quad _3$

23. $C_6H_2BrBr.CH_3.NH.(C_2H_3O)$.   Fus. p. 154°.
    $_{2\ 6}\quad\quad _3$

24. $C_6H_2BrBr.NH_2.CH_3$.   Fus. p. 73°.
    $_{3\ 5}\quad _4$

25. $C_6H_2BrBrNH_2CH_3$.   Fus. p. 82°.
    $_{1\ 5}\quad _?$

26. $C_6H_2BrBrNH_2.CH_3$.   Fus. p. 95°.
    $_{3\ 5}\quad _?$

27. $C_6H_2BrBrNH_2.CH_3$.   Fus. p. 83°.
    $_{2\ 5}\quad _?$

28. $C_6H_2BrBrNH_2.CH_3$.   Fus. p. 50°.
    $_{3\ 5}\quad _2$

29. $C_6H_3BrNH_2.CH_3.HNO_3$.
    $_3\ _6$

30. $C_6H_3BrNH_3CH_3.HCl$.
    $_3\ _6$

31. $(C_6H_3BrNH_2.CH_3)_2H_2SO_4 + 1\frac{1}{2}$ aq.
    $_3\ _6$

32. $C_6H_3BrNH_2.CH_3$.   Fus. p. 57°.
    $_3\ _6$

33. $C_6H_3Br.CH_3.NH.C_2H_3O$.   Fus. p. 156 .
    $_3\quad _6$

34. $\beta\ (C_6H_3Br(NO_2)CH_3SO_3)_2Ba + 7$ aq.

35. $\beta\ (C_6H_3Br(NO_2)CH_3SO_3)_2Pb + 3$ aq.

36. $\beta\ (C_6H_2Br(NO_2)CH_3SO_3)_2Ca + 9$ aq.

37. $\beta\ (C_6H_3Br.SO_3.CH_3)_2Pb + 3$ aq.
    $_3\quad _6$

38. $\beta\ C_6H_3Br.SO_3.CH_3.K$.
    $_3\quad _6$

39. $(C_6H_3Cl.CH_3.SO_3)_2Ba + 2$ aq.
    $_4\quad _2$

40. $(C_6H_2Cl(NO_2).CH_3.SO_3)_2Ba.$
    ${}_4{}_?{}_?$

41. $C_6H_2BrBrI.CH_3.$ Fus. p. 86° ; boil. p. 270°.
    ${}_3{}_5{}_4$

42. $C_6H_4Br.CH_3.$ Boil. p. 182°.

43. $C_6H_4Cl.CH_3.$ Boil. p. 157°.
    ${}_3$

44. $C_6H_2Br_3.CH_3.$ Fus. p. 70^c.

45. $C_6H_3BrBr.CH_3.$ Boil. p. 238°
    ${}_2{}_5$

46. $C_6H_3BrBr.CH_3.$ Boil. p. 237°.
    ${}_3{}_4$

47. $C_6H_3BrBr.CH_3.$ Boil. p. 246^c.
    ${}_2{}_6$

48. $C_6H_3BrBr.CH_3.$ Fus. p. 60°.
    ${}_3{}_5$

49. $C_6H_3BrBr.CH_3.$ Fus. p. 42·5°.
    ${}_2{}_3$

50. $C_6H_3(CH_3)_2C_2H_5.$ Boil. p. 185°.
    ${}_{1,3}{}_5$

51. $C_6H_4.CH_3.C_2H_5.$ Boil. p. 157°.
    ${}_3$

52. $C_6H_3Cl.(NH.C_2H_3O).CH_3.$ Fus. 99°.
    ${}_3{}_4$

53. $C_6H_3Cl.CH_3.NH_2.HCl.$
    ${}_3{}_4$

54. $C_6H_3Cl.CH_3.NH_2.HNO_3.$
    ${}_3{}_4$

55. $(C_6H_3Cl.CH_3.NH_2)_2.H_2SO_4.$
    ${}_3{}_4$

56. $(C_6H_3Cl.CH_3.NH_2)_2.C_2H_2O_4.$
    ${}_3{}_4$

57. $C_6H_3Br.NH(C_2H_3O).CH_3.$ Fus. p. 117·5°.
    ${}_3{}_4$

58. $C_6H_3Br(CH_3)NH_2.HCl.$
    ${}_3{}_4$

59. $C_6H_3Br(CH_3)NH_2.HNO_3.$
    ${}_3{}_4$

60. $(C_6H_3Br(CH_3).NH_2)_2.H_2SO_4.$
    ${}_3{}_4$

61. $(C_6H_3Br(CH_3).NH_2)_2C_2H_2O_4.$
    ${}_3{}_4$

62. $C_6H_2Br(NO_2)CH_3.NH(C_2H_3O).$ Fus. p. 210·5°.
    ${}_3{}_5{}_4$

63. $C_6H_2Br(NO_2)CH_3.NH_2.$ Fus. p. 64·5°.
    ${}_3{}_5{}_4$

64. $C_6H_3Br(NO_2)CH_3.$ Fus. p. 86° ; boil. p. 279°.
    ${}_3{}_5$

65. $C_6H_3Br.CH_3NH_2.$ Boil. p. 265°.
    ${}_3{}_5$

66. $C_6H_3Br.CH_3.NH_2.HNO_3.$
    ${}_3{}_5$

67. $C_6H_3Br.CH_3.NH_2.HCl.$
    ${}_3{}_5$

68. $(C_6H_2Br.CH_3.NH_2)_2H_2SO_4.$
    ${}_3{}_5$

69. $C_6H_4Br.CO_2H.$ Fus. p. 154°.
   5   1

70. $\beta\ C_6H_3Cl.NH_2.CH_3.$ Fus. p. 83°; boil. p. 241°.
   4   ?

71. $\beta\ C_6H_3Cl.NH_2.CH_3.HCl.$
   4   ?

72. $\beta\ C_6H_3Cl.NH_2.CH_3.HNO_3^*.$
   4   ?

73. $\beta\ C_6H_3Cl(NO_2)CH_3.$ Boil. p. 253°.
   4   ?

74. $\beta\ C_6H_3Cl.I.CH_3.$ Boil. p. 240°.
   4

75. $\alpha\ C_6H_3Cl.CH_3.NH_2.$ Boil. p. 238°.
   4   ?

76. $\alpha\ (C_6H_3Cl.CH_3.NH_2)HCl.$
   4   ?

77. $\alpha\ C_6H_3Cl.CH_3.NH_2)HNO_3.$
   4   ?

78. $\alpha\ C_6H_3ClCH_3.NO_2.$ Boil. p. 243°.
   4

79. $C_6H_2Cl_2(NO_2)CH_3.$ Boil. p. 274°.
   3,4   ?

80. $C_6H_2Cl_2(NH_2).CH_3.$ Fus. p. 88°; boil. p. 259°.
   3,4   ?

Wroblevsky.

1. Orthonitrocinnamates.
2. Orthonitrocinnamic ethers.
3. Orthonitrobenzoic ether and Orthonitrobenzoic amide.
4. Trichlorbenzoate of barium.
5. Trichlorbenzoate of ethyle.
6. Trichlorbenzoate of ammonium.
7. Trichlorbenzoate of calcium.
8. Trichlorbenzoic acid.
9. Trichlorbenzoic amide.
10. Trichlorbenzoic aldehyde.
11.
12. } No. 4–12 of Liebig's Ann. 152.
13–28. Chlorine derivatives of Toluene (Liebig's Ann. 146,150,152).
   (1.) $C_6H_4Cl.CH_3.$
   (2.) $C_6H_3Cl_2.CH_2Cl.$
   (3.) $C_6H_2Cl_3.CH_3.$
   (4.) $C_6H_3Cl_2.CHCl_2.$
   (5.) $C_6H_4Cl.CH_3.$
   (6.) $C_6H_2Cl_3.CH_2Cl.$
   (7.) $C_6H_4Cl.CCl_3.$
   (8.) $C_6Cl_5CH_3.$
   (9.) $C_6H_3Cl_2.CCl_3.$
   (10.) $C_6H_2Cl_3.CHCl_2.$
   (11.) $C_6HCl_4.CH_2Cl.$
   (12.) $C_6H_2Cl_3.CCl_3.$
   (13.) $C_6Cl_5.CH_2Cl.$
   (14.) $C_6HCl_4.CHCl_2.$
   (15.) $C_6HCl_4.CCl_3.$
   (16.) $C_6Cl_5.CHCl_2.$

29–34. Substituted Benzylic alcohols (Liebig's Ann. 147).
   (1.) $C_6H_4Cl.CH_2.OH.$
   (2.) $C_6H_3Cl_2.CH_2OH.$
   (3.) $C_6H_2Cl_3.CH_2.OH.$
   (4.) $C_6HCl_4.CH_2.OH.$
   (5.) $C_6Cl_5.CH_2.OH.$
   (6.) $C_6H_3(NO_2)_2.CH_2.OH.$
35. α Nitronaphtalidine.
36. Trinitronaphtalene (35 and 36 comp. Liebig's Ann. 169).
37–39. Toluylendiamine (Liebig's Ann. 158).
   (1.) $C_6H_3(NH_2)p(NH_2)m.CH_3.$
   (2.) Sulphate.
   (3.) Stannous chloride double salt.
40–53. Derivatives of Toluene and Nitrotoluene (Liebig's Ann. 155).
   (1.) m Nitro–o Acettoluid.
   (2.) m Nitro–o Toluidine.
   (3.) m Nitro–p Acettoluid.
   (4.) m Nitro–p Toluidine.
   (5.) o Nitro–m Acettoluid.
   (6.) o Nitro–m Toluidine.
   (7.) Dinitro–p Acettoluid.
   (8.) Dinitro–p Toluidine.
   (9.) o Nitro–p Toluidine.
   (10.) m Nitrotoluene.
   (11.) m Nitro–p Bromtoluene.

(12.) Dinitrotoluene (Fus. p. 60·5°).

(13.) Trinotrotoluene (Fus. p. 77°–80°).

Beilstein and Kuhlberg.

54. α Dichlorbenzoic acid.
55. Calcium-salt.
56. Barium-salt.
57. α Dichlorbenzoic amide.
58. β Dichlorbenzoic aqid.
59. Calcium·salt.
60. Barium-salt.
61. β Dichlorbenzoic amide (No. 54–61 of Liebig's Ann. 179).

Beilstein.

62–82. Derivatives of Aniline (Liebig's Ann. 176).
(1.) $C_6H_3Cl.NH_2.NO_2$
　　　1　4　3
(2.) $C_6H_3Cl.NH_2.NO_2$.
　　　3　1　5
(3.) Acetylic compound of No. 2.
(4, 5.) $C_6H_3Cl.NO_2.NH_2$.
　　　　2　5　4
(6.) $C_6H_3.NO_2.NH_2.Cl$.
　　　1　　4　2
(7.) Acetylic compound of No. 6.
(8.) $C_6H_2ClCl.NO_2.NH_2$.
　　　3　4　6　1
(9.) Dichloraniline (1, 4—2).
(10.) Dichloranilin (1, 2—4).
(11.) Acetylic derivative of No. 10.
(12.) Acetylic compound of Dichloranilin (1, 3—4).
(13.) Picrate of *p* Chloranilin.
(14.) Picrate of Chloranilin.
(15.) *m* Chloracetanilid.
(16.) *o* Chloracentailid.
(17.) Trichloranilin (1, 2, 3—4).
(18.) Trichloranilin (2, 4, 6—1).
(19.) Trichloranilin (1, 2, 3—5).
(20.) Trichloranilin (1, 2, 5—4).

83–87. *o* Dichlorbenzolsulphonic acid.

88. $C_6H_3ClCl(NO_2)$.
　　　1　2　4

89. *m* Dichlorbenzol.
90. $C_6H_3ClCl(NO_2)$.
　　　1　3　4
91. *m* Dichlorbenzolsulphonic acid.
92. *o* Chlornitrobenzol.
93. $C_6H_3Cl_3$ (1, 3, 5).
94. $C_6H_3Cl_5$ (1, 2, 3).
95. $C_6H_2ClClCl(NO_2)$.
　　　1　2　3　5
96. $C_6H_2Cl_4$ (1, 3, 4, 5).
97. $C_6HClClClCl(NO_2)$.
　　　1　3　4　5　2
98–101. Trichlorphenylguanidin and Salts.

Beilstein and Kurbatoff.

102. Dinitroacetanilid.
103. Dinitrocitraconanil.
104. Citraconanil.
105. Dinitroanilin.
106. Parasulfocinnamic acid.
107. Parasulfocinnamate of Barium.
108. Acid metasulfocinnamate of Barium.

Rudnew.

109. Barium-sulfobutyrate.
110. Calcium-sulfobutyrate.
111. Silver-sulfobutyrate.
112. Copper-sulfobutyrate.
113. Lead-sulfobutyrate.
114. Zinc-sulfobutyrate.
115. Triphenylmethan.
116. Triphenylmethan with Benzol.
117. Diphenyltolylmethan.
118. Naphtyldiphenylmethan.
119. Diphenyl-phenylenmethan.
120. Triphenylcarbinol.
121. Triphenylcarbinol-ethylic ether.
122. Benzhydrolbenzoic acid.
123. Barium salt.
124. Trinitrotriphenylmethan.

Hemilian.

125. Magnesium-cuminate $(C_{10}H_{11}O_2)_2Mg.6H_2O$.
126. Calcium - cuminate $(C_{10}H_{11}O_2)_2Ca.3H_2O$.
127. Barium - cuminate $(C_{10}H_{11}O_2)_2Ba.2H_2O$.

Beilstein and Kupffer.

2536h. Pyroxyline. Detection of poisons in medico-legal investigations. *N. W. Sokoloff.*

2536h. Chemical Laboratory of the Imp. Medic. Surg. Academy. 45 Flasks containing organic bodies remaining after the researches made by Professor A. Borodin and by others under his direction.

1. Acetic ether of isocapric alcohol, $(C_2H_3O)(C_{10}H_{21})O$. (Jahresbericht für Chemie, 1864, 338.)
2. Aldehyde of isocapric alcohol, $C_{10}H_{20}O$. (Jahresb. f. Ch. 1870, 680.)
3. Condensation product of the valeral (divaleral), $C_{10}H_{20}O_2$. (Berichte der Deutsch. Chem. Gesellschaft, V. 480.)
4. Hydrate of divaleral, $(C_{10}H_{20}O_2)_2H_2O$. (*Ibidem*, VI. 982.)
5. Condensation product of valeral-aldehyde, $C_{10}H_{18}O$. (*Ibid*. V. 480.)
6. Acid from the aldehyde, $C_{10}H_{18}O$. $(C_{10}H_{18}O_2.)$ (*Ibid*.)
7. Condensation product of the valeral, $(C_{20}H_{38}O_2$. (*Ibid*.)
8. Base, isomeric with the cuminhydramide $(C_{30}H_{36}N_2)$, crystallized from alcohol. (Ibid. VI. 1,253.)
9. The same crystallized from benzine. (*Ibid*.)
10. Hydrochlorate of the base. $(C_{30}H_{36}N_2 . ClH.)$
11. Sulphate of the base, crystallized from alcohol, $(C_{30}H_{36}N_2)_2H_2SO_4$.
12. The same crystallized from alcohol acidulated with sulfuric acid.
13. Nitrate of the base. $(C_{30}H_{36}N_2 . HNO_3.)$ (*Ibid*.)
14. Nitrosoamarin. $(C_{21}H_{12}(NO)N_2)$. (*Ibid*. VIII. 933.)
15. Sulphocarbobenzidide, $(C_{12}H_{10}(CS)N_2)$. (Jaresb. f. Chem. 1860, 356.)
A. Borodin.
16. Cerotic acid (from the bees-wax), fusing at $91°$ $(C_{34}H_{68}O_2 ?)$. (Bericht der Deutsch. Chem. Gesell. IX. 278).
17. Pelargonamide, $(C_9H_{14}O)N_7H_2$. (*Ibid*. VI. 1252.)
Dr. M. Schalfejeff.
18. Amido-desoxybenzoine $(C_{14}H_{11}(NH_2)O)$ crystallized from alcohol.
19. The same from its hydrochlorate.
20. Hydrochlorate of amido-desoxybenzoine. $(C_{14}H_{11}(NH_2)O.HCl)$.
21. Oxalate of amido-desoxybenzoine, $(C_{14}H_{11}(NH_2)O)_2C_2H_2O_4$. (*Ibid*. VI. 1253.)
Dr. P. Goloubeff.
22. Succinyl-dibenzoine, $(C_{14}H_{11}O_2)_2C_4H_4O_2$. (*Ibid*. V. 331.)
Mrs. Adelaida Lukanin.
23. Potassium-sulphocetenate, $(C_{16}H_{31}SO_3)K$.
24. Magnesium-sulphocetenate, $(C_{16}H_{31}SO_3)_2$ Mg.
25. Ferrous sulphocetenate, $(C_{16}H_{31}SO_3)_2Fe$.
26. Zinc-sulphocetenate, $(C_{16}H_{31}SO_3)_2$ Zn.
27. Cupric sulphocetenate, $(C_{16}H_{21}SO_3)_2Cu$.
28. Argentic sulphocetenate, $(C_{16}H_{31}SO_3)Ag$.
29. Uranic sulphocetenate. (*Ibid*. VII. 125.)
Dr. P. Lazorenco.
30. Mono-iodophenol $(C_6H_4I(HO))$, fusing at $64°$–$66°$.
31. Isomeric mono-iodophenol, fusing at $89°$.
32. Tri-iodophenol, $C_6H_2I_3(HO)$.
33. Isomeric fluid mono-iodophenol. (*Ibid*. VI. 1251.)
Dr. C. Lobanoff.
34. α Dinaphtol, $C_{20}H_{12}(HO)_2$. (*Ibid*. VII. 125, VI. 1252.)
35. Dibenzoyl-α-dinaphtol, $C_{20}H_{12}(C_7H_5O . O)_2$. (*Ibid*. VII. 487.)
36. β Dinaphtol, $(C_{20}H_{12}(HO)_2$. (*Ibid*. VI. 1252.)
37. Mono-benzoyl-β-dinaphtol. $(C_{20}H_{12}(C_7H_5O . O)(HO).)$ (*Ibid*. VII. 125.)
38. Dibenzoyl-β-dinaphtol, $C_{20}H_{12}(C_7H_5O . O)_2$. (*Ibid*.)
39. β Oxydinaphtylene. $(C_{20}H_{12}O.)$ (*Ibid*. VIII. 166.)
40. Picrate of β oxydinaphtylene. $(C_{20}H_{12}O . C_6H_2(NO_2)_3HO.)$ (*Ibid*.)
41. Dithymol, $C_{20}H_{24}(HO)_2H_2O$. (*Ibib*. VIII. 166.)
42. Dibenzoyl-dithymol, $C_{20}H_{24}(C_7H_5O . O)_2$. (*Ibid*.)
Mr. A. Dianin.

43. Dinitro-diphenol, $C_{12}H_6(NO_2)_2(HO)_2$. (Jahresbericht für Chemie, 1874, 466.) Mr. M. Goldstein.
44. Nitrobenzoylanilide, $(C_7H_5(NO_2))C_6H_5N$.
45. Benzoylnitranilide, $(C_7H_6)C_6H_4(NO_2)N$. (Jahresb. f. Ch. 1870, 759.)
Dr. P. Lazorenco.
46 Apparatus for a new and simplified method of quantitative estimation of urea, by means of sodium hypobromite. (Will be published.)
Prof. A. Borodin.

2536i. Collection of Chemicals (39).
The Imperial Moscow University, Moscow.

1. Ac. isobutyric.
   (The first isomeric acid discovered among the monobasic acids (synthetic).
2. Isobutyrate of calcium.
3. Isobutyr. of plumb.
4. Isobutyr. of argent.
5. Isobutyr. of ethyl.
5a. Chlorure of isobutyrile.
6. Ether ethyl pseudopropylic.

⎫ Annalen der Chemie und Pharmacie, B. 138, par V. Markovnikoff.

7. Ac. bromisobutyric.
8. Oxyisobutyrate of calcium.
9. Oxyisobutyrate of zinc.
10. Chlorure of butyrylechlory.
11. Oxybutyrate of zinc.
12. Bromhydrine of pseudopropyleneglycol.

⎫ Annalen der Chemie und Pharmacie, B. 153, par V. Markovnikoff.

13. Ac. oxyisocaprylic $\left(\begin{smallmatrix}CH_3\\CH_3\end{smallmatrix}> CH\right)^2 C(OH)CO_2H$. Journale de la Société Chimique Russe, v. II. (1870).

14. Ether oxyisocaprylic.

15. Pseudoheptylene $\begin{smallmatrix}CH_3\\CH_3\end{smallmatrix}> CH - CH = C < \begin{smallmatrix}CH_3\\CH_3\end{smallmatrix}$. Journ. de la Soc. Chim. Russe, III. (1871), par V. Markovnikoff.

16. Dichloracetane $CH_2ClCOCH_2Cl$ Journ. de la Soc. Ch. Russe, V., par V. Markovnikoff.

17. Ac. dibromisocaproic
   $\begin{smallmatrix}C_2H_5\\CH_3CHBr\end{smallmatrix}> CBrCO_2H$
18. Ac. isocaproic (diethylacetic)
   $\begin{smallmatrix}C_2H_5\\C_2H_5\end{smallmatrix}> CHCO_2H$

Journ. de la Soc. Chim. Russe, V.; Berichte der Deutsch. Chem. Geselsch., VI., par V. Markovnikoff et Drobjazgine.

19. Ac. oxypyrotartric normal (glutanic).
20. Glutanate of zinc.
21. Glutanate of magnesium.
22. Ac. pyrotartric normal (synthetic) $(CH_2)^2(CO_2H)^2$.
23. Ac. pyrotartric normal (by reduction).
24. Pyrotartrate of magnesium (normal).
25. Ac. ethylmalonic (α pyrotartric).
26. α Pyrotartrate of zinc.
27. α Pyrotartrate of ethyl.
28. Ac. dimethylmalonic (β pyrotartric) $\begin{smallmatrix}CH_3\\CH_3\end{smallmatrix}> C < \begin{smallmatrix}CO_2H\\CO_2H\end{smallmatrix}$
29. β Pyrotartrate of plumb.
30. Ether bromisobutyric.
31. Iodure of pseudoprapyle. Zeitsch für Chemie, 1865, Ann. d. Ch. u. Ph., B. 138, par V. Markovnikoff.

32. Iodure of acetylene $C_2H_2I_2$.
33. Bibromure of acetylene.   } Ann. der Ch. u. Ph., B. 178, par
34. Tetrabromure of acetylene.   A. Sabanejeff.
35. Acetylene triphenyltriamine.
36. Oxychlorure of antimony } Journ. de la Soc. Russe, par A. Sabanejeff.
         SbOCl
37. Isodibromantracene $C_6H_4C_2H_2C_6H_2Br_2$.

2536r. Chemical Reagents, &c. A series of 835 specimens of those employed in scientific research. Not specially
prepared for exhibition.   *Hopkins & Williams.*

2678. Digester, made of cast-steel, for digesting under
pressure. It holds 4 litres, is silvered inside and coppered outside.
Manufactured by Fr. Krupp, of Essen.
                         Prof. H. Landolt, Aix-la-Chapelle.

2588. Diagram, showing elevation, section, and plan of a
" digestorium " (iron closet constructed for effecting dangerous
digesting operations).
      *Chemical Institute of the University of Strassburg,
      Elsass.*

2584. Instrument for **Calculating Atomic and Volu-
metrical combinations,** to 4 places of logarithms.
                         Rudolph Weber, Aschaffenburg.

2456s. Bunsen's explosion Eudiometer, with the wire
cut off, and ground flush with the external surface of the tube in
order to prevent straining of the glass by the bending of the wires.
                         Prof. H. McLeod.

When an explosion has to be made the wires are connected to an induction
coil by the help of a wooden paper-clip with strips of platinum foil fastened
inside it, each strip being provided with a wire loop to which the conducting
wires from the coil may be attached.

24561. Regnault's Eudiometer.
         Golaz, 22 Rue des Fosses, St. Jacques, Paris.

This apparatus is made of a triangular iron stand, and is furnished with
a set screw for fixing to the upright a slide moved by rackwork ; a vessel of
mercury is fixed upon the slide, and can consequently be raised or lowered to
the height required. On the opposite side of the iron upright is fixed a plate
provided with an iron stop-cock and two tubes. In one of these tubes is
cemented a straight tube, open at its ends, and divided into millimetres
throughout its length. In the other tube is cemented a tube, terminating at
its upper extremity by a capillary glass tube, and bent horizontally ; this tube
is divided into millimetres, and is traversed at its upper part by two platinum
wires, through which the electric spark is made to pass. The whole of
this manometer apparatus is surrounded with a glass cylinder 15 centimetres
in diameter to be filled with water, in order to keep the manometer at a
uniform temperature. The second portion of the apparatus consists of the
experimental tube in which the gases are submitted to the various absorbents ;
this eudiometer is thus composed of two distinct parts:—1. The measuring

tube, in which the gases are measured under determined conditions of temperature and moisture. 2. The experimental tube. The upper part of the iron upright is furnished with a copper carrier, to which is fixed a steel stopcock, which is connected with another similar one by means of a collar. The stop-cock fixed to the carrier is cemented to the measuring tube which dips into the cylinder ; the second steel stop-cock is cemented to the experimental tube, and dips into the vessel of mercury. It can be seen by this arrangement that it is easy to connect or disconnect the experimental tube and the measuring tube, to make the gas pass from the experimental tube into the measuring tube, or to make the mercury run into either of the two tubes of the manometer apparatus, as is explained by M. Regnault, vol. 4, p. 73, " Cours de Chemie Élementaire."

2456m. Regnault's Apparatus for measuring the Coefficients of expansion of Gas.

Golaz, 22 Rue des Fosses, St. Jacques, Paris.

This apparatus consists of two distinct parts, a boiler and a manometer The boiler is a brass vessel in which either water or ice is placed, it is closed by a cover provided with a tube by which the vapour may escape, and is supported by a cast-iron stand on which it can be raised or lowered ; it is also capable of motion in a horizontal groove, which allows of its being advanced or withdrawn.

A large alcohol lamp, which is also capable of being raised or lowered, is used for heating the water,

The manometer consists of a mahogany board fixed to an iron stand supporting an iron framework furnished with two vertical tubes connected by a three-way cock ; by this means communication can be established between the two tubes or with the outside air. Thus the mercury can be made either to pass from one tube into the other, or to flow away from either into the outside air. Into one of the tubes is cemented a glass tube with millimetre graduations, and ending at its upper part in a capillary glass tube ; the second tube is graduated like the first, and terminates in an opening, which constitutes an open manometer. The capillary ending of the first tube is closed by a steel three-way cock.

A glass globe of about three-quarter litre capacity is fixed in the heater, and into it is introduced the gas to be investigated. The globe is attached to the stop-cock of the manometer by means of a copper capillary tube, the flexibility of which renders the adjustment of the apparatus easy. In the first apparatus of this description constructed the tube was of glass and cemented to the globe, but this method was found to render the formation of a proper joint very difficult.

The details of the experiments made with the apparatus are to be found in the "Mémoires de l'Académie des Sciences," tome 21, page 96, and plate 1, fig. 7."

For this second method the same boiler and globe are employed. The manometer alone is different. It is composed of two tubes, the first having a large diameter for the greater part of its length, and ending in a capillary tube attached to a steel stop-cock, the second tube is open at its upper end. This part of the apparatus is enclosed in a case of strong canvas, provided with a glass side, thus permitting the manometer to be kept at a constant temperature, and the progress the experiment at the same time to be watched.

The capillary tube of the globe in the boiler is connected with the steel stop-cock of this apparatus.

An air-pump or pneumatic apparatus is put into communication with the third tube of the two stop-cocks, and serves to dry the globe and the tube of the manometer, by taking care to turn the lower stop-cock in such a way as to render impossible the rise of the mercury in the tube of the manometer.

2456p. Regnault's Apparatus for determining the Tension of Steam at high pressure.

Golaz, 24 Rue des Fosses, St. Jacques, Paris.

This apparatus consists of four parts, 1, the boiler; 2, the reservoir of compressed air; 3, the manometer; 4, the force pump.

1. The boiler is a cylinder with hemispherical ends. The upper portion is traversed by two tubes fixed in it, and which are prolonged into the cylinder. In these tubes are placed the thermometers indicating the temperature of the steam.

The cylinder is connected with a horizontal tube surrounded by another cylinder which forms a refrigerator for the tubes, and is so placed as to be able to furnish a constant current of water. At the extremity of this tube is an arrangement for connecting the apparatus to the reservoir of compressed air.

2. The compressed air reservoir is of a cylindrical shape; at the lower part there is a junction for the attachment of a leaden tube which connects it with the boiler. At the upper part are two stop-cocks connecting it respectively with the manometer and the force pump. This pressure reservoir is surrounded by a cylinder filled with water to keep it at an equable temperature.

3. The manometer is fitted with a cast-iron reservoir, and a force pump with which the mercury is forced into the tubes; a stop-cock is closed when the mercury has risen to the height desired. A three-way cock allows the compressed air to escape from the reservoir.

The upper part of the pressure tube of the manometer is put in connection with the compressed air reservoir by a small flexible leaden tube provided with the necessary stop-cocks.

4. The compression pump is worked by a shaft provided with a fly-wheel and winch handle, and with it one man can readily obtain a pressure of 15 or 20 atmospheres. The valves are of leather, and are thus easily replaced in case of derangement.

For details of the apparatus and of the experiments made with it, see "Mémoire de l'Académie", t. 21 (vol. 21). This apparatus was made by the exhibitor for M. Regnault at the time of his first experiments.

2456q. Regnault's Apparatus for determining the Tension of Vapour from 0° to 100°.

Golaz, 24 Rue des Fosses, St. Jacques, Paris.

This apparatus consists of a sheet iron vessel furnished with an agitator, one side of it being formed of a pane of thick glass, so that all that is going on inside may be seen. A cylinder is supported on an iron frame, upon which are placed a mercury vessel, a screen, and an alcohol lamp; two barometrical tubes enter the vessel, one being a standard barometer, and the other terminating at its upper end in a horizontal tube, whilst to the other end a glass globe can be readily attached by means of a suitable mounting. In the middle of the horizontal tube is placed a stop-cock by means of which the globe and the barometrical tube can be emptied. In the globe is placed the substance, the vapour tension of which is to be investigated. See "Mém. de l'Acad." t. 21, vol. 21.

2456r. Apparatus for compression of Liquids by Regnault's Piezometer.

Golaz, 24 Rue des Fosses, St. Jacques, Paris.

This apparatus, constructed by the exhibitor for M. Regnault, consists of a thick copper cylinder closed at its upper part by a copper plate ⸂bolted to the circumference and provided with three tubes. To the central tube is cemented the glass stem of the piezometer. The right hand tube has a stop-cock which opens into the atmosphere and serves for relieving the pressure ; the left-hand tube has also a stop-cock which is attached by a leaden tube to a block fitted with two other stop-cocks at right angles to it, and which is cemented at the other end to the stem of the piezometer ; the other branch is connected with a compression pump or a pressure reservoir. The cylinder is first filled with water by means of the two upper stop-cocks, and pressure can then be applied under all the conditions indicated by M. Regnault. "Mém. de l'Acad." t. 21.

III. APPARATUS USED IN ANALYSIS.

2456. Apparatus used in the Analysis of Drinking Water. *E. Frankland, F.R.S.*

For full particulars, see Journ. of Chem. Soc. for March 1868 and June 1876.

1. Determination of total solid constituents or total solid impurity.

Platinum capsule in which half a litre of the water is evaporāted to dry ness, and the residue weighed.

2. Determination of organic carbon and organic nitrogen.

A. Apparatus in which one litre of the water is first heated to boiling with sulphurous acid, and then evaporated to dryness.

B. Steel spatula for scraping out the dry residue from the glass capsule.

C. Cupric oxide for the combustion of the organic matter in the dry residue.

D. Piece of hard glass tube ready to receive the mixture of dry residue, cupric oxide, and metallic copper for combustion.

E. Combustion tube charged and drawn out ready to be placed in furnace.

F. Sprengel mercurial pump for extracting air from combustion tube before the ignition of the mixture of dry residue and cupric oxide, and for extracting the gases from the same tube after ignition.

G. Hofmann's gas furnace, containing a pair of charged tubes ready for combustion.

H. Pair of inverted glass tubes for the collection of the gases produced by the combustion of the organic matter and ammonia in the solid residue.

I. Apparatus for the analysis of the gases collected in the tubes H.

3. Determination of nitrogen in the form of nitrates and nitrites.

This is effected by dissolving, in a small quantity of distilled water, the solid constituents obtained in determination No. 1, and adding a slight excess of argentic sulphate to convert the chlorides into sulphates. The filtered liquid is then concentrated to a small bulk, mixed with rather more than its own volume of concentrated sulphuric acid, and then introduced over mercury into the tube K, through the cup and stop-cock. The lower end of the tube is now firmly closed by the thumb, and the contents violently agitated by a simultaneous vertical and lateral movement, in such a manner as always to leave an unbroken column of mercury, at least an inch long, between the acid liquid and the thumb. In about a minute from the commencement of the agitation a strong pressure begins to be felt against the thumb of the operator, and mercury spurts out in minute streams as nitric oxide gas is evolved. In from three to five minutes the re-action is completed, and the nitric oxide evolved must then be transferred for measurement to the apparatus No. 2 I. The volume of the nitric oxide so measured is exactly equal to that of the nitrogen as nitrates and nitrites contained in one litre of the water.

2455. Apparatus for the Analysis of Gases.
E. Frankland, F.R.S.

Several of the mechanical arrangements of this apparatus were first employed in an instrument invented by Messrs. Regnault and Reiset, but the principles involved in the determination of the gaseous volumes are different in the two instruments.

The apparatus, which is represented by the accompanying figure, consists of the tripod A furnished with the usual levelling screws, and carrying the vertical pillar B, B, to which is attached, on the one side, the movable mercury trough C, of gutta percha, with its rack and pinion a, a, and on the other, the glass cylinder D, D, with its contents. This cylinder is 36 inches long and 3 inches in diameter, its lower extremity is cemented into an iron collar, c, the under surface of which can be screwed perfectly water-tight upon the bracket-plate, d, by the interposition of a vulcanised caoutchouc ring. The circular iron plate d is perforated with two large apertures for the passage of the tubes E and F, besides a smaller one into which the brass cock b is screwed. E is a glass tube of uniform bore, and about 18 mm. internal diameter, and marked with 10 divisions equidistant from each other. It serves to contain the gas during measurement. Its upper extremity terminates in the capillary tube e, which carries a glass stop-cock, and is cemented into the steel cap f in such a way as to exclude any air space. For the passage of the electric spark, two platinum wires are fused into E at g At its lower extremity the tube E, after passing water-tight through a caoutchouc cork in the plate d, is connected with the junction piece h.

F is a second glass tube only 5 mm. internal diameter, graduated with a millimetre scale throughout its entire length, and reaching to a height of about 8 inches above the glass cylinder. It is furnished at top with a small funnel, i, into the neck of which a glass stopper, about 2 mm. in diameter is carefully ground. In this tube the pressure of the gas in E is measured. Like E, it also passes water-tight through the plate d, and is connected with the junction piece h, which is furnished with a stop-cock, and is continued to the floor, where it is connected by a long caoutchouc tube with the glass mercury reservoir G. The caoutchouc tube is 3 mm. internal diameter, and is strengthened by a jacket of tape to enable it to withstand the pressure of mercury. The reservoir G can be raised or lowered by the cord and pulley k, l. H is an absorption tube, in which the gases are brought into contact with various liquid or solid re-agents. It is supported on the clamp o, and connected with the capillary tube e, by the cap and junction clamp p.

The tubes E, F, are held firmly in position by the clamp m, which is screwed to the under side of the plate d.

Before using the apparatus, the cylinder D, D is filled with water through the cock b, and a slow stream is afterwards established through the syphon n, one leg of which encloses the upper part of the tube F. The internal walls of the tubes E and F are, once for all, moistened with distilled water by the introduction of a few drops into each, through the stop-cock on e and the stoppered funnel i. The reservoir, previously filled with mercury, is now placed in communication with the tubes E and F, and is raised until the mercury flows from the cock on e, which is then shut, and rises into the cup i, the stopper of which is then firmly closed. The absorption tube I; being now filled with quicksilver, and attached to e by the screw clamp p, the instrument is ready for use.

With a proper supply of water the temperature of the gas remains constant during the entire analysis, and therefore no correction on that score has to be made. The atmospheric pressure being altogether excluded from exercising any influence upon the volumes or pressures, no barometrical observations are requisite; and as the tension of aqueous vapour in E is exactly balanced by that in F, the instrument is in this respect self correcting.

The readings of pressures in this instrument give, without any calculation or correction, the true volumes of the gases, and as the manipulations are very simple, the analysis of even a complex gaseous mixture is made very rapidly. Thus the determination of the separate constituents of a mixture of carbonic anhydride, oxygen, hydrogen, carbonic oxide, marsh gas, and nitrogen can be made in about two hours.

2456a. Apparatus for Analysis of Gases. Modification of that described by Frankland and Ward. For full description see Journ. of Chem. Soc. 1869 [2], vii., 313. *Prof. H. McLeod.*

2456n. Doyèré's Apparatus for Gas Analysis.
Golaz, 24 Rue des Fosses, St. Jacques, Paris.

This apparatus consists of a vessel for water and mercury, a telescope, a sheet of plate glass with levelling screws, a transfusing pipe, a detonating eudiometer, and a mercury cistern which is used for the transfusing of the gas.

The vessel for water and mercury is formed of four plates of glass set in a brass frame and resting on a cast-iron slab. Through this latter is an aperture leading to a cistern which projects beyond the vessel, and rises five centimeters above the slab. The latter rests on four feet and stands on a plate levelled by screws. The mercury is poured in by the exterior aperture of the cistern until it has risen one centimeter from the bottom of the vessel; the vessel is then filled with water from above, up to ¾ of its height. The mercury bearing the mass of water rises about 1½ centimeters through the exterior aperture of the cistern, and the levels of the two liquids are established permanently. The graduated tube is put in its place, by making its lower aperture rest in a small iron spoon filled with mercury. By this arrangement the spoon and the tube can traverse the whole mass of water and reach the mercury in the basin, without any of the water penetrating into the interior of the tube. The tube is held by brass pincers, which allow of the reading being made along the whole length of the tube, and consequently of the measurement of even the very smallest volume of gas. The pincers are supported by a rod clamped to a gallows which can be moved vertically by a rack and pinion outside of the vessel. The tube can thus be adjusted to any required height.

The stand of the telescope is formed of two tubes sliding one within the other, with a rack and pinion; at the top of this sliding tube is fixed horizontally a telescope with a micrometer eyepiece. This micrometer allows of the fractions of divisions being measured with great precision. The stand of the telescope is fixed to a tripod furnished with levelling screws. In order to increase the

R r

precision of the readings of the instrument and to be able in various cases to advance or draw back the eye-glass in the same optical axis, it rests upon a glass plate mounted in a narrow-brimmed basin provided with wedging screws. The regulator is formed of a glass cylinder, from whose lower extremity issues a tube of small diameter which is bent upwards and rises by the side of the cylinder, is bent at the top, descends paralled to itself, bends once more, and finally ends in a glass reservoir, open at the top.

This tube is fixed to a strip of plate glass on which is engraved an arbitrary scale to serve as a datum for the air-bubble in the tube. The apparatus is clamped to the brass frame at the top of the water vessel, in which it hangs vertically. By its means all calculations of the corrections for temperature, barometric pressure, and the tension of the aqueous vapour, are avoided.

The detonator is a mercurial eudiometer constructed like those usually employed in chemistry.

The transfuser or gas pipette is made of two glass bulbs connected together by a bent tube ; one of these bulbs is surmounted by a tube which is bent at an angle of about 35°. It is by this tube that the aspiration or compression is effected by means of the mouth. The other "bulb" is surmounted by a capillary tube, which rises vertically, is bent horizontally, then descends vertically, is bent once more, and rises again parallel to the descending tube. Its extremity is drawn out to a point. It is with this branch that the gases are exhausted from or forced into the tube in which the measurements are taken in the water-vessel. The whole instrument is mounted on a wooden stand. The cistern for the manipulation is like that on which is placed the water-vessel, but it differs inasmuch as the rim of the surface is lower.

For complete details of experiments see " Les Annales de Physique et de Chimie," vol. XXVIII., third series, 1850.

2456o. Doyéré's Pipette for Gas Analysis Apparatus.
Golaz, 24 Rue des Fosses, St. Jacques, Paris.

2456t. Automatic Eudiometer with alarum.
D. Monnier, Geneva.

2456b. Two Bischof's Evaporating Apparatus to use for Frankland's apparatus. One graduated nitric acid tube. Enamel back. *E. Cetti & Co.*

2455a. Frankland's Water Analysis Apparatus.
E. Cetti & Co.

2456b. McLeod's modification of **Frankland's Apparatus** for the **Analysis** of **Gases.** *E. Cetti & Co.*

2456i. Ure's Eudiometer. *E. Cetti & Co.*

2456k. Mitscherlich's Eudiometer. *E. Cetti & Co.*

2534. Apparatus for boiling off the air from sea-water.
Prof. Oscar Jacobsen, Rostock.

2573. Bunsen's Cathetometer. *Julius Schober, Berlin.*

2660. Apparatus for the **determination** of **Water, Carbonic Acid,** &c., principally for elementary organic analysis, made by Dr. Geissler in Bonn.

Dr. Drevermann, Hoerde, Westphalia.

2556. Geissler's Apparatus for determining carbonic acid.

Ch. F. Geissler & Son, Berlin.

2580. Collection of small **Glass Apparatus** for the **Estimation** of **Carbonic Acid.** *Dr. H. Geissler, Bonn.*

2402. Plattner's Diamond Mortar, in steel.

No. 1, with brass capsule.
No. 2, without „
No. 3, „ „ *A. Herbst, Berlin.*

2403. Freyberg's Diamond Mortar, in steel.

No. 1, with brass capsule.
No. 2, without „
No. 3, „ „ *A. Herbst, Berlin.*

2574. Habermann's Air-bath. *Julius Schober, Berlin.*

2574a. Water and Air Bath, with regulator.

P. Waage, Professor of Chemistry at the University of Christiania, Norway.

(NOTE.—For reference letters see photograph.)

(A) is a water-bath heated by two Bunsen lamps, a compartment (*b*) in which serves as an air-bath heated by a copper tube (*c*) coiled in the water and opening into the air-bath.

A tube (*d*) keeps the air-bath in communication with the outer air.

The regulator (B) consists of a **U** tube (*e*) half filled with mercury, and the float (*f*) which rising or falling with the surface of the mercury, shuts or opens the glass cock (*h*) by means of the lever (*g*). This movement can be adjusted by the regulating screw (*i*).

(*l*) is the drying tube filled, for instance, with chloro-calcium.

(*p*) is connected with an apparatus for exhaustion, for instance, a Bunsen's pump.

If the cock *o* and *n* are shut, whilst the cock *m* is open, and the air is exhausted at *p*, the mercury rises in the long arm of the glass-tube, whilst it sinks in its short arm. In this way the floating piece sinks and the cock *h* is opened. Through *i*, the regulating screw, the floating piece may be placed higher and lower in such a way that the cock *h* admits the air into the apparatus, in a greater or less state of rarefaction.

If the cock *m* is shut, whilst the cocks *n* and *o* are open, and when these cocks have been connected with *c* and *d*, through the india-rubber tubes *r* and *s*, whilst the other end of the tube *l* ending in the air-bath, is connected with the interior opening of the copper-tube *d*,—then the air that enters through the drying tube by the cock at a given rarefaction, will first go into the copper coil at *c*, and then, having been heated to 100° C., will enter the air-bath to be from thence exhausted by the pump through the cock *o*.

The air-bath can thus be filled with warm dry air at any pressure that may be desired.

Besides, with the object just named, the regulator may also be used under filtration with Bunsen's pump. In this case one of the india-rubber tubes, for instance *s*, is brought into connection with the filtering apparatus, the

R r 2

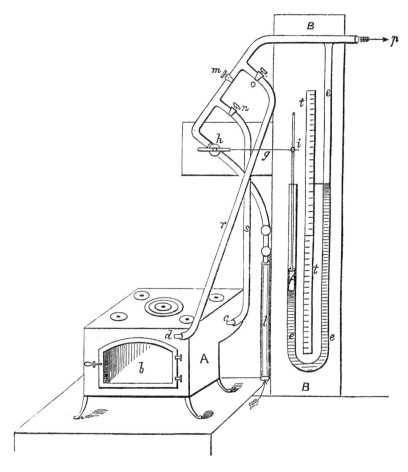

cock *o* is opened, but *n* and *m* are shut, and the floating piece is, by the help of the millimeter-scale *t*, placed so as to give the desired rarefaction.

The regulator may also be employed with a distilling apparatus in connection with the rarefied air. In this case, one india-rubber tube is put in connection with one end of the air-tight apparatus of distillation, and the other india-rubber tube is put in connection with its other end.

2578. Collection of Burettes and Burette Stands.
Dr. H. Geissler, Bonn.

2579. Collection of Pipettes, with stand.
Dr. H. Geissler, Bonn.

2585. Drying Closet, provided with mica-plates, for watching the drying of precipitates obtained in quantitative chemical analysis (*see* Mineralogy). *Max. Raphael, Breslau.*

2586. Mitscherlich's Apparatus for organic analysis by means of mercuric oxide. *Prof. A. Mitscherlich, Münden.*

The apparatus is described in Mitscherlich's paper entitled "Elementar analyse vermittelst Quecksilber oxyd," published by E. S. Mittler and Son, Berlin, copies of which accompany the apparatus.

2587. Mitscherlich's Metal Drying Closet, with accessories. *Prof. A. Mitscherlich, Münden.*

A description of this apparatus will be found in Mitscherlich's papers, entitled "Chemische Abhandlungen," published by E. S. Mittler and Son, Berlin, 1859–65.

2530. Phosphorus Eudiometer.
University of Munich, Prof. von Jolly.

2443. Apparatus for Volumetric Analysis.
James How & Co.

2563. Hofmann's Filtering Stand, the metal rings of which are lined with wooden rings slanting at an angle of 60°.
Julius Schober, Berlin.

2567. Hofmann's Combustion Furnace.
Julius Schober, Berlin.

2568. Bunsen's Combustion Furnace.
Julius Schober, Berlin.

2569. Finkener's Combustion Furnace.
Julius Schober, Berlin.

2456d. Set of **Russell** and **West's Apparatus** for estimating the urea in urine. *E. Cetti & Co.*

2456e. Six **Burettes** and **Scale Pipettes,** with registered enamel back to show up the divisions. *E. Cetti & Co.*

2456g. Nessler's Tubes for **Colour Test.** *E. Cetti & Co.*

2592. Self-filling Burette, used for oxidizable liquids, together with filling bottle, wash-bottle, stand, and india-rubber tubes.
Berggewerkschaftskasse, Bochum, Rhenish Prussia.

The working of this burette is described in Fresenius' Zeitschr. f. anal. Chem. XIV., 3 and 4.

IV.—APPARATUS IN GENERAL USE IN LABORATORIES.

2525. Apparatus for regulating the Pressure of Gas.

Prof. A. Crum Brown.

The apparatus consists of a gasometer, the pressure of the gas in which depends upon the weight of the inverted vessel and that of the counterpoises. The peculiarity of the arrangement lies in the mode in which the gas is introduced into the gasometer. The gas is led from the main or meter to a fixed vertical tube, and thence by a short piece of india-rubber tube to a metal tube fixed to one of the counterpoising weights, and from that tube to the gasometer. When the pressure in the main increases the vessel rises, and the counterpoise falls; the short india-rubber tube is thus bent, and the entrance of gas into the gasometer hindered. This short india rubber tube is thus a self-acting valve, always admitting exactly the same quantity of gas as is drawn off from the gasometer, and as the motion of the gasometer is very small, the pressure is practically constant.

2524a. Oxygen Generator.   *John Craig.*

The gas (generated from chlorate of potash and manganese, in equal quantities) passes from the retort C to the bottom of the gas holder B (which is filled with water) by the tube F. The water is displaced through the tube H into the reservoir A until the pressure has been reached for which the lever Q has been loaded. When that point has been reached, the pressure acting on the bellows K raises the lever Q, and so diminishing the source of heat; the other end of the lever regulating the air to the amount of gas admitted to the Bunsen burner. (By changing the fulcrum of the lever to U the sliding tube is reversed in its action, and adapted for diminishing the flame of a spirit or other lamp where gas is not available.) When the pressure is again reduced below the required limit, the lever, falls, opens the stop-cock O, and more gas is generated. The object of the second retort is that the one may be re-charged while the other is in use, thereby securing continuous action. A convenient working pressure is found with 3 inches of water in the reservoir. The regulator J controls the pressure of gas through the exit pipe I, fulfilling the same office as the regulator in the hydrogen generator. By this means both the production and working pressure of the gas is regulated.

The *advantages* of the combined apparatus are, 1st, a continuous or intermittent supply of oxygen and hydrogen gases may be obtained in any quantity and at any pressure, as the rate of production depends solely on the rate at which the gases are consumed; 2nd, being self-acting, little superintendence is required; 3rd, great economy (as compared with the use of bags), all the gas made being available; 4th, danger from explosion is reduced to a minimum (*a.*) from the small quantity of gas in stock, and (*b.*) undue tampering with the levers would only result in bursting the cloth bellows of the regulator.

At G there is a small valve, which remains shut until the heat is withdrawn from the retort; when, to prevent the water rising to fill the partial vacuum from condensation in the retort, communication is at once established by the valve rising and allowing gas to restore the equilibrium.

The apparatus is suitable for lighthouses, signal lights, ships' lights, &c.

The three copper tubes are adopted to the hydrogen generator, the three brass tubes to oxygen.

2524b. Hydrogen Generator.   *John Craig*

To generate the gas pieces of zinc are placed in the lower vessel at E. Dilute acid, added to the upper vessel A, comes in contact with the zinc

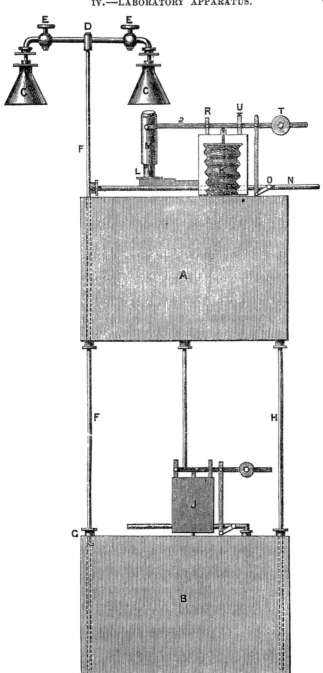

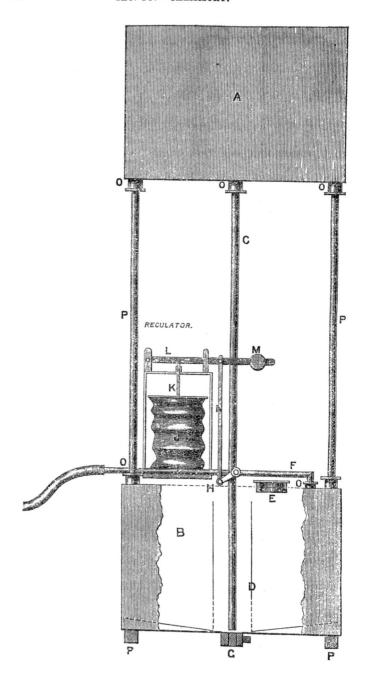

through the pipe C, when hydrogen is at once produced. If no gas is taken away the acid liquid is all forced back into the upper vessel, when the production ceases. When gas is taken away at F the lessened resistance allows as much acid to return to the lower vessel as will supply the demand. The fountain head pressure is determined by the distance between the upper and lower vessels, combined with the quantity of liquid in the upper chamber. The burning pressure is determined by the weight M on the lever L, which when once fixed remains constant and is independent of changes produced by the rising and falling of the liquid in the pipe C.

D is a perforated fence to prevent the zinc closing the bottom of tube C F the tube through which the gas passes to the burner. G the stop-cock to withdraw spent acid. H the stop-cock crank, at an angle of 45° when open. I where the gas enters the bellows of the regulator. J bellows. K stud. L lever. M weight. N connecting rod. O screw couplings. P supports.

2572. Collection of Gas Lamps of novel Construction.

(1.) Gas lamp provided with one regulator for air and gas.
(2.) Gas lamp with separate regulator for air.
(3.) Gas lamp with star support and chimney.
(4.) Bunsen's gas lamp, single burner.
(5.) Bunsen's gas lamp, with regulation for air supply.
(6.) Bunsen's gas lamp, with star support and chimney.
(7.) Bunsen's gas lamp, consisting of 3 burners.
(8.) Bunsen's gas lamp, with regulation for air supply.
(9.) Bunsen's gas lamp, consisting of 6 burners.
(10.) Bunsen's gas lamp, provided with one regulator for both gas and air.
(11.) Hofmann's gas lamp.
(12.) Hofmann's gas lamp, provided with an air regulator, star support, and chimney.
(13.) Hofmann's gas lamp, with a bent burner tube.
(14.) Hofmann's gas lamp, with a bent burner tube, and provided with an air regulator.
(15.) Griffin's rose burner.
(16.) Griffin's small gas lamp.
(17.) Griffin's large gas-lamp.
(18.) Microchemical lamp.
(19.) Iserlohn gas lamp, small size.
(20.) Iserlohn gas lamp, middle size.
(21.) Iserlohn gas lamp, larger size.
(22.) A fusion gas lamp, consisting of 11 single burners, and fixed on a stand.
(23.) A gas lamp, consisting of 3 burners, each possessing a separate regulator for a simultaneous supply of air and gas.
(24.) A gas lamp, containing 4 burner tubes.

Julius Schober, Berlin.

2446. One-inch Solid Flame Bunsen Burner and two Perforated Caps, the fine one for ordinary use, and the coarse

one to exhibit the green beads above each perforation more distinctly.   *John Wallace, Messrs. Tangye Bros., and Rake.*

2447. Two-inch Solid Flame Bunsen Burner, with a piece of sheet platinum suspended across the top of the cap in order to show by its incandescence that the flame is solid to the centre. These burners will burn safely a mixture of $4\frac{6}{10}$ volume of air per volume of gas.

John Wallace, Messrs. Tangye Bros., and Rake.

By the addition of a perforated metallic cap, adjustable to various heights on the burner tube, the ordinary Bunsen burner may be increased to $2''$ or more in diameter, while at the same time the combustion is greatly improved and the disposition to light within prevented. In the ordinary half-inch burner one and a quarter volumes is about the maximum quantity of air which may safely be mixed with the gas previous to combustion without running the risk of lighting within. In the new burner $4\cdot6$ volumes of air may be safely mixed previous to combustion.

By raising or lowering the cap the proportion of air drawn in at the base of the burner may be varied at pleasure, and when the maximum amount of air is used a flame of marvellous beauty is obtained, which differs in many respects from the ordinary atmospheric flame. The hollow conical space within the flame entirely disappears, leaving a perfectly solid flame, while above each perforation of the cap is seen an intensely brilliant green bead, indicating the beginning of combustion. These beads are hollow, and are filled with the mixture of air and gas passing through the cap. Their colour is entirely on the surface, and they fade and merge into each other on a slight interruption of the air supply below. They cannot be regarded as separate flames, as they have no separate air supply, and their colour is the same whether the cap be of copper or iron.

As the pre-admixture of air increases the flame becomes smaller and more intense; the heat is produced in less space, and is thus better adapted for boiling liquids or heating metals. The absence of the hollow cone points to some interesting peculiarities in the flame. In a flame $3''$ diameter from a $2''$ burner the green beads (those most sensitive indices of any change in the physical condition of the flame) are of the same size and colour from the outer edge to the middle of the cap. They are all filled with the same mixture of gas and air, and their uniformity can only be accounted for by supposing that each one gets its complement of air with equal facility.

About $6\frac{1}{3}$ volumes of air are necessary to consume London gas, and one half of this amount will produce a carbonic oxide flame. But when $4\frac{6}{10}$ volumes are mixed previous to combusion there is more than the amount required to produce a carbonic oxide flame ; this fact, together with its uniform appearance, offers grounds for supposing that the flame is a carbonic acid one, and that the oxygen which combines during combustion reaches every part of it with equal facility. The penetrating power of the oxygen is here so remarkable that the question naturally arises, " does it drag its associated nitrogen with " it into the heart of the flame, parting only at the time of combination, or, " does dissociation take place before the oxygen combines with the carbon " and hydrogen of the gas ? "

2447a. Wallace Bunsen Burner, Tangent pattern.

John Wallace, Messrs. Tangye Bros., & Rake.

This apparatus consists of a hemispherical chamber or vessel into which a jet of coal gas is discharged at a tangent carrying with it a certain proportion of air which by eddying round within the chamber is ultimately mixed with the gas. The mixture escapes by a short tube in the middle of the chamber, and burns with an amber flame containing a brilliant green cone within it. This cone is a hollow space in the flame in which the head of a lucifer match, if dexterously inserted, will not ignite. The flame above will readily melt gold, silver or copper, indicating a temperature equal to about 3000° Fahr.

2740. Chemical Apparatus, Photometrical and **Gasometrical Apparatus, Lamps, &c.,** invented by Professor Bunsen, of Heidelberg, and described by him in his " Gasometrical Methods," and other publications.   *C. Desaga, Heidelberg.*

Spirometer.
Goniometer.
Galvanic battery, consisting of four elements.
Table blowpipe lamp, with 2 jets.
A 6-Bunsen burner blowpipe table lamp.
An 18-Bunsen burner blowpipe table lamp.
Oxy-hydrogen blowpipe.
Iron tripods, four different sizes.
Small tripods for fixing on lamps.
Small tripods for spreading out the flame.
Electrometer.
Bunsen's water vacuum pump.
Air-pump plate.
Spark-producer.
24 compression clamps of different sizes.
Potash apparatus.
4 supports for flasks, &c.
Apparatus for cooling (condenser).
Bunsen gas burners, 18 different sorts.
Babo's lamps, five different kinds.
Breitenlohner's lamp.
Lamps with double and treble air draught.
A 7-Bunsen burner lamp.
Sevenfold lamp.
Gas blowpipe, provided with stopcock.
Gas blowpipe, mounted.
Furnace, containing 25 burners.
Circular gas furnace, three different forms.
Absorptiometer.
Diffusion apparatus.
Dividing machine for glass gas apparatus.
Cathetometer.
Oxy-hydrogen generating apparatus.
Mercury trough.

System of convergent lines, etched on glass, for making divisions.

Hydrogen generating apparatus.

Apparatus for measuring the rapidity with which gases issue through a narrow aperture.

Spectroscope.

Spectroscope, provided with micrometer movement for the two telescopic tubes, an eccentric movement for the adjustment of the observation telescope, comparison prisms, cross-thread, &c.

Pocket spectroscope.

Universal clamp.

Stand for holding flasks, &c.

Stand for holding Geissler's tubes.

Stand for holding carbon points.

2 stands provided with adjusting screws.

2 stands provided with three legs.

3 holders for burettes.

6 rings of different sizes.

6 filtering rings of different sizes.

2 forks.

7 clamps.

Photometer.

Gas-clock.

Test-tube holder.

2 water baths, 2 sizes.

Apparatus for regulating the level of water.

Bunsen's gas regulator.

Drying apparatus.

Apparatus for showing the reversal of the sodium spectrum.

Several cases containing chemical substances for practising spectroscopic analysis.

Chimneys for the various lamps, 15 different sizes.

2574b. Doberawer's or Philosopher's Lamp.

Parkinson & Frodsham.

This lamp when charged with dilute sulphuric acid is self-igniting; it suffices to raise the arms of the small figure, and the spongy platinum then brought into view, becomes incandescent. The explanation of this is that the hydrogen generated in the body of the lamp, which contains zinc, is drawn into such close contact with the oxygen of the air, by reason of the extreme fineness of the spaces between the particles of the platinum, that the two gases combine chemically, or in other words burn, whereby sufficient heat is generated to raise the temperature of the platinum to a red heat, sufficient to ignite paper, &c.

2572a. Heating Apparatus for Laboratories.

M. Wiesnegg, Paris.

1. (?) Arched Bunsen burner.
2. Plain Bunsen burner, with support.

3. Bunsen burner of 8 millimetres.
4. Three tye-pieces for gas burners.
5. Treble Bunsen burner.
6. Wiesnegg burner, with stand.
7. Berzélius burner, for gas and air.
8. Stove "à couronne," No. 1.
9. Do.   do.   No. 2.
12. Laboratory blow-pipe, for petroleum.
13. Bellows, with handle and treddle.
14. Leclerc oven.
15. Cloës universal stand, with hook.
16. A hand blow-pipe.
17. Complete Schlœsing blow-pipe.
18. Two Schlœsing ovens, No. 1.
19. One   do.   No. 2.
20. Tube oven, and model of oven for melting platinum.
21. Perrot apparatus, No. 1.
22. Clothing framework of Perrot apparatus.
23. Perrot apparatus, No. 2.
24. Perrot coppel, No. 5, with treble clothing.
25. Oxygen retort.
26. Complete oxygen bladder.
27. Two Perrot cocks and light strikers combined.
28. Cloës " étuve."
29. Ranvier " étuve."
30. Wiesnegg " étuve."
31. Fremy " étuve."
32. Two gridirons for analysis (18 and 14 burners).
33. Stove "à couronne," No. 1, for petroleum.
34. Stove "à couronne," No. 2, for petroleum.
35. Gas stove for incineration, with two jackets.
36. ? Candelabra for lecture theatre.
37. Complete still.
38. Perrot burner, No. (?), for petroleum.
39. 150 burners of various sorts, copper.
40. Pressure and gas caoutchouc.
41. Twelve Wiesnegg briquettes, No. 2, and 8 quarter briquettes.
42. Twenty Wiesnegg briquettes, No. 1 (?), and 8 quarter briquettes.
43. Eight Perrot plots.
44. Twelve manometer glasses.
45. A small ingot mould and a large one.
46. Two charcoal and crucible tongs.

2524. Laboratory Forge, with double draught, fitted with a cerfeu for the fusion of metals.

With this forge the pressure can be increased by reducing the quantity of air when soldering is required, either with gas, or with Sainte Claire Deville's forge lamp, or with Schlösing's blow-pipe.

2560. Apparatus for producing a **Blast of Air** by the fall of water. (Catalonian "Trompa".) *Dr. Otto Bach, Leipzig.*

Founded on the principle of the hydraulic ram; with the sole difference that the caoutchouc bag (or conduit) takes the place of the valve. It is especially fitted for the use of glassblowers on account of its quiet and perfectly uniform blast of air, and by the introduction of Bunsen valves (ventiles) it can be adapted for all the purposes of a water air-pump, as it then corresponds to the "Pulsirpumpe" designed by Faguo in Moscow, and described in the report of the German Chemical Society, 5.328.

2598. Blowing Apparatus for working a blow-pipe.
C. Osterland, Freiberg, Saxony.

This blowing apparatus has two wind-chambers, whereby a perfectly regular stream of air may be obtained, even without any previous practice. It has a movable jet, constructed by the exhibitor, with an air-tight ball-and-socket movement. It may be screwed fast to a table, and will continue blowing for nearly a minute before the air-chambers get exhausted. It can also be worked by the foot, so as to leave both hands free. The stream of air can be regulated by a tap, in order to suit every requirement.

2570. Hofmann's Table Blowpipe Lamp.
Julius Schober, Berlin.

2571. Bunsen's Table Blowpipe Lamp.
Julius Schober, Berlin.

2571a. Apparatus for developing the new system of **Blowpipe Chemistry,** called "pyrology," a "spectrum lorgnette," pyrological candles, and other novelties; together with some of the results, as the sublimate of gold on aluminium plate, &c.
Major W. A. Ross.

2561. Small Copper Steam Generator.
Julius Schober, Berlin.

2596. High-pressure Steam-boiler, together with a distilling and condensing apparatus, as well as a drying closet, capable of being heated by steam.   *E. A. Lentz, Berlin.*

The apparatus is heated by 10 Bunsen burners; when steam is once got up only half the number is required to keep it up; the other half may then be turned off.

The three openings in the apparatus for evaporations are separated from the steam chamber by copper casings, and the steam is let into the latter when required for use by means of small valves. It is thus possible to maintain a certain pressure of steam in the boiler, which is of especial importance when steam has to be used for distillations and for conducting to a distance, as, for instance, in the heating of distant drying-closets.

2596a. Apparatus for Drying in Vacuo.
Prof. Markovnikoff, Professor of Chemistry at the University of Moscow.

2596b. Laboratory Furnace; magnesia.
Prof. Schloesing, Conservatoire des Arts et Métiers, Paris.

2596c. Laboratory Furnace; brick.
Prof. Schloesing, Conservatoire des Arts et Métiers, Paris.

2596d. Distilling Apparatus, for laboratory.
Prof. Schloesing, Conservatoire des Arts et Métiers, Paris.

2456b. Apparatus for simultaneous production of hydrogen, carbonic acid, and sulphuretted hydrogen.

P. Waage, Professor of Chemistry and Director of the Chemical Laboratory at the Royal University of Christiania, Norway.

The apparatus consists of a reservoir for hydrochloric acid of 10–15 per cent. strength. At the bottom of this reservoir are three apertures through which it is connected in the usual manner by glass tubes and stopcocks with three cylindrical jars, one filled with zinc, the second with marble, and the third with sulphide of iron. Beneath each of these, and communicating with the sewers, is another reservoir into which the solutions of chlorides are gathered. The three gases generated are conducted through wash bottles furnished with stopcocks of glass ; when these are turned off the several gases and the solutions of chlorides will pass through the lower reservoirs into the sewers.

The apparatus once fitted up is always ready for use, and all superfluous gas is completely led away without producing any smell in the room where the apparatus is fixed.

2523. Stand used in chemistry for the fusion of metals by means of a gas apparatus.

2593. Stand provided with pincers and air-bath.

Ball, Freiburg, Baden.

2595. Heat Regulator.　　*C. Kramer, Freiburg, Baden.*

2550. Frerich's filtering Apparatus, made of platinum.

F. Sartorius, Göttingen.

2551. Frerich's filtering Apparatus, made of gutta percha.　　*F. Sartorius, Göttingen.*

2552. Frerich's filtering Apparatus, made of porcelain.

F. Sartorius, Göttingen.

2553. Frerich's Apparatus, for determining specific gravities in scientific investigations.　　*F. Sartorius, Göttingen.*

2456c. Water Quick Filter Pump for use with a fall of 40 feet.　　*E. Cetti & Co.*

2576. Water Vacuum Pump, constructed for a fall of water of at least 20 feet.　　*F. A. Wolff & Sons, Heilbronn and Vienna.*

This pump is of simple construction and great solidity. Any desired vacuum may be obtained by making use of a regulator.

2663. " Water Jet " Pneumatic Pump, for laboratories, &c. ; designed by Professors Arzberger and Zulkowsky, and constructed by P. Böhme, Mechanician, Brünn.

Prof. Fred. Arzberger and Chas. Zulkowsky, of the Polytechnic Institute, Brünn, Austria.

The effect of this pump is based on the injection principle ; it requires for working it a water supply conveyed by high pressure, but no fall below ;

hence the difficulties are avoided which frequently impeded the employment of earlier designed pneumatic water pumps. The rarefaction of air produced by it, which is entirely dependent on the amount of pressure caused by the influx of the water, may be increased with a single pressure of about one atmosphere to a degree of tension equal to that of steam. Described in Liebig's Annalen der Chemie, Vol. 176.

2558. Apparatus for producing a **Vacuum** for purposes of **Crystallization** and **Filtration.** *Hermann Fischer, Hanover.*

2559. Spring-Vacuometer, belonging to the above apparatus. *Hermann Fischer, Hanover.*

The apparatus, which is principally intended to replace Bunsen's filtering pumps, is constructed on the principle of the injection pump, producing a vacuum by means of a powerful jet of water. The little vacuometer contains a Schinz' tube. If the tube, marked water, be joined to the water supply pipe by means of an india-rubber tube, and the tube which in the drawing is not shown at all be joined to the waste pipe, and connexion be made with the vessel which is to be exhausted, a corresponding vacuum is readily produced. With a fall of water of 11 m. a vacuum will be formed which at most will only fall short by one cm. from the absolute barometric height. If the water used be allowed to flow freely into a vessel, the level of the water must be deducted from the height of the water column above the apparatus. The vacuometer is made small in order to render the apparatus more handy. The apparatus may also be used without the vacuometer.

The advantages of the whole apparatus are as follows :—(1.) It is very handy; (2.) It may be used on every work-bench which is provided with a supply of water ; and lastly (3.) It is very cheap.

2554. A. Müller's Subsiding and Washing Apparatus, provided with a set of sieves and a second indicator, for agricultural analysis. *Franz Schmidt and Haensch, Berlin.*

This apparatus is accompanied by a pamphlet descriptive of its application.

2565. Wolf's improved **set of Sieves** with **Brushes.**
Franz Hugershoff, Leipzig.

1959. Two Hydrostatic Rotary Engines or Precipitating Machines, with table to which eccentric motion is conveyed. The machines were invented by the exhibitor for the purpose of facilitating the solution, aggregation, or precipitation of chemical compounds, which they do as effectually in half an hour as if the solutions were allowed to stand for 24 hours (tested by quantitative experiments). The table is open in the centre, so that a beaker or flask may be heated by a Bunsen burner, and it is furnished with double-sliding clamps so as to securely hold the vessel in its place.

A, engine intended for delicate quantitative experiments.
B, engine for ordinary purposes.
Joseph William Thomas, Cardiff.

2589. Mica-plates used for effecting fusions and for evaporations (*see* Mineralogy). *Max. Raphael, Breslau.*

Mica is not affected by most acids. This material may also be employed as a cheap substitute for platinum in many chemical operations.

2590. Mica Powder, for use in filtrations.
Max. Raphael, Breslau.

2591. Mica-spectacles, for use in laboratories, to protect the eyes in all operations when dangers from spitting or from explosions have to be guarded against. *Max. Raphael, Breslau.*

2594. Ozone generating Tube, provided with drying apparatus. *C. Kramer, Freiburg, Baden.*

2547a. Fairley's simple form of Ozone Generator in Glass. *Harvey, Reynolds, and Co.*

2577. Hydrofluoric Acid Apparatus, in lead.
F. A. Wolff & Sons, Heilbronn and Vienna.

For the use of public laboratories.

2557. Collection of Glass Stop-cocks of various sizes.
Ch. F. Geissler & Son, Berlin.

2581. Collection of Glass Stop-cocks.
Dr. H. Geissler, Bonn.

2602. A. Mitscherlich's Apparatus for the continuous evolution of chlorine gas, together with a water bath having a constant water level. *Prof. A. Mitscherlich, Münden.*

This apparatus is described in A. Mitscherlich's chemical publications, part 2, p. 6.

1377c. Apparatus for the **Production** of **Hydrogen Gas.**
Alvergniat Frères, Paris.

V. MODELS, DIAGRAMS, APPARATUS, AND CHEMICALS EMPLOYED IN OR RESULTING FROM CHEMICAL MANUFACTURES; ANALYTICAL APPARATUS USED IN MANUFACTORIES.

2566b. Patent Platinum Apparatus.
Johnson, Matthey, and Co.

Newest form for the concentration of sulphuric acid, securing great strength, productive power, safety and economy in working, and highest degree of purity of acid, with a minimum of platinum : —

40075. 8 s

Boiler.—By the corrugated form of bottom (Prentice's patent) the greatest possible amount of strength, surface, and consequent evaporating power is obtained, and a considerable saving in fuel is effected.

Pans.—By means of these vessels the large and costly leaden tanks for the previous concentration of the chamber acid, which require constant repair and renewal and more or less contaminate the acid, are entirely done away with.

The setting of these boilers and open pans is of the simplest kind ; they are placed upon an iron frame over a straight flue, and they may be multiplied or enlarged to any desired capacity of production. without sacrifice of existing plant.

Cooler.—An improved economical and convenient form, securing great cooling power with a minimum of water and space.

The total cost of such apparatus composed entirely of platinum, and; therefore, always of great intrinsic value in comparison with the first outlay, is less than that of any other form yet introduced.

Briefly, the chief advantages of this construction are :—

Great economy in first outlay and daily expense of working.

Great intrinsic value in a realisable form in proportion to the cost.

Purity of acid, and freedom from danger.

2689. Common Salt. From the salt works of Northwich, Cheshire, used in the manufacture of salt cake or sulphate of soda.
Sullivan & Co.

2690. Salt Cake, or commercial sulphate of soda, containing about 98 per cent. of real sulphate, made by decomposing common salt with vitriol in an iron pan heated externally. When about three fourths of the decomposition has been effected, the charge is pushed upon the bed of a "muffle" furnace, maintained at a bright red heat till the decomposition is completed. The gaseous hydrochloric acid evolved during the process is conducted away through pipes or flues to condensing towers, where, being absorbed by water, it assumes the form of liquid acid.   *Sullivan & Co.*

2691. Limestone, or native carbonate of lime from Derbyshire, used in combination with slack or small coal for converting salt cake or sulphate of soda into carbonate of soda.   *Sullivan & Co.*

2692. Slack, or small coal from the Lancashire coalfield, used in combination with limestone for converting salt cake or sulphate of soda into carbonate of soda.   *Sullivan & Co.*

2693. Black Ash, or ball soda, containing about 24 per cent. of soda, produced by fluxing together in a suitable furnace a mixture of salt cake or sulphate of soda, limestone, and slack, by which means the sulphate of soda is converted into carbonate of soda, the carbonate of lime of the limestone into calcium sulphide; and the slack into coke.   *Sullivan & Co*

2694. Vat Liquor. A saturated solution of soda produced by lixiviating " black ash " or " ball soda " with warm water in iron tanks or vats ; when the solution is saturated it is run off into iron pans to be boiled down, the calcium sulphide and unburnt coke of the " black ash " being left behind in the vat as " tank " or " vat waste." *Sullivan & Co.*

2695. Salts, or crude carbonate of soda, obtained by boiling down " vat liquors " in iron pans, heated either externally or by passing the flame over the surface of the liquor in the pan. *Sullivan & Co.*

2696. Soda Ash, or finished carbonate of soda of commerce, containing 58·5 per cent. of real alkali, produced by gradually heating to redness in a reverberatory furnace " salts " or crude carbonate of soda. *Sullivan & Co.*

2697. Salt, as used in the " Hargreaves " process. The salt is placed in a moist state on drying floors, and when dried into hard flat pieces is broken by machinery. *Sullivan & Co.*

2698. Salt Cake, or sulphate of soda containing 99 per cent. of real sulphate, made without the use of sulphuric acid by the " Hargreaves " direct action process. Salt maintained at a red heat in iron cylinders is exposed to the direct action of sulphurous acid, air, and steam. The hydrochloric acid from the salt is condensed as in the ordinary process. *Sullivan & Co.*

2469. Drawings (4) of Hargreaves and Robinson's Sulphate of Soda Apparatus. 1. General plan. 2. Enlarged sectional plan. 3. Transverse sectional plan. 4. Condensing apparatus. *J. Hargreaves & T. Robinson.*

2470. Samples of : A. Salt prepared for the converting cylinders above. B. Sulphate of soda. C. Hydrochloric acid. *J. Hargreaves & T. Robinson.*

Sulphurous acid, steam and air, are made to react upon salt directly, producing sulphate of soda and hydrochloric acid, without the preliminary manufacture of sulphuric acid. The salt is placed in the iron cylinders, and the mixture of sulphurous acid, air, and steam passed through it. The evolved hydrochloric acid is taken out at the bottom of the cylinders, and conveyed to the condensers.

2738a. Drawing of Root's Patent (special) Blower for Hargreaves' process in the manufacture of sulphates. *Thwaites & Carbutt.*

2738b. Model of Root's Patent Pressure Blower.
Thwaites & Carbutt.

2448. Models of Hydrochloric Acid Condensing Towers, as erected at Messrs. Hutchinson's Alkali Works at Widnes.   *Prof. Roscoe, F.R.S.*

2449. Model of Salt-Cake Furnace, with open roaster.
Prof. Roscoe, F.R.S.

2450. Model of Finishing (Alkali) Furnace.
Prof. Roscoe, F.R.S.

2451. Model of " Weldon " plant for the regeneration of Black Oxide of Manganese in the Manufacture of Chlorine.
Prof. Roscoe, F.R.S.

2717. Soda Crystals, or **Washing Soda.** To obtain these crystals, soda ash is dissolved in hot water, and the clear solution is run into iron vessels, where a large part of the soda crystallizes out.   *Gaskell, Deacon, & Co.*

Composition :—

$$Na_2CO_3 + 10 \ Aq.$$

2708. Mother Liquor. When the crystallization of the soda is complete, the remaining liquid, or "mother liquor," is run off. This contains nearly all the soluble impurities in the "soda ash."
Gaskell, Deacon, & Co.

2709. Refined Alkali. To obtain this, soda ash is dissolved in hot water, the clear solution is evaporated, when a mono-hydrated carbonate of soda is precipitated, which is then calcined and ground.   *Gaskell, Deacon, & Co.*

2710. Red Liquor. In some cases the vat liquor is not evaporated to dryness, the residual liquor (called, from its colour, " red liquor ") containing the greater part of the caustic soda, sodium chloride, sulphate, sulphites, sulphide, and other impurities, is separated, and subsequently treated for the production of caustic soda.   *Gaskell, Deacon, & Co.*

2711. Oxidized Red Liquor. The sulphides, sulphites, &c. in the red liquor are wholly or partially oxidized by blowing air through the red liquor in deep iron vessels.
Gaskell, Deacon, & Co.

2712. Causticised Liquor. After oxidation the red liquor is diluted to about 1·080 specific gravity, and rendered caustic by agitation in contact with caustic lime.   *Gaskell, Deacon, & Co.*

2713. Cream Caustic. The causticised liquor is concentrated by evaporation until the solution sets hard on cooling, and tests 60 per cent. of alkali. During this concentration some of the sodium sulphate and a little carbonate are precipitated, and are ladled out, and any further oxidation required is effected by the addition of nitrate of soda. *Gaskell, Deacon, & Co.*

2714. White 60 per cent. Caustic Soda. To obtain this the causticised liquor is concentrated until the unevaporated portion fuses, when the iron (which has probably been held in solution by combination with cyanogen), and the greater part of the alumina and silica, are precipitated and allowed to settle ; the clear fused caustic soda, containing some sodium sulphate and chloride, is packed in iron drums. *Gaskell, Deacon, & Co.*

2715. White 70 per cent. Caustic. This is obtained in the same manner as white 60 per cent., except that care is taken during the concentration to remove sufficient sodium sulphate and chloride to enable the resultant caustic to be of the required strength. *Gaskell, Deacon, & Co.*

2716. Chloride of Calcium. This is obtained as a bye product from the manufacture of bicarbonate of soda or from the " Weldon " process. The solution of chloride of calcium is purified and settled, and concentrated by evaporation.

Gaskell, Deacon, & Co.

Composition :—

$CaCl_2$.
Aq.

2472. Samples of Bicarbonate of Soda unground ; also ground and dressed. *John Hutchinson & Co.*

2471. Samples illustrative of Mond's process for the recovery of sulphur from alkali (vat or tank) waste :

1. Alkali (vat or tank) waste before oxidation.
2. Alkali (vat or tank) waste after oxidation.
3. Sulphur liquor.
4. Precipitated sulphur.
5. Sulphur in bulk.
6. Roll sulphur.

John Hutchinson & Co.

2477. Soda Waste, Alkali Waste, Black Ash Waste, or Tank Waste. Insoluble residue formed in the second or black ash stage of Leblanc's process for obtaining soda from common salt. *John Hutchinson & Co.*

From the appended analysis by Kopp of a sample it will be seen that a large quantity of sulphur is present in the waste. A part of this can be profitably extracted by Mond's and other processes.

Sulphide of sodium	2,880
Carbonate of lime	13,636
Silicate of lime	5,680
Hydrate of lime	8,588
Monosulphide of calcium	22,162
Alumina, magnesia	1,466
Sulphide of iron	2,670
Carbon	1,800
Sand	2,000
Water	36,700
Combined water and loss	2,418

2478. Oxidized Soda, Alkali, or Vat Waste.

John Hutchinson & Co.

In Mond's process a current of air at the ordinary temperature is forced through the waste as it lies in the vats, directly the last soda liquors are drained away. The waste becomes hot, oxydizing, and forming polysulphides of calcium, hyposulphite and hydrosulphate of calcium. The waste is then lixiviated, and the liquid removed, the residue being again treated as above several times.

2479. Sulphur Liquor, the soluble portion of **Oxidized Alkali,** or **Tank Waste.**   *John Hutchinson & Co.*

The oxydized waste, lixiviated in the vats with warm water, gives a yellow solution containing polysulphides of calcium, hyposulphite of calcium, and hydrosulphate of calcium.

Liquor generally contains equal to 5·0 of sulphur, distributed as follows:—

         2·0 hyposulphite of calcium.
         2·0 polysulphide   do.
         1·0 hydrosulphate do.

2480. Precipitated Sulphur, formed when sulphur liquor and muriatic acid are mixed.   *John Hutchinson & Co.*

Sulphur liquor is mixed with common hydrochloric acid from the condensers, the mixed liquors being heated to 140° Fahr. and well stirred.

The muriatic acid forming chloride of calcium in the liquor, sulphur is deposited as a yellow flocculent precipitate.

In a good sample of liquor all the sulphide of hydrogen disengaged by decomposition of the sulphides reacts on the sulphurous acid liberated at the same time from the hyposulphite, sulphur and water being the products.

2481. Lump Sulphur, prepared by melting the moist sulphur of a previous stage of the process in a strong cast-iron vessel.   *John Hutchinson & Co.*

After washing out the chloride of calcium from the moist sulphur, the drained product is put into a strong cast-iron cylinder, the filling-in aperture screwed down, and steam of a pressure of 35 lbs. admitted through a coil of cast-iron pipe. The sulphur rapidly melts, and at the expiration of a certain time, found out by experience, the whole charge is forced by the pressure of the steam through the discharge pipe into tight wooden wagons, and when cool broken up for sale.

2482. Roll Sulphur. The roll brimstone of commerce, used in medicine. *John Hutchinson & Co.*

Melted sulphur run into round wooden moulds and allowed to cool.

2483. Bicarbonate of Soda (lump), prepared by exposing crystals of carbonate of soda to a current of carbonic acid gas till saturated, and drying the product. *John Hutchinson & Co.*

Soda crystals, or carbonate of soda, made by dissolving the soda-ash of Leblanc's process in hot water, and, after settling, allowing to cool and crystallize, are put into an air-tight iron chamber, and carbonic acid gas (prepared by decomposing limestone with waste muriatic acid from the condensers) turned in by a pipe on the roof of the chamber. After some hours the crystals are changed into bicarbonate of soda, part of the water of crystallization escaping during the process.

The moist bicarbonate is dried in kilns heated not over 100° Fahrenheit.

2484. Bicarbonate of Soda. Carbonate of soda of the shops, used in medicine and for making effervescing drinks.
John Hutchinson & Co.

The dry lump bicarbonate ground in a mill and dressed to separate coarse particles, as in grinding and preparing flour from wheat.

2462. Deacon's Apparatus for exposing porous materials and currents of gases to mutual action. Sectional working model, illustrating the application of one form of the apparatus to Deacon's process for producing chlorine. *Henry Deacon.*

In this example, the layer or "wall" is vertical and circular, and forms a section of a cylinder. The frames resemble those of venetian blinds, with the laths inclined at an angle of 45°, and so far apart, that an imaginary line joining the upper edge of each lath and the lower edge of the one above it is more horizontal than the natural angle of repose of the porous material itself, which is thus retained and supported by each lath in succession. A "wall" of this kind on being raised in height adds to the pressure on the bottom layers only so long as the height is less than that of a cone whose base is the width of the "wall," and whose sides are at the same angle as the natural angle of repose of the material. This increase of pressure diminishes in inverse proportion to the height, and ceases when the height of the imaginary cone is reached, all the weight of additional height above that point being borne by the retaining frames. In the model, the "wall" and framework are in a cast-iron vessel, which is heated in a brick furnace. The porous material is distributed to, and gathered from, the wall, by covering and inverted cones from, and to, central pipes. The gaseous current passes from the outside of, and through, the "walls" to the space they and the covering cone enclose, and is withdrawn from this space, or the direction of the current is reversed at pleasure.

2699. Native Peroxide of Manganese (Spanish), containing 80 per cent. of peroxide. *N. Mathieson & Co.*

2700. Hydrochloric Acid, obtained by absorbing the gaseous acid in water. *N. Mathieson & Co.*

2701. Chloride of Manganese from native " manganese " stills *after* neutralization with carbonate of lime.

N. Mathieson & Co.

2702. Limestone Dust. Crushed carbonate of lime.

N. Mathieson & Co.

2703. Manganese Mud, as precipitated *before* blowing with air. *N. Mathieson & Co.*

2704. Manganese Mud, *after* oxidation by blowing.

N. Mathieson & Co.

2705. Burnt Lime, from Buxton, Derbyshire. This lime is unslaked. *N. Mathieson & Co.*

2706. Hydrate of Lime, sifted and ready for the bleaching powder chambers. *N. Mathieson & Co.*

2707. Bleaching Powder, containing 38 per cent. of chlorine.

N. Mathieson & Co.

2724. No. 1. Cupreous Iron Pyrites, employed in the manufacture of sulphuric acid. *Widnes Metal Company.*

Water	·70
Sulphur	49·00
Arsenic	·47
Iron	43·55
Copper	3·20
Zinc	·35
Lead	·93
Lime	·10
Siliceous residue	·63
Oxygen, and traces of various metals	1·07
	100·00

2725. No. 2. Burnt Ore, or cinder remaining after the almost complete elimination of the sulphur from the pyrites as sulphurous anhydride. *Widnes Metal Company.*

Water	3·85
Sulphur	3·76
Arsenic	·25
Iron	58·25—83 per cent. Fe_2O_3.
Copper	4·14
Zinc	·37
Cobalt	traces
Silver	traces
Lead	1·14
Lime	·25
Oxygen and loss	26·93
Insoluble residue	1·06
	100·00

2726. No. 3. **Mixture** of **Burnt Ore** and **Salt** (the latter being about 12 per cent. of the whole), ground and passed through sieve of about 16 holes per square inch.

Widnes Metal Company.

This mixture is furnaced during a period of 5¾ hours, at the expiration of which time the copper has usually been almost entirely converted into a soluble chloride of that metal.

2727. No. 4. **Mixture** of **Salt** and **Burnt Ore** after calcination during 5¾ hours. *Widnes Metal Company.*

2728. No. 5. **Purple Ore,** or residue left from No. 4 after the extraction of the copper by lixiviation with hot water and dilute hydrochloric acid. *Widnes Metal Company.*

This residue is chiefly employed for the " fettling " of puddling furnaces.

Ferric oxide	-	- 96·20 = iron 67·35 per cent.
Lead	- -	- ·86
Copper	- -	- ·18
Cobalt	- -	· trace
Alumina	- -	- ·45
Lime	- -	- ·46
Soda	- -	- ·10
Phosphoric anhydride		- none
Arsenic	- -	- trace
Sulphuric anhydride		- ·49
Sulphur	- -	- ·16
Chlorine	- -	- ·03
Silica	- -	- 1·22
		100·15

2729. No. 6. **Purple Ore** compressed into blocks for use in the blast furnaces (the brick exhibited was made by Messrs. N. Mathieson & Co., of Widnes). *Widnes Metal Company.*

2730. No. 7. **Solution of Copper** as drawn from the lixiviators, containing, in addition to copper, &c., from three to four grains of silver per gallon, with traces of gold.

A quantity of iodide of potassium (or zinc) sufficient to precipitate the silver present is added to this solution, and the precipitate formed is allowed to subside, the supernatant liquor being passed into vessels in which the copper is precipitated by means of metallic iron.

Widnes Metal Company.

ANALYSIS of COPPER SOLUTION of Sp. Gr. 1·24.

Contents per gallon.
Grains.

Sodium sulphate - - - -	- 10,092
Sodium chloride - - - -	- 4,474
Chlorine (combined with metals) - -	- 4,630
Copper - - - - -	- 3,700
Zinc - - - - -	- 480
Lead - - - - -	- 40
Iron - - - - -	- 32
Lime - - - - -	- 52
Silver - - - - -	- 3·06

Arsenic, antimony, bismuth, &c. not estimated.

2731. No. 8. Precipitate caused by addition of a soluble iodide to solution of copper, &c. (Claudet's process), consisting of iodide of silver mixed with various salts of lead, chiefly sulphate.

Widnes Metal Company.

2732. No. 8A. Silver Precipitate, obtained by the reduction of the iodide by means of metallic zinc.

Widnes Metal Company.

Silver - - - - -	- 5·95
Gold - - - - -	- ·06
Lead - - - - -	- 62·28
Copper - - - - -	- ·60
Zinc oxide - - - - -	- 15·46
Ferric „ - - - - -	- 1·50
Lime - - - - -	- 1·10
Sulphuric anhydride - - -	- 7·68
Insoluble residue - - - -	- 1·75
Oxygen and loss - - - -	- 3·62
	100·00

2733. No. 8B. Iodide of Zinc, obtained by reduction of iodide of silver by metallic zinc, and used for the treatment of a further quantity of argentiferous copper solution.

Widnes Metal Company.

2734. No. 9. Copper Precipitate thrown down by the immersion of thin scrap iron in the cupreous solutions.

Widnes Metal Company.

(Contains on an average 75 to 80 per cent. of metallic copper.)

2735. No. 10. Spent Liquor, or liquid remaining after complete precipitation of copper from solution. This liquor is run to waste. *Widnes Metal Company.*

2459. Spence's Process for the Manufacture of Alum.
Specimens to illustrate the process, consisting of :—
(1.) Shale of the coal measures before calcination (source of alumina).
(1a.) Shale after calcination.
(2.) Pyrites or bi-sulphide of iron (source of sulphur for the manufacture of sulphuric acid).
(2a.) Nitrate of soda, for oxidising the sulphurous acid produced from the pyrites into sulphuric acid.
(2b.) Sulphuric acid, specific gravity 1·6, as used in the manufacture of alum.
(3.) Gas liquor (source of ammonia).
(3a.) Sulphate of ammonia.
(4.) Sulphate of alumina.
(5.) First crystals of ammonia alum (a double salt or compound of sulphate of ammonia and sulphate of alumina).
(6.) Ammonia alum. Second and final crystallisation.
Peter Spence.

2473. Specimens of Alum Crystals.
3 Crystals of potash-alum.
3 ,, ,, ,,
10 ,, ,, ,,
4 Crystals of chrome-alum.
2 ,, ,, ,, covered with potash-alum.
1 Crystal of potash-alum of 5 kilo.
1 ,, ,, ,, 3½ kilo.
H. B. J. van Rijn, Venlo, Netherlands.

2679. Chlorate of Potash. Muriate of potash containing about 90 per cent. chloride of potassium, and some impurities consisting of chloride of sodium, sulphates of potash, soda, lime, and magnesia.
James Muspratt & Sons.

2680. Chlorate of Potash. Milk of lime of about 1·08 specific gravity, made by well stirring up about 18 ctr. slacked lime in about 6,800 litres water in a cast-iron vessel. The sample contains a small quantity of chlorate and of chloride of calcium from the previous solution made in the same vessel.
James Muspratt & Sons.

2681. Chlorate of Potash. Solution of chlorate of lime of about 1·15 specific gravity, containing at the same time chloride of calcium in the molecular proportion of about 1 5. Obtained by saturating milk of lime with chlorine gas. Towards the end of the process the temperature rises, when permanganate of lime is formed owing to the small quantity of manganese contained in the lime, by which the liquor obtains its pink colour.   One litre contains from 45–48 grammes of chlorate of lime.

James Muspratt & Sons.

2682. Chlorate of Potash.   Finished ground chlorate of potash, containing about 0·022 per cent. chloride of calcium, and about 0·125 per cent. moisture.   *James Muspratt & Sons.*

2683. Chlorate of Potash.   First crystals of chlorate of potash, containing about 85 per cent. chlorate of potash, and 3–4 per cent. chloride of calcium.   *James Muspratt & Sons.*

2686. Chlorate of Potash.   First mother liquor of chlorate of potash, of about 1·3 specific gravity.   The solution of chlorate of lime is boiled down with a sufficient quantity of muriate of potash, and allowed to cool and crystallize for 8–10 days.

James Muspratt & Sons.

2684. Chlorate of Potash.   Finished crystals of chlorate of potash, containing about 0·0502 per cent. chloride of calcium.

James Muspratt & Sons.

2685. Chlorate of Potash.   Last mother liquor of chlorate of potash, of 1·35–1·40 specific gravity.   The first mother liquor is boiled down still further, and allowed to cool and crystallize again for 8–10 days.   It contains 30–40 grammes chlorate of potash, about 520 grammes chloride of calcium, and 3–5 chloride of potassium in the litre.   *James Muspratt & Sons.*

2687. Caustic Soda Ash, containing about 12 per cent. caustic soda, 76–80 per cent. carbonate of soda, total strength about 54–56 per cent. Na^2O.   Black ash is lixiviated in iron tanks, the liquor thus obtained is boiled down in open wrought-iron pans, and, as the crystals fall, they are fished out and calcined in an open furnace.   *James Muspratt & Sons.*

2688. Nitre Cake, obtained in the manufacture of sulphuric acid by decomposing nitrate of soda with sulphuric acid.   It is composed of sulphate of soda, together with small quantities of sulphate of iron, lime, and magnesia, of chloride of sodium, nitrate of soda undecomposed nitrate of soda, and free sulphuric acid.

James Muspratt & Sons.

2718. No. 1. **Alkaline Silicate of Soda** (glass), suitable for soap making.   *William Gossage & Sons, Lancashire.*

Composition :—

$$100 \text{ parts, } Na_2O.$$
$$180 \quad , \quad SiO_2.$$

2719. No. 2. **Silicate** of **Soda** (glass), suitable for calico printing. *William Gossage & Sons.*

Composition :—

    100 parts, Na_2O.
    310 ,, SiO_2.

2720. No. 3. **Silicate** of **Potash** (glass).

William Gossage & Sons.

Composition :—

    100 parts, K_2O.
    310 ,, SiO_2.

2721. No. 4. **Solution** of **Alkaline Silicate** of **Soda,** made from No. 1. 1,500 specific gravity. *William Gossage & Sons.*

2722. No. 5. **Solution** of **Silicate** of **Soda,** 1,375 specific gravity. Made from No. 2. *William Gossage & Sons.*

2723. No. 6. **Solution** of **Silicate** of **Potash,** 1,350 specific gravity. Made from No. 3. *William Gossage & Sons.*

2738f. Examples of Technological Diagrams.

Diagram 6. Retort for the manufacture of coal gas.
   ,,   7.   ,,   ,,   ,,
   ,,   28. Extraction of zinc (Silesian furnace).
   ,,   31. Lime-kiln (continuous process).
   ,,   32. Lead smelting (German cupellation hearth).
   ,,   36. Lead smelting blast furnace.
   ,,   37. Extraction of salt by evaporation of brine.
   ,,   38.   ,,   ,,   Graduation house.
   ,,   50. Manufacture of steel by cementation.

The Council of the Yorkshire College of Science, Leeds; Prof. Thorpe.

2460. Oxalic Acid. Two specimens. One crystallised from a solution containing sulphuric acid, the other from an aqueous solution. *Roberts, Dale, & Co.*

This product is made by the action of caustic potash, or a mixture of caustic potash and caustic soda on woody fibre (sawdust). The result of this action is oxalate of potash. The oxalic acid is isolated by precipitation as oxalate of lime, and the subsequent decomposition of this latter product by sulphuric acid. Specimen No. 2 is crystallised from water. A comparison will show the marked difference in crystalline form due to the presence of sulphuric acid.

2461. Binoxalate of Potash. *Roberts, Dale, & Co.*

2463. Carbolic Acid, chemically pure, free from taste and smell of tar, and fusing at 108° Fahrenheit. Specially prepared for internal medicinal use. *F. C. Calvert & Co.*

2464. Carbolic Acid, crystallised, fusing at 95° Fahrenheit. Specially prepared for external medical application.
F. C. Calvert & Co.

2465. Cressylic Acid, used for disinfecting purposes.
F. C. Calvert & Co.

2466. Sulpho-Carbolates. A series of pharmaceutical products, comprising sulpho-carbolates of potash, soda, ammonia, lime, iron, copper, and zinc. *F. C. Calvert & Co.*

2467. Carbolic Acid Preparations, comprising carbolic acid soap, carbolic acid disinfecting powder, and carbolised tow.
F. C. Calvert & Co.

2468. Picric Acid Crystals and paste, Aurine (rosolic acid). Used in the arts. *F. C. Calvert & Co.*

2475. Chemicals for use in various Manufactures.
Liquid archill, made from Orchella weed, for dyers and printers.
Sulphate of alumina, for sugar refiners, paper makers, and dyers.
Aluminous cake, for paper makers.
Cudbear, made from Orchella weed, for dyers.
Bichromate of potash, with samples of chrome ore, limestone, and sulphate of potash, from which it is made.
W. J. Norris and Brother.

2476. Aniline and other Chemical Products used for Dyeing. *Brooke, Simpson, and Spiller.*

		Commercial Name.
Rosaniline base - - -	$C_{20}H_{19}N_3, H_2O$	
Rosaniline hydrochlorate - -	$C_{20}H_{19}N_3HCl$	Roseine crystals.
Crysaniline - - - -	$C_{20}H_{17}N_3$	Phosphine.
Methyl-rosaniline acetate -	$\left\{\begin{array}{l}C_{20}H_{18}\\ CH_3\end{array}\right\} N_3 \left.\right\} C_2H_3O_2$	Hofmann violet.
Tri-methyl rosaniline acetate -	$\left\{\begin{array}{l}C_{20}H_{16}\\ CH_3\\ CH_3\\ CH_3\end{array}\right\} N_3 \left.\right\} C_2H_3O_2$	Eclipse violet.
Dimethyl-phenyl-rosaniline acetate - - - - -	$\left\{\begin{array}{l}C_{20}H_{16}\\ C_6H_5\\ CH_3\\ CH_3\end{array}\right\} N_3 \left.\right\} C_2H_3O_2$	Spiller purple.
Phenyl-rosaniline hydrochlorate -	$\left\{\begin{array}{l}C_{20}H_{18}\\ C_6H_5\end{array}\right\} N_3 \left.\right\} HCl$	Imperial violet.
Tri-phenyl-rosaniline hydrochlorate - - - -	$\left\{\begin{array}{l}C_{20}H_{16}\\ C_6H_5\\ C_6H_5\\ C_6H_5\end{array}\right\} N_3 \left.\right\} HCl$	Pure opal blue.

Commercial Name.

				Commercial Name.
Sodium tri-phenyl-rosaniline sul-phonate - - - -	$\left\{\begin{array}{l}C_{20}H_{16}\\C_6H_5\\C_6H_5\\C_6H_3\end{array}\right.$	N_3	$\left.\right\}N_aSO_3$	Fast blue, 6B.
Tri-phenyl-rosaniline sulphonic acid - - - - -	$\left\{\begin{array}{l}C_{20}H_{16}\\C_6H_5\\C_6H_5\\C_6H_5\end{array}\right.$	N_3	$\left.\right\}HSO_3$	—
Sodium phenyl-rosaniline sul-phonate - - - -	$\left\{\begin{array}{l}C_{20}H_{18}\\C_6H_5\end{array}\right.$	N_3	$\left.\right\}N_aSO_3$	Alkali violet.
Di-phenyl-rosaniline acetate -	$\left\{\begin{array}{l}C_{20}H_{17}\\C_6H_5\\C_6H_5\end{array}\right.$	N_3	$\left.\right\}C_2H_3O_2$	Atlas blue.
Calcium tri-phenyl-rosaniline di-sulphonate - - - -	$\left\{\begin{array}{l}C_{20}H_{16}\\C_6H_5\\C_6H_5\\C_6H_5\end{array}\right.$	N_3	$\left.\right\}C_a(SO_3)_2$	Soluble blue.
Methyl-rosaniline hydrochlorate (fused) - - - -	$\left\{\begin{array}{l}C_{20}H_{18}\\CH_3\end{array}\right.$	N_3	$\left.\right\}HCl$	Hofmann violet cake R.
Di-methyl-rosaniline hydrochlo-rate (fused) - - -	$\left\{\begin{array}{l}C_{20}H_{17}\\CH_3\\CH_3\end{array}\right.$	N_3	$\left.\right\}HCl$	Hofmann violet cakeBB.
Rosaniline hydrochlorate (fused) -	$C_{20}H_{19}N_3HCl$			Roseine cake.
Tri-methyl-rosaniline methyl io-dide - - - -	$\left\{\begin{array}{l}C_{20}H_{16}\\CH_3\\CH_3\\CH_3\end{array}\right.$	N_3	$\left.\right\}(CH_3I)^3$	Iodine green crystals.
Anthracene - - - -	$C_{14}H_{10}$			Anthracene.
Alizarin - - - -	$C_{14}H_8O_4$			Alizarin (blue shade).
Anthrapurpurin - - -	$C_{14}H_8O_5$			Alizarin (red shade).

2476a. Chrysoidine, a new basic orange dye for silk, wool, cotton, leather, &c. Manufactured by the exhibitors.

Williams, Thomas, and Dower.

Chrysoidine is, according to quality, the sulphate or hydrochlorate of a new bi-acid base, belonging to the benzene series. It is a valuable orange-yellow, and owing to the readiness with which it crystallizes, can be obtained in a state of perfect purity. Except phosphine, it is the only basic yellow dye known, and as it is the result of a direct process (not a secondary product) it can be obtained at a much lower price than phosphine, which it equals, and in some of its applications surpasses in strength and beauty of shade. It dyes in neutral or slightly basic baths, and has a great affinity for silk, wool, and even for unmordanted cotton. Mordanted cotton is dyed a yellow or orange shade according to the mordant employed. It combines readily with magenta and other red colours, giving rise to very clear and beautiful scarlets. The chrysoidine was discovered quite recently in our laboratory by Dr. Otto N. Witt.

2450a. Models of **Stoneware Distilling Apparatus** as used by **Manufacturing Chemists,** and others, and comprising still, still head, false arm, dip arm, condensing worm, three receivers with connecting pipes and taps, mixing pan, and stone jar. Made of improved vitrified stoneware, and warranted to resist the action of the strongest acids. *James Stiff & Sons.*

2450b. Model of Hofmann's Patent Kiln.

Hermann Wedekind.

2450c. Drawing of Bull's Patent Kiln.

Hermann Wedekind.

2566a. Drawing of Price's Patent Retort Furnace.

W. Price.

DESCRIPTION OF DRAWING.

Fig. 1.—Longitudinal section through centre of furnace.
Fig. 2. — Plan through the "retort," "combustion," and "heating" chambers.
Fig. 3.—Cross section through combustion chamber.
Fig. 4.—Elevation, part in section, of the retort chamber.
Fig. 5.—Sectional plan through retort chamber at E E.

Fig. 1.—A is a combustion chamber filled with grate bars in the ordinary way. B, a heating chamber, separated from A by the usual bridge. C, is the neck descending into an underground flue D, leading into an upcast or retort chamber, as it has been designated, E. In the centre of the chamber (E) is a firebrick circular pillar F, with spaces around marked in Fig. 5, E E E E, and on which is placed a cast-iron cylindrical air-vessel G, which is protected by fire-brick.

On this air vessel (G) is built a retort H, partly of fire-brick, partly of cast iron. The top of the cast-iron part of the retort is fitted with a hopper, I, in the throat of which is a damper, J, worked by a rocking shaft and lever, K, from the ground. The lower portion of the retort made of fire-brick has two necks, L L, the one leading to the combustion chamber for the passage of fuel, the other to the outside of the furnace for the insertion of stoking tools, to force the fuel forward into the combustion chamber. The entrance of the outer neck is closed by an air-tight door M. The upcast or retort chamber (E) extends to near the top of the retort, where it is closed by brickwork, but is opened at the side by the flue N, leading to the stack O.

Near the bottom of the chamber E, and in a line with the centre of the circular air vessel G, are pipes P P, inserted in the walls of the chamber (E), passing all round the chamber as shown in Fig. 5. In front of the inner side of the circuit of pipes, and opening into the chamber E, are a number of portholes Q Q Q (see Fig. 5), leading to the space around the pipes (P P), which space affords scope for expansion and a free circulation of heat. These pipes (P P) are connected with the blast as shown at E, Fig. 4, and pass into the central chamber G, as shown at F, Fig. 5, the outlet, R, from the air vessel leads into the ash-pit S.

The practice in working is to light a fire on the grate-bars, and generate heat in the usual manner, until the furnace is well heated. The retort is then filled with fuel, and the firing commences from the retort, and by the time the fuel at the top descends to the bottom of the retort, it is well heated, and a continuous supply of heated fuel is then kept up. All raw fuel is from this time supplied to the hopper (I) only, and let into the "retort" by the damper without the access of air.

The gases so generated in the combustion chamber (A) pass over the bridge into heating chamber (B), down the neck (C), into the underground flue (D), into the upcast or "retort" chamber (E), filling the spaces around, and giving up their heat to the circular air chamber (G), the retort (H), and the air pipes (P P), and their residue passing off by way of the flue (N) into the stack (O), the heat so stored being carried back into the furnace by the heated fuel. Combustion is supported by air under pressure from a fan. The air entering in as shown at (E), Fig. 4, traverses the entire circuit of pipes, passing into the central air-vessel (G), out through the outlet (R), into the ash-pit (S), and so up through the grate-bars.

2664. Description and drawing of an **Apparatus** for the **Determination** of **Sulphur** (see the accompanying treatise : "Aphorismen über Steinkohlen von Dr. E. Muck.")
Berggewerkschaftskasse, Bochum, Rhenish Prussia.

2665. Reich's Apparatus for **estimating Sulphuric Acid.** *Royal Mining Academy, Freiberg, Saxony.*

2661. Collapsing Aspirator, of vulcanised rubber, for rapid gas analysis. Arranged to measure the gas to be examined and to contain the absorbent. On agitation, the gas and absorbent are brought into contact. Used by H.M. Inspectors of Alkali Works. *Alfred E. Fletcher.*

2566. Winkler's Apparatus for gas analysis.
Franz Hugershoff, Leipzig.

One is intended for relatively small quantities of gases, carbonic acid, &c. Its advantage consists in the considerably narrowed part of the measuring tube above the lower three-way stop-cock, whereby finer divisions are rendered possible.

The other is constructed with a second three-way stop-cock, in the upper part of the measuring tube, which allows for the easy connexion with any other apparatus.

2666. Winkler's Apparatus for technico-chemical **Gas Analysis.** *Royal Mining Academy, Freiberg, Saxony.*

2736. Max Liebig's Apparatus for determining, for technical purposes, the quantity of oxygen and carbonic acid contained in various gaseous mixtures.
Rhenania Alkali Works, Stolberg, near Aix-la-Chapelle.

A description of this apparatus will be found in Ding. Polyt. Journ., Jan. 1873, of which copies may be had.

2597. Dittrich's Apparatus for the **Volumetric Determination** of **Gases,** accompanied by a printed description, together with a drawing and 5 tables. *J. H. Büchler, Breslau.*

This apparatus may be used for scientific as well as technical determinations of the carbonic acid contained in chalk, bone charcoal, &c., and is extensively used in the sugar refineries of Germany and Russia.

2520. Fixed appliance for the analysis of gases with three reagents. *M. Orsat, Paris.*

2521. Fixed appliance for the analysis of gases with two reagents. *M. Orsat, Paris.*

2522. Portable Appliance for the analysis of gases, giving the quantity of gases, giving the quantity of the hydrogen and its carburets. *M. Orsat, Paris.*

40075. T t

2454.   Apparatus for Gas Analysis.   By M. Orsat, of
Paris.   *Robert Galloway.*

The apparatus is so constructed as to be readily available for the use of
ordinarily intelligent workmen, and to furnish not only ready but comparatively trustworthy indications.

2740a. Rudorff's Apparatus for determining the car-bonic acid in illuminating gas.

*W. J. Rohrbeck and F. Luhme and Co., Berlin, Dr. Herm.
Rohrbeck.*

This apparatus consists of a three-necked Woulff s jar, of about 1,000 cc.
capacity.

Into one of the necks is ground a tube (*a*), reaching nearly to the bottom,
which can be closed by a stop-cock, whilst the second neck is closed by a
glass-cock burette, divided into cc., and the third by a manometer. In the
lower part a cock is fixed, perforated in such a manner that the interior of
the bottle can be brought into connexion either with the open air by closing
the manometer, or with the manometer alone.

The apparatus, the volume capacity of which has been previously ascertained by weighing with distilled water, is thereupon filled with the gas to be
examined in such a manner as to allow the gas to flow in through the pipe
reaching nearly to the bottom of the jar, by means of a tube, whilst at the
same time the air in the apparatus escapes through the cock attached to the
manometer. When it is supposed that the jar has been completely filled
with the gas, first the outside air is shut off by means of the manometer cock
by which the same is brought into connexion with the interior of the jar, and
next the gas supply pipe. The difference of pressure occurring at the moment
the connexion is being shut off, which can be observed on the manometer,
must be equalised by carefully opening the manometer cock, whereupon the
gas can be tested as to the per-centage of carbonic acid contained in it.

It is advisable, in order to remove the disturbing influence of the changes
in the temperature, to surround the glass jar with water of the same temperature as that of the room. After a lapse of three or four minutes at the
utmost, the interior of the jar with the surrounding water will have assumed
the same temperature, as can be easily ascertained by thermometers attached
within and without.

The principle of determining the carbonic acid is, that the carbonic acid is
allowed to be absorbed by potash-lye, and then replaced by an equal volume
of potash-lye.

By opening the glass-cock of the burette, some drops of potash-lye are put
into the jar. At the first moment the pressure will be increased, which,
however, is soon lessened by the absorption of the carbonic acid. Thereupon potash-lye is dropped in in the same quantity as the absorption of the
carbonic acid is consumed, so that the position of the indigo solution in the
manometer remains the same.

Where the absorption of the carbonic acid is completed, it will be necessary
to wait for a few minutes, and in case there is still a difference of pressure, to
drop in so much of potash-lye that the manometer show again exactly its
first position.

The volume of carbonic acid having been replaced by potash-lye, this
volume can be read on the burette.

As the capacity of the jar is known,—at Rudorff's experiments with this apparatus it contained 8·80 cc.,—there will be necessary of potash-lye :—

at 1 per cent. volume of carbonic acid   8·80 cc.
at 2   „   „   „   „   2·88 =
at 3   „   „   „   „   3·88 cc.

In order to form a judgment on the exactness of this method, Rudorff filled the jar with hydrogen and carbonic acid of known strength, and thus he obtained :—

		Found.	Calculated.
1 per cent.	CO_2	8·7 cc.	8·8
2 „	„	17·8	17·6
3 „	„	26·8	26·4
4 „	„	35·1	35·2

Some consecutive experiments made with illuminating gas, resulted in consumption of potash-lye :—

12·0 cc. = 1·36 per cent.
12·1 cc. = 1·37   „   „
11·8 cc. = 1·34   „   „

The apparatus, of course, is only intended for ascertaining small quantities of carbonic acid, such as is the case with illuminating gas. The other component parts of gas are of no perceptible influence on this method.

The analyses mentioned prove this sufficiently in order to secure for a method a certain adoption in practice, which recommends itself just as well by its trustworthy results, as by easy and speedy execution.

2456f. Tube with **Beads** to collect the **Ammonia** in **Coal Gas** as used at the testing stations.   *E. Cetti & Co.*

2737. Demby's Amylometer (potato-tester) accompanied by a description and case.   *J. H. Büchler, Breslau.*

This instrument is quite new (1875). It is remarkable for its simplicity and accuracy, as well as for the rapidity with which the starchy matter in potatoes can be estimated. Rain water may be employed.

2738. Stammer's Testing Apparatus applicable to all the processes met with in distilleries, accompanied by a descriptive treatise.   *J. H. Büchler, Breslau.*

This apparatus is quite new. It comprises absolutely true standard hydrometers for the determination of specific gravities, and replaces (for the most part) the purely empirical methods of testing which have hitherto been in use, by rational chemical methods.

2667. Soleil-Scheibler improved **Polarising Apparatus** (or optical saccharoetmer) for determining sugars.
   Dr. C. Scheibler, Berlin.

All the instruments exhibited by Dr. C. Scheibler are for the use of chemical laboratories in beet root sugar refineries. They are described in the "Journal " of the Society for the Advancement of the Beet-root Sugar Industry of the " German Empire," edited by the exhibitor, and are used in the chemical laboratory of the society of which Dr. C. Scheibler is the director.

The Soleil-Scheibler Polariscope, serves for the determination of the optical rotary power of sugar solutions, besides ascertaining quantitatively the amount of the sugar contained therein. The improvements made by the exhibitor in this instrument (and described in the Zeitschrift for 1870, p. 609) are as follows:—

(1.) In the place of the two quartz-wedges sliding against one another in the earlier instruments, and which carry the scale and vernier, these instruments have the wedges so arranged that only one of them carrying the scale is movable, the other bearing the vernier is fixed. By this means a much firmer motion of the first wedge is secured, and the so-called dead-motion of the setting-screw does not easily occur. In consequence of this the zero point of both wedges remains unaltered, which is a matter of considerable importance.

(2.) The above-mentioned vernier resting on a firmly fixed quartz-wedge is movable within definite dimensions, by means of a key, without the position of the wedge itself being changed. In this way an accurate adjustment to the zero point is easily accomplished.

(3.) For the more convenient reading off of the experimental numbers obtained, a small round magnifying mirror is fixed above the scale.

(4.) To prevent any side rays interfering with the readings, all light is kept out by a completely light-tight sliding casing fitted over the observation tubes.

2668. Mitscherlich's simple Polarising Apparatus for use in sugar refineries.   *Dr. C. Scheibler, Berlin.*

2669. Scheibler's Apparatus for the **estimation** of **Carbonate** of **Lime** in bone charcoal, as well as for the quantitative volumetric analysis of carbonates generally.
Dr. C. Scheibler, Berlin.

This apparatus was described by Scheibler in the above-mentioned Zeitschrift for 1859, p. 285, and for 1861, p. 525.

2670. Scheibler's Apparatus for the **volumetric estimation** of **Carbonic Acid** contained in the gases of saturated solutions.   *Dr. C. Scheibler, Berlin.*
Described in Zeitschrift for 1866, p. 644.

2671. Scheibler's Apparatus for estimating the **Refinery Value of raw Sugar.**   *Dr. C. Scheibler, Berlin.*
Described in Zeitschrift for 1872, p. 297, and 1873, p. 304.

2672. Stammer's Colorimeter for estimating the **Colour** of **Liquids.**   *Dr. C. Scheibler, Berlin.*

2673. Hydrostatic Balance for estimating the **Density of Sugar** solutions and other liquids.   *Dr. C. Scheibler, Berlin.*
Described in Zeitschrift for 1870, p. 264.

2674. Brix's Standard Saccharometer, consisting of three hydrometer tubes and a thermometer packed in a case.
Dr. C. Scheibler, Berlin.

2675. Gerlach's Standard Saccharometer.
Dr. C. Scheibler, Berlin.

2676. Scheibler's Muffle of platinum, together with platinum Scales and Stand used for the estimation of the ash in raw sugars.

Dr. C. Scheibler, Berlin.

Described in Zeitschrift for 1867, p. 338.

2476b. Card with three patterns of dyed cloth.

Messrs. J. Marshall, Son, & Co.

Orchella weed.	Bottle of Cudbear, red shade.
Moss Teneriffe.	Do. Orchill liquor.
Orchella weed, Ceylon.	Do. do. paste.
Do. do. Zanzibar.	Do. Cudbear, violet shade.
Do. do. Cape de Verde.	

2450c. Two Bink's Alkalimeters with registered enamel back. *E. Cetti & Co.*

2450d. Nitric Acid Tube with graduated enamel back.

E. Cetti & Co.

2738e. Artificial Fruit Essences, being solutions of various organic ethers in rect. spirit of wine. *Hirst, Brooke, and Hirst.*

They are nearly identical in flavour and chemical composition with the flavouring principles of the fruits they are intended to imitate :—

Jargonelle pear.	Raspberry.
Pine apple.	Greengage.
Apple.	Peach.
Strawberry.	Cherry.

2665a. Collection of Chemical Substances.

Messrs. Bouch & Co.

Sulphate of copper; copper precipitate; copper oxide; agricultural sulphate of copper; nitrate of lead; sulphate of ammonia; sulphate of zinc; red prussiate of potash; yellow prussiate of potash; crude sulphur; roll sulphur; flour sulphur; crude carbolic acid; coal tar solvent naphtha; coal tar creosote; benzole $90°/_0$; benzole $50°/_0$; anthracene; anthracene oil; naphthaline; pitch.

VI. MODELS, DIAGRAMS, AND APPARATUS ILLUSTRATING PROCESSES AND RESEARCHES IN AGRICULTURAL CHEMISTRY ; APPARATUS, DIAGRAMS, &c., ILLUSTRATING THE CHEMISTRY OF THE GRAPE VINE.

2741. Specially made Balance, and other appliances, used in an investigation by Messrs. Lawes and Gilbert, to determine the amount of water given off by plants during their growth. See Journal of the Horticultural Society of London, vol. v., p. 38,

1850, and vol. vi., p. 227, 1851.   The experiments were, however, continued to 1858, inclusive.   *John Bennet Lawes.*

The balance, which was made by Mr. Oertling, of London, was constructed to turn with less than one grain, when loaded with 50 lbs., or even more, on each side. This it accomplished, but it was found that the quantity of water given off by the plants during their growth was so great that such accurate weighing was not necessary. In fact, during the whole period of growth, as much as from 15 to 20 lbs. of water was in some cases given off from a single jar of plants, and during the most active periods of growth as much as from 1,500 to 2,000 grains per day. In the earlier experiments the vessels in which the plants were grown were made of glass, but afterwards of zinc. There was no opening at the bottom for drainage. The top was closed by a glass plate, firmly cemented to the rim, but having a hole in the centre for the plants to grow through, and another, smaller one, nearer the side by which to supply weighed quantities of water as needed, but which was, at other times, closed by a cork. To prevent, as far as possible, evaporation from the soil other than through the plant itself, small pieces of glass were laid over the centre hole, close up to the stems of the plants as they grew. Each jar held about 42 lbs. of soil. A standard leaden counterpoise was kept in the weight pan, and only the deviations above or below its weight were determined ; a set of weights, from 10,000 grains down to one-tenth of a grain, being provided for the purpose. The weighings were generally taken at intervals of 10 days; but sometimes at shorter periods.

The list of plants experimented upon included wheat, barley, beans, peas, clover, mangold wurzel, turnips, and various evergreen and deciduous trees.

2742. Case of Casts of White Silesian Sugar-beet, illustrating the influence of different manures on the amount of produce, and on the per-centages of dry matter and of sugar in the roots. First season of the experiments, 1871.   *John Bennet Lawes.*

2743. Table of **Average Results** obtained on growing the crop (sugar-beet) five years in succession on the same land.   *John Bennet Lawes.*

The experiments were conducted on the farm of John Bennet Lawes, Esq., Rothamsted, near St. Albans.

2744. Apparatus used in an investigation by Messrs. Lawes, Gilbert, and Pugh, to determine whether plants assimilate free or uncombined nitrogen ; with drawings of some of the plants grown. See Philosophical Transactions, Part 2, p. 493, 1859; and Journal of the Chemical Society, new series, vol. i. ; Entire Series, vol. xvi., 1863.   *John Bennet Lawes.*

The tap being opened, and water allowed to flow from a raised reservoir into the large stoneware Woulfe's bottle, air passes from it by the small leaden exit tube, through two glass Woulfe's bottles containing sulphuric acid, then through the long tube filled with fragments of pumice saturated with sulphuric acid, and, lastly, through a Woulfe's bottle containing a saturated solution of ignited carbonate of soda ; and, after being so washed, the air enters the glass shade, from which it passes by an exit tube through an eight-bulbed apparatus containing sulphuric acid, by which communication with the unwashed external air is prevented. Entering the shade at the side opposite to this exit tube is a tube for the supply of water or solutions to the soil, but which is at other times closed. In front of the shade is a bottle connected by a tube with the bottom of the earthenware lute-vessel, for the collection of the condensed water, which is from time to time withdrawn from the bottle by suction, and returned to the soil. The shade enclosing the pot and plant stands in the groove of a specially made, hard-baked, glazed, stoneware lute-vessel, mercury being the luting material. Carbonic acid is supplied as occasion may require, by adding a measured quantity of hydrochloric acid to the bottle containing fragments of marble, the evolved gas being, as will be seen, washed through one of the bottles of sulphuric acid, through the long tube, and through the carbonate of soda solution, before entering the shade. The short leaden pipe, bent and opening downwards externally to the large stoneware bottle, passes nearly to the bottom of it inside, and is a safety tube for the overflow of the water when the vessel is full, and so to prevent it passing into the wash bottles, &c. When full, the cork near the bottom of the stoneware vessel is withdrawn, and the water flows by means of a drain back into a tank, from which it is pumped into the raised reservoir for re-use. It will be observed that, by the arrangement described, the washed air is forced, not aspirated, through the shade, and the pressure being thus the greater within the vessel, the danger of leakage of unwashed air from without inwards is lessened. In 1857 twelve sets of such apparatus were employed, in 1858 a larger number, some with larger lute-vessels and shades, in 1859 six, and in 1860 also six. The whole were arranged side by side in the open air, on stands of brickwork, as indicated.

2744a. Apparatus for Analysing Soil.
Herr Bela Gonda, Magyar, Ovar, Hungary.

2744b. Eight Bottles of Specimens illustrating constituents **of Cotton Fibre** and case containing 14 glass bottles.
Dr. Schunck, Owen's Coll., Manchester.
Film of colouring matter mounted between two pieces of glass.
Pamphlet on constituents of cotton fibre.

2534a. Azotometer arranged for the easy and exact determination of the nitrogen contained in the manures employed in agriculture. *M. Honzeau, Paris.*

2564. Knop's Improved Azotometer.
Franz Hugershoff, Leipzig.

2583. Knop's Azotometer, as modified by Dr. Wagner.
Ehrhardt and Metzger, Darmstadt.
Described in "Zeitschrift für anylitiche Chemie von Fresenius," 1874, 4 Heft; 1876, 3 Heft.

2760a. Ebullioscope, for weighing alcohol in wines, Vidal's system, improved. *M. Malligand fils, Paris.*

The improved Ebullioscope of M. Malligand fils, is an instrument for ascertaining easily and correctly, in a few moments, without distillling, and merely by ebullition, the quantity of alcohol contained in dry or sweet wines. It is now used by the Syndical Chamber of the Wine Trade of Paris, who have found the Ebullioscope to be the most practical and most correct instrument of all those used hitherto for ascertaining the alcoholic properties of wines. (Sittings of 7th July and 6th October 1874.)
Declared by the Institute of France to be the best process known hitherto for weighing alcohol in wines. (Report of the Academy of Science, vol. 80, No. 17. Sittings of 3rd May 1875.)

Set of Tephrylometers.   *Dr. Herbert Major.*

APPARATUS, DIAGRAMS, &C., ILLUSTRATING THE CHEMISTRY OF
THE GRAPE VINE, EXHIBITED BY PROF. DR. LEONARD
ROESLER, KLOSTERNEUBURG.

2745. Model of Apparatus, designed to impregnate divers
fluids, *e.g.*, must, wort, &c., with air, before fermentation.

2746. Diagram, exhibiting the results of analysis of must,
grapes, and trial-manurings, 10 tables.

To render the proportions of the various substances more conspicuous,
their quantities are represented by curves.

2747. Grape-growing Map of Tyrol, with results of
analysis of Tyrolean wines at the station.

2748. Grape-growing Map of Portugal, with results of
chemical analysis of Portuguese wines.

2749. Results of Wine Analysis, with specimens of
chemicals.

2750. Wine Ashes, prepared by means of electricity, by Dr.
Roesler.

2751. Apparatus for Calcination, after a design by Dr.
Roesler.

2752. Apparatus for Wine Analysis, in box.

**2753. Apparatus for studying the Influence of Elec-
tricity upon Wine.**

2754. Cheap Electric Battery, adapted for electrolytical
experiments.

2755. Pneumatic Pump, with arrangement by which the sap
of plants can be extracted, devoid of air, for studying the influence
of oxygen upon fermentation.

2756. Apparatus for the Filtration of Fruit-saps, without oxygen.

2757. Diagram, showing the results of water analysis, with
special reference to apparatus for determining nitric acid.

2758. Diagrams (5), illustrating apparatus employed in experiments on the growth and respiration of grape vines and fruit.

2759. Two Portfolios, containing diagrams referring to
wine analysis, microscopical research, the laboratory fittings for
fermentation, &c., at the experimental station.

2760. Soil Thermometer of peculiar construction.

VII.—METALLURGY.

2772. Old Cupellation Furnace, supposed to have been the one used by Sir Isaac Newton, when Master of the Mint, in some experiments on the cupellation of silver. *The Master of the Mint.*

In general construction it is precisely similar to those now in use, the only difference being that, in modern forms, more perfect means are adopted for regulating the draught.

2773. Touchstone for the Assay of Gold, formerly used in the Royal Mint. *The Master of the Mint.*

The method is based on the fact that the greater the amount of gold contained in an alloy, the brighter is the gold yellow colour of a streak drawn with it on a black ground, and the less is it attacked by pure nitric acid or by a "test" acid. In ascertaining the richness of the alloy under examination its streak is compared with marks drawn with alloys whose richness is accurately known.

2770. Case showing successive processes of Gold Assaying.

1. Tray on which assays are placed when ready for the furnace.
2. Muffle containing cupels.
3. Tray for annealing the buttons after being rolled or flatted.
4. Tray of platinum cups for "parting" the assays.
5. Platinum boiler in which the assays are treated with strong nitric acid.
6. Assay balance, capable of indicating the $\frac{1}{3000}$th of a grain when loaded with 7·5 grains in each pan.
7. Mould for making cupels, each of which holds four assays.
8. Cupel tongs.
9. Assay tongs. *The Master of the Mint.*

The process of gold assaying comprises six distinct operations.

1st Process.—The portion of metal to be assayed is adjusted to an exact weight by cutting and filing (*see* specimen A. in Case). Such accurately weighed portions of alloy are wrapped, together with definite weights of pure silver (B), in capsules of lead foil (C C¹), and placed in order on a tray, 1.

2nd Process.—The packets are transferred to porous cups or "cupels" of phosphate of lime, which are arranged in rows, corresponding to those on the tray, in a muffle or small oven, 2, which is fixed in a suitable furnace and maintained at a bright red heat. The lead oxidizes and is absorbed by the cupel, together with the copper and other oxidizable metals present, and the silver and gold remain behind in the form of a button (D), which may also contain platinum, iridium, or other metals possessing similar properties.

3rd Process.—These buttons are hammered out into discs (E), which, after being annealed, are rolled into thin strips (F), and these are again annealed and bent into loose coils or "cornets" (G). The annealing takes place in an iron tray, 3.

4th Process.—The cornets are placed in small perforated cups of platinum arranged in a perforated tray of the same metal, 4. The whole is then introduced into a platinum boiler, 5, which contains boiling nitric acid of specific gravity 1·26, in which it is allowed to remain for 15 minutes; it is then transferred to a similar boiler with acid of specific gravity 1·31. The silver is removed by the action of the acid, and the gold remains in a spongy state (H).

5th Process.—The gold sponge (which retains the original form of the cornet) is rendered coherent by annealing at a dull red heat. This is accomplished by introducing the platinum tray into the muffle. The cornets then assume the appearance of (I).

6th Process.—The final operation consists in weighing the cornets. This is done in a specially constructed "assay" balance, 6, which is capable of indicating about $\frac{1}{3000}$th of a grain when loaded with 7·5 grains in each pan. The weights employed bear a decimal relation to the original weight of the piece of metal operated upon. The percentage of gold, therefore, present in the alloy, is at once indicated without calculation. The weighing in the first process was of course conducted on the same or a similar balance.

2771. Appliances used in the Assay of Silver.

1. Pipette for "standard" salt solution used in silver assaying.
2. Pipette for "decimal" salt solution.
3. Riemsdijk's apparatus for adding drops of decimal solution.

The Master of the Mint.

The assay of silver can be conducted by cupellation, an operation similar to that already described in the 2nd process of the assay of gold, the only differences being that no pure silver is added to the assay piece, and the operation terminates when the button (D) has been obtained and weighed. The wet method of Gay Lussac is, however, now usually employed for the assay of silver when the "standard" is approximately known, and the alloy contains not less than 50 per cent. of silver. It consists in precipitating the precious metal from the solution of a known weight of the alloy to be assayed, the weight being so adjusted that sufficient silver is present to neutralize a given volume of the solution employed as a precipitant. This solution is usually one of common salt, but hydrochloric acid or hydrobromic acid may sometimes be used with advantage. The assay pieces, having been carefully weighed, are placed in numbered bottles, and a definite amount of moderately dilute nitric acid is added to each, the bottles being then moderately heated to assist solution. The standard solution is then carefully introduced by means of a pipette, and the bottles are vigorously shaken until the precipitate coheres and the solution becomes clear. A cubic centimetre of "decimal" solution, which is $\frac{1}{10}$th as strong as the "standard" solution, is then added to each bottle, and they are again shaken. This is repeated until the decimal solution produces either no cloud or a very slight one. This indicates the conclusion of the operation, as the amount of silver present can be calculated when the weight of salt which is required to saturate it is known.

From the above description it will be seen that the only special apparatus required in assaying silver by the method of Gay Lussac is a pipette for measuring out the "standard" solution, and one for adding "decimal" solution. 1 and 2 are the forms of these used in the Royal Mint. The pipette for standard solution, 1, is fixed in a vertical position and filled by an india-rubber tube from below. The opening at the upper end of the pipette is closed by the finger, the india-rubber tube is removed, and the solution thus accurately measured is added to the contents of a bottle. No. 2 is divided into cubic centimetres, and the additions made by means of it as already described. When great accuracy is required the decimal solution may be added drop by drop by means of the apparatus No. 3, designed by Chevalier Van Riemsdijk, of the Utrecht Mint.

2415e. Appliances for testing Gold and Silver.

All these objects date from the time of the first quarter of the 18th century, and are the property of His Highness Prince Pless, of Fürstenstein Castle. *The Breslau Committee.*

2458. Gay Lussac's Apparatus for assaying Silver by the Wet Way. *Aug. Bel and Co.*

Copper cistern lined with resinous cement to contain the normal solution of salt, 100 standard measures of which correspond with 896-thousandth of fine silver in the assay. Funnel for use in filling the cistern. Pipette of 100 cc, for use with normal solution. Pipette for use with decimal salt solution. Pipette for use with the decimal silver solution. Cage for conveying the bottles to the shaking apparatus. Shaking apparatus, filled with 10 assay bottles in position for use. Dark case for bottle, containing the decimal silver solution. Case in which the assay bottle is placed during the addition of the normal solution to the assay. Hollow pillar, with sponge for absorbing the excess of liquid from the front pipette during the adjustment of the level to gauge mark. Water bath used for heating assay bottles during the solution of assay pieces. Bellows used for removing the nitrous fumes from the assay bottles. Whisk used for stirring the salt solution in the cistern.

2761. Chart with Photographs of an Assay-Balance for Buttons of a Weighing-out Assay-Balance and of a Blowing Apparatus. *C. Osterland, Freiberg, Saxony.*

The balances represented by the photographs have the columnar lifting apparatus constructed by the exhibitor, which not only allows of the beam being raised from the outside, but which renders the displacement of so light a beam impossible. Assay balances of this kind have been made by the exhibitor, which weigh to the 20th or 40th of a milligramme.

2762. Improved Furnace for Puddling Iron. *Jeremiah Head, M.Inst.C.E.*

The object is to utilize a portion of the waste heat which ordinarily is discharged from the chimney, by causing it to heat air to be afterwards supplied for the combustion of the fuel. Part of the chimney is enlarged into a chamber, having a vertical partition extending nearly to the top. One half of the chamber thus divided contains a cast-iron stove pipe, and the other half is provided with a damper.

When the damper is withdrawn the heated products of combustion take the nearest route to the chimney, but when it is closed they are obliged to pass by the more circuitous route, heating the stove pipe on the way.

The air for combustion is injected, by means of a steam jet, into a funnel connected with one side of a divided box, upon which the stove pipe stands. Moistened with the steam it becomes a powerful absorber and radiator of heat. It passes through the heated stove pipe, and afterwards through the back of the furnace into a closed ash-pit, and a portion through tuyeres into the space above the fuel.

It has then attained a temperature of about 650 deg. Fahr. The consumption of coal of this furnace has averaged 12 cwt. 2 qrs. 11 lbs. per ton of puddled bar over two months of ordinary work, including lighting up and lost heats. This is about one half of the usual consumption of fuel. The iron (refined) used per ton of puddled bar in the same time averaged 20 cwt. 2 qrs. 26 lbs. The heating chamber is surmounted by a boiler, intended still further to utilize the waste heat. This, however, is not essential, and is hardly worth the extra expense. An ordinary iron-cased chimney is preferable.

1762a. Description of Revolving Puddling Furnace.

Thomas Russell Crampton.

The peculiarity of this revolving puddling furnace is that one chamber forms the gas producing, combustion, and working chamber, no separate fire-place, fire-bricks, or fire-bars being employed. The heat being produced without smoke by the automatic injection of powdered fuel and air into the chamber. The whole furnace is protected by water circulating between a double casing.

Puddle balls of wrought iron up to 30 cwt. can be produced in one mass without fatigue to the men; the puddling being effected by revolving the furnace mechanically.

2763. Models to illustrate Dr. C. William Siemens' processes for the production of wrought iron from iron ore, and of cast steel, in large quantities.

No. 1. Regenerative Gas Rotative Furnace, from which wrought iron is obtained from iron ores mixed with fluxing and carbon-aceous materials by a direct process. The puddled balls thus made may either be shingled and treated for the production of wrought iron, or be transferred to a steel melting furnace for the production of cast steel.

2. Regenerative Gas Furnace for the production of cast steel in large quantities on the open hearth, from pig iron, puddled blooms, iron or steel scrap, and iron ore.

Geological Museum, Jermyn Street.

2764a. Whitwell's Fire-brick Hot Blast Stove or Oven, as specially designed for heating the blast for blast furnaces.

Thomas Whitwell.

This model, to the scale of 1 inch to 1 foot, represents a stove 22 feet diameter × 28' 6" high, capable of heating 8,000 cubic feet of air per minute to a temperature of 1,400–1,450° Fah. during 60 consecutive minutes, after which it is again re-heated by the furnace gases, the combustion and absorption being so perfect that the products of combustion pass off to the chimney at a temperature of 250° Fah. only. These stoves are largely adapted to furnaces making Bessemer pig iron direct, also for anthracite fuel, and the various qualities of charcoal iron, Cleveland, spathic, spiegeleisen, &c.

The stoves are made of different dimensions to suit situation, but cost from 350l. upwards, according to size. Four stoves to the scale of the model make 500 tons a week of Bessemer iron with 19 cwt. of coke to the ton of iron. There is no loss of pressure by friction or loss of blast by leakage, and they require only two-thirds the quantity of gas that the ordinary cast-iron pipe system demands.

2764b. Set of Drawings of two Blast Furnaces,

erected at Middlesbrough by Messrs. B. Samuelson and Co., together with the requisite heating stoves, kilns for calcining ironstone, blowing engines, &c. &c.   *Bernhard Samuelson, M.P.*

The peculiarity is in the large dimensions of the blast furnaces (height from bottom of hearth to charging plate, 85 feet; diameter of bosh, 28 feet), resulting in a great economy of fuel; so that after they have been in blast nearly six years the quantity of coke required to produce a ton of No. 1 or 3 foundry pig iron is on the average less than 22 cwt.

2767. Sections of **Steel Ingots,** one cast in the ordinary way, the other compressed while in a fluid state.

Sir Joseph Whitworth & Co., Limited.

By the ordinary method of manufacture, it is found to be impossible to produce sound ductile steel suitable for constructive purposes, owing to the presence of honey-combed air-cells, which are altogether uncertain in their size and situation, and undiscoverable until laid bare by fracture or sections.

By compressing the metal while in a fluid state this defect is overcome, and a sound trustworthy material produced. This is shown by the two ingots, one cast in the ordinary way, and the other compressed while fluid.

2765. Tubes tested with **Gunpowder,** to show the strength and ductility of Whitworth fluid compressed steel.

Sir Joseph Whitworth & Co., Limited.

These tubes were tested to ascertain the strength and ductility of fluid compressed steel, as made for guns, torpedoes, &c.

The ductility is shown by the metal bulging under the strain, instead of flying in pieces.

2766. Sample Pieces of **Metal,** used for testing, to ascertain the strength and ductility of metal.

Sir Joseph Whitworth & Co., Limited.

There is no scientific line of demarcation between iron and steel. Sir Joseph Whitworth proposes that such a line should be established, and that the quality of a metal should be represented by two numbers, showing its strength and ductility.

These test pieces are similar to those in use by Sir Joseph Whitworth in testing his fluid compressed steel, to ascertain the proportions of strength and ductility which is required for different purposes.

The greater the strength, and the greater the ductility, the higher the quality of the metal.

2767a. Sample of Iron melted by means of compressed air.

M. Enfir fils, Paris.

2769. Set of cubical specimens of **Coal, Ironstone, Limestone,** and **Cold Blast Iron.** Illustrating the exact proportions, both in weight and bulk, of the minerals, coal, ironstone, and limestone, consumed in the blast furnaces for the production one cubic inch of cast iron at the Bowling Ironworks.

The Bowling Iron Company, Limited.

The coal is coked and the ironstone calcined preparatory to their introduction into the blast furnace.

2774. Diagram, illustrative of a Westphalian blast furnace of most recent construction, for the use of lecturers on metallurgy.

Prof. Dr. Dürre, Aix-la-Chapelle.

2566c. Furnace for Melting Platinum, by Deville's method, with cover and jets. *Johnson, Matthey, and Co.*

Size in which 3,250 ounces troy (about 100 kilos.) is melted in one charge.

2775a. Saturn Steel. *M. Breguet, Paris.*

2775. Diagram, illustrative of a lead smelting furnace on Pilz's principle, for the use of lecturers on metallurgy.

Prof. Dr. Dürre, Aix-la-Chapelle.

These diagrams are drawn to scale, and can also be employed for teaching constructive drawing and as designs for smelting works.

2415d. Seven different metal Wires of equal Strength and Weight, for determining the Specific Gravity from their length, by means of a scale.

The Breslau Committee.

VIII.–MODELS, DIAGRAMS, APPARATUS, AND CHEMICALS USED IN TEACHING CHEMISTRY.

2444. Series of $\frac{1}{2}$-inch cardboard cubes, with different colour for each chemical element on which its symbol, atomicity, and combining weight are marked, illustrating the laws of chemical combination and the binary theory of Salts.

Rev. Nicholas Brady, M.A.

Only a comparatively small number are sent, as they will sufficiently show their method of use, thus $\boxed{\underset{1}{\text{H}} \quad \underset{1}{\text{H}} \quad \underset{1}{\text{O}}}$ represents the molecule of water, the dyad oxygen requiring two monads to satisfy it, the chemical equivalent of the compound being $1 + 1 + 16 = 18$.

2445. Lecture-Room Diagrams of Alkali Works and Sulphuric Acid Plant. In 4 sheets.   *Prof. Roscoe, F.R.S.*

I.—*Sulphuric Acid Plant.*

Sheet No. 1. General plan, showing the arrangement of the most modern form of sulphuric acid plant, drawn to scale, with descriptions.

Sheet No. 2 contains a longitudinal section of the apparatus and chambers on the line C D, sheet No. 1.

Sheet No. 3 contains a sectional elevation of the same apparatus on the line A B, sheet No. 1.

Details on sheet No. 2 and sheet No. 3 show various points of importance on an enlarged scale.

No less than 850,000 tons of sulphuric acid were made last year in Great Britain. The diagrams show the most approved method for the manufacture of this important acid, together with the best arrangement of the plant, and are so fully numbered and explained that further description appears unnecessary.

II. *Alkali Works.*

Sheet No. 4 is an enlarged plan of a model alkali works, showing the best arrangement of the various portions of the works; also fully lettered and numbered. As in the above case a description appears unnecessary.

2442. Drawings, two, illustrating the making of.coal-gas, used for teaching chemistry in the Secondary Schools.
　I. Coal-gas factory.
　II. Regulator of coal-gas factory.
　　Dr. D. de Loos, Director of the Secondary Town School, Leyden.

2452. Model of the Chemical Laboratory of the Secondary Town School, Leyden. Constructed by J. Noest, custodian of the building.
　　Dr. D. de Loos, Lecturer on Chemistry, and Director of the Secondary Town School, Leyden.

A. Forcing-pump. The pump fills a large cistern in the upper part of the building. This cistern discharges its contents into the *white* pipes, and supplies the laboratory and the other parts of the building with water. The *red* pipes carry off the drainage from the laboratory into a sink constructed below the surface. The large *black* pipes, as also the small ones over the pupils' work-tables, are the gaspipes.
　B. Register-school.
　C. Chimney with three passages ; one to conduct off the smoke of the stove, the others to carry off the bad gases disengaged in chemical operations.
　D. Pump for rain-water and well-water.
　E. Cupboards for twenty-two pupils, in which they keep the re-agents ; the largest, which is open, is for the use of the lecturer on chemistry.
　F. Drawer for small apparatus, as spoons, glass tubes, &c.
　G. Small cupboards for the use of the pupils, in which to keep chemical preparations still in progress.
　H. Bunsen's pump to filter under a lesser pressure. The water, running through the red pipe from the top to the bottom of the laboratory, carries the air from the bottle along with it.
　I. Entrance to the balance-room.
　K. Collection of implements for performing chemical experiments.
　L. Large cupboard for chemical preparations and apparatus.
　M. Table with apparatus for blowing glass.

2455b. Case containing five stands and a collection of spheres made to **demonstrate** the **structure of chemical combinations** according to the theories of A. W. Hofmann.
　　Manuel Gonzalez, Madrid.

2453. Diagram of Carre's Ice-making Machine, by Mr. Hubertus Sattler. 　　　　*Dr. Alexander Bauer, Vienna.*

Adapted to be used for making experiments at lectures on technical chemistry to small audiences.

2408. Apparatus for exhibiting in the lecture room the contraction which takes place when oxygen is changed into ozone. It is a modified Siemens' tube : the contraction being shown by the ascent of the sulphuric acid in which the open end of the tube is immersed. 　　　　　　　　*Dr. Andrews, F.R.S*

2575. Filtering Apparatus, provided with a glass receiver and 2 india-rubber rings and supports.

F. A. Wolff & Sons, Heilbronn and Vienna.

This filtering apparatus is intended for the use of schools.

2456h. Cavendish's Eudiometer.   *E. Cetti & Co.*

2575a. Model and Section of Spongy Iron Filter, with filtering materials employed.   *Gustav Bischof.*

Spongy iron is metallic iron in a spongy or porous state, obtained by the reduction of an oxide without fusion.   The powerful property of metallic iron to purify water has been known for a long time, but its application in the spongy state only renders the purification sufficiently rapid to be of practical use.   The water in passing through the spongy iron dissolves a minute quantity of iron, which is completely retained by a mixture of sand and pyrolusite underneath.

Pyrolusite has the property of converting protosalts of iron into persalts. The latter are again decomposed, probably by the agency of calcic carbonate in water, and the ferric hydrate formed is mechanically retained by the sand. Ammonia and other substances are likewise oxidised by pyrolusite.

The hardness of water is very considerably reduced, and every trace of lead removed by filtration through the spongy iron filter.   (See VI. Report of the Royal Commission on Rivers Pollution, 1875, pp. 220, 221 ; Report of the Registrar General, 8th January 1876 ; Journal of the Royal Agricultural Society of England, Vol. XI., part I., 1875, p. 158.)

A small lateral opening in the regulator tube forms the only communication between the upper part of the filter and the reservoir for filtered water. The flow of the water is thus completely controlled by the size of such opening to ensure a sufficient contact between the purifying medium and the water.

2455a. Apparatus for showing the **Decomposition** of **Steam** by the **Heat** of **Electric Sparks.**

Prof. Frankland, F.R.S.

SECTION 14.—METEOROLOGY.

WEST GALLERY, GROUND FLOOR, ROOM L.

I.—SPECIAL COLLECTIONS.

METEOROLOGICAL INSTRUMENTS, A SET AS LENT TO CAPTAINS
OF MERCHANT SHIPS, COMPRISING :—

2776. Board of Trade Marine Barometer, Kew pattern ;
the best barometer yet devised for use at sea, serviceable also as
a station barometer; by Adie, marked B.T. 12 ; certificate of
Kew verification in box.

2777. Board of Trade Marine Thermometers. Set of
six, Kew pattern ; one for sea-water, in copper case, one for air,
and one for evaporation temperatures, with three spare to replace
breakages.; by Casella, marked B.T. 1720 to 1725 ; certificate of
Kew verifications in box.

2778. Board of Trade Thermometer Screen, for " dry-
bulb " and " wet-bulb " thermometers, with water-cup. For use
on ship board.

2779. Board of Trade Marine Hydrometers. Set of
four; by Casella, marked B.T. 610, 620, 630 ; for testing the
specific gravity of sea-water; by Hicks, A. 170, for the denser
water of the Suez Canal; certificate of Kew verifications in box.

2780. Board of Trade Prismatic Azimuth Compass, in
box. This instrument is lent only in a few special cases.

2781. Meteorological Register, for observers using the
above instruments at sea.

2782. Rough Register, for use at sea. The observations
should be entered at the time of making them, and afterwards
copied into the preceding register.
Meteorological Committee of the Royal Society.

2783. Equipment of a Second Order Station, approved
by the METEOROLOGICAL SOCIETY, consisting of the following
instruments :—

STANDARD BAROMETER.
THERMOMETER STAND, containing DRY BULB THERMOMETER,
WET BULB THERMOMETER, MAXIMUM THERMOMETER, and MINI-
MUM THERMOMETER.

40075.   U u

RAIN GAUGE.

BLACK and BRIGHT BULB THERMOMETERS *in vacuo*, for determining the intensity of solar radiation.

MINIMUM THERMOMETER for determining the intensity of terrestrial radiation.

ANEMOMETER.

With specimen forms, instructions for taking, and tables for reducing, the observations.

Publications of the Meteorological Society :—
Quarterly Journal, Vols. I. and II., 1872–1875.
List of Fellows, January 31st, 1876.
Catalogue of the Library, revised to December 31st, 1875.
Instructions for the observation of phenological phenomena,
    charter, and byelaws.   *Meteorological Society.*

2784. Instruments used at the Russian Meteorological Stations. A specimen set.

Wild's adjustable syphon barometer, No. 45, with screw for suspension and key for valve.

Three thermometers by Geisler, dry bulb and wet bulb, No. 244, minimum No. 138.

Thermometer-screen, No. 141, made of zinc, to revolve round its vertical axis, carrying the above thermometers.

Saussure's hair hygrometer, No. 9, in box.

Russian rain and snow gauge, with glass measure, &c., No. 118.

Wild's anemometer on iron swinging plate, for observing direction and pressure of wind.

Meteorological Committee of the Royal Society.

2786. Meteorological Instruments, made under the direction of the late Sir John Leslie, and used by him.
Scottish Meteorological Society.

1. Portable Hygrometer (with specimen of Sir J. Leslie's writing).
2. Stationary Hygrometer.
The covered ball in both cases being wetted and exposed to evaporation, the liquor soon marks, by its descent in the opposite stem, the dryness of the air.
3. Leslie's Atmometer. It consists of a ball of thin porous earthenware, to which is cemented a wide glass tube, bearing divisions which correspond each to the measure of a film of water that would cover the external surface to the thickness of the thousandth part of an inch.

1. Two volumes of the Society's Quarterly Reports from 1856 to 1863.
2. Vols. I., II., III., and IV. of the Society's Journal from 1863 to present date.
3. Three Papers by A. Buchan, Secretary, from Trans. Roy. Soc. Edin.   *Scottish Meteorological Society.*

2787. Meteorological Instruments, set, for the use of colleges and public schools. *Francis Pastorelli.*

These instruments when compared with the standards at the Kew Observatory will be found to be within the error permitted by that institution. They consist of a comparative standard barometer, maximum and minimum thermometer, wet and dry bulb hygrometer, terrestrial radiation thermometer, solar radiation (in vacuo) thermometer, rain gauge and graduated measure.

2788a. Special set of Observatory Standard Meteorological Instruments. Negretti and Zambra's special selection of standard meteorological instruments for a first class observatory station, consisting of :—Standard barometer, Negretti and Zambra's patent maximum thermometer, Rutherford's minimum thermometer, Negretti and Zambra's patent mercurial minimum thermometer, hygrometer, wet and dry bulb, Negretti and Zambra's patent solar radiation thermometer, terrestrial radiation thermometer, Robinson's anemometer, rain-gauge (Glaisher's pattern). *Negretti and Zambra.*

2788. Specimens of Meteorological Instruments, supplied by the Meteorological Office to the Royal Navy :—

Kew-pattern Marine Barometer, iron cistern, tube protected with India-rubber packing against damage by concussion from gun-firing, &c. ; by Adie, marked A 430; certificate of Kew verification in box. Glass tube and packing, as fitted to the above barometer.

Pair of Thermometers, maximum ⋀ A 369, and minimum⋀ A 394; constructed specially for use at sea in chronometer-rooms, by Hicks. Certificates of verification in box.

Aneroid, compensated for temperature, marked Elliott, A 421.

N.B. The thermometers, screen, and hydrometers supplied to Her Majesty's ships are of the same pattern as those lent to merchant ships, Nos. 2777–79.

Portable Instruments used by Admiralty Surveyors : Pocket Aneroid, compensated, Casella,⋀ 6 (548).

Pocket Thermometer, mounted in boxwood, in leather case, Casella, No. 13.

Portable thermometer (thermometre fronde), in padded tin case, Hicks, B. T. 499; to give the temperature of the air when whirled round at the end of a string. *Meteorological Committee.*

2789. Specimens of Instruments used at Stations of the Meteorological Office :—

Barometer, used at Coast Stations as a weather-glass, by Hicks, M. O. 143.

Stevenson's Thermometer Screen, by Hicks.
Maximum Thermometer, by Negretti and Zambra, B. T. 215
(5577).
Minimum Ditto, B. T. 216 (10,435) ; certificates of verification
in boxes.
Copper Rain, and Snow Gauge, Can, and Glass, by Casella,
M. O. 150.
Solar Radiation Thermometer, in vacuo ; Negretti's Maximum,
with platinum points for testing, Hicks, M. O. 10.
Terrestrial Radiation Thermometer, with hollow bulb for sensi-
tiveness, Hicks, 18,429.   *Meteorological Committee.*

2789a. Milne's Barograph, by West, clockwork by Schoof.
Meteorological Office.

2789b. Portable Barometer, graduated on the glass tube,
with a sliding vernier, invented and used by Sir John Richardson, M.D., F.R.S., in his expeditions in North America.
Meteorological Office.

2789c. French Station Barometer, by Tonnelot, J.G. 48,
cistern of large diameter to diminish capacity error.
Meteorological Office.

2789d. Von Lamont's Atmometer. *Meteorological Office.*

2789e. Wall Screen, for registering thermometers.
Meteorological Office.

2789f. Tube, Lath, and Thermometer, by Hicks A 164,
for taking earth temperatures.   *Meteorological Office.*

2789g. Hypsometer Apparatus, as improved by Dr. G.
Henderson, with two maximum thermometers, by Hicks, 39621,
39622, in leather sling-case, with certificates.
Meteorological Office.

2789h. Johnson's Deep-sea Pressure Gauge, ⋀ 1, with
box.   *Meteorological Office.*

2789i. Deep-sea Thermometers :—
Johnson's metallic, ⋀ 6.
Negretti's protected bulb, ⋀ 9.
Casella's protected bulb, 21260, with certificate.
Elliott's unprotected, with certificate, ⋀ 36.
Pastorelli's unprotected, No. 5, out of order, used by Sir J. C.
Ross. Notwithstanding the thickness of the glass a similar instru-
ment to this rose 9·2°, under a pressure of two tons on the square
inch.   *Meteorological Office.*

**2789k. Quarterly Journal of the Meteorological
Society,** Vols. I. and II., 1872 to 1875. *Meteorological Society.*

SET OF INSTRUMENTS made on the KEW PATTERN, by P. ADIE, for the PHYSICAL OBSERVATORY, at POTSDAM.

27891. Magnetometer. *Dr. H. W. Vogel.*

2789m. Declination Magnetometer. *Dr. H. W. Vogel.*

2789n. Vertical Force Magnetometer.
Dr. H. W. Vogel.

2789o. Automatic recording Apparatus.
Dr. H. W. Vogel.

II.—BAROMETERS.

a. MERCURIAL.

2790. Large Standard Barometer, $\frac{7}{10}''$ tube, with English and metrical scales, glass plunger cistern, and pipette in tube for preservation of vacuum. *Patrick Adie.*

2791. Medium Standard Barometer, $\frac{5}{10}''$ tube, with glass plunger cistern, and pipette in tube for preservation of vacuum. As used in Government meteorological observations.
Patrick Adie.

2792. Small Standard Barometer, with glass plunger cistern, and pipette in tube for preservation of vacuum. Medical Department, War Office pattern. *Patrick Adie.*

2793. Large Standard Barometer, with single readings. Kew Observatory verification, and pipette in tube for preservation of vacuum. Kew pattern. *Patrick Adie.*

2794. Medium Standard Barometer, with single readings, Kew Observatory verification, and pipette in tube for preservation of vacuum. Kew pattern. *Patrick Adie.*

2795. Mountain Barometer, Fortin's, on tripod, with pipette in tube for preservation of vacuum. *Patrick Adie.*

2796. Gay-Lussac's Portable Barometer, with pipette in tube for preservation of vacuum. *Patrick Adie.*

2796a. Standard Barometer with Electrical Adjust-ment. *Negretti and Zambra.*

This is a tube dipping into a glass cistern of mercury, fitted with a vertical adjustment screw. Through the top of the tube a platina wire is passed and hermetically sealed. The cistern has a metallic connexion, so that by means of copper wires (in the frame) a galvanic circuit is established; another connexion also exists by a metallic point dipping into the cistern. The circuit, however, can be cut off from this by means of a switch placed about midway up the frame. On one side of the tube is placed a scale of inches, with a circular vernier, divided into 100 parts, connected with the dipping point and working at right angles with the scale.

2796b. Negretti and Zambra's Standard Barometer, constructed on Fortin's principle, proved to be the most reliable arrangement yet introduced. The level of the mercury in the cistern being adjusted previous to each observation to a fixed ivory zero-point, loss of mercury from leakage or oxidation is of little or no importance, and does not affect the accuracy of the readings. The tubes are of varying internal diameter, and are filled with pure mercury, very carefully boiled in the tube to perfectly expel all air or moisture. *Negretti and Zambra.*

2796c. The Gun Marine Barometer, constructed by Negretti and Zambra for special use in Her Majesty's navy, and now the adopted Admiralty pattern. It differs from barometers hitherto made by having its tube packed with vulcanised india-rubber, which checks vibration from concussion, thereby doing away with the necessity of unshipping the barometer during gun firing.

See Admiral Fitzroy's report, 9th number of the Meteorological Papers, issued by the Board of Trade. *Negretti and Zambra.*

2796d. Negretti and Zambra's Self-registering Mercurial Barometer. *Negretti and Zambra.*

In this instrument the various parts of the mechanism have been so arranged that the recording is effected by means of a clock, which causes a drum to revolve, carrying round it a paper. On this paper is traced the barometric curve by a pencil attached to a float, and placed on the mercury of a syphon barometer tube.

See Negretti and Zambra's Treatise on Meteorological Instruments.

2796e. Howson's Patent Long Range Barometer.
Negretti and Zambra.

In this barometer the tube is fixed, but its cistern is sustained by the upward pressure of the atmosphere. Looking at the instrument it appears as though the cistern with mercury in it must fall to the ground. The bore of the tube is about an inch across. A long glass rod is fixed to the bottom of the cistern. The tube being filled with mercury, the glass rod is plunged into the tube as it is held upside down, and when it is inverted the mercury partly falls and forms an ordinary barometric column. The cistern and glass rod, instead of falling away, remains suspended. There is no material support to the cistern; the tube only is fixed,.the cistern hangs to it. The glass rod, being so much lighter than mercury, floats and sustains the additional weight of the cistern by its buoyancy, and the almosphere acting on the mercury keeps up the barometric column.

2796f. Long Range Barometer. *Negretti and Zambra.*

This instrument consists of a mercurial tube on the syphon principle, one side of the syphon, the closed end, being about 33½ inches long, and the other only a few inches in length. To this short end or leg is joined a length of glass tubing of a much smaller (internal) diameter, both legs being of equal length; the smaller tube is filled with a fluid many times lighter in specific gravity than mercury, the rising and falling of the mercurial column in the large tube having a lighter fluid to balance, and that dispersed over a larger space by reason of the difference in the diameter of the two tubes, a longer range is obtained due both to the unequal capacity of the two tubes and the difference in the specific gravity of the mercury and the second fluid employed. The range of these barometers is from 6 to 10 inches to the inch of the ordinary mercurial barometer; the $\frac{1}{100}$th of an inch can easily be observed without the use of a vernier.

It is a most interesting instrument, as from the extremely extended scale the slightest variation is plainly visible.

Large Household Barometer, for ordinary use.
Negretti and Zambra.

Domestic Standard Barometer, of the same construction as 2796*b*, but of smaller size. *Negretti and Zambra.*

2797. Mercurial Barometer, an old Dutch instrument by Reballio, combining syphon and long range barometer, thermometer, and hygrometer. *Pillischer.*

2798. Drawing of a " Balance " Barometer, of which a model was executed, submitted to the Royal Irish Academy, and tested during some months. *Jos. P. O'Reilly.*

The column is inclined from the vertical, and suspended by a knife edge, as the beam of a balance, whence the proposed name. The displacement of the mercury in the column causes this to incline more or less from the vertical, the amplitude of movement showing itself on a graduated limb by means of an index. The mode of action and the degree of sensitiveness of the instrument are therefore comparable to those of a beam balance, and the indications given without the intervention of wheel work.

2799. Barometer of De Luc, formerly belonging to **H. B. de Saussure,** and carried with him in his Alpine excursions.
H. de Saussure, Geneva.

2800. Meteorological Barometer by Wild, used in all meteorological stations throughout Russia.
Geneva Association for the Construction of Scientific Instruments.

This barometer is a combination of Fortin's barometer, and Gay-Lussac's old siphon barometer. For minute description, see " Mélanges physiques et chimiques, tirés du Bulletin de l'Académie Impériale des Sciences de St. Pétersbourg," vol. ix, 23rd September to 5th October 1875.

2800a. Barometer of a New System of Construction.
W. Gloukhoff, Ministry of Finance, St. Petersburg.

For taking an observation with this barometer, the mercury of the cistern is forced to pass by means of the screw *A* through the hole *a* into the open space *m n*, formed by a glass ring, and to cover its bottom. Then the movable scale of the barometer is lowered so as to bring the steel end C of the scale near the surface of the mercury. Finally, by the screw A the surface of the mercury is made to touch the end C of the scale. The last part of the observation is made in the usual way.

2800b. Drawing of a Normal Barometer and Manometer of the Central Physical Laboratory, St. Petersburg, after the design of Wild, constructed by Brauer, St. Petersburg.

Dr. H. Wild, Director of Central Physical Observatory, St. Petersburg.

2800c. Drawing of a Balance Barograph, with temperature compensation according to a design by Wild, constructed by Hasler, Berne (Switzerland), registering by electricity every ten minutes.

Dr. H. Wild, Director of Central Physical Observatory, St. Petersburg.

2801. Stevenson's Portable Iron Barometer, with Mr. Sang's improvements.   *Scottish Meteorological Society.*

All the stop-cocks being open, and the plug at lower limb out, fill in mercury till it begins to escape at this opening, shut stop-cock of lower limb, and fill to above upper stop cock of upper limb; shut both stop-cocks of longer limb, put in the plug and open stop-cock at lower limb. The float should then show the true reading as compared with a standard instrument. In this way a reading can always be obtained at the most inaccessible stations as accurately as when the instrument left the maker's hand. Designed by T. Stevenson, C.E., F.R.S.E., Honorary Secretary, and described in the Society's Journal, Vol. iv, p. 265.

2801a. Capt. George's Improved Patent Mercurial Barometer for travelling.   *Henry Porter.*

This process of filling by the spiral cord enables the traveller to fill the tube, and produce a perfect vacuum in about 15 minutes, and when the observation is registered the mercury is returned to its iron bottle and the *empty* tube returned to its packing of india-rubber and brass tube, so that it can be carried over the roughest country in perfect safety.

Hicks's Patent Flexible Mountain Barometer.
          J. J. Hicks.

It consists of a flat bulb of flexible glass filled with mercury exhausted of all air and hermetically sealed at both ends. The inches are divided on the glass tube itself, which is mounted on a metal scale with small attached thermometer and sliding scale to compensate for temperature.

2802. Mercurial Barometer; may be rendered entirely void of air in half an hour without boiling.
          Prof. Dr. Bohn, Aschaffenburg.

The instrument is easily constructed, even by persons without special training, at the place of observation itself. It is thus well adapted for transport. Permits controlling of the vacuum and avoids the errors of capillarity. No boiling of the mercury is required, and thus no loss of the graduated tubes is risked. The expense of constructing the instrument is very small.

2803. Barometer with movable bottom, Kupfer's method, improved by Köppen ; constructed by Fuess, Berlin.
Imperial Admiralty Hydrographical Bureau, Berlin, and Deutsche Seewarte, Hamburg.

Instruments of this kind will in future be used as normal barometers in the meteorological system of the Naval Observatory.

2804. Glycerine Barometer.   *Jas. B. Jordan.*

This instrument is designed for the purpose of affording a delicate " weather glass," indicating small changes of pressure by large oscillations of a fluid column, at the same time preserving all the accuracy of the mercurial barometer. The fluid used is glycerine, in a maximum state of purity, which has a specific gravity of $1\cdot26$, or about one tenth that of mercury. It has the advantage of giving a vapour of very low tension in the Torricellian vacuum from its high boiling point, and is therefore free from the masking effect of back pressure which interferes with the indications of a water barometer. The fluctuations of the column are observed in a glass tube of 1 inch sectional area, or 100th that of the cistern. The tube forming the body of the instrument is an ordinary composition gas pipe, $\frac{5}{8}$ inches diameter, 27 feet long, placed in the well of the staircase, between the upper and lower galleries. The exposed surface of the glycerine in the cistern is protected by a layer of paraffin oil, in order to prevent absorption of moisture.

The divided scale on the right hand side is in inches and tenths in absolute measure, while that on the left shows the equivalent values reduced to a column of mercury.

Instruments of this class can be constructed in the most simple form and at a moderate cost, and for museums and public institutions would be of great interest.

2805. Compensated Barometer.
Prof. Dr. A. Krueger, Helsingfors.

The upper part of the tube is enlarged to a retort, the volume of which corresponds to a length of one or two metres of the tube. A quantity of air is introduced into the tube, the pressure of which is equal to about $34\cdot5^{mm}$ mercury. The scale is divided with due regard to the effect of that depression. The zero having been adjusted by comparison with a standard barometer the reading will immediately give the barometer height reduced to normal temperature of mercury and scale. One millimetre of the scale

$$1^{mm} = \frac{P+p}{P} + \frac{34\cdot5}{\lambda}, p$$ being the area of the tube, P of the cistern, and $p.\lambda$ the volume of the upper part above $725\cdot5^{mm}$ from the level in the cistern. An instrument of this construction has been in use some years at the Helsingfors' Observatory.

2806. Standard Barometer, mounted in metal frame, with glass cistern, and pointed for adjusting 'the mercury before an observation is taken.   *Elliott Brothers.*

2807. Diagonal Barometer invented by Sir Samuel Moreland, and made by T. Whitehurst, of Derby, 1772.
The Committee, Royal Museum, Peel Park, Salford

The action of this form of barometer is explained in Rees' Cyclopedia, vol. 3., 1st edit., 1819.

2807a. Anerora or Mercurial Barometer, suitable for public buildings, at seaports, &c. Diameter of the dial, 1½ metre.

M. Redier, Paris.

2807b. Antique Baroscope.   *G. J. Symons.*

2807c. Two Mountain Barometers (old forms).

G. J. Symons.

2808. Standard Metal Marine Barometer, Board of Trade pattern, as supplied to H.M. ships of war.

Francis Pastorelli.

The frame and cistern are of metal, bronzed, suspended by gymbals and a spring metal arm ; it has a rotary motion to obtain the best light for observation. The barometer scale is divided to inches, tenths and 0·05 of an inch, the vernier, by means of a rack and pinion, works between two longitudinal openings ; it reads direct to 0·002 of an inch, and by estimation to 0·001. The divided portion of the brass tube is protected from dust and moisture by a glass shield ; the barometer tube is surrounded and packed by india-rubber to resist breakage by the discharge of heavy guns.

The barometer tube is made with a glass air-trap (a small portion of air ascending to the top of the tube would cause a great and variable error) ; this prevents the air from passing up to the top of the tube, which might occasionally happen with the barometer in careless usage. The interior diameter of the tube is about 0·35 of an inch. The mercury is carefully boiled in the tube to expel all particles of air and moisture. It is so contracted that an inch fall of mercury occupies four minutes of time ; this is to prevent the oscillation of the mercury by the ship's motion. The scale divisions are corrected so that the error arising from the displacement of the zero by a rise or fall of the mercury in the cistern does not cause an error (in a well made barometer) of more than 0·008 of an inch.

2808a. Pastorelli's Mountain Barometer, in metal frame, similar in form to the Comparative Standard Barometer, specially designed for the use of civil engineers and scientific travellers.

F. Pastorelli.

Its great portability may be judged from the fact that its weight does not exceed 1¼ lbs. Another very great advantage is that it cannot be deranged by careless use. The greatest error in this instrument rarely exceeds ·008 in. throughout its scale. It can be confidently recommended to engineers for a preliminary survey where greater accuracy is required than can be attained by use of the aneroid.

2809. Standard Barometer, upon Fortin's principle.

Francis Pastorelli.

The barometer tube is enclosed in a brass frame ; connected with the tube is a glass cistern, which is fixed by three pillars, the ends of which have screws passing through an upper and lower brass plate, by means of which the necessary pressure can be applied to make it mercury-tight ; at the bottom of the cistern is a leather bag, which is raised or lowered by an adjusting screw, permitting the surface of the mercury to be brought into perfect contact with a piece of ivory which forms the zero of the scale; this point is seen through the glass cistern. The vernier works between two longitudinal openings ; it is moved by a rack and pinion, so that it may be adjusted to the apex of the surface of the mercurial column, and it is divided to read to the 0·002 of an inch, and by estimation to the ·001 of an inch; it can be divided to read directly to ·001 of an inch, but in that case the use of a magnifying glass or microscope is indispensable. The thermometer has the divisions etched upon the stem ; it is fixed on the brass frame nearly in contact with the mercury tube. The barometer is suspended from a bracket fixed to a mahogany board, having a lower bracket with adjusting screws to fix it in a truly vertical position. The instrument permits of a rotary motion, in order to obtain the best light for observation. The internal diameter of the tube of this barometer is ·44 of an inch ; the mercury is carefully boiled in the tube in order to insure the expulsion of all particles of air or moisture.

2809a. Carved Oak Barometer.   *E. Cetti & Co.*

2809b. Small Pocket Standard Marriotte Barometer, Macneill's Patent, combining in itself nearly equal portability with the aneroid, and the constant correctness of the standard barometer, its action depending on the well-known laws of Marriotte and Boyle applied to the expansion and compression of air.   *L. Casella.*

2809c. First Barometer, with weights, by Conté. Used in the expedition of Egypt.
Conservatoire des Arts et Métiers, Paris.

2809d. Barometer, with overfall. Constructed by Meynie for Lavoisier.   *Conservatoire des Arts et Métiers, Paris.*

2809e. Skeleton of Construction of the Largest Barometers for Public Buildings. (For demonstrating purposes.)
M. Richard, Paris.

2809e. Metal Barometer.
Dr. Wilhelm Tinter, Professor of Practical Geometry at the I. R. Polytechnic Institute, Vienna.

2809f. Metal Barometer, of great sensitiveness (diameter, 0·20m), of which the index describes a complete circle, under a differential pressure of one millimetre of mercury.
M. Richard, Paris.

2809g. Metal Barometer, for measuring heights (diameter, 0·14m), of which the index describes a complete circle, by a difference of pressure of one centimetre of mercury.
M. Richard, Paris.

2809h. Aerostatic Barometer, with equal divisions, the index of which describes a complete circle, by a difference of pressure of one decimetre of mercury.   *M. Richard, Paris.*

b. ANEROIDS.

2810. Aneroid Barometer. The ends of the axle which carries the index hand are jewelled like the pivots of a watch, and the hand works under the cap. By this means greater sensitiveness and especially greater definiteness of the indications are obtained. This aneroid will show a difference in height of 2 feet.   *The Hon. Ralph Abercromby.*

2811. Aneroid Barometer, capable of measuring up to 5,000 metres from the level of the sea. With case, tables of comparisons, instructions, &c.   *J. Goldschmid, Zürich.*

2812. Aneroid Barometer, capable of measuring from 9,000 to 10,000 metres from the sea level. With case, tables, and instructions, &c.   *J. Goldschmid, Zürich.*

2813. Pocket Aneroid Barometer, of German silver. With case and instructions, &c.   *J. Goldschmid, Zürich.*

2814. Aneroid Barometer, Weilenmann system. With tables.   *J. Goldschmid, Zürich.*

2815. Aneroid Barometer, capable of measuring up to 5,000 metres. With tables, &c.   *J. Goldschmid, Zürich.*

The faces of the above aneroid barometers are of German silver. The variations of reading are measured by a fine micrometer. A table, specially prepared, accompanies each instrument, and gives the height of the barometer. The correction for temperature, given in a second table, is founded on the observation of a small thermometer applied to the instrument.

2816. Miniature Aneroid Barometer, the dial measuring $\frac{5}{8}$ inch, the case $\frac{6}{8}$ inch in diameter, the bearings set in jewelled centres, compensated for temperature.   *Pillischer.*

This is believed by the maker to be the smallest instrument of the kind ever constructed.

2816a. Pocket and Watch-sized Aneroid Barometers.
Negretti and Zambra.

The patent for the aneroid having expired, the late Admiral Fitzroy urged upon Negretti and Zambra the desirability of reducing its size as well as improving its mechanical arrangement, and compensating it for temperature. They after great labour and numerous experiments, succeeded in reducing the dimensions to less than three inches in diameter and two inches in thickness. These instruments have since been further reduced by Negretti and Zambra, until they have at last reached the dimensions of an ordinary watch. These very small instruments are found to act quite as correctly as the largest, and are much more convenient; they may be had with a range sufficient to measure heights of 20,000 feet, with a scale of elevation in feet, as well as of pressure in inches.

2817. Two Aneroid Barometers, exhibitor's construction, with visible movement.   *R. Deutschbein, Hamburg.*

2818. Two Metal Barometers, exhibitor's construction, with visible movement. *R. Deutschbein, Hamburg.*

2819. Two Spring Barometers, exhibitor's construction, with visible movement. *R. Deutschbein, Hamburg.*

The first four are house barometers, distinguished by good workmanship, shape, and cheapness. The last two instruments are spring barometers, specially adapted to meteorological observations, determination of heights, &c.

The aneroid barometers of the Reitz system are metal barometers with a vacuum box according to Vidi. The movements of the box are read off by means of a microscope on a scale which is divided into hundredths of milli-mètres. Each degree can easily be further subdivided by estimation into ten parts. One such tenth corresponds to about $\frac{1}{30}$ mm. of the mercurial column, or a difference in height of $\frac{1}{4}$ meter.

The instrument is chiefly destined for the preparatory tracings of railway lines and similar engineering works, and also for the observation of minute oscillations in the atmospheric pressure. Its construction is considerably simpler than usual; optical means serve to magnify the movements of the vacuum box, instead of the customary transmission by lever; the instrument is easily handled in the field. Care has been taken in the correct applica-tion of the thermometer, which is to indicate the temperature of the most important parts of the instrument.

2820a. Field's Engineering Aneroid, with extra com-pensation for the temperature of the air. *L. Casella.*

2820b. Open Range Aneroid, with registering indices, most portable and convenient for showing the maximum and minimum pressure of the air. *L. Casella.*

2820c. Aneroid with Gold Band, for moist climates and coal mines. *L. Casella.*

2820d. Cary's Improved hardy Aneroid Barometer made strong expressly for the use of travellers, in improved wood and leather case and sling. *Henry Porter.*

2820e. Wheel Aneroid Barometer on stand. *G. Washington Moon.*

c. SYMPIESOMETERS.

2821. Sympiesometer, a sensitive instrument for sea use. *Francis Pastorelli.*

This consists of a syphon tube, containing a volume of air and a fixed fluid that partly fills the tube, also a thermometer. Its principle of action is upon Mariotte or Boyle's law. By an increase or decrease of the weight of the atmosphere the fluid is raised or lowered (arising from the elasticity of the enclosed air), through equal distances for each barometric inch if the con-fined air were unaffected by varying heats, but as it is affected by temperature this error is allowed for by a temperature scale. To take a reading note first the temperature of the thermometer; now set the pointer attached to the movable scale of inches, by means of its rack motion, to the corresponding degrees of temperature of the syphon tube, the position of the fluid indicates the height of the barometer. This instrument is considered by many a most valuable instrument; it is more sensitive than the barometer, and when accurately constructed should give good results.

III.—SPECIAL THERMOMETERS.

2823. Old Floating Thermometer, by Gay Lussac and Collardeau, of Paris.   *G. J. Symons.*

2824. Thermometer of Translation or Integrator of variations of temperature.   *Scottish Meteorological Society.*

The bar of zinc is fixed at its lower end during expansion by the needle points catching in the teeth of the rack below, so as to produce lengthening upwards, while during contraction the bar is held by the needle at the top, so that the shrinking is upwards. In this way the centre of gravity is moved upwards. The total annual march or creep of the bar will measure the total amount of fluctuation of temperature. Designed by Thomas Stevenson, C.E., Honorary Secretary.

2825. Von Lamont's Terrestrial Thermometer, for determining the temperature of soil from one to four feet deep.
       Prof. Ebermayer, Aschaffenburg.

2826. Three Vacuum Thermometers for studying solar radiation.

2826a. Six's Self-registering Thermometer, for registering the degree of heat and cold.   *Francis Pastorelli.*

The thermometer is continuous; in form it appears as three parallel limbs, the interior of the central and shorter one is filled with a fluid, the other two limbs partly with the same fluid and mercury. Floating in the fluid in each of the outside limbs above the mercury are two registering indices; they consist of pieces of fine steel enclosed in delicate glass tubes. Attached to them are hair springs, which retain them in a fixed position, unless acted upon by a force. The force used to set them is a magnet.

2826c. Standard Maximum Thermometer with divisions on the tube, protected by a glass shield covering from the rain and action of the atmosphere; by this contrivance it retains its legible appearance.   *Francis Pastorelli.*

2826d. Standard Minimum Thermometer. Wet and Dry Bulb Hygrometer; mounted on a vulcanite scale.
       Francis Pastorelli.

The maximum is constructed upon the valve principle patented by Negretti and Zambra. This thermometer has in its bend near the ball a small piece of solid enamel glass partly fused; the ball and part of the stem are filled with mercury, the upper part is a vacuum, the piece of enamel glass acts as a valve, for by applying heat to the mercury ball, the mercury in expanding passes the enamel glass (but it cannot return on cooling), it registers the amount of heat applied, which is read at the upper end of the column.

To set this thermometer for a new observation hold the ball downwards in the hand, with a gentle shake the mercury in the tube will unite with that in the ball. The minimum is upon Rutherford's principle. A small glass index floats in the spirit; its end farthest from the ball is flattened. To set the thermometer gently incline it with the ball uppermost, the index will then fall to the end of the film of spirit; place it in a horizontal position, as the temperature falls the index will be drawn back by the last film of spirit by attraction and there remain, the lowest point to which the spirit has receded; on an increase of temperature the spirit passes the index, it cannot move it.

2826e. Minimum Thermometer, with flat bulb.
Dring and Fage.

Constructed with a view to overcome the great drawback to the use of spirit thermometers, sluggishness. The bulb is made flat, so as to expose as large a surface as possible, while the glass is made as thin as is consistent with a non-barometric action.

2826f. Symons' Earth Thermometer. A portable arrangement for showing temperature at any depth, divested of the risk and difficulty of using the long and awkward thermometers hitherto employed for this purpose. *L. Casella.*

2822. Thermometers. The set of fourteen employed by the Exhibitor in experiments on the sensitiveness of thermometers. (Quarterly Journal Meteor. Soc., Vol. ii, p. 123.) *G. J. Symons.*

2827a. Plain Thermometer, Thermometer with Enamel Tube. *Negretti and Zambra.*

A plain and an enamel thermometer placed side by side, showing the immense advantage of the enamel over the plain. The extremely delicate investigations of medical and scientific men could not be carried on by the aid of such sensitive thermometers as are now manufactured had the process of enamelling not been introduced.
The enamel tube was *invented* by Negretti and Zambra.

2827b. Negretti and Zambra's Patent Mercurial Minimum Thermometer. *Negretti and Zambra.*

This thermometer has a plug of platina wire inserted in a small supplementary tube. When the thermometer is inclined, the mercury flows from the end of the supplementary tube until it reaches the platina plug, then by affinity of the mercury for the platina the column is maintained at the existing temperature; on a decrease of temperature the mercury recedes in the long or indicating tube, but on an increase of temperature it rises in the short tube, leaving the column of mercury in the thermometer indicating the minimum temperature.

2827c. Negretti and Zambra's Patent Recording Thermometer is upon the same principle as the deep-sea thermometer, but without the protected bulb.
Negretti and Zambra.

In this case the instrument is turned over by a simple clock movement, which can be set to any hour it may be desirable; the thermometer is fixed on the clock, and when the hand arrives at the hour determined upon, and to which the clock is set, as in setting an alarum clock, a spring is released, and the thermometer turns over (as in the case of the patent deep sea thermometer), transferring the mercury from the thermometer to the auxiliary or recording tube.

2827d. Board of Trade Thermometer. Original instrument, as used formerly in Her Majesty's service.

Negretti and Zambra.

The scale is brass, which, after constant use in salt water, soon becomes corroded and the figures obliterated.

2827e. Board of Trade Thermometers, with porcelain scales as patented by Negretti and Zambra, and now universally adopted. *Negretti and Zambra.*

2827f. Negretti and Zambra's Patent Self-registering Standard Maximum Thermometer.

Negretti and Zambra.

This instrument consists of a tube of mercury fitted on an engraved scale. The tube above the mercury is entirely free from air, and at a point in the bend above the ball is inserted and fixed with the blowpipe a small piece of glass, which acts as a valve, allowing mercury to pass on one side of it when heat is applied ; but not allowing it to return when the thermometer cools. When mercury has been once made to pass the valve (which nothing but heat can effect) and has risen in the tube, the upper end of the column registers the maximum temperature.

2827g. Negretti and Zambra's Standard Minimum Registering Thermometer, on Rutherford's principle ; an alcohol thermometer for recording the lowest temperature during any given period of time. *Negretti and Zambra.*

2827h. Negretti and Zambra's Patent Recording and Deep Sea Thermometer. *Negretti and Zambra.*

In shape this instrument is like a syphon with parallel legs. At the bottom on the left hand side there is a small glass plug or contraction on the plan of Negretti and Zambra's patent maximum thermometer. The mercury rises or falls as in an ordinary thermometer, but at the moment the temperature is desired to be recorded, the thermometer by a simple contrivance (which may be termed a vertical propeller) is made to pivot on its centre, causing the mercury to break off at the plug, and to pass into the right hand leg, where it remains fixed, indicating the exact temperature. The bulb is protected from the water pressure by an outer covering of thick glass, the intervening space being nearly filled with mercury.

2827i. Vacuum Solar Radiation Thermometer.

Negretti and Zambra.

The instrument consists of Negretti and Zambra's patent maximum thermometer with a blackened bulb enclosed in a glass tube and globe from which all air has been exhausted. This form of instrument was first made by Negretti and Zambra. An important addition (patented by Negretti and Zambra) to the instrument has since been made by inserting with the thermometer a small mercurial vacuum gauge, thereby showing whether a perfect vacuum has been obtained.

2827j. Negretti and Zambra's Thermometer for the determination of Terrestrial Radiation.
Negretti and Zambra.

The bulb of this instrument is transparent with divisions engraved on its stem. In use, to be placed with the bulb fully exposed to the sky, resting on grass with its stem supported by forks of wood. The divisions and figures are protected by an outer glass tube.

2826o. Original Self-registering Maximum Thermometer. Made by Prof. Phillips. *Hicks.*

Metallic Thermometer, Maximum and Minimum. By Hermann and Pfister, of Berne. *R. H. Scott, F.R.S.*

Wet and Dry Bulb Hygrometer. Same principle as that used with the thermometer No. 2827e. *Negretti and Zambra.*

Large Six's Self-registering Thermometer, with black scale on transparent glass. *Pastorelli.*

Small Six's Self-registering Thermometer, with two arms to tube. *Pastorelli.*

Delicate Thermometer, with gridiron bulb. The only instrument of the kind ever constructed. *Negretti and Zambra.*

Standard Meteorological Thermometer, with elongated bulb. *Casella.*

Standard Thermometer, with elongated bulb.
Elliott Bros.

Portable Wet and Dry Bulb Thermometers (hygrometer), with elongated bulbs, on brass stand. *Pastorelli.*

Small Portable Wet and Dry Bulb Thermometers (hygrometer), on black scale, with white figures and graduations, mounted on its own case. *Pastorelli.*

Wet and Dry Bulb Thermometers (hygrometer), with maximum and minimum thermometers, on one support.
S. G. Denton.

Wet and Dry Bulb Thermometers (hygrometer).
J. Hicks.

Wet and Dry Bulb Thermometers (hygrometer).
Negretti and Zambra.

Wet and Dry Bulb Thermometers (hygrometer), with elongated bulbs. *Casella.*

40075. X x

Six's Self-registering Maximum and Minimum Thermometer.   *S. G. Denton.*

Six's Self-registering Thermometer, maximum and minimum.   *Dring and Fage.*

Terrestrial Radiation Thermometer, on brass stand.
Pastorelli.

Portable Maximum and Minimum Thermometers, mounted on their own case.   *Pastorelli.*

Portable Maximum and Minimum Thermometers, on metal scales.   *Pastorelli.*

2826o. Hicks's Patent Thermometer Shield.
J. J. Hicks.

This is an arrangement for effectually protecting the figures and divisions of thermometers and other scales from the corrosive action of air and moisture, and consists of an outer glass tube hermetically sealed on to the thermometer stem whereby the scale is rendered absolutely indestructible. It his here exhibited, applied to maximum, minimum, and other thermometers.

2826p. Delicate Thermometers, with gridiron and spirally formed bulbs.   *J. J. Hicks.*

By the diffusion of the mercury in the bulb over a large surface, and selecting tubes of small and uniform bore, very great sensibility is attained.

2826q. Six's Thermometer with Patent Cylinder Jacket Bulb.   *J. J. Hicks.*

In this instrument by diffusing the spirit over an extended area, a degree of sensibility is attained equal to mercury.

2826y. Mercurial Standard Thermometer, by Troughton and Simms.   *Royal Society.*

2826z. Spirit Thermometer, for very low temperatures, by Newman.   *Royal Society.*

Hicks's Patent Terrestrial Radiation Thermometer.
J. J. Hicks.

The difficulty arising from the condensation of moisture in the outer cylinder of ordinary terrestrial minimum thermometers, is, in this instrument, entirely obviated by simple but efficient means.

Highly sensitive Thermometers, with open scales in sets of six, one scale being continuous with the other, and so open as to admit of readings being taken to one-tenth of a degree.
J. J. Hicks.

Hicks's Patent Solar Radiation Thermometer with electrical test.   *J. J. Hicks.*

In constructing this improved radiation thermometer, special precautions are adopted to ensure perfection of vacuum, and the enclosing tube is furnished with platinum wire terminals to admit of the vacuum being tested by the electric current from a powerful Ruhmkorff coil by the aid of which the vacuum can be ascertained to exist to within one-tenth of an inch of pressure.

2827k. Thermometers of extreme sensitiveness.
Negretti & Zambra.

Negretti & Zambra's Instantaneous Thermometer, with gridiron, spiral, or coiled form of bulb, and divided upon the stem, as shown in the International Exhibition of 1862, and used by Mr. Glaisher in his balloon ascents to obtain very rapid thermometric readings.

2826g. Standard Thermometer, graduated on the stem, in maroon case, with Kew certificate. *L. Casella.*

2826h. Standard Maximum Thermometer, graduated on the stem (Kew certificate). *L. Casella.*

2826i. Sensitive Minimum Thermometer, with forked bulb (Kew certificate). *L. Casella.*

2826j. Thermometer, 2 meters long, set in wood and zinc, divided into $\frac{1}{5}°$, for investigations of terrestrial heat.
Warmbrunn, Quilitz, & Co., Berlin.

2826k. Two Thermometers, maximum, in case.
Will. Haak, Neuhaus, Thüringen.

2826l. Two Thermometers. Maximum on brass scale.
Will. Haak, Neuhaus, Thüringen.

2826m. Continuous Self-registering Thermometer.
W. Harrison Cripps, F.R.C.S.

The object of the instrument is to obtain a continuous registration of heat. The instrument is in two portions : 1st, the thermometer for indicating the temperature ; 2ndly, the clock-work for registering the hours and minutes. The thermometer consists of six coils of glass tubing wound concentrically round an axis in such a manner as to form a spiral glass wheel 4 inches in diameter. The last coil is moved slightly away from the others, so that it shall form the circumference of a circle 5 inches in diameter. To each end of the axis a fine needle-pointed-pivot is attached. These pivots rest in minute depressions between two parallel metal uprights. By this arrangement the glass wheel can rotate freely between the uprights. The spirit in the thermometer fills the spiral portions of the tube, and also 3 or 4 inches of the last coil (the one forming the circle). The spirit then comes into contact with a column of mercury 4 inches in length. Beyond the mercury are a few drops of spirit to moisten the glass. The remaining portion of the tube is hermetically sealed, enclosing a small quantity of air. On the spirit expand. ing with heat, the column of mercury is driven forwards. This immediately alters the centre of gravity, and the wheel revolves in a direction contrary to that of the moving mercury. When the spirit contracts on cooling the enclosed air acting as an elastic spring keeps the mercury in contact with it, and the wheel regains its original position. By this arrangement the two forces, heat and gravity, acting in contrary directions, generate a steady rotatory motion.

The method by which this movement is made serviceable is by a grooved wheel 2 inches in diameter fixed to one of the pivots, and therefore revolving

X x 2

with the thermometer. Fixed to and passing over this wheel is a fine thread, from which is suspended a pencil holder, moving up and down on a vertical slide. The pencil will be raised or lowered according to the direction in which the wheel is moving. The other portions of the clock-work are arranged in a manner similar to that employed in the barograph.

In the present instrument a cylinder $4\frac{1}{2}$ inches, both in width and diameter, is made to revolve once in seven days. Around this cylinder is placed a paper, on which the days and hours are indicated by vertical lines. The cylinder is so placed that the surface of the paper is $\frac{1}{10}$th of an inch away from the pencil point, moving at right angles to its surface. A small striker is connected with the clock-work in such a manner that at every quarter of an hour it gives the pencil a tap, striking its point against the paper. The registration of the temperature is thus indicated be a dotted line which can be read off by a prepared scale supplied with the instrument.

The scale for reading off the register is prepared by observing the extent to which the pencil rises for every 10°, as indicated by a standard thermometer. In this particular instrument it is found that for every 10° the pencil is raised $\frac{3}{4}$ of an inch ; this gives $\frac{3}{40}$ths for each degree, and the index is accordingly divided to this scale.

2826n. Fluctuation Thermometer. *Prof. Balfour Stewart.*

IV.—ANEMOMETERS.

2830. Static Anemometer, for measuring the force of the horizontal component of the wind, especially of gusts.

Scottish Meteorological Society.

Two sets of Robinson's cup anemometers are placed one above the other on one vertical spindle, so that the couple tending to turn the spindle depends on the force of the horizontal component of the wind, not on its direction. To the spindle is attached a spring, so that the magnitude of the couple (and therefore the force of the horizontal component) is measured by the angle through which the spindle is turned. This is recorded by a pencil which is raised and lowered by a screw cut on the spindle. The clockwork and paper for recording have not been sent. Designed by Professor Crum Brown, M.D., F.R.S.E., Member of Council.

2831. Anemometer, for ascertaining pressure of wind.

Scottish Meteorological Society.

This anemometer acts by lengthening (not compressing). The maximum result is recorded by the thread which is fixed to the rod and pulled through a hole in the brass plate fixed to the side of the box. To ascertain the maximum elongation which takes place, press the thread against the plate, then push in the disc until the part of the thread which has been drawn through the hole is again tightened, and read off the result from the graduated tube. The small disc is for high winds, the large for light. Designed by Thomas Stevenson, C.E., F.R.S.E., Honorary Secretary, and described in the Society's Journal, vol. iv., p. 266.

2832. R. Ballingall's Anemometer, for continually registering the pressure of the wind. *Scottish Meteorological Society.*

The principle of this anemometer consists in a cistern of mercury in the left hand chamber, with a wooden plunger, which acts in connexion with the pressure plate. There is an arrangement by which the accuracy may be tested at any time. Designed by the late R. Ballingall, and described at the general meeting of the Society, 2nd July 1874.

2834a. Anemometer, constructed by P. Schultze (Dorpat).
Prof. A. von Oettingen (Dorpat).

After 50 revolutions of the Robinson's cups an electrical contact is made, but only for a small fraction of a second, because after the movement of the differential wheel the electrical communication is broken, but at the same time prepared for the next operation.

2835. Anemometer (statical), in case.
Dr. G. Recknagels, Kaiserslautern (Physical Collection of the Royal School of Industry).

The indications of the instrument are proportional to the pressure of the wind. The small size of the instrument, and the ease with which its indications may be read off, make it specially advantageous for investigating ventilation. The velocities of the first revolution are marked on the outer scale; otherwise the constants are given with the apparatus. A small weight is added by way of a check; it balances the spring suspended at a mark of the hand.

2835a. Five Anemometers. *F. Darton and Co.*

2835b. Howlett's Anemometer. *Elliott Brothers.*

Consists of a copper sphere of such a diameter that the pressure of the wind on its hemisphere shall be equal to the whole or any required portion of a square foot; the sphere is mounted on a vertical rod that is suspended on knife edges like a balance, and registers the force and direction of the wind on a slate slab or on printed forms placed in the instrument for recording daily observations.

2836. Improved Anemometer of Combe.
Herm. Recke, Freiberg, Saxony.

All axle bearings in these anemometers are of stone. The setting and suspending is accomplished by pulling at one and the same knob. To facilitate the readings, dials with hands are provided. The fans are of trapezoidal shape and reach nearly down to the axle; they are of mica and capable of adjustment, in order to ascertain the most sensitive position.

2836a. Improved Anemometer of Combe.
Herm. Recke, Freiberg, Saxony.

In the preceding the axle of the fans remains unchanged during suspension inasmuch as only the dial work is moved; in this the axle is lifted parallel to itself.

2837. Pendulum Anemometer (drawing), with explanation.
Prof. Prestel, Emden.

2838. Tangent Scale, for determining the mean direction of the wind. *Prof. Prestel, Emden.*

2839. Balance Anemometer, constructed by Mr. Francis Ronalds at the Kew Observatory, in 1843, for the purpose of measuring the force of the wind.

Kew Committee of the Royal Society.

It consists of a light board, 1 foot square, fixed transversely to a cross of wood, suspended by a brass axis passing through its centre, and turning in glass tubes in such a way that the system can partially rotate in a vertical plane. The lower end of the bar carrying the board is counterpoised, so as to keep the surface of the board vertical, and a scale pan, hung to one end of the horizontal bar of the cross, serves to receive the weights, which are necessary to counterbalance the force of the wind, pressing on the board opposed to it, at any time. A small box, covering the scale pan, serves to shield it from the action of the wind. The instrument was, at the time of observation, placed so that the surface of the pressure plate should stand at right angles with the direction of the wind, as indicated by a vane.

2841a. Anemometer, large. As supplied to the Government for registering on a diagram the velocity and direction of the wind. *L. Casella.*

2842. Electrical Anemometer, by which the velocity of the wind in miles, &c., can be shown on dials in an observatory or study. *Yeates & Sons.*

2843. Robinson's Anemometer, for measuring the wind's velocity reading from $\frac{1}{10}$ up to 1,000 miles. *Francis Pastorelli.*

This instrument consists of four hemispherical cups fixed to four strong metal arms, that radiate from a central boss at a distance of 90° apart; at right angles to the plane of the cups is attached the vertical axis, its lower end terminates with an endless screw; this works two wheels which differ in the number of teeth, so that by their common revolution one has 100 times less velocity than the other.

The front wheel has two divided circles, the interior denoting 10 miles; each mile is figured and sub-divided into 10 parts; the outer circle is divided into 100 parts; each of these divisions represent 10 miles, every fifth division 50 miles, and they are numbered 50, 100, &c.; therefore readings from $\frac{1}{10}$ up to 1,000 miles can be taken by this instrument.

2844. F. Pastorelli's Electric Anemometer, to indicate from $\frac{1}{10}$ up to 10,000 miles. By the use of this instrument the velocity of the wind at any distant station may be seen by inspection in the observatory. *Francis Pastorelli.*

It consists: (1.) of Dr. Robinson's cup arrangement attached to a vertical axis; as it revolves the motion is conveyed by its endless screw to a wheel with a cam mounted in a rectangular metal box; here the angular velocity is reduced, so that each contact has the value of $\frac{1}{10}$ of a mile; (2.) the receiving instrument with its dial is mounted in a polished cabinet; it is worked by an electro-magnet and lever, so when the revolutions of the cups have indicated $\frac{1}{10}$ of a mile, the action of the lever works a wheel, and motion is communicated to a series. On the face of the instrument are divided circles and indexes which register from $\frac{1}{10}$ to to 10,000 miles of velocity. (3.) A Leclanché battery of four No. 2 cells is connected with the above, so that each $\frac{1}{10}$ of a mile, indicated by the revolutions of the cups, may be transmitted electrically and consecutively to the receiving instrument.

This instrument has the minimum amount of friction. The cup arrangement can be placed in any distant and convenient position, and the receiving instrument in the observatory. It is portable, being of small dimensions.

2845. New Portable Anemometer, by Francis Pastorelli, for measuring accurately the velocity of the air or wind, specially adapted for scientific travellers and explorers. *Franeis Pastorelli.*

It consists of four small hollow hemispherical cups (Dr. Robinson's form) fixed to a vertical axis ; the lower end has an endless screw. This works a wheel, the complete revolution of which is equal to $\frac{1}{10}$ of a mile. Its action is conveyed to others that carry indexes over divided circles on the face of the dial, which is mounted on a circular box fixed on a metal base ; readings can be taken from $\frac{1}{100}$ up to 1,000 miles. Its weight, $1\frac{1}{2}$ lbs., and it packs in a mahogany case about $4\frac{1}{4}$ inches square.

This is a most sensitive instrument; its indications are as accurate as those of the large kind.

2845a. Anemometer, with shaft whose length can be adjusted to suit different stations, and with dial reading to 10,000 miles, constructed for stations of the 2nd and 3rd order connected with the Canadian Meteorological service.

G. T. Kingston, M.A., Toronto, Canada.

This instrument is designed to reconcile adequate exposure of the hemispheres with accessibility of the dial.

A short spindle bearing a small set of Robinson's cups is connected with a horizontal cogged wheel resting on friction rollers, and with a long shaft suspended from its centre, as in the anemographs of the British observatories. It is contrived so that the shaft may make 101 complete turns for 200 miles of wind.

The mode of recording the miles is as follows :—

At the lower end of the shaft is an endless screw which acts on the circumferences of two toothed wheels of equal diameter, turning in vertical planes about a common axis, and having 100 and 101 teeth respectively.

From the centre of the *back* wheel (that of 100 teeth) projects forwards a short hollow pin which incloses and works on a solid pin fixed to a support behind the back wheel.

The *front* wheel (that of 101 teeth) which turns on the above named hollow pin, and slides closely on the face of the back wheel, has a graduated ring on its face, containing 100 divisions. The *outer* ends of the lines of graduation indicate miles, while the *inner* ends of the same lines, reckoned in the reverse order, indicate hundreds.

The endless screw, at every turn, causes both wheels to advance *two* teeth, so that for 100 miles the front wheel makes one complete turn, or 100 divisions, while the back wheel make one complete turn and one tooth, and thus advances one division with respect to the front wheel. The miles up to 100 are shown by a fixed pointer, and the hundreds by a pointer attached to the end of the hollow pin.

For fractions of a mile there is a contrivance which needs adjustment at each observation.

A correction of 1 per cent. nearly should be subtracted from the fractional parts.

If the anemometer be two distant from the observer's office to allow the dial to be read conveniently with the required frequency (if, for instance, it be on an adjacent hill or tower), it may be connected electrically with a dial or with a self-recording apparatus.

In such a case, the dial above described should be read periodically as a check on the electrical dial.

To adopt the length of the shaft to the circumstances of various stations, the shafting is supplied in lengths of four feet, with one of two feet, one of one foot, and one telescopic piece.

2845b. Windmill Vane with electrical arrangement whereby the direction of the wind at any instant may be known by sound, constructed for stations of the second and third order in connection with the Meteorological Office of the Dominion of Canada.   *G.T. Kingston, M.A., Toronto, Canada.*

This instrument is designed to meet the case when a vane, if suitably exposed, is too distant to admit of being connected by a shaft with a dial in an accessible position.

Surrounding the step of the vane and attached to the bottom of the box is a flat brass ring divided by radial lines into four equal parts corresponding to, but not necessarily in the direction of the four cardinal points, and separated by small equal intervals.

The direction arcs (as they may be termed) are insulated, except as regards connection with their screw cups.

Clamped to the spindle, and capable of adjustment in azimuth, is a circuit maker, which consists of an arm bearing a brass arc, which is made to press on the flat brass ring. The length of this arc is 45°+ interval between the fixed arcs. The step of the vane is connected by wire with one pole of a battery, and the direction arcs with four screw cups in near proximity to a brass plate fastened to the wall of the office, and so contrived that, by aid of a plug, metallic connection may be made at will between the brass plate and any one of the four wires.

Finally, two wires from the poles of a small telegraph sounder are attached to the brass plate and to the other pole of the battery, and the apparatus is complete.

To ascertain the direction of the wind, notice by the sounder which direction arcs or pair of arcs is placed in circuit as the brass plate is connected by the plug with the four wires in succession.

If, for instance, N. only sounds, the direction is nearer to N. than either to N.W. or N.E., unless it be N.N.W. or N.N.E. exactly, but if N. and E. *both* sound the direction is nearer to N.E. than either to N. or E., unless it be N.N.E. or E.N.E. exactly.

The azimuth of the vane box need be governed only by appearance and the position of the door.

To adjust the vane, unclamp the circuit maker, and insert a pin (provided for the purpose) through holes at the middle point of the circuit maker and the arc at the left of the box, which arc may represent any one of the four cardinal points. Turn the vane to that point, clamp the circuit maker, and remove the pin, when the adjustment will be complete.

The apparatus works well with a single gravity cell, when the vane is 60 feet or more from the battery and sounder. With a stronger battery it works well at the distance of a mile or more.

3388a. Signor Bianchi's Air Meter.

Conservatoire des Arts et Métiers, Paris.

V.—RAIN GAUGES.

2847. Rain Gauge. In use at meteorological stations belonging to the Norway Meteorological Institute.

Prof. H. Mohn, Christiania.

Square surface, 15 x 15 centimetres, height 50 centimetres, for catching snow; the lower part protected against evaporation. The rain (or melted snow) water is to be poured out of the gauge through one of its upper corners, into a measuring cylindrical glass, divided to show the height of fallen rain in millimetres. The gauge is made of plate iron, after design made by Professor H. Mohn. The measuring glass was calibrated at the Meteorological Institute, in Christiania.

2848. Rain Gauge, No. I. *Scottish Meteorological Society.*

Designed to obviate errors due to out-splashing and in-splashing of rain drops. Designed by Thomas Stevenson, C.E., F.R.S.E., Honorary Secretary, and described in the Edinburgh Philosophical Journal, 1842.

2849. Jagga's Rain Gauge, No. II.

Scottish Meteorological Society.

The principle of this gauge consists in making the diameter of the funnel equal to 4·697 inches, so that a fluid ounce of rain-water collected equals one-tenth of an inch of rain. Designed by G. V. Jagga, Rao of Vizagapatam, and introduced to the Society by Sir Walter Elliot, Member of Council, and described in the Society's Report, quarter ending June 1861, p. 9.

2850. Marine Rain Gauge.

Scottish Meteorological Society.

This instrument is composed of a cylinder poised on an upright pivot projecting from the bottom of a square box, enclosing the whole, having the top open to the atmosphere. The cylinder is divided into two parts, an upper one having a diaphragm to collect the rain, and the lower, the receiver for the same having a conical floor inside in which works the pivot below. The horizontal bearing of the gauge is thus maintained in all cases of the rolling and pitching of the vessel by the motion on the pivot.

2851. Ronalds' Rain and **Vapour Gauge,** erected at the Kew Observatory in 1843.

Kew Committee of the Royal Society.

An instrument constructed at the Kew Observatory in 1843, by Mr. Francis Ronalds, for indicating a mean result from the quantity of water which may have fallen between any two given periods, *minus* the quantity of vapour which has evaporated in the same time, on and from a circular plane of one foot diameter. It is described in the British Association Report for 1844. It consists of two cylindrical vessels, connected by a tube, the one being one foot in diameter, and open at the top, whilst the other is 3 inches, and (with the exception of a small hole) entirely closed by a cover, which carries a frame, holding a circular divided arc, with an index moving over it. The index is attached to a small pulley, over which a cord passes, having its end fixed to the float in the cylinder. This, rising and falling with the changes of water level, indicates the amount of rain or evaporation on the metal scale.

2851a. Stutter's Self-recording Rain Gauge.

J. J. Hicks.

This is a step in the direction of cheapness without sacrificing efficiency. An eight-day clock rotates a funnel, the tubular end of which passes successively over the openings of 12 or 24 glass jars numbered to correspond with the hours of the day. The funnel-shaped openings over each jar being brought to a knife edge, a well defined separation is effected, so that each jar receives the rain for each hour, and the sum of their contents gives the rainfall for 12 or 24 hours.

2851b. Beckley's Self-recording Rain Gauge, as used at the observatories in connection with the Meteorological Office.

The Meteorological Committee of the Royal Society.

This instrument is described in the report of the Meteorological Committee for 1869.

The rain is delivered into a receiver floating in mercury. As soon as 0·2 in. of rain has been collected a syphon arrangement is brought into play, and the receiver is emptied and rises to its original position.

The motion of the receiver is recorded on paper by a pencil.

2852. Glaisher's Rain Gauge.   *Francis Pastorelli.*

This instrument has a greater internal depth, the coned part being 3 inches from the surface; in this respect it more resembles the Admiralty pattern; it prevents heavy rain splashing over, and consequent loss.

2852a. Mountain Rain Gauge. Capacity 48 inches. Pattern employed in the English lake district and at very wet mountain stations in Wales and Scotland.   *G. J. Symons.*

2852b. Engineer's Rain Gauge. Capacity 12 inches. Adapted for rough observations in ordinary hilly districts.

G. J. Symons.

2852c. Glaisher's 8-inch Rain Gauge. Adapted for, and largely used by, private observers.   *G. J. Symons.*

2852d. Snowdon Pattern Rain Gauge, originally designed for use in North Wales, the deep cylinder being added to secure accurate observations during snow. This gauge is now in general use in all parts of the British Isles.   *G. J. Symons.*

2852e. Indestructible Monthly Rain Gauge for private observers.   *G. J. Symons.*

3852f. Copper Rain Gauge, Glaisher's form.

E. Cetti and Co.

2852g. Electrical Self-registering Rain Gauge.

Yeates & Sons.

The peculiarity of the above is a novel form of rocking bucket, the partition of which is so constructed that it will register correctly, no matter at what rate the rain may fall.

2852h. Self-registering Rain Gauge. *Elliott Brothers.*

This instrument consists of an upright square metal case, with funnel or receiver 10 inches square, with a set of counting wheels with dials registering from $\frac{1}{100}$ to 100 inches of rain fallen. The water falls into a trough with a division in the centre, having a motion on an axis, and being put at an angle, the rain overbalances it alternately, and is registered by the motion being communicated to the set of wheels, and consequently there is no evaporation to be deducted.

Funnel, Bottle, Copper Pan, and Glass Measure for Rain Gauge. *F. Darton and Co.*

VI.—HYGROMETERS.

Mostra Umidaria (Hygrometer). An invention of Folli da Poppi, perfected by the Accademia del Cimento; the hygroscopical body is a band of paper. *The Accademia del Cimento.*

Hygrometer, at condensation, invented by the Grand Duke Ferdinand II. dei Medici. *The Accademia del Cimento.*

Folli's hygrometer, made in 1664, consists of a band of paper fixed at both ends, and having in the middle a corresponding weight, which in its rising or falling (caused by the moisture shortening or lengthening the paper) transmits its motion by means of a cord and pulley, to a hand which points out the degrees of humidity on a quadrant. The members of the Accademia del Cimento substituted parchment for paper, whilst Torricelli used oats as an hygrometric substance.

The hygrometer No. 23, that of the Grand Duke, is founded on the principle of the condensation of the aqueous vapour of the atmosphere, by means of a cold substance. It consists of a truncated cone, made of a sheet of tinned iron, covered on the inside with a layer of cork, and supported on a tripod. Below the smaller aperture, turned downwards, there is suspended a hollow glass cone, ending in a closed point likewise turned downwards, and provided towards the upper part with an escape pipe. The upper cone is then filled with snow or ice, which in melting runs into the glass cone, which remains at the temperature of 0°, whilst the excess of water runs through the pipe into a separate vessel. The moisture of the air coming in contact with the cold side, condenses itself and covers it with dew, which collecting by degrees into drops, runs towards the point whence it falls into a graduated vessel. If the time the experiment has lasted be taken into consideration, the quantity of water thus collected will be found to be in proportion to the humidity of the air.

Experimenting in this manner, the members of the Accademia del Cimento, found that the south winds are so charged with moisture, that in one minute the hygrometer has given as much as 35, 50, and even 80 drops of water; whilst the north wind leaves the glass perfectly dry.

2853a. Capt. Kater's Hygrometer, by Robinson.
Royal Society.

2853b. Jones' Hygrometer, with stem bent at an acute angle. *Royal Society.*

2863a. Chameleon Hygrometer, based on the property of cobalt of changing colour with the dampness or dryness of the air.
Walter B. Woodbury.

2871g. Slide Rule for hygrometrical calculations by John Welsh, F.R.S. *Surgeon-Major F. de Chaumont.*

2971h. Mason's Hygrometer. *E. Cetti & Co.*

2853. Saussure's Hygrometer; an old specimen by V. F. Hausman. *G. J. Symons.*

2854. Whalebone Hygrometer, by Thos. Jones, of Oxendon Street. *G. J. Symons.*

2855. Daniell's Hygrometer. No maker's name, but formerly belonging to Sir James South. *G. J. Symons.*

2856. Balance Hygrometer (drawing and essential part of the instrument). *Prof. Buys-Ballot, Utrecht.*

At one arm of a balance is hung a wide glass tube filled with chloride of calcium or any other hygroscopic substance. This tube is closed at its upper part by a cork stop, bearing two ∩∪ bent glass tubes plunging in oil baths, but care is taken not to submerge their open ends. Two glass bells, ending in tubes at their upper parts and plunging for nearly half an inch in the same oil, render it possible to aspirate air through the chloride of calcium tube without affecting its movability. One of the bells is joined by an india-rubber tube with the spot the air of which is to be examined; the other with a gas meter and aspirator. In a given time the water contained in a quantity of air indicated by the gas meter can in this manner be weighed, and its humidity ascertained. It is obvious that the instrument can easily be made self-registering.

2857. Collection of Hygrometers and Psychrometers.
Dr. H. Geissler, Bonn.

2857a. Standard Hygrometer, with divisions and figures, protected by an outer glass tube, as in Negretti and Zambra's terrestrial radiation thermometer. *Negretti and Zambra.*

2857b. Negretti and Zambra's Standard Dry and Wet Bulb Hygrometer or Psychrometer. *Negretti and Zambra.*

Two thermometers as nearly identical as possible are placed side by side, one marked DRY and the other WET. The bulb of the wet thermometer is covered with thin muslin, and has twisted round the neck conducting threads of cotton, passing into a vessel containing water placed on one side so that the water may not affect the reading of the dry bulb thermometer. The temperature of the air and of evaporations is given by the reading of the two thermometers, from which can be calculated the dew point.

2858. Klinkerfues' Bifilar Hygrometer, with reduction disc, executed by W. Lambrecht, Göttingen.

Prof. Klinkerfues, Göttingen.

The bifilar hygrometer shows the relative dampness without further reduction, upon a stereotyped scale of equal divisions, and also the dew point by means of the reduction disc.

2859. Reduction Discs for Psychrometers.

Prof. Klinkerfues, Göttingen.

The reduction discs for the psychrometer give the dew point according to the following rule :—

The outer disc is turned round the inner one in such a manner that the two places of the evaporation temperatures, read off from the moist thermometer, coincide; with the place of the air temperatures upon the one will then also coincide the place of the dew point temperature upon the other. The one disc has a second division, which comes into use in the case of the evaporation temperature falling below zero.

The barometric pressure is assumed to be 750 mm. ; for any other pressure, b, the quantity $\frac{4}{50}$ ($b-750$), taken in nearest round numbers, can be easily multiplied in the head by the thermometric difference, likewise taken in round numbers. The product expresses the number of hundredths of a degree, and has to be added to the air temperature, in order to obtain, after the setting of the disc, the dew point with greater precision. This correction is, however, seldom required in practice.

2861. Psychrometer Scale, for determining the relative and absolute moisture of the air, as well as the dew-point, without calculation. Model with explanation. *Prof. Prestel, Emden.*

2862. August's Psychrometer, with two thermometers divided into · 1° on a stand. *Warmbrunn, Quilitz, & Co., Berlin.*

2863. Catgut Hygrometer, dating from the first quarter of the 18th century. Property of His Highness the Prince of Pless.
Breslau Sub-Committee (Prof. Poleck).
Interesting on account of its age.

2864. Eight-Haired Saussure's Hygrometer, by Richer of Paris, formerly the property of Mr. Francis Ronalds, and used by him at the Kew Observatory in 1843.
Kew Committee of the Royal Society.

2865. Three Hair Hygrometers, by H. B. de Saussure.
M. Henri de Saussure, Geneva.
Original models, originally the property of M. H. B. de Saussure, accompanied by tables drawn up by him and his son Theodore de Saussure.

2867. Hair Hygrometer of De Saussure, with two sets of graduations, one being fractional of relative moisture.
Geneva Association for the Construction of Scientific Instruments.

The faults found with the hair hygrometer are caused, generally, by the very great imperfection of the manufacture of those usually sold by the trade. The hair deteriorates, and from time to time its indications alter, because, as a rule, the weight of tension is too great. Hair properly prepared, and subject to due tension only, altered so little that M. Regnault tells of having found an old hygrometer made by Paul as correct as any modern instrument with which he has compared it. Another cause of irregularity proceeds from the careless choice of the hair. All hair that has been pulled about, and of which the limit of elasticity has been exceeded, should be avoided. The most isolated hamlets have now to be searched in order to obtain hair uncombed. The object of the Geneva Association for the Construction of Scientific Instruments is to revive the hygrometer of De Saussure in an improved form. The general construction of the De Saussure hygrometer has been maintained as being the best, but the following modifications have been introduced:—

1st. The marking needles on the axis of the pulley are made of aluminium bronze, thus making the pulley lighter, consequently more moveable, and lessening the friction of the axle.

2nd. The weight of tension of the hair is replaced by a gold spiral, which makes the instrument more portable and avoids the twitching of the hair by oscillation and accidental displacement of the weight.

3rd. The hygrometer has two graduations (this is the chief modification); the first is an arbitrary division in equal parts of 0 to 100; the second marked out on a moveable arc, is superposed to the first and registers in hundredths the relative moisture, or the fraction of saturation. Thus, when the hygrometer registers 50, it is certain that the air contains half the quantity of water that it can contain in the state of saturation.

The hygrometer is graduated according to "Regnault's" method.

2868. Hygrometer, modified by Dr. Geissler, with a delicate thermometer on a stand.

Will. Haak, Neuhaus am Rennweg, Thüringen.

2869. August's Psychrometer, on stand. The thermometer divided into tenths of a degree from −30 to +45° C.

Will. Haak, Neuhaus am Rennweg, Thüringen.

2870. Psychrometer, August system, with stand, in a case.

Ch. F. Geissler & Son, Berlin.

2870a. Psychrometer, August system, with stand, in a case, for travelling.   *Ch. F. Geissler & Son, Berlin.*

2871. Early Hygrometer.

The Council of King's College, London.

2871a. Oatbeard Hygrometer.   *G. J. Symons.*

2871b. Old Travelling Hygrometer.   *G. J. Symons.*

2871c. Modern Saussure's Hygrometer.   *G. J. Symons.*

2871d. Mason's Wet and Dry Bulb Hygrometer for Observatories.   *L. Casella.*

2871e. Pocket Hygrometer, in maroon case, for travellers.
L. Casella.

2871f. Dine's Sensitive Hygrometer, for taking rapid indications of the dew point.   *L. Casella.*

VII.—SELF-RECORDING INSTRUMENTS.

Barometrograph of Fontana.
The Royal Institute of " Studii Superiori," Florence.

Felice Fontana, a native of Roveredo in the Tyrol, was the first director of the Royal Museum of Physical Science and Natural History founded in Florence by the Grand Duke Pietro Leopoldo. Towards the close of last century he constructed several registering meteorological instruments, and among them the present Barometrograph. A float on the surface of the mercury of a large barometer transmits its motion to a section of a cylinder about 70 mm. in diameter, and covered with paper. Every hour an impression is made upon this paper by a steel point set in motion by a clock. The point itself advances a certain distance at each impression, so that its indications end by drawing on the paper, a curve of the barometical oscillations.

2872. Barometrograph, or Self-Recording Aneroid Barometer.   *Pillischer.*

The construction of this instrument differs materially from others of a similar nature, by having the entire mechanism placed in a vertical line, whereby friction is reduced to a minimum.

The aneroid barometer, 18 inches in diameter, has two vacuum chambers; below it is placed the cylinder, carrying a ruled paper coinciding with the scale of the barometer, and driven by a powerful 8-day clock. The pencil point is moved up and down upon a metal rod by the action of the barometer, and, by a simple mechanical arrangement connected with the clock, imprints the changes which occur from hour to hour on the ruled paper. Thus a black dotted undulated line is produced, showing the rise and fall of the barometer.

2873. Self-recording Barometer and Thermometer for use on board ships.
Dr. Franz Paugger, Director of the I. R. Commercial and Nautical Academy, Trieste.

This apparatus, which is enclosed in a small box, consists of three principal parts :
A thermometer, a barometer, and a contrivance for registering.
The thermometer is composed of a system of 10 zinc tubes, one foot (English) in length, placed side by side in the form of a cylinder, each one of which, commencing from the first, transmits, by means of a lever, to the next tube, in a somewhat augmented degree, its linear expansion produced by increase of heat. This system of tubes is suspended in the open air and surrounded by a shell or cover in the form of venetian blinds at the back or the parallel tubular box. From the end of the last tube a somewhat longer lever extends into the interior of the box, as far as the writing cylinder, by

which the variation of the temperature, caused by the total expansion or contraction of the tubular system, is automatically recorded from time to time. The motor for the measurement of the atmospheric pressure is composed of 10 aneroid cases, which are joined together in the shape of a column. Their expansion or contraction, produced by the changes in the atmospheric pressure, is transmitted to the writing cylinder by means of a lever.

The arrangement for registering consists of a cylinder placed vertically, which is turned on its axis every 24 hours by a clock movement in the front part of the box. The two levers, both extending horizontally from the thermometer and the barometer, as far as the same edge of the cylinder, are at a mean temperature and a mean atmospheric pressure in a position the one in the centre of the lower, and the other in the centre of the upper half of the writing cylinder. Every 15 minutes, the ends of the levers, which are provided with pencils, are pressed against the cylinder, and by these means the variations of the atmospheric pressure and of the temperature are regularly registered. The instrument is in all its parts so contrived as not to require a fixed position, and being easily transportable, it can be used without difficulty on board ship.

2873a. Self-registering and Signalling Vessel Barometer.   *Dr. Friedr. C. G. Müller, Osnabrück.*

This self-registering and signalling barometer is new, both as a whole, and in its details. The points of novelty are:

I. The bulb barometer, which is entirely independent of the registering and signalling mechanism, and which possesses the following properties:—(a.)The reading takes place at the *lower* level by unchanged position of the upper. (b.) The galvanoscopic adjustments, accomplished by means of the platinum point fused into the Torricellian vacuum, requires no sight, is very simple, quickly performed, and free from errors. (c.) The construction of the cistern with the barometric tube passing through it, whereby the level of mercury in the cistern is kept invariable. (d.) No boiling of the tube is required. (e.) By the introduction of a small quantity of dry hydrogen into the vacuum, the influences of temperature may easily be compensated.

II. The uninterrupted automatic adjustment, which makes all changes in the atmospheric pressure, even the smallest oscillations, *audible*. Specially deserving of attention are:—(a.) The two platinum points in connexion with (b.) the two sliding relays, working in opposite directions, and (c.) the two little electro-motor machines, which by their beats give direct indication of all variations in the atmospheric pressure, and announce it indirectly by setting an electric clock-work into action. (d.) By connecting the two relays with the same point, the utmost sensitiveness of the automatic adjustment is reached, whilst the machines are in continuous action.

III. The registering mechanism, which is connected with the self-regulating barometer cistern, and moving in which the mercury in the barometer has no work to perform. Attention is called to the following points:—(a.) The force moving the marking pencil is a considerable one, in consequence of the multiplying gear; the wheels have to revolve 80 times in order that the cistern may move 1 mm. (b.) The registering mechanism allows, without losing anything of its simplicity and reliability, a magnifying of the movement of the barometer cistern. (The present apparatus doubles the movement.) (c.) The instrument continues working for a month, and even longer, without interruption. (d.) The introduction of a new slip is easily accomplished in about one minute, and no new correction is then required. (e.) The second fixed pencil marks along the margin of the paper slip a line, which by being broken every three hours marks the time. This line determines also for the subsequent measuring of the curve the adjusting of a transparent scale, with the co-ordinate lines of time and height.

IV. Finally, the mechanical execution of the apparatus, which comes from the establishment of Herr *Wanke*, is deserving of attention.

2873b. Photographic Thermograph, as employed at the seven observatories of the Meteorological Committee. Described, Report of Meteorological Committee of Royal Society for 1867.

The Meteorological Committee of the Royal Society.

2874. Electric Registering Anemometer with momentary contact contrivance.

Prof. Osnaghi, Imperial Central Meteorological Institute, Vienna.

The apparatus registers each kilometre by marks impressed on a slip of paper moved by clockwork.

2875. Wind-Current Autograph, or Registering Apparatus.   *John G. Schoen.*

This "Wind-Current Autograph" marks, or registers, continuously and correctly, on a strip of paper moved by clockwork, the motion or direction of the currents of wind in such a manner that the time is indicated as the abscissa, and the angle of elongation of the weather-vane towards the north shown at every particular moment, as the ordinate.

2876. Electrical self-recording Anemometer and Printing Apparatus, invented by the Exhibitor.

J. E. H. Gordon.

The figures on the left hand side of the paper give the hours, those on the right the direction of the wind at each quarter of an hour, while for every mile of wind that passes over the cups on the roof a dot is made in the centre of the paper. The number of dots between any two consecutive figures is the velocity in miles per hour for that hour. The communication with the roof being made by electricity no shaft is required.

2877. Automatic Light Registering Apparatus.

A. Salted paper.
B. Silver nitrate solution.
C. Trough for silvering paper.
D. Drying reel.
E. Dark box.

F. Insolator.
G. Cover for insolator.
H. Clock.
K. Battery.
L. Punch.

M. Reading-off apparatus :—

    a. Drum.
    b. Graduated strip.
    c. Stand.
    d. Sodium carbonate.
    e. e. Platinum wires.

    f.f. Stands for platinum wires.
    g. Bunsen burner.
    h. Lens.
    k. Spirit lamp.

Prof. H. E. Roscoe, F.R.S.

This method depends on the fact that the depth of colour produced on a properly prepared chloride of silver paper is directly proportional to the intensity of the light multiplied by the time of exposure.

40075.   Y y

The apparatus consists therefore essentially of two parts, one in which the prepared paper can be exposed for definite periods of time to the action of the light, and a second part in which the intensity of the tint obtained can be determined.

The paper (A), previously salted by immersion for five minutes in a 3 per cent. solution of sodium chloride, and cut into strips, is silvered by floating for two minutes in a 12 per cent. solution of silver nitrate (B), contained in the long trough (C), and afterwards dried on the reel (D).

The prepared paper may be preserved either before or after exposure in the dark box (E). It is next wound on to the bobbin of the insulator (F), which is placed in electric communication with the clock (H) by means of the battery (K); the free end of the paper, passing over the large wheel, being held in position by means of a small pin on the inside circumference of the wheel.

When a current of electricity passes, the magnet attracts the armature, and the wheel moves through a small space; the circuit is immediately broken and the armature released, and this slight movement of the wheel is repeated every time a current of electricity passes through the apparatus. The insulator is provided with a cover (G), in the top of which is a small circular hole, against which the prepared paper is pressed by means of a spring, and as the wheel revolves, fresh portions of the paper are successively brought under this hole, and thus exposed to the action of the light. The mechanism of the clock is so arranged that discs of prepared paper shall be exposed to the action of the light each hour for 10 different periods of time, which have been exactly determined, varying from 2 to 30 seconds, the object of this being to obtain, either with the feeble light of the morning, or the strong light of mid-day, a tint neither too light nor too dark to be read off This is accomplished by means of a large metal disc in the clock, which revolves once in two minutes, and is in metallic connexion with one pole of the battery. On the face of the disc are placed 11 platinum pegs, arranged at equal distances from the centre of the disc, but at such different distances from one another that the first 10 intervals correspond as closely as possible to 2, 3, 4, 5, 7, 10, 12, 17, 20, 30 seconds respectively (their values being afterwards experimentally determined to one-fifth of a second), whilst the value of the last interval is of no importance. The other pole of the battery is connected with a metallic lever tipped with platinum, the insulator forming a part of the circuit.

Each hour this lever is lowered mechanically so that it comes in contact with the first platinum peg; the circuit is completed, the magnet in the insulator attracts the armature, causing the wheel to make a small fraction of a revolution, and a fresh surface of paper is exposed to the light. Contact is immediately broken, the peg passing away from under the lever, to be again made when the next peg passes by, and so on through one revolution of the disc. After the 11th peg has passed by, the lever is automatically raised, and remains out of contact till the next hour. The last exposure, therefore, corresponds to the interval, the value of which has not been determined, this portion of the paper remaining exposed until the next hour.

After the apparatus has been in work for 24 hours the strip of paper is exhausted, and must be replaced by a new one.

On the paper thus exposed will be found repetitions of a series of 10 discs, of a tint which in each series gradually increases in intensity, separated by one black disc when the paper has been exposed for the whole hour. One half of two or three of the discs in each series is cut away by the semi-circular punch (L), and the intensity of the tint in each case read off. For this purpose the apparatus under the glass case (M) is employed. Round the drum (a) is pasted the graduated strip (b), the intensity of the tint on any point on which may be found with reference to a table prepared for the purpose. The drum revolves freely on the stand (c). The paper to be examined is placed over the graduated strip, being held in position by the two clamps attached to the stand.

By means of the semi-circular hole, which has been punched into the exposed paper, the tint of the paper and of the graduated strip are brought side by side, and by revolving the drum with the hand various portions of the graduated strip are brought successively into juxtaposition with the tint of the exposed paper, and when the two tints are seen to be identical, the value of the exposed tint is known, corresponding of course to the one for that point on the graduated strip, and ascertained by reference to the table of intensities. A number of such readings for each tint are obtained and the mean taken as correct. These comparisons must be made by the light of the monochromatic soda flame.

Beads of sodium carbonate (d) on the platinum wires (e, e,) are held by means of the stands (f, f,) in the flame of the Bunsen burner (g), and the rays of light are concentrated on to the strip by the lens (h). When gas is not available the spirit lamp (k) is employed.

It now remains alone to divide the value of the tint thus obtained by the number of seconds which the tint has been exposed, to determine the intensity of the light for one second of time, which is the standard adopted. Two or three of the discs obtained each hour are thus read off, and the mean taken as representing the correct intensity.

2877a. Insulator and Whirling Apparatus.
Capt. Abney, R.E., F.R.S.

This instrument is based on the same principles as that of Professor Roscoe, but it was constructed with the idea that its portability may cause it to be employed in localities where his larger apparatus might be too bulky.

It consists of a revolving cylinder (round which is wound sensitive paper), driven by clockwork enclosed in a cover. In the cover is a narrow slit, extending across the cylinder, over which is pressed a black glass wedge of known graduation. The light falls through the wedge on to the paper in different gradations, according to the intensity of the light. The integration of the darking is obtained by placing the cylinder in a whirling machine, and the position of the standard tint is noted on a scale which depends on the graduation of the wedge. The total insolation is thus arrived at. When Professor Roscoe's and this instrument are worked together the results are accordant.

2879a. Full sized Lever Anemometer for registering the pressure of the wind.
C. O. F. Cator, M.A.

The pressure-plate is circular, has an area equal to 1 square foot, and is kept constantly face to wind by a vane (or by windmill fans) ; as it is driven forward by the wind, its motion is conveyed by means of a wire to a pencil, which continuously records every movement on paper revolving by clockwork below.

The peculiarity of this instrument is, that, instead of employing springs as the resisting medium, which from their proximity to the pressure plate and consequent continual exposure to the weather, cannot always preserve the same strength and elasticity, the resistance is furnished by a system of leverage. This consists of two eccentric curves of different sizes rigidly connected together and revolving on the same axle, of which the curvatures decrease respectively in opposite directions, so that the effect is doubled. Round the larger one a cord is carried, from which hangs a fixed weight. Round the smaller one a chain is passed, which is connected directly with the pressure plate, and also with the recording pencil. As the motion is direct, the spaces moved through by the pencil and by the pressure plate are exactly equal to another one and also to the primeter of the smaller curve.

One great advantage of this apparatus consists in its close proximity to the recording pencil, and in fact that the resistance is always the same, as every part of the instrument except the pressure plate is under cover and free from exposure to the weather. Another peculiarity of this instrument is that the plate is furnished with a conical back, so as to diminish the error arising from the formation of a partial vacuum behind it in strong winds.

The direction of the wind is also continuously recorded by Beckley's method, on the same paper as the pressure.

2880. Whewell's Anemometer.   *Elliott Brothers.*

Consists of a delicate wheel with vanes and endless screw, working a series of cog wheels, which communicate with an ordinary lead pencil, that registers the force and direction of the wind on a vertical japanned cylinder. The cylinder is enclosed in a wooden case to prevent it from being injured by exposure to the weather.

2882. Registering Aneroid Barometer, for showing at a glance the various fluctuations that have taken place in the barometer.   *Elliott Brothers.*

2883. Howlett's Portable Anemograph, an instrument which records the varying direction and force of the wind in the form of a map.   *Elliott Brothers.*

2885. Self-recording Aneroid for Hall or Library ; a graphic delineation of the change in pressure of the air for each week can be seen at a glance.   *Francis Pastorelli.*

Fixed nearly in line, in a case, is an aneroid, an eight-day clock, and a revolving cylinder that occupies the central position; it is covered with metallic paper ruled for the days of one week, with the barometric scale in inches and tenths; upon this is marked every hour the pressure of the atmosphere ; the markings have the appearance of a curved line, and the rise and fall of the barometer is seen at a glance for every hour of each day for one week, when a new paper has to be placed on the cylinder. The index hand on the dial of the aneroid indicates the pressure of the air at the time of observation. Having already described the construction of the aneroid, it is only necessary to explain the method by which it is made self-registering. Connected with the lever that carries the chain round the arbor is a long watch chain that passes through the top of the aneroid over a pulley ; it terminates with a metallic point, which is kept in working position by means of a vertical bar ; this is capable of revolving upon its points. Behind the metallic point or pencil is the vertical revolving cylinder covered with the metallic paper; this is moved by the clock, which also presses the pencil point down upon the paper, and leaves a mark at each hour.

2887. A. Von Oettingen's " Self-recording Wind Components Integrator." Constructed by P. Schultze.

Dr. Arthur Von Oettingen, Professor at the Imperial University, Dorpat (Russia).

The wind moves a system of hemispherical cups, like Robinson's, (which motion is replaced by clockwork for exhibition) acting on a circular plate, whose velocity is ordinarily proportional to the velocity of the wind. Four systems of sliding-rollers rest on this plate, whose bearings can be moved round a vertical axis, the principal planes of which imitate all variations of a wind vane. Each sliding roller can rotate about a horizontal axis, but only in one direction, and after a half rotation an electrical contact is made. A mechanism limits the contact to a fraction of a second. When the contact is made, one of four wheels, with number-types, is moved. Every half an hour the position of these four wheels is shown by printing numbers on a strip of paper. The differences of those readings represent the mean velocity of the wind from N., E., S., and W. They are converted into absolute values by means of a table. Different mechanism adjusts the portion of every sliding roller.

2888. Complete Meteorograph.
F. Van Rysselberghe, Ostend.

4553. Drawings, on a large scale, **of the Meteorograph** of Padre Secchi, with printed description annexed.
Observatory of the Collegio Romano, Rome; Director, Padre Secchi.

AUTOMATIC REGISTERING APPARATUS.

2889. Photographic Proofs, obtained at various depths and at different seasons, to ascertain the penetration of the solar rays in the waters of the Lake of Geneva.
Prof. Dr. F. A. Forel, Morges, Switzerland.

From these observations it has been found that the extreme limit of penetration at which the solar rays are capable of acting on chloride of silver is—
1. In the summer, when the water is, relatively speaking, not very clear, from 40 to 50 metres deep.
2. In winter, when the water is less clear, 90 to 100 in depth.

To enable a comparison to be made with the clearness of the water in other lakes and in the ocean, and to give an idea of the degree of transparency of the waters of Lake Leman, it should be mentioned that on the 10th March 1875, at noon, the sun being at 39° 10′, a white plate or plaque 25 centimetres in diameter became invisible at a depth of 17·0 metres.

2890. Drawings of the principal parts of a meteorological registering apparatus, viz.:—
  a. Scale-barometer, natural size.
  b. Instrumental thermometer, natural size.
  c. Air thermometer, natural size.
  d. Clock-work and registering contrivances for only one of these instruments, since the mechanisms of all of them are identical; natural size. *Dr. P. Schreiber, Chemnitz.*

The apparatus represented in the drawings is intended for registering, at intervals of 10 minutes, atmospheric pressure, atmospheric temperature, and the temperature of the registering instruments. Only drawings could be sent in, as the construction of the apparatus is not yet finished. The instrument consists of four parts, viz., the barometer, the thermometer for measuring the temperature of the instruments, the thermometer for measuring the temperature of the air, and the propelling and registering mechanism.

The barometer is a balance-barometer of simple solid structure, with supporting pillar and beam of iron.

The thermometer for indicating the temperature of the instrument is a tube, closed at the top and open below, filled with air, dipping into mercury, and balanced by a weight on a string passing over a pulley. This tube moves 2 mm. for every degree C. of change in the temperature of the instrument.

The other thermometer, intended for the measuring of the atmospheric temperature, is likewise an air thermometer, consisting of a tube similar to the above described, but which contains an iron tube, joined to a narrow lead tube (2 to 3 mm. clear width), which again is connected with the thermometer bulb. This bulb is freely exposed to the atmosphere, though protected in a proper manner against rain and sunshine. It is made of copper, has a capacity of 5 litres, and a corresponding surface area. Here, too, a change of 1° C. gives a motion of 3 mm.

Let b, t, τ, signify pressure and temperature of the atmosphere and of the instrument, x, y, z, the respective positions of the tubes I., II., and III.; then $x = f_1 (b, \tau)$, $y = f_2 (b, \tau)$, $z = f_3 (b, t, \tau)$, and hence $b = \phi_1 (x, y)$, $\tau = \phi_2 (x, y)$, $t = \phi_3 (x, y, z)$, that is to say, the three quantities sought are determined by the positions of the three tubes.

The movements of the tubes are communicated to rods which move vertically in front of cylinders set revolving by means of a clockwork. This same clockwork moves also hammers, which hit every 10 minutes upon the rods, whereby marks are produced upon the cylinders, and thus the positions of the tubes noted. The clockwork is specially constructed for the instrument, and the largest toothed wheel, with 120 teeth, has a diameter of 120 mm.

Before each marking the tubes are slightly moved forward by steel cylinders which dip into the mercury. All movements are produced by a horizontal axis, which completes one revolution every 10 minutes, and which is provided with the necessary wheels, &c.

Electro-magnetism is used—

(1.) To mark the hours upon the cylinders.

(2.) To work signals, which will indicate a possible stoppage of the clock-work, or other interruptions in the working of the apparatus.

(3.) To sound, by means of the clock and a bell, the hours, thus giving evidence of the right condition of all the agents employed.

The advantages of the apparatus here described are :—

(1.) The constituent agents, the balance barometer, air thermometer, and registering clockwork, are all well tested instruments. The movements to be registered are so large that it is not necessary to magnify them. They can be calculated beforehand by means of precise formulæ from constants that are easily determined with great exactness. The apparatus has, therefore, nothing of the nature of an interpolating instrument, a character belonging at present to all registering mechanisms, but must be viewed as an instrument of precision. Checking observations, *daily* required by all self-registering instruments, become superfluous with this apparatus.

(2.) The registering instruments can be put up in a dry place, protected against changes of temperature.

(3.) All parts of the apparatus are of metal and glass, wood is excluded. The instrument is, therefore, more durable.

(4.) The construction is extremely simple ; any disorder is easily repaired. The purely mechanical motors are free from the disturbances to which electro-magnetic motors are so frequently subject. If moderately well executed, and tolerably well handled, the instrument will, practically, never refuse to work.

(5.) The registering of the temperature of the instrument might prove useful in other observations, since it gives the temperature of a closed space.

(6.) Not the least of the advantages is the cheapness of the instrument.

The theory of this apparatus is given in full in Volume XI. of Carl's "Repertorium für Experimentalphysik."

G. Lorenz, Chemnitz (Lachsen), has undertaken the construction of the apparatus.

2891. Kreils' Barograph, formerly in use at the Kew Observatory, for registering the movement of the barometer.

Kew Committee of the Royal Society.

An instrument employed at the Kew Observatory in 1845, for the purpose of registering automatically the height of the barometer. It consists of a syphon barometer, having a float resting upon the surface of the mercury, in the open end of the tube. Immediately above the tube a lever is fixed horizontally, and a cord, wrapped round a sector on the short arm, passes down and is attached to the float. The other end of the lever carries an ordinary pencil, which, being struck every five minutes by a hammer moved by a clock, makes a dot upon a sheet of paper fixed to a frame drawn in front of it by clockwork.

2892. Ronalds' Photo-Barometrograph, for registering photographically the changes in the height of the barometer, formerly erected at the Kew Observatory.

Kew Committee of the Royal Society.

An instrument for registering the variation in the height of the barometer upon a daguerreotype plate; constructed in 1847 by Mr. Francis Ronalds, afterwards erected at the Kew Observatory, and described by him in the British Association Report for 1851.

The light from an argand lamp, after passing through a condensing lens, falls on a narrow slit cut in a metal plate attached. A barometer tube, the mercury in which, by rising or falling, varies the length of the slit illuminated.

An achromatic combination of lenses, by Voigtlander, throws a magnified image of the bright slit upon an aperture in the case, past which a daguerreotype plate is moved slowly by clockwork, and so registers the changes in the height of the barometer.

The barometer itself, together with its cistern, which is of large area, is suspended from an arrangement of levers and zinc rods, on the principle of the gridiron pendulum, in such a manner as to render the indications unaffected by fluctuations of temperature.

An improved form of this instrument, in which the photographic image is impressed upon paper, is now in use at Kew, and at many other observatories.

2892aa. Photographic Barograph, as employed at the observatories in connection with the Meteorological Office.

Meteorological Committee of the Royal Society.

This instrument is described in the Report of the Meteorological Committee of the Royal Society for 1867. It is the improved form of Ronald's barograph, No. 2892.

2892a. Barometrograph.   *M. Bréguet, Paris.*

2893. Barometer adapted to automatic registration by Photography.   *Chas. Brooke, M.A., F.R.S.*

A vertical cylinder serves (as at Greenwich) for the registration of the barometer and the balanced magnetometer. *See* Magnetism.

2894. Barograph.   *M. Redier, Paris.*

2895. Registering Barometer, mercurial or aneroid, showing enlarged curves without the aid of electricity or photography.
                                                                 M. Redier, Paris.

2895a. Registering Mercurial Thermometer, after the plan of M. Hervé Mangon.   *M. Redier, Paris.*

2896. Barograph, balance barometer; executed by Greiner and Geissler, Joint Stock Company for manufacturing meteorological instruments at Berlin.

> *Imperial Admiralty Hydrographical Office, Berlin, and German Naval Observatory, Hamburg.*

This instrument is put up and kept working at each of the normal observatory stations on the German coast. The stations are subordinate to the German Naval Observatory (Seewarte).

VIII.—EVAPOROMETERS.

2897. Evaporometer, in the form of a spring steel-yard.
        Prof. F. Osnaghi, Imperial Central Meteorological Institute, Vienna.

This instrument shows on a sector the number of millimeters evaporated from a certain quantity of water in a given time. As its action is produced by gravity, it is also useful in winter when ice is formed on the scale. It differs from other steel-yards in the weight acting on the inner end of a spiral spring.

2898. Apparatus for determining the **Evaporation** from different soils.
                        Sydney B. J. Skertchly, F.G.S., H.M. Geological Survey.

The apparatus consists essentially of an evaporimeter composed of two vessels, the innermost of which receives the material to be experimented upon, and the external one supplies water to compensate for evaporation. Over this is a glass vessel which receives the vapour given off by the material. The temperature, &c., are registered by a hygrometer and barometer in the glass receiver, and the temperature of the soil by a ground thermometer. Any given temperature can be obtained by means of a platinum spiral heated by a galvanic battery. The evaporimeter maintains the material in a natural condition so far as regards temperature and moisture. Dry air is admitted into the glass receiver, and the air with the evaporated water passes from the top of the receiver into a train of drying tubes; the current of air being produced by an aspirator containing oil. By means of this apparatus various soils, &c.,

can be brought under similar conditions of temperature, &c., and the evaporations compared for any temperature. The apparatus was especially designed to determine the proper amount of water which should be discharged by the artificial drainage system of the Fen Land.

2899. Integrator of Sun's Heat.
Scottish Meteorological Society.

When the water in the globe expands, some of it passes out at the bent upper tube. The level of the water is kept constant by a supply from the cistern which communicates with the globe by the india-rubber ball which acts as a valve.

Note.—The instrument, which is in principle a weight thermometer, may also be used for ascertaining the mean temperature of the air. Designed by Thomas Stevenson, C.E., F.R.S.E., Honorary Secretary.

2900. Instrument, designed to ascertain the temperatures at which visible vapour is found. *Scottish Meteorological Society.*

Water is heated in the main chamber by a lamp beneath, and its temperature is read at the point when vapour appears on one of the pieces of glass which are made to revolve slowly above the open end branch tube, and again read as the water cools, at the instant when condensation ceases to be observed. Designed by Thomas Stevenson, C.E., F.R.S.E., Honorary Secretary.

2901. Ebermayer's Evaporation Apparatus, for determining the degree of evaporation of different kinds of soil.
Prof. Ebermayer, Aschaffenburg.

The evaporating apparatus and the earth thermometer are described at greater extent in "Die physikalischen Einwirkungen des Waldes auf Luft " und Boden," von E. Ebermayer, Aschaffenburg, 1873.

2902. Morgenstern's Atmometer. *W. Apel, Göttingen.*

Morgenstern's atmometer differs from every other by its being founded on the principles of capillarity and of Mariotte's bottle.

The evaporating vessel is filled with siliceous sand, below which there may be placed a flat stone. This sand is saturated with water by capillarity; any loss of water by evaporation is at once replaced by a corresponding volume of water from a burette. This burette forms a Mariotte's bottle, the upper part of which is closed against the outer air by means of mercury. A tube, bent in the shape of a horseshoe, of which one branch enters the burette from below, conducts air into the latter in proportion as water is lost through evaporation. When a large portion of the burette has become filled with air, the danger arises that the air column on expanding, by a possible rise of temperature, would exert a pressure upon the water below it in the burette, and thus lead to an over-saturation of the sand. To prevent this the branch of the last-mentioned tube which is placed outside the basin is provided with a small globular vessel, into which the water, pushed on by the expansion of the air column, enters. With progressing evaporation this water returns again into the burette, or can later be drawn into the burette. This globular vessel is further intended for the filling of the burette with water, which purpose is accomplished by fixing to the open end an india-rubber tube, dipping into water, and sucking at the upper end of the burette. Before the burette is completely filled the india-rubber tube is removed, and the sucking at the upper end of the burette resumed, until, in consequence, the globular vessel is emptied. The connexion of the burette with the sand is closed during the operation of filling.

The evaporating vessel has a surface area of one square decimètre.

To exclude, as much as possible, the influence of the temperature, the evaporating vessel is enveloped in some bad conducting material.

The sensitiveness of the instrument is so great that a little dry sand, or a piece of blotting paper, or the fraction of a drop of water, put upon the surface of the evaporating vessel is immediately indicated by the water column in the burette.

2903. Atmometer or Evaporometer, for determining the quantity of water evaporating from the surfaces of waters as well as from different sorts of soil.   *Prof. Prestel, Emden.*

Atmidometer with Hicks' Patent enamel Stem for measuring the rate of evaporation from water, ice, or snow.

J. J. Hicks.

The enamel stem has the divisions and figures in black on white enamel which renders the reading easy and the scale incorrosible.

2904. Apparatus for the direct determination of the tension of aqueous vapour in the atmosphere, constructed in the year 1868 by Dr. Geissler, of Bonn, according to the instructions of the late Prof. Schulze, professor of chemistry in Rostock.

Prof. Matthiessen, Rostock.

The U-shaped tubes serve for the reception of the mercury. The absolute vapour tension of the atmospheric air enclosed in the flask-shaped vessel is directly determined, at the differential barometer, through absorption of the vapour by concentrated sulphuric acid, which is introduced for that purpose.

IX.—OZONOMETERS.

2904a. Smyth's Ozonometer, made of brass, lined with sealing-wax, and provided with brass stop-cocks, invented by contributor previous to ozonometer No. 2905.   *John Smyth, jun.*

2904b. Smyth's Ozonometer, made of brass, lined with glass, and provided with glass stopper to admit air, invented by contributor previous to ozonometers Nos. 2904 and 2905.

John Smyth, jun.

2904c. Diagram showing the rough form of *ozonometer* and *aspirator* used by contributor in his first experiments, August 1865.   *John Smyth, jun.*

2904d. Diagram showing the contributor's *ozonometer* connected to Dr. Andrew's form of aspirator by means of which his later experiments have been made.   *John Smyth, jun.*

2905. Smyth's Ozonometer, for the determination of the amount of ozone in a measured volume of air by means of an aspirator, invented by the contributor and described by him in a paper read at the meeting of the British Association in Birmingham in 1865. *John Smyth, jun.*

It consists of a box-wood tube or cylindrical box, about two inches long and two inches in diameter, one end of which is closed, except in the centre, where it is pierced by a quarter-inch tube communicating with the aspirator; the open end is covered by a lid or second box of the same material, which is so large as to slide over the first, and is also pierced by a quarter-inch tube, which, when the ozonometer is arranged for an experiment, directs the air against the centre of the test paper stretched across the open end of the inner or first box, and is secured there by an india-rubber band lying in a groove.

2905a. Schönbein's Ozonometer, rendered self-recording. An instrument for exposing each hour a fresh piece of Schönbein's ozone test-paper to the influence of the atmosphere.

R. C. Cann Lippincott.

Two cylinders (one large, the other small) are enclosed in boxes, the openings of which are guarded by india-rubber lips. The boxes are $2\frac{1}{4}$ inches apart. The large cylinder is moved round $2\frac{1}{4}$ inches each hour by means of a driving-shaft attached to the clock. A strip of test-paper, about 5 ft. long and $\frac{5}{8}$ of an inch wide, is rolled round the small cylinder, the free end of it being fastened to the large cylinder; $2\frac{1}{4}$ inches of paper are thus exposed to the influence of the atmosphere.

The clock, which goes 8 days, indicates the minutes only, and shifts the paper by causing the large cylinder to rotate $2\frac{1}{4}$ inches of its circumference exactly at the hour, thus removing the portion of paper ($2\frac{1}{4}$ inches long) exposed during the past hour, and exposing a fresh portion, to be in its turn removed at the end of another hour and succeeded by a new piece, and so on. At the end of 24 hours the whole strip of test-paper is unrolled by the observer from the large cylinder, dipped, and read by the scale (Schönbein's) ; a fresh strip is then rolled round the small cylinder, and its free end is attached to the large cylinder as before.

The instrument was made for me by Mr. Casella, in 1868, after a plan of my own suggestion.

The following are the results of some observations made with the instrument taken from the " Proceedings of the Meteorological Society," Vol. V., p. 51.³

MEAN HOURLY AMOUNTS of OZONE deduced from Observations taken every Hour (with a few omissions when the clock stopped) from Feb. 20 to Nov. 18, 1869. Test paper, Schönbein's ; scale, 0 to 10.

Hours.	Midnight to 1 A.M.	1 A.M. to 2 A.M.	2 A.M. to 3 A.M.	3 A.M. to 4 A.M.	4 A.M. to 5 A.M.	5 A.M. to 6 A.M.
Mean amount of ozone - -	2·8	2·8	2·7	2·7	2·8	2·8

Hours.	6 A.M. to 7 A.M.	7 A.M. to 8 A.M.	8 A.M. to 9 A.M.	9 A.M. to 10 A.M.	10 A.M. to 11 A.M.	11 A.M. to Noon.
Mean amount of ozone - -	2·8	2·7	2·4	2·0	2·2	2·4

Hours.	Noon to 1 P.M.	1 P.M. to 2 P.M.	2 P.M. to 3 P.M.	3 P.M. to 4 P.M.	4 P.M. to 5 P.M.	5 P.M. to 6 P.M.
Mean amount of ozone - -	2·4	2·5	2·7	2·7	2·9	2·8

Hours.	6 P.M. to 7 P.M.	7 P.M. to 8 P.M.	8 P.M. to 9 P.M.	9 P.M. to 10 P.M.	10 P.M. to 11 P.M.	11 P.M. to Midnight.
Mean amount of ozone - -	2·7	2·5	2·6	2·7	2·8	2·8

Mean of all the hours = 2·63.

2906. Aspirator, with gutta percha tubing and wooden box for tests, used by the Ozone Committee during their experiments in 1869.   *Scottish Meteorological Society.*

The small tin box was made by a native Arab for Dr. Arthur Mitchell, in connexion with ozone experiments made by him in Algiers in 1855.

2907. First Ozone Generator.   *Dr. Werner Siemens.*

The air or oxygen to be ozonised is caused to pass between two concentric cylinders coated with tinfoil, and electrified by an induction apparatus.

2907a. Tisley's Ozone Generator. The central tube being used as a water channel, a uniform low temperature can be maintained, thereby greatly improving the per-centage of ozone.

Tisley and Spiller.

See "Nature," 19th June 1873.

X.—MISCELLANEOUS.

2909. Plan and View of the Observatory of the Central Imperial Institute for Meteorology and Terrestrial Magnetism, Vienna.

I. R. Central Institute for Meteorology and Terrestrial Magnetism, Vienna.

The building was erected, under the direction of the Imperial Minister of Education, by the architect Ferstel, and finished in April 1872. The Observatory is provided with a great number of self-registering instruments for all observations, among which may be particularly noticed Dr. Theorell's printing meteorograph for meteorological observations, and Adie's photographical magnetometer (Kew model), for the observation of magnetic variations.

2910. Meteorological Photographs, specimens as ordinarily produced at the Radcliffe Observatory, Oxford, containing the unreduced observations made with the barograph, thermograph, hygrograph, and anemograph, by the use of the waxed paper process, from November 19th to December 6th, 1873.

Rev. Robert Main.

2911. The Storm Atlas of the Meteorological Institute of Norway, by H. Mohn. *Prof. H. Mohn, Christiania.*

"Atlas des Tempêtes de l'Institut Météorologique de Norvége, publié avec "le concours de la Société scientifique de Christiania, par H. Mohn."

2911a. Goddard's Cloud Mirror. *G. J. Symons.*

2911b. Daily Bulletin of the Signal Service, U.S.A., May 1873. *J. Norman Lockyer, F.R.S.*

2911c. Atlas Météorologique de l'Observatoire de Paris, 1872-3-4. *J. Norman Lockyer, F.R.S.*

2912. Pane of Glass pierced by a **Hailstone** during the storm at Geneva in the night of the 7th and 8th July 1875.

Messrs. Ramboz and Schuchardt, Geneva.

This pane formed part of the roof of the printing office, in a court of the Rue de la Pelisserie, at Geneva. No thunder-bolt fell in the neighbourhood, so that this strange oval opening, the upper edge of which, it may be noticed, is blunted, cannot be attributed to the action of lightning.

2812a. Large Globe in which H. B. de Saussure collected air on the summit of Mont Blanc. With portable case or shoulder basket, capable of holding two such globes.

M. H. de Saussure, Geneva.

2913b. Dietheroscope.

Prof. Luvini, through the Meteorological Office.

2913c. Seven pictures of Clouds, according to Howard's nomenclature. *Meteorological Office.*

2913d. Meteorological Diagrams. *Meteorological Office.*

7 storm charts.

List of stations.

Colliery explosions and weather.

2914. Map of Scotland, showing the Society's stations, the prevailing winds, and the annual rainfall.

Scottish Meteorological Society.

The present stations are indicated by a *black* circle, and stations at which observations are no longer made by a *circle and a cross.* Rainfall stations by a *red* circle. Constructed by Alexander Buchan, M.A., F.R.S.E., Secretary.

2915. Temperature of the British Islands, for each month and for the year.   *Scottish Meteorological Society.*

On all the maps the isothermals up to 44° are coloured *black;* those from 45° to 54° are coloured *blue;* and those from 55° and upwards are coloured *red.* Prepared by Alexander Buchan, F.R.S.E., Secretary, and published in the Society's Journal, vol. iii. p. 102.

2916. Seasonal Distribution of the Rainfall of Europe, showing the month of *greatest* rainfall and the month of *least* rainfall of its different regions.   *Scottish Meteorological Society.*

The months from November to February are coloured *blue;* from March to May, *green;* from June to August, *red;* and September and October, *black.* Prepared by Alexander Buchan, Secretary.

2917. Charts, showing the mean monthly and annual amount of the diurnal oscillation of the barometer over the globe, from the a.m. maximum to the p.m. minimum, by lines of 10, 20, 40, 60, 80, and 100 (and upwards) thousandths of an inch (0·010, 0·020 inches, &c.).   *Scottish Meteorological Society.*

The portions shaded *red* indicate an oscillation of 0·100 inch and upwards. Prepared by Alexander Buchan, Esq., F.R.S.E., Secretary of the Society, and published in Trans. Roy. Soc. Edin., vol. xxvii. p. 397.

2918. Charts, showing by isobaric lines the mean pressure of the atmosphere, and by *arrows* the prevailing winds over the globe for each month.   *Scottish Meteorological Society.*

Isobarics of 30 inches and upwards are coloured *red,* and those under 30 inches *blue.* Thus the red isobarics indicate where pressure is in excess, and the blue isobarics where the pressure is in defect.

Calms are marked with a circle, and variable winds with an asterisk. The arrows fly with the wind.

Constructed by Alexander Buchan, F.R.S.E., and published in Trans. Roy. Soc. Edin., vol. xxv., p. 575.

2919. Diagrams, showing for London the influence of weather on mortality from different diseases and at different ages.   No. I.

Scottish Meteorological Society.

Mortality classed according to—

A. Age (average of 5 years).—Under 5 years coloured *red;* 5 to 40, *blue;* 40 and upwards, *black.*

B. Sex (average of 30 years).—Males coloured *red;* females, *blue.* The weekly deviations from the mean line are given in per-centages above and below the average.

C. Groups of diseases (average of 30 years).—The upper *thin* line represents weekly averages from all causes of death whatever; and the *thick black* line, from all causes minus violent deaths. The *red* space embraces deaths from all bowel complaints; and it will be observed that the subtraction of these diseases takes away the summer maximum from the general curve. The *blue* space embraces all diseases of the respiratory organs; and the *dark* space, diseases of the brain and consumption.

The mean weekly temperature is represented by the *red* line.

Prepared by Alexander Buchan, Secretary, and Dr. Arthur Mitchell, Chairman of the Medico-climatological Committee, and published in the Society's Journal, vol. iv., p. 187.

2920. Diagrams, showing the influence of weather on mortality from different diseases and at different ages. Nos. II. and III.
Scottish Meteorological Society.

D. Showing mortality curves from special diseases, viz., small-pox, measles, whooping-cough, scarlet fever, typhus and typhoid fevers, erysipelas, metria, rheumatism, pleurisy, and pericarditis.

E. Showing mortality from bowel complaints, viz., dysentery, British cholera, cholera, and diarrhœa.

F. Mortality from diseases of the respiratory organs, viz., asthma, bronchitis, pneumonia, and laryngitis.

The curves under D, E, and F are the averages of 30 years, and the deviations from the mean lines are given in per-centages above and below the average.

Prepared by Alexander Buchan, Secretary, and Dr. Arthur Mitchell, Chairman of the Medico-climatological Committee, and published in the Society's Journal, vol. iv., p. 187.

2921. Diagrams, showing the mortality of British large towns. No. I. shows the weekly mortality from *all causes* for large towns of England, and No. II. for large towns in Scotland and Ireland. The averages are in most cases 10 years. The weekly averages are calculated at the annual rate of mortality per 1,000 of the population. *Scottish Meteorological Society.*

The mean annual mortality per 1,000 of the population of each town is given in the left-hand margin. The *red* lines show the deviations from this average for each week of the year.

No. III. shows the weekly mortality from *diarrhœa* for large towns of England, and for Edinburgh. The averages are in most cases for 10 years. The weekly averages are calculated at the annual rate of mortality per 1,000 of population. The breadth of space coloured red shows, for each week, the rate of fatality for each town.

Prepared by Alexander Buchan, Secretary, and published in the Society's Journal, vol. iv., p. 337.

2922. Diagram, showing the *steadiness* of the mortality curve of scarlet fever in each of the six epidemics which have occurred in London from 1840 to 1874. *Scottish Meteorological Society.*

A. represents the per-cent. of deviation from the mean line of each week's average of each of the six epidemics. The duration of each epidemic is given on the left-hand margin, and its average weekly death-rate on the right-hand margin.

B. shows the gross mortality for each year of the period, each epidemic being marked by a different colour.

By Arthur Mitchell, M.D., F.R.S.E., Chairman of the Medico-climatological Committee. Published in the Society's Journal, vol. iv., p. 340.

2923. Diagram, showing for London the relation of diarrhœa to temperature. *Scottish Meteorological Society.*

A. represents the mean temperature of 1859, having the hottest summer in London from 1845 to 1874 ; of 1860, having the coldest summer during the same period ; and of 1861, having an average summer temperature.

B. represents the mean weekly death-rate in London from diarrhœa during each of these same years, the colours of the diarrhœa curves corresponding with those of the temperature curve for the same years. The weekly death-rate has been calculated in each case at the annual rate of 1,000 of the population.

By Arthur Mitchell, M.D., F.R.S.E., Chairman of the Medico-climatological Committee.

2924. Barometric Gradients.
Scottish Meteorological Society.

Model illustrative of the principle of the barometric gradient, which is calculated by dividing the distance between any two barometers by their difference in reading, both being reduced to 32° and sea level (or the same level). Designed by Thomas Stevenson, C.E., Honorary Secretary, and described in the Society's Journal, 1867, when the proposal was first made.

2925. Drawing of an instrument for ascertaining sea and river temperature by thermometers continuously immersed.
Scottish Meteorological Society.

The case containing the instruments is suspended from the pier, or lightship or in the river, and is drawn up bringing with it some water in the cistern at the bottom, in which the thermometer bulbs are placed. Designed by Thomas Stevenson, C.E., F.R.S.E., Honorary Secretary, and described in the Society's Journal, vol. iv., p. 44.

2926. Cloud Reflecting Compasses. For ascertaining direction of higher currents of air.
Scottish Meteorological Society.

When a cloud is to be observed the compass should be turned round till one of the hues on the mirror coincides with a well defined edge of the cloud, and the compass is then made to revolve, gradually keeping the hue constantly on the edge of the cloud. The angle indicated by the magnetic needle being afterwards read off, the direction of the cloud's motion in azimuth is at once ascertained. Designed by Thomas Stevenson, C.E., F.R.S.E., Honorary Secretary, and described in the Edinburgh Philosophical Journal, 1855.

2926a. Hypsometric Map of the Caucasus, recently published by the Topographical Office at Tiflis.
The Pulkowa Observatory.

2926b. Diagrams Illustrative of Underground Temperature.
G. J. Symons.

2927. Graphic Representations of Underground Temperatures in Fifeshire, by Sir J. Leslie. *University of Edinburgh.*

2928. Graphic Representations of Underground Temperatures near Edinburgh, by Principal Forbes.
See Trans. Roy. Soc. Edin., 1846, for a full account of each.
University of Edinburgh.

2928a. Map of the **British Isles,** showing sites at which rainfall observations are being made. *G. J. Symons.*

2928b. Engravings of various apparatus employed in rainfall experiments. *G. J. Symons.*

2928d. Diagram showing the fluctuations of rainfall in central England, from 1726 to 1869. *G. J. Symons.*

2928e. British Rainfall. On the distribution of rain over the British Isles during the year 1874, as observed at about 1,700 stations in Great Britain and Ireland, with maps and illustrations.
G. J. Symons.

2928f. Monthly Meteorological Magazine.
G. J. Symons.

2928g. Rain ; How, When, Where, and Why it is measured? Being a popular account of rainfall investigations.
G. J. Symons.

2928h. Series of Blocks to illustrate the production of *weather charts* in *newspapers,* as carried out by the Patent Type Founding Company.
The Meteorological Committee of the Royal Society.
1. Plaster block, with outline of land.
2. Do. engraved by drill pantograph.
3. Cast of No. 2 in fusible metal.
4. Mould from No. 3 in papier mâché.
5. Cast from No. 4 in type metal.
6. Proof from No. 5.

2929. Meteorological Stand, with Psychrometer case, as used in the Imperial Navy. *C. Bamberg, Berlin*

2930. Case for the exhibition of telegrams and weather reports of the Naval Observatory at the signal stations on the German coast.
Imperial Admiralty Hydrographical Office, Berlin, and German Naval Observatory, Hamburg.

2931. Meteorological Window-stand, Reinert's system for the stations of the German Naval Observatory.
Imperial Admiralty Hydrographical Office, Berlin, and German Naval Observatory, Hamburg.

40075.   Z z

The case contains a psychrometer and a thermograph (minimum and maximum thermometer) ; it is kept at a proper distance from the window, and when in this situation the sliding door is closed ; on bringing it nearer to the window, for the purpose of reading off the indications, the door opens automatically.

2932. Collection of Weather Maps of the German Naval Observatory, for the month of March 1876.

Imperial Admiralty Hydrographical Office, Berlin, and German Naval Observatory, Hamburg.

2933. Storm Warner and Weather Indicator, original instrument, 0·85 in. long and 0·50 in. broad.

Annexed to the same is a treatise entitled : "Der Sturmwarner " und Wetteranzeiger, ein nach wissenschaftlichen Grundsätzen " ausgeführtes und durch Beobachtung und Erfahrung bewährtes " Instrument zur Vorherbestimmung von Sturm und Wetter."

Prof. Prestel, Emden.

2934. Graphical Representation of the changes of the weather in the years 1857, 1858, and 1859.

Prof. Prestel, Emden.

2935. Diagram of Solar Spots and their connexion with the variations of magnetic declination.

Prof. Rod. Wolf, Zürich.

The upper black curve gives the monthly relative numbers, introduced by him, for the years 1831–1875 ; the red one their means from twelves to twelves. The lower black curve gives the corresponding yearly relative numbers for the years 1745–1875 ; the red curve their reduction to the scale of the variation of magnetic declination at Prague. To the last curve are added (in black colour) those obtained from observation of the variation of magnetic declination at Mannheim, Paris, London, Göttingen, and Prague, demonstrating the connexion between the frequency of the solar spots and the variation of magnetic declination.

2937. Shortrede's Barometric Slide Rule, as arranged by Major-General A. De Lisle, R.E., for barometric readings and boiling points. *Elliott Brothers.*

The barometric scale is calculated from Bailey's formula. The range extends to about 15,000 feet of altitude, occupying the three faces of the slides. The back has logarithmic lines for computing the corrections, scales of which are given on the edges. The method of using the lines is given on the face of the rule. The corrections are : 1st, for temperature of mercury ; 2nd, for temperature of air ; and, 3rd, for latitude. · Aneroids only require the two last. This rule was invented by the late Major-General Shortrede, formerly of the Great Trigonometrical Survey in India.

The logarithmic slide solves the following equations :—

$$x = ab \; ; \; x = \frac{a}{b} \; ; \; x = \frac{ab}{c} \; ; \; x = \frac{c}{ab}$$

$$x = a^2 \; ; \; x = \frac{a^2}{c} \; ; \; x = \frac{c}{a^2}$$

2938. Thermometer Screen for meteorological stations of the 2nd order. In use at the stations of the Norway Meteorological Institute.   *Prof. H. Mohn, Christiania.*

Made of plate iron. To be mounted outside a window and kept in shade. The screen contains one psychrometer (dry and wet bulb), one minimum thermometer (both instruments made by R. Grave in Stockholm), and one hair-hygrometer (made by Herman and Pfister in Bern). The screen is constructed after the designs of Prof. H. Mohn.

2939. Thermometer Screen for one single thermometer for obtaining the temperature of the air. In use at stations belonging to the Norway Meteorological Institute.
Prof. H. Mohn, Christiania.

Made of plate iron. To be mounted outside a window or on a wall, and kept in shade. With thermometer, made by R. Grave, in Stockholm. Screen constructed after the design of Prof. H. Mohn.

2940. Thermometer Screen for one minimum thermometer. In use at stations belonging to the Norway Meteorological Institute.   *Prof. H. Mohn, Christiania.*

Made of plate iron, with double walls. Suspended on a cylindrical rod intended to pass through the window-frame. The rod has a handle inside, so that the screen with the thermometer can be turned for "setting" in the same way as a thermograph. The double walls prevent the rising of the thermometer, even in direct sunshine, to more than a few degrees above the temperature of the air. The minimum thermometer is made by R. Grave, in Stockholm. The screen is executed after designs of Prof. H. Mohn.

2941. Stevenson's Box for Thermometers.
Scottish Meteorological Society.

The box is louvre-boarded, and painted white inside and outside, and screwed to four stout posts, also painted white, firmly fixed in the ground. The posts must be of such a length that when the thermometers are hung in position the bulbs of the minimum thermometer, and of the dry and wet bulb thermometers, will be exactly at the same height of 4 feet above the ground, the maximum thermometer being hung immediately above the minimum thermometer. The thermometer box is to be placed over a plot of grass, and in a free open space to which the sun's rays have free access during as much of the day as surrounding conditions enable the observer to secure. The thermometers are suspended on cross-laths in the centre of the box, and face the door, which should open to the north.
Designed by Thomas Stevenson, C.E., F.R.S.E., Honorary Secretary, and described in the Society's Journal, vol. i., p. 122.

2942. Plunging Apparatus, by "Baudin," for observations at sea, with maxima air-bubble thermometer of "Walferden," and minima vertical thermometer of "Baudin," withdrawn from external pressure.   *M. Baudin, Paris.*

2942a. Model, illustrative of Meteorological Sections of the atmosphere.   *Scottish Meteorological Society.*

Showing vertical gradient for temperature, pressure, humidity, &c., ascertained by dividing the difference of readings between the instruments at the two heights by the height between them. Designed by Thomas Stevenson, C.E., F.R.S.E., Honorary Secretary, described at a general meeting of the Society, and published in "Nature."

2942b. Maps of the **Distribution** of **Temperature** over the **British Islands.**   *Prof. Hennessy, F.R.S.*

Maps derived from the results of observation and showing the distribution of temperature in the British Islands as illustrative of the law of distribution in islands whose coasts are bathed by heat-bearing currents. See the "Atlantis" for June 1858, and the "Philosophical Magazine," 1858.

2942b. Proceedings and Transactions of the Meteorological Society of Mauritius. Vol. vi.

The Royal Commissioners for the International Exhibition of 1851.

Monthly notices of same Society.

Contributions to the Meteorology and Hydrography of the Indian Ocean, Part I.

1 Vol. of Weather Charts, Indian Ocean, 1861.

2 Charts showing the number and tracks of the cyclones in the Southern Indian Ocean.

SECTION 15.—GEOGRAPHY.

I.—SURVEYING INSTRUMENTS.

COLLECTION OF INSTRUMENTS LENT BY THE ORD-
NANCE SURVEY.—MAJOR-GENERAL CAMERON.
R.E., C.B., F.R.S., DIRECTOR-GENERAL.

2943. Colby's Compensation Bars.

The first operation in an extensive survey, viz., the measurement of a base
line, is beset with many difficulties, one of the greatest of which arises from
the fact that the lengths of the measuring rods, bars, or chains vary with the
temperature. This could be got over if one could ascertain at any moment
the precise temperature of the rods, bars, or chains, but it is not easy to be
assured on this point. Hence arose the idea of a compensation apparatus
which was carried out by Major-General Colby, at the commencement of the
survey of Ireland, in the following manner. Two bars of different metals
and different rates of expansion are laid parallel and close to each other as
AB, CD—

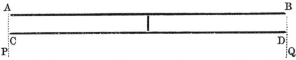

They are firmly connected at the centre, from or to which point they are
free to expand or contract. At a given temperature they are taken of the
same length, and in this state suppose lines ACP BDQ to be drawn through
their extremities, so that AP=BQ, make also the ratio of AP to CP and
of BQ to DQ, equal to the ratio of the expansion of the bar AB to that of
the bar CD. Now, if we suppose both bars to receive an equal increment of
temperature, and in this position lines to be drawn through their extremities,
these lines will pass through the points P, Q.
The compensation bar, then, consists of a bar AB of brass, united at its
middle to an equal bar CD of iron; at each extremity of the bars is a metal
tongue, connected with them by pivots. These pieces are about six inches in
length, and on a silver plug at the extremity of each is a finely engraved dot.
The dots are 10 feet apart at all temperatures.
When used in measurement of a line, the dots on two adjacent bars are
brought to the precise distance of six inches apart by means of a compensation microscope. The compensation microscope is formed by uniting two
ordinary microscopes by parallel bars of iron and brass, in such a manner
that their outer foci are "compensated," and measure six inches whatever
may be the temperature. The whole apparatus consists of six bars and
seven microscopes, of which two bars and three microscopes are now
exhibited.

Two base lines have been measured with this apparatus, one in the north of Ireland, in 1827, and the other on Salisbury Plain, in 1849. Upon these two lines depend all the results of the trigonometrical survey.

2944. Ramsden's Theodolite.

This instrument was made by the celebrated Ramsden, and first used in 1792. Since that time it has been in use at a vast number of stations, including the highest mountains in England and Scotland, the Hebrides, Orkney, and Shetland. It is now in as good condition as when it was first made, having never in all its travels met with any accident or ill-usage. Subsequent to the year 1843, the observations with it have been made by non-commissioned officers of the Royal Engineers. This instrument is a slight improvement on another made by Ramsden,—the 3-ft. theodolite of the Royal Society,—which was completed and first used in July 1787, but the two are almost identical. The divisions of the circle on the Ordnance Theodolite are rather closer than in the Royal Society's. The focal length of the telescope is 36 inches, and the aperture 2·5 inches, the magnifying power ordinarily used 54. The probable error of a single observation of a fine object under favourable circumstances is about 0·18″. The astronomical determination of the direction of the meridian has been effected at many stations with this instrument. In 1862 it was used at Fairlight, near Hastings, and St. Peter's Church, Isle of Thanet, in the extension of the English Triangulation into France and Belgium.

Many of the observations made with this theodolite have been to points upwards of 100 miles off (seen by means of a heliostat, a looking-glass reflecting the sun's light). The greatest distance ever observed was 120 miles, from Ben More in South Uist to a hill in Sutherlandshire.

The diameter of the circle of this instrument is 36 inches. The weight of the whole is 200 lbs.

2945. Ramsden's Small Theodolite.

This instrument is very similar in appearance and construction to the large instrument, but is only 18 inches in diameter. The focal length of the telescope is 19·5 inches, the aperture of the object glass 2 inches, and the ordinarily used magnifying power 30. The horizontal circle is read by three microscopes.

It has been used at a very large number of stations, many of them being church towers and spires. At Thaxted Church, of which the tower is 79 feet high, surmounted by a spire of 93 feet, the instrument was 178 feet above the ground. It was also used at St. Paul's Cathedral, over the cross. At one of the stations in the north of France, some 14 miles east of Boulogne, this theodolite was mounted on a scaffolding 75 feet high.

2946. Troughton and Simms' 24-Inch Theodolite.

This theodolite, of a totally different construction from Ramsden's, was made by Messrs. Troughton and Simms, at the commencement of the Irish Survey. In Ramsden's theodolites, the divided horizontal circle revolves with the telescope, while the microscopes which read it are fixtures. In Simms', the microscopes are connected with the telescope, and rotate with it, while the circle is fixed.

The focal length of the telescope is 27 inches, and the aperture 2·125 inches. The instrument has a repeating table, and is, strictly speaking, an altazimuth. The horizontal circle is read by five micrometer microscopes.

It has been in continual use, at a very large number of stations, from 1829 until 1862, and, like Ramsden's theodolites, has been so fortunate as to escape accidents.

2947. Troughton and Simms' 14-inch Theodolite.

This instrument is one of excellent construction. The telescope has 2 inches aperture and 18 inches focal length, the ordinary magnifying power 40. The horizontal circle is read by three micrometer microscopes : the vertical circle by two micrometer microscopes. It is admirably suited for astronomical observations. It is, comparatively with the other theodolites exhibited, a new instrument.

2948. Airy's Zenith Sector.

This instrument was used between the years 1842 and 1850 in the determination of the latitudes of 27 stations of the principal triangulation of the kingdom. On the destruction of Ramsden's Zenith Sector in the great fire at the Tower of London this instrument was invented and constructed under the direction of the Astronomer Royal, Sir George Airy. The first principle in the instrument is the arrangement for making successive observations in two positions of the instrument, face east and face west, at the same transit of a star. The second principle was the substitution of a level or system of levels for the previously used plumb-line. The third principle was the casting in one piece, as far as practicable, of each of the different parts of the instrument, in order to avoid the great number of screws and fastenings with which most instruments are hampered, and to secure if possible perfect rigidity.

The focal length of the telescope is 46 inches, the aperture 3·75, and the magnifying power 70. The vertical arc is read by four micrometer microscopes.

The weight of the entire instrument is rather more than half a ton.

2949. Zenith Telescope.

This instrument has done much service in North America and in Scotland in the determination of latitudes. It is most simple in its construction, consisting of little more than a good telescope capable of being set to any zenith distance, and rotating round a long and very firm vertical axis. In the focus of the eyepiece, besides five ordinary transit wires, is a micrometer wire, adapted to measurement of zenith distance, to the extent of 45′. The mode of using the instrument is this : suppose two stars pass the meridian at nearly equal zenith distances, and within a few minutes of one another in point of time, the one star being to the north, the other to the south of the zenith ; the observer sets the telescope for the first star (of course in the plane of the meridian), reads the micrometer wire, rotates the instrument 180° round the vertical axis, and looks for the second star : as it passes through the field he bisects it with the micrometer wire. The difference of the reading of the micrometer on the two stars leads to the immediate knowledge of the latitude ; the indications of the level being of course taken into account. This instrument was made by Wurdemann of Washington, United States, America, for the North American Boundary Commission : it is a great favourite with observers, and certainly leads to the most excellent results.

2950. Portable Transit Instrument.

The small transit instrument exhibited has a telescope of 21 inches focal length and 1·7 inches aperture. The uprights are of mahogany. The telescope is provided with a reversing apparatus.

2951. Clinometer.

For determining the values of levels, such as are used in the zenith sector, zenith telescope, and other astronomical instruments. The two micrometer screws at the right hand of the instrument are moved simultaneously. The value of one division of these screws corresponds to a change of inclination of the upper bar, of one second of angle.

2952. Standard Toise.

This bar of cast steel was formed to be a connecting link between the English yard and the "toise," which is the geodesic unit of length of so many countries of Europe. Its length in terms of the standard yard (about 76·74 in.) is known, from some 4,000 observations, with a probable error of one ten-millionth part of itself. It has been compared with the standard toise of Belgium, with that of Prussia, with the standard double toises of the Russian Geodetic Survey, with a standard toise of Vienna, and with the Spanish four-metre bar.

Connecting the survey of so many countries, this standard bar takes a place of the utmost importance in the determination of the figure of the earth.

The continental toises are generally measures "by contact;" that is, the measure is not indicated by lines on the bar, but by the entire length of the bar between its extremities.

2953. Thermometer Calibration Apparatus.

It is necessary in the comparisons of standards that the errors of the thermometers should be known to two or three hundredths of a degree Fahrenheit. The calibration errors of the best standard thermometers (such as those made by Casella for the Ordnance Survey) are extremely small, but it is necessary to determine what they actually are in order to correct the observations made with them. The apparatus consists of a cast-iron frame, having two parallel rods above, on which slides a microscope. The thermometer lying in a carriage below is read by the microscope, and by means of the micrometer screw on the right a small movement is communicated to the carriage and thermometer. Now, if we break off from the column of mercury a piece, say of 30° in length, and cause that piece to take different positions in the tube, it will be seen that we can by means of the micrometer and the microscope detect errors among the division lines; and finally, by using columns of various lengths, determine as many errors as we choose.

2954. Isometric Drawing of Expansion Apparatus.

This drawing shows the interior of the room in which the comparisons of bars are made; it is a room 20 feet by 11, half sunken below the surface of the ground, and entirely surrounded and covered over by an outer building. In the drawing are shown the three stone piers for carrying microscopes. In the experiments depicted, two microscopes are used and stand upon the outer piers. The long boxes contain bars, one hot, the other cold; they are alternately brought under the microscopes. In this manner the absolute expansions of two Indian standards, two English standards, and two American standards have been determined. The drawing shows the flexible tubes which form supply and waste pipes for the currents of water, hot and cold, which maintain the bars at steady fixed temperatures.

Surveying Instruments exhibited by the Ordnance Survey
    Department of the Royal Prussian General Staff,
    Berlin.

2955. Distance and Angle Measuring Telescope Ruler
(Kippregel), old construction, by Breithaupt, No. 71.

2956. Distance and Angle Measuring Telescope Ruler (Kippregel), construction of 1874, by Breithaupt, No. 97.

2957. Distance and Angle Measuring Telescope Ruler (Kippregel), construction of 1875, by Sprenger, No. 108.

2958. Surveyor's Plane Table, old construction, by Baumann, No. 53.

2959. Surveyor's Plane Table, improved construction, by Baumann, No. 82.

2960. Surveyor's Plane Table, construction of 1874, by Breithaupt, No. 91.

2961. Surveyor's Plane Table, construction of 1875, by Sprenger, No. 113.

2962. Two Surveyor's Plane Tables. Prepared for surveying.

2963. Two Surveyor's Plane Tables.

2264. Distance Staff.

2965. Universal Instrument, latest construction.
With such instruments as these the topographical survey of Prussia is carried on. The Trigonometrical Department of the Royal Survey lays down the network of triangles, and gives the geographical longitude, latitude, and elevation of each individual trigonometrical point to the Topographical Department, in a similar form to that shown in No. 2962. This sheet (section Kalisch) contains the trigonometrical points which are marked thus " ⊕," the station being at the point of intersection of the two lines. The minute network drawn upon the sheet is on the scale of 1:25,000, to which the topographical surveys are executed.

The surface covered by each section between 54° and 55° N. lat. amounts to about 2·2 square (German) miles.

The position of the square in the sections in relation to the degree net is shown by the two general tables appended to No. 2962. Here the section Kalisch is coloured red.

The *Universal Instrument* (2965) is one of the instruments used for measuring the angles of the triangles of the *first* order by the Trigonometrical Department of the Royal Land Survey. The one here exhibited was taken out to Kerguelen's Land by the expedition for observing the transit of Venus.

SECRETARY OF STATE FOR INDIA.

PHOTOGRAPHS OF INSTRUMENTS USED FOR THE GREAT TRIGONOMETRICAL SURVEY OF INDIA.

2966. Photographs of the Great Theodolite for the great trigonometrical survey of India. Designed by Lieut.-Colonel A. Strange, F.R.S.

Constructed by Messrs. Troughton and Simms. Telescope, aperture $3\frac{1}{4}$ inches, focus $36\frac{1}{4}$ inches. Horizontal circle, diameter 3 feet. Vertical circle. 2 feet. (Photograph about $\frac{1}{8}$ real size.) No. 1. General perspective view. No. 2. Elevation in plane of pillars. No. 3. Side elevation, showing vertical circle and microscopes. No. 4. Side elevation, showing counterpoises of vertical circle and microscopes. No. 5. Horizontal circle (plan of upper surface). Vertical axis socket. Horizontal microscope arms, &c. (plan of under surface). No. 6. Horizontal circle (plan of under surface). Table, pillars, horizontal microscope arms, &c. (plan of upper surface).

2967. Photographs of the Zenith Sector for the great trigonometrical survey of India. Designed by Lieut.-Colonel A. Strange, F.R.S.

Constructed by Messrs. Troughton and Simms. (Telescope, aperture 4 inches, focus 4 feet. Sector, diameter 3 feet.) (Photograph about $\frac{1}{8}$ real size.) No. 1. Front elevation, complete. No. 2. Front elevation, telescope and sector removed. No. 3. Back elevation, complete. No. 4. Side elevation, complete. No. 5. Telescope and sector (side elevation). Plinth (plan of inner surface). Vertical axis. Cradle, with microscopes and levels (plan).

2968. Photographs of the Electro-Chronograph for the great trigonometrical survey of India.

Constructed by M. Eichens and M. Hardy, of Paris, under the superintendence of Lieut.-Colonel A. Strange, F.R.S., assisted by M. Leon Foucault, member of the Institute. (Photograph about $\frac{1}{8}$ real size.) No. 1. Front elevation. No. 2. Back elevation. No. 3. End elevation, showing clockwork. No. 4. End elevation, showing connexion of pointer slide with barrel. No. 5. Plan of upper surface-barrel removed.

2969. Photographs of the Transit Instrument for the great trigonometrical survey of India. Designed by Lieut.-Colonel A. Strange, F.R.S.

Constructed by Messrs. T. Cooke and Sons. Telescope, aperture 5 inches, focus 5 feet. (Photograph about $\frac{1}{12}$ real size.) No. 1. General perspective view. No. 2. Elevation in plane of telescope. No. 3. End view of transit axis and levels. No. 4. Front view of transit axis and levels. Bearings and their foundation plates (plan of upper surfaces). No. 5. Bearings and their foundation plates (side and inner elevations).

2970. Photographs of the Diagonal Transit Instrument for the great trigonometrical survey of India.

Constructed by Messrs. T. Cooke and Sons, under the superintendence of Lieut.-Colonel A. Strange, F.R.S. Telescope, aperture 2·64 inches, focus 34·6 inches. (Photograph about $\frac{1}{8}$ real size.) No. 1. Elevation in plane of telescope. No. 2. Telescope raised for reversal.

2971. Photographs of the Vertical Circle for the great trigonometrical survey of India.

Constructed by M. M. Repsold, of Hamburg, under the superintendence of Lieut.-Colonel A. Strange, F.R.S. Telescope, aperture 1·9 inch, focus 18·8 inches. Vertical circle, diameter 12 inches. (Photograph about $\frac{1}{8}$ real size.) Front and side elevations.

II.—HYDROGRAPHY.

ADMIRALTY—HYDROGRAPHIC DEPARTMENT.

DEEP-SEA SOUNDING AND DREDGING APPARATUS, AND SURVEYING INSTRUMENTS, USED IN H.M.S. "CHALLENGER," 1872-6. EXHIBITED BY THE HYDROGRAPHICAL DEPARTMENT, ADMIRALTY.

2971a. Altitude and Azimuth instrument, 8-inch.

Repeating circle.

Theodolite, 6½-inch.

Sextant, observing.

Sextant, sounding.

Sextant, pocket, or box.

Micrometer telescope, 24-inch.

Station pointer, 6-inch.

Protractor, circular. (Bullock.)

Protractor, semi-circular.

Protractor, rectangular, 12-inch. (Metal.)

Protractors, rectangular, 8-inch. Case of three.

Scale, brass, graduated, 3 feet.

Straight-edge, steel, 6 feet.

Parallel rulers, 12, 18, and 24 inch. Case.

Sector. (Metal.)

Mathematical instruments. Case.

Baillie weight detaching apparatus, date 1872. By Nav. Lieutenant C. W. Baillie, R.N.

Hydra weight detaching apparatus, date 1868. By Gibbs, Artificer, H.M.S. " Hydra."

Tube lead, 1 cwt. For sounding in depths not exceeding 1,000 fathoms, and procuring bottom specimen, date 1872.

Cup lead, 1 cwt. For sounding in depths not exceeding 1,000 fathoms, and for procuring bottom specimen, date about 1858.

Sounding ring and rod, fitted with wire.

Sounding weights, ½ cwt.

Dredge, medium size, complete with net.

Trawl, with beam and irons.

Dredge rope, 5 fathoms of 3-inch.
>>   >>   2½-inch.
>>   >>   2-inch.

Sounding line, 5 fathoms.

Accumulators, one set.   For easing the strain, upon dredge rope or sounding line, caused by the motion of the ship.

Gin block, 9-inch, patent.   For attaching to accumulators.

Slip Water Bottle, for obtaining specimen of bottom water, is a brass rod, with three radiating plates to strengthen it and act as a guide for the brass cylinder which encloses the water. At the bottom and half-way down the radiating plates are finely-ground sections of cones, upon which the upper and lower surfaces of the cylinder accurately fit.   Over the notch of the tumbler at the top of the brass rod is placed the bight of a line which suspends the cylinder over the cones while the bottle is descending, and which is released by the tumbler falling over, when the strain is removed, by the instrument reaching the bottom, thus allowing the cylinder to fall on the cones and enclose the bottom water.   A tap at the bottom of the bottle facilitates withdrawing the water.

Stop-cock Water Bottle, for obtaining specimens of water at different depths, consists of a brass tube, with stop-cock and bell-shaped mouth at either end.   Each stop-cock has a lever attached to the brass side-rod in such a manner that the movement of the rod opens or closes the cocks simultaneously. The brass plate, hinged on a stud near the top of the side-rod, is so arranged that, while the bottle is descending and the water passing freely through, it remains in a vertical position.   On the sounding line, to which the bottle is attached, being checked, the plate assumes a horizontal direction, and on heaving up, the pressure of the water on the plate closes both stop-cocks; a still greater pressure is sufficient to release the plate from the spring, which retains it in position when it falls down in a line with the side-rod.   A small tube, with spiral spring, in the upper part of the bottle, allows the water to expand.

Tow Net.

Set of Sieves, for sifting mud dredged from the bottom.

INSTRUMENTS COMPRISED IN THE ORDINARY OUTFIT OF ONE OF H.M. SHIPS EMPLOYED ON SURVEYING SERVICE.

2972. Azimuth and Altitude Instrument.

2973. Theodolite, 5-inch.

2974. Sextant, observing, and stand.

2975. Sextant, used for measuring angles between terrestrial objects, to fix the position of a ship or boat when engaged in sounding.

2976. Sextant, pocket or box.

2977. Artificial Horizon, roof.

2978. Artificial Horizon, black glass, portable. Used in travelling.

2979. Micrometer Telescope. Rochon. Invented by the Abbé Rochon. Date 1812. Used (in surveying) for the measurement of distances by means of a base a few feet in length.

It is formed of two prisms of rock crystal, the one cut parallel to the axis of the crystal, and the other parallel to one of the faces of the pyramid; these are placed one on the other in contrary directions, and cemented together in a cell which slides in the body of the telescope, and by means of a slit in the direction of its length, so that the prism may be moved along the tube by means of the index, and a milled head screw, which is connected with the prism-box inside.

The scale on the telescope tube is divided into minutes of arc, and by means of a vernier, reading to seconds.

A table accompanies the instrument, giving the distances for the angular measurement of each foot of base.

2980. Heliostat. A mirror for reflecting the sun's rays, used in trigonometrical surveys to indicate, from great distances, the position of a station.

2981. Station Pointer, 6-inch. For placing the observer's position on the chart from angles taken between three objects, the relative positions of which are known.

2982. Protractor, circular, with arms. For projecting the angles of a survey.

2983. Protractor, circular. Bullock.

2984. Protractor, semi-circular.

2985. Protractor, rectangular.

2986. Protractor, horn.

2987. Parallel Rulers.

2988. Scale, brass, graduated.

2989. Scales, ivory ; set of 6.

2990. Beam Compass, wood. For the accurate measurement of long sides in projecting triangulation.

2991. Drawing Instruments ; magazine set.

2992. Drawing Instruments, for service in boats.

INSTRUMENTS OCCASIONALLY FURNISHED TO OFFICERS ENGAGED IN NAVAL SURVEYING EXPEDITIONS.

2993. Reflecting Circle. Troughton.

2994. Sextant, Travelling, small.

2995. Sextant, Observing, with micrometer and indicators. Davis.

2996. Sextant, Sounding, Double. Beechey.

2997. Sextant, Pocket, Double. George.

2998. Dip Sector. By Troughton. For measuring the angle between the apparent and true · horizons, from an elevation above the surface of the earth.

2999. Artificial Horizon, portable. George.

3000. Stadiometer. Blakey. An ordinary telescope fitted with a sliding measure and linear arrangement on object-glass ; designed for the purpose of determining the distance apart of two ships, the height of the masthead of the second ship being known.

3001. Optical Opposite. Raper. An instrument for determining and preserving a position in a direct line between two objects.

3002. Beam Compasses, tubular.

3003. Beam Compasses, bar, metal.

SOUNDING MACHINES AND APPARATUS USED BY H.M. SHIPS IN DEEP SEA EXPLORATION.

3004. Clam, for bringing specimens of the sea bottom. Sir John Ross ; date 1818. Designed and used by Sir John Ross in H.M.S. "Isabella" in Baffin Bay.

The claw is kept open by hinged arms within ; whilst thus open, the outer case is kept up, and a spike connected with the arms projects below the claw. On striking bottom the spike forces the arms up ; this releases the outer case ; as the claw closes, the case slides down, and the contents in claw are secured.

3005. Brooke's Rod, for bringing up specimens of the sea bottom, and weight detaching apparatus (about 1856). By Mr. Brooke, Midshipman, United States Navy. Used, with various modifications, to 1868.

In the first instrument employed in the United States Navy the valve securing the specimen of the sea bottom was not applied, quills being inserted in the tube for that purpose. The hook, with the wire and ring, supports the sinker weight as long as there is strain on the sounding line ; the moment that strain ceases by the rod touching sea bottom, the sinker weights turn the suspending hook and the wire being released, the sinkers slide off as the rod is drawn upwards.

3006. Brooke's Rod, another mode of detaching sinker weights ; date 1857. Used in sounding North Atlantic.

3007. Skead's Weight Detaching Apparatus ; date 1857. By Mr. F. Skead, Master R.N., H.M.S. "Tartarus." Used in Mediterranean Sea survey.

A 68-lb. shot (represented by the wooden ball, exact size) is slung with wire, and the detaching apparatus attached. On reaching sea bottom, the small weight on the apparatus, which had till this time been kept above the hook by the weight of the shot, is freed, and, falling, reverses the hook ; the shot is thus left behind. The small weight is roughed to secure a small portion of the sea bottom.

3008. Bonnici's Weight Detaching Clam ; date 1857. By Mr. C. Bonnici. Used in Mediterranean Sea survey.

A 68 lb. shot (represented by the wooden ball) is slung with wire, and the claw attached, the weight of the shot preventing the claw opening. On reaching sea bottom and the line slackened, the arms of the claw fall by their own weight, and the sinker is released.

3009. Bulldog Clam, for bringing up specimens of the sea bottom ; date 1860. Used in H.M. ships "Bulldog" and "Porcupine," North Atlantic Ocean.

The claw is kept open by a tubular sinker weight, resting on the four horns. When the weight is detached on reaching sea bottom, the india-rubber bands contract, closing the claw, and thus securing the sea bottom contained within it.

3010. Fitzgerald Sounding Machine, combining apparatus for detaching the weight and procuring specimens of the sea bottom ; date 1867. By Lieutenant Fitzgerald, R.N., employed in H.M.S. "Cordelia," in deep sea sounding between Jamaica and Cuba.

An iron sinker weight (represented by wood model) is hooked to the side of the sounding bar, and the hook or lever, to which the sounding line is attached, is inserted in the hole at the upper end of the bar. When suspended, the weight of the sinker keeps the lever in the hole and the bar nearly vertical. On reaching sea bottom, the scoop is driven into the ground, and the hook being freed by the slackening of the sounding line the weight falls over ; the action of hauling in the sounding line reverses the bar, unhooks the weight, and the scoop closes with its contents.

3011. Hydra Weight Detaching Apparatus ; date 1868.
By Gibbs, Artificer, H.M.S. " Hydra." Used in sounding Atlantic and Indian Oceans.

The wooden models represent (exact size) iron sinkers, each of one hundred-weight ; they are suspended on the sounding tube by the iron ring and wire, to a button which protrudes through a steel spring on the sliding rod above, the spring being kept back by the weight of the sinkers. On reaching sea bottom and the sounding line slackening, the rod slides down, the sinker weights then resting, the steel spring throws the suspending wire off, and the weights are left behind as the tube is drawn through them. A butterfly valve within the tube at the bottom secures a portion of the sea bottom.

3012. Baillie Weight Detaching Apparatus ; date 1872.
By Navigating Lieut. C. W. Baillie, R.N. In general use in H.M.S. " Challenger."

Iron sinkers of half a hundredweight each (exhibited in wood models) in numbers sufficient for varying ocean depths, are placed on the sounding tube and suspended by the ring and wire to two shoulders that project from the sides of a sliding rod working in the upper part of the sounding tube. On touching sea bottom and the sounding line slackening, the sinker weights draw the sliding rod downwards, and the shoulders passing within the sounding tube, the wire is thrown off, and the weights released. The lower portion of the tube, to which a valve is attached, receives the specimen of the sea bottom.

3013. Burt's Bag and Nipper, for sounding in moderate depths without stopping the ship's way. Invented by Mr. Gould, an American (about 1812).

The bag is first soaked in water to render it air-tight, and when used is in-flated ; a wooden tube and peg is affixed for the purpose. The sounding line is placed in the nipper or snatch attached to the bag, and on the sounding lead being cast from the ship, the whole is thrown overboard, the inflated bag floating as the line runs through the attached nipper, the bag keeping immediately over the descending lead.

On the lead reaching the bottom, indicated by the spring of the bag, no more sounding line will pass through the nipper, and the place where it is thus nipped is the vertical depth of water as marked on the line.

3014. Massey's self-registering Sounding Machine ; date 1800. Adapted for moderate depths. In use in H.M. Navy.

3015. Ericsson's self-registering Sounding Machine ; date 1836. For sounding in moderate depths.

This instrument records the vertical depth irrespective of the amount of sounding line thrown out from the ship.

The depth is ascertained by the compression of air within a glass tube, the value of the compression, in accordance with depth, being recorded by the quantity of water passing into the tube as the air is compressed, and a scale adapted to it.

3016. Cup Lead. For sounding in depths not exceeding 1,000 fathoms, and to procure specimens of the sea bottom (about 1858) ; model. Originally used in sounding North Atlantic.

3017. Tube Lead. For sounding in depths not exceeding 1,000 fathoms, and procuring specimens of the sea bottom ; date 1872 ; model. In use in H.M.S. "Challenger."

3018. Specimens of Sounding Line. Used in H.M. ships. No. 1. used in "Challenger" ; No. 2. medium.

These lines are constructed of the best Italian hemp.

No. 1 line is one inch in circumference ; 100 fathoms (or 600 feet) weighs 18 lbs. 9 oz. When wet this line breaks at a minimum strain of about 14 hundredweight.

No. 2 line is 0·8 inches in circumference ; 100 fathoms weighs 12 lbs. 8 oz., and it bears a strain of about 10 hundredweight when wet.

THERMOMETERS AND APPARATUS FOR PROCURING SEA WATER, USED IN DEEP SEA EXPLORATION.

3019. Six's Thermometer, with protected bulb. Negretti. Designed by Fitz Roy and Glaisher (about 1860) ; used in H.M. ships "Bulldog" and "Porcupine" in North Atlantic.

3020. Six's Thermometer. Hydrographic Office pattern ; bulb not protected. Casella ; date 1867. Used in H.M. ships "Lightning" and "Porcupine" in North Atlantic.

This instrument (the full bulb of which was unprotected by an outer glass casing) was found to be affected by pressure, and gave erroneous results at great ocean depths ; and this led to the construction of the protected bulb thermometer (3021).

3021. Six's Thermometer. Miller pattern ; bulb protected as proposed by Dr. Miller. Casella ; date 1869. In use in H.M.S. "Challenger :" generally adopted, after being tested to three tons on the square inch.

The full bulb is protected by an outer glass casing. The errors of these thermometers are determined under pressure, and are found to be reduced to the limits of about 1·2 for depths of 2,500 or 3,000 fathoms.

Two protected bulb thermometers which were broken by pressure at a depth of 3,875 fathoms in the Atlantic Ocean.

3022. Johnson's Metallic Self-registering Thermometer. Invented by Mr. Henry Johnson (about 1858). Not in general use.

3023. Saxton's Metallic Self-registering Thermometer (about 1855). Used on the United States Coast Survey. Not in general use.

3024. Water Bottle. In ordinary use for bringing up water from ocean depths. [Superseded in H.M.S. "Challenger" by Buchanan's design.]

3025. Water Bottle. Buchanan's; date 1872. A drawing. Invented by Mr. Buchanan, one of the scientific civilian staff attached to " Challenger."

3026. Barometer, Diagonal (about 1750). Watkins and Smith, London.

3026a. Barometer with Thermometer.
Paul Greiner, Hamburg.

The baro-thermometer, an instrument for measuring the depth of the sea, is also intended for recording the temperatures and the salinity of the sea water.

III.—SURVEYING AND OTHER INSTRUMENTS LENT BY VARIOUS CONTRIBUTORS.

3026b. Kater's Reversible Pendulum.
The Royal Society.

The original pendulum employed by Kater (Phil. Trans. for 1818), with the tail-pieces subsequently modified by Sir Edward Sabine (Phil. Trans. for 1829).

The pendulum is made to vibrate with one of the steel wedges called "knife-edges," resting on firmly-supported and carefully-levelled agate planes, and the vibrations are referred to a clock by the method of coincidences. The pendulum is then inverted so as to vibrate about the other knife-edge, and the movable weights are adjusted by a combination of calculation and trial, till the times of vibration about the two knife-edges are found to be the same. In this condition the length of a simple pendulum vibrating in the same time as the Kater is equal to the distance between the knife-edges, which can be accurately measured; and the length of the seconds pendulum will then be got by dividing by the square of the time of vibration expressed in seconds.

This pendulum has just been employed in a re-determination of the length of the seconds pendulum, particulars of which are nearly ready for publication.

3132. Pendulum Apparatus. Great Trigonometrical Survey of India. *Kew Committee of the Royal Society.*

The pendulum vibration apparatus used by Captain Basevi, R.E., and Captain Heaviside, R.E., in their pendulum operations in connexion with the Great Trigonometrical Survey of India, 1865 to 1874.

It consists of a vacuum chamber, of stout copper, firmly supported by a strong wooden stand; the chamber contains in its upper part a pair of agate plates, upon which the knife edge of the pendulum rests, and it has at the bottom an apparatus for starting and stopping the pendulum, as well as a graduated arc for determining its extent of motion. In the chamber there is also a fixed pendulum containing thermometers.

The Shelton clock exhibited with the other pendulum apparatus was used for timing the vibrations of one pendulum in the receiver, being firmly fixed to a wall erected for the purpose behind it.

The air pump served to exhaust the receiver, the vibrations being made in a vacuum.

The coincidences were observed by the small telescope erected at a distance.

3134. Pendulum Apparatus. Sabine.

Kew Committee of the Royal Society

Pendulum vibration apparatus used by Sir E. Sabine, and similar to that now in use by the Arctic Expedition.under Captain Nares, R.N.

It consists of a massive iron triangular frame, supporting a pair of agate planes, upon which the invariable pendulum is swung; a strong wooden frame, carrying a Shelton clock, and a telescope, with which the coincidences of beat of the detached pendulum and that of the clock are observed.

3133. Photograph of a Reversion Pendulum Apparatus, distance of knife-edges 1 metre from each other, executed for the Central Bureau of the European Measurement of Degrees, Berlin. *A. Repsold and Sons, Hamburg.*

The apparatus consists of the *stand* made of brass tubing, the *pendulum*, the *scale*, and the *comparator*. The pendulum is a brass tube provided with knife edges. The scale is of the same tube as the pendulum, and contains a steel tube, and this again a zinc tube, both of which are united with each other and with the brass tube in such a manner that for the determination of the temperature a small divided scale which moves by the different expansion of steel and zinc, may be read off near one of the end divisions of the main (brass) tube. The divided surfaces are in the middle line of the tube. The scale may stand with either end upwards. By rotation of the comparator round its foot the two microscopes may be alternately directed upon the scale and upon the suspended pendulum.

3100. Photograph of a Base-measuring Apparatus, in course of construction for the United States Lake Survey, Detroit. *A. Repsold and Sons, Hamburg.*

A measuring rod 4 mètres in length is used, consisting of a steel and a zinc rod, enclosed together between rollers in a wide iron tube, and with divisions upon platinum at the ends. The measuring rod rests on its extremities (below the divisions) upon its supports, by which the advantage is attained that each support remains in place for two positions, and, therefore, only one has to be arranged for each new position. Before making a measurement microscopes are placed vertically upon separate supports above the rod-supports, and beneath these the rod is carried forward. To fix the daily termination a cylinder with a level is arranged perpendicularly under the last microscope on the small hemisphere of a terminal plate on the ground, and by the microscope a division on the upper terminal surface of the cylinder is read off in the two positions differing by 180° The apparatus is accompanied by a comparator for the comparison—(1) of the measuring rod with a standard mètre; (2) of the measuring rod with a spare measuring rod. In the first case the standard mètre is laid, together with the measuring rods, upon a carriage 4 mètres in length, and compared one by one with the intervals of the measuring rod divided into mètres. For this purpose the carriage is moved to and fro upon the base piece under two microscopes placed upon a common bearer. This bearer may be pushed along the whole length upon the base piece (like the standard mètre upon the carriage). In the second case the two measuring rods are laid side by side upon the carriage, and compared under the microscopes, one of which is attached to the common bearer, and the other to a special arm of the base piece at a distance of 4 mètres from the former. The comparator is also arranged, if desired, for the comparison of the standard metre line measure (*à traits*) with a yard end measure (*à bout*). For this purpose small cylinders with scales are applied to the standard yard in order to compare suitable lines of these scales with the mètre ; then the sum of the lengths of the cylinders up to the lines of the scale employed is found by putting the two directly opposite one another.   The whole comparator is to be enclosed in a wooden case, in order to keep the measuring rods at an equable temperature.

3093. Experimenting Model (quarter of the natural size) for the preparation of bars for a new *base-measuring apparatus.*
Carl Bamberg, Berlin.

The steel tube is bored through cylindrically in an axial direction, closed at the ends and filled with mercury, which also passes into a glass tube, closed above, and communicating with the bore.   The height of the column of mercury in the glass tube will be a measure of the alteration of the length of the scale by change of temperature.   The proportions of the cavity of the passage in the steel and glass tubes are so arranged that every change of length in the scale is increased 100 times in the glass tube.

This instrument has been made as an experiment by order of the Royal Prussian Land Survey.

3093a. Drawing of a Base-measuring Apparatus, with employment of the measuring bars mentioned with the above apparatus.   *Carl Bamberg, Berlin.*

1831a. Telescope, by Dollond.

Sextant by Ramsden.   *Prof. Winnecke, Strasburg.*

Universal Instrument and Goniometer, by Robinson.
Prof. Winnecke, Strasburg.

The telescope is the " Dollond," so often mentioned in Humboldt's account of his travels in America.

The sextant and the universal instruments was taken by Humboldt in all his journeys in Asia and America, and on the former depend nearly all the determinations of position.   It contains an inscription to this effect, and in the case may still be found the paper in Humboldt's handwriting with instructions for the engraver.

These instruments form a portion of the Humboldt collection, with which friends in Berlin enriched the new observatory of Strasburg, erected in the year 1873.

3106a. Quadrant used by Captain Cook during his several voyages.   *Richard Caulfield.*

3027. Instruments used by the late **Dr. Livingstone** in his *last* journey.   *The Royal Geographical Society of London.*

Pocket chronometer, by Jas. McCabe, 194, Royal Exchange, London.
Sextant, by John Dalton, of Hartlepool.
Hypsometrical boiling apparatus, by Casella.
2 boiling point thermometers, by Casella.
1 ordinary thermometer, by Casella.

3153e. Two Cases of Instruments used by the late Captain Speke in Africa, with map of his discovery of the region of the Nile sources. *William Speke, jun.*

3027a. The First Traces ever found of the Franklin Expedition. *Admiral Ommanney.*

a. THEODOLITES.

3039. Five-Inch Theodolite by Adams. *Royal Society.*

3047. Theodolite with 3-inch circle.
Messrs. Troughton & Simms.

The telescope is of the bent form, having a small rectangular prism in the axis. The light received by this prism is turned through the axis of the instrument, and an image of the object is formed outside the pivot. By this construction the instrument may be kept low, and possess at the same time considerable optical power.

3047a. Five-Inch Theodolite. *Joseph Casartelli.*

In making this theodolite the object has been to reduce the number of parts and simplify the construction. It has double conical bearings to the axes, with a ball-and-socket adjustment, and spring verniers to the horizontal circle, producing an easy and smooth motion. The whole is made of gun-metal.

3047b. Travellers' Transit Theodolite and Stand. Improved by the late Lieut.-Col. Strange, F.R.S., and adapted for alpine and military surveys or occasional astronomical observations. *L. Casella.*

3047c. Travellers' Transit Theodolite and Stand, with telescope in centre. Improved by the late Lieut.-Col. Strange, F.R.S., and adapted for alpine and military surveys or occasional astronomical observations. *L. Casella.*

3047d. Cary's Improved 5-in. Theodolite, with improved rack adjustment to cross webs of eyepiece, and improved form of tripod foot. *Henry Porter.*

3048. Theodolite (10·5 in.), in which the vertical arc is jointed on an ordinary sector, and covered with a variety of scales. Made by Sissons, London. *Major M. L. Taylor, R.A.*

3049. Repeating Theodolite, horizontal circle 14 cm. in diameter, divided to 20′ reading to 30″ by vernier. The divided limb is completely covered and the verniers are protected by glass plates. Telescope, 12 lines free aperture, magnifying 25 times, vertical circle divided to 30′ reading to 1′ by vernier. Level to arrange the telescope in two positions. *A. and R. Hahn, Cassel.*

Besides the addition of the movable level to the telescope, this instrument possesses a new arrangement for its vertical movement. A small female screw with right and left threads on opposite sides is inserted near the base of the bearer. Corresponding to these threads are male screws tightly fastened to the bearer, and, by turning the female screw to the inner or outer side. the height of the telescope is increased or diminished.

3050. Universal Theodolite (Bende's construction), horizontal circle 12 m. diameter, divided to 20' and reading by two verniers to 30''. The divided limb is completely covered, and the verniers are protected by glass plates. Telescope, 15 lines aperture, magnifying 35 times, adapted for reversal in the bearings, level for setting; tangent screw exactly 1 mm. range. The vertical are is replaced by a long tangent screw with micrometer head, by which vertical angles up to 45° can be measured.

A. and R. Hahn, Cassel.

By means of this instrument elevations can be measured directly with the tangent instead of by the arc or its chord as hitherto. Its advantage over other instruments consists in its peculiar construction, which admits of using the micrometer screw in the measurement of angles up to 45°, whilst hitherto angles up to about 10° only have been so measured. The relation between the reading on the screw and the height or distance to be measured is a simple geometrical one.

3051. Repeating Theodolite (No. 148), with stand and leather case. *Otto Fennel, Cassel.*

Covered horizontal circle, 14 cm. in diameter, graduated on silver to 20', and reading by vernier to 20 seconds. Achromatic erecting telescope Vertical circle with vernier reading to minutes.

3052. Metford's Improved Theodolite, made by F. Pastorelli, under the direction of the inventor. *Francis Pastorelli.*

DESCRIPTION OF THE INSTRUMENT.

Levelling Gear.—It consists of three left-handed screws with the balls fitting closely to their beds on the under surface of the traversing box; they are secured by an elastic three-cornered plate, having boxes at their ends to protect them from dust and grit. The lower ends of the three screws pass through a triangular plate with broad ends, with sufficient spring to permit the female screws to be slightly twisted.

Traversing Stage.—The main hollow centre of the instrument carries a circular disc $3\frac{1}{2}$ inches in diameter and $\frac{2}{10}$ of an inch thick; the traversing stage is a flat plate $5\frac{1}{2}$ inches in diameter; in its bottom the levelling screws are seated as previously explained. The upper surface has a ring round its edge the depth of the circular disc. There is a 2-inch hole in the stage to let the plumb cord traverse with the instrument; the disc will traverse 1 inch in any direction from the centre. To secure the instrument there is an upper plate screwed to the ring, so that the stage becomes a very shallow box; there is also a washer that keeps out dust and grit. A three-arm pinching screw running on the hollow centre secures the disc.

Horizontal Limb and Vernier Plate.—The horizontal limb and vernier plate are solid; the latter has mounted upon it a compass, circular bubble, and a memorandum plate; also attached to it is the tangent motion. The limb divided to read 20 minutes, the vernier to 30 seconds.

Bracket Support of Vertical Limb and Telescope.—On the side of the main pivot is attached a strong curved bracket' with two arms at the top. This bracket has a T section throughout; to it is fixed the vertical circle and female centre. The telescope and vernier circle, with its tangent motions, are fixed on the male centre. The main spirit bubble is fixed at the back of the vertical circle; there is a screw in the bracket for perfecting its adjustment. The axis of the telescope is suspended over the axis of the instrument, and admits of a transit motion; this is an important improvement.

Tripod Staff Head.—This is an adoption of W. Froude, Esq., C.E., F.R.S., by which steadiness is obtained, which is of great importance in taking angular measurements. The cheeks are set wider apart, the leg joints being similar to an inverted mortar with strong trunnions which can be tightened in their bisected cylindrical bearings by means of capstan-headed screws. As wear goes on they can always be kept perfect; this cannot be done with the ordinary staff head.

3052a. Metford's 7-inch Theodolite, with curved arm instead of the usual structure for carrying the vertical circle, and telescope, mounted on his traversing stage, with legs, invented by Mr. Froude. *H. Husbands.*

3052b. Metford's 4-inch Theodolite. Instead of being mounted on the traverser is placed on the ball movement, and is more especially recommended for very steep and hilly ground, and for preliminary surveys. *H. Husbands.*

3052c. Theodolite. An extremely old, if not unique, instrument, purchased by the late Sir James South as a rarity. In the original oak box. *G. J. Symons*

3053. Small Theodolite, for rapid operations.
Geneva Association for the Construction of Scientific Instruments.

This instrument presents the following advantages :—
The telescope is reversible to give the correction of the zero of the vertical circle.
The position of the level is symmetrical.
The divisions are strongly marked for rapid observations.
The vertical circle has two graduations; one to show degrees and minutes, the other tangents. This latter division shows the reduction for inclination without computation.

3054. Gambey's Theodolite.
Conservatoire des Arts et Métiers.

3059. Large repeating Theodolite, with arc and altitude circle, divided to 10'; with verniers for direct reading to 10'
Dennert and Pape, Altona.

3060. Theodolite with circles of six and four inches.
Julius Wanschaff, Berlin.

This theodolite has been constructed (of greater durability than usual) for making very accurate measurements. The division is not the exhibitor's;

only those on the reflecting circle, and on the universal instrument, were made with the exhibitor's new machinery.

The circles on both these instruments have no errors which could be detrimental, or which can be ascertained by the instruments themselves, since the exhibitor's dividing machine (of 90 cm. diameter) is only arranged for copying, and the normal-division itself being correct as far as 1·2 seconds.

The horizontal circle on the theodolite can be turned, and is provided with a contrivance for centering; the verniers indicate at a division of—10 to 10 minutes—10 seconds. Those on the altitude circle indicate at a division of —20 to 20 minutes—30 seconds.

3062. Universal Instrument with eight-inch Circles, and reading microscope.   *Julius Wanschaff, Berlin.*

The circles on the universal instrument read to four seconds, but they have only figures of half the value, in order to obtain at once the mean by the addition of the two readings of the opposite microscopes.

3066. Small Universal Instrument for measuring heights, &c., with stand.   *Zimmer Brothers, Stuttgart.*

Small levelling instrument and theodolite in one, for measuring heights, horizontal angles, &c., and for levelling, and setting out right angles. Useful to foresters and agriculturists, on account of its cheapness, and capability of being easily handled.

3073. Theodolite with two Vertical Circles.
Geodetic Institute of the Royal Polytechnic School, Munich, Prof. Dr. von Bauernfeind.

This instrument was made by Ertel & Son, of Munich, from the exhibitor's design, and is suited for investigations as to the relative value of the vernier and microscope readings of five divisions, and, secondly, for observations upon the magnitude of terrestrial refractions and their variation with temperature and the pressure of the air. The exhibitor has not yet described the instrument, because the number of observations made with it is still too small to allow certain results to be deduced from them. The instrument, however, perfectly fulfils its purpose.

3074. Repeating Theodolite, with 15 cm. horizontal circle, and 12 cm. covered altitude circle.   *A. Bonsack, Berlin.*

Both circles are divided on silver ; the first to 20′ with two verniers reading to half minutes ; the second to 30′ reading to one minute by one vernier. The telescope is adapted for distance measurements by a method invented by the exhibitor, and which has been strongly recommended by Herr S. Woyike, who used it in operations on the Ostbahn.

3075. Theodolite with Micrometers on the horizontal circle.   *F. W. Breithaupt and Son, Cassel.*

Theodolite for Geodetic purposes.—Horizontal circle 20 centimètres in diameter, with two graduations. the coarse to 20′, the fine to 5′, reading by micrometers, to two seconds directly ; the fine division is protected by a cover ; the verniers of the vertical circle show 10 seconds ; orthoscopic telescope for transit (*Durchschlagen*) and *reversal* in the Y's with arrangement for distance-measurement, reversing spirit-level and level on the tube of the telescope. The microscopes have parallels and reversion scales on glass. This theodolite belongs to the Royal Forest Academy at Münden.

3076. Pocket Theodolite.

F. W. Breithaupt and Son, Cassel.

Capable of employment, on account of its small dimensions, in very confined spaces, mines, &c.

3077. Francis' Patent Pocket Theodolite, made by Negretti and Zambra. *George Francis, C.E.*

The improvements claimed for this invention are, 1st, its portability and cheapness; 2nd, its simpleness and easy application; 3rd, the proportional increase of length between the sights for dialling; 4th, it is less liable to error in taking horizontal angles; 5th, its adaptability for taking the angle of an underlie or gradient; 6th, it requires little or no adjusting; it combines with the theodolite a clinometer, protractor, and plotting scales for reference, from 1 to 60 feet to the inch.

At *each* setting of the instrument, the following observations can be taken: —1st, the horizontal angle of the back and forward sights to $\frac{1}{60}$th part of a degree; 2nd, the magnetic bearing to $\frac{1}{4}°$; and 3rdly, the rise or fall in the sight to 1' and also in inches to the fathom for perpendicular and base.

3080. Repeating Theodolite, with reading microscopes.

Dennert and C. W. Pape, Altona.

The tripod stands on screws which turn upon separate supports. The instrument is fixed to the stand by a brass plate pressed up against the lower surface of the head by a screw and spiral spring. The body rests upon an annular horizontal surface and can be turned accurately round its axis, an arrangment which enables the repetition movement to be accurately made. The vertical axis of hardened steel and the horizontal circle are firmly united with the body and round the vertical axis turns the upper part, having two double arms and serving as a support for the bearers of the axis of rotation of the telescope, which is arranged for reversal and carries the vertical circle, with an arrangement for vertical repetition. On the other end of the axis is the counterpoise with a graduated circle. The telescope has an achromatic object glass and orthoscopic eyepiece upon which a prism and sun glass can be placed. The cross wires consist of spiders' threads and are held by adjusting screws. Provision is made for illuminating the cross threads for observations at night. The two circles are divided into sixths of degrees, and by means of two micrometers accurate readings to 10 seconds may be obtained.

3081. Theodolite for Horizontal Angles with reading microscopes. *Imperial German Navy (August Lingke and Co.).*

3082. Theodolite, with reading microscopes.

Ed. Sprenger, Berlin.

Theodolite, with horizontal and vertical circles 0·17 meter in diameter.— Both circles reading by microscopes to 2 seconds. Focal length of telescope, 0·32 m.; magnifying power, 36 times. The tripod has a complete circle, and a third microscope for the exact and easy examination and determination of the errors of division. The substructure and axis bearer are of a single piece of cast steel, coated with chloride of platinum for protection from rust. The circles are solid and coned, by which means unequal expansion is avoided; this is not the case with spoked circles. The object-glasses are by Fraunhofer's successors at Munich.

The Topographical Department of the Royal Land Survey has exhibited a telescope ruler, which is now introduced by the Department. It has double verniers and a reversing telescope, and is also furnished with a double bubble-tube, rendering the instrument very useful for levelling. The distance-measure is removable, but so contrived that no screw is visible outside, so that nothing can act upon it.

3082b. 6-inch Transit Theodolite. *G. W. Strawson.*

3082c. 5-inch Transit Theodolites (2).

G. W. Strawson.

3106. Repeating Theodolite.

Meissner, Berlin. (H. Müller and F. Reinecke).

3072. Distance and Altitude Measuring Theodolite, by Dennert and Pape, Altona, with stand.

Prof. Helmert, Aix-la-Chapelle.

The theodolite is intended for topographical surveys. The division of the circles is so fine that frequently the reading of one index will suffice, its position being appreciable to 1 minute with the lens, without the employment of the vernier. As the stand has great stability, the vernier arm is not movable together with a spirit level, but both are fixed.

3079. Tachymeter (theodolite), with stands.

Bau Deputation, Hamburg.

3079a. Theodolite (Tachymeter), from the mechanical workshops of Dennert and Pape, Altona.

Royal Prussian Ordnance Survey Department.

This was specially constructed by Dennert and Pape, Altona, according to the directions of the Survey Department for the measurement of distances and heights, as well as for geodetical purposes. The arrangement has been such that, besides lightness, sufficient stability is afforded for work of such a nature. The circles are divided to 10', so that they can be read without the aid of the vernier to 5', and that the operator is not obliged to leave his place during the work for the purpose of reading at the circles.

For distance and height measurement, after the centre horizontal thread has been adjusted to any point of the levelling staff, and the readings taken, and the difference of the readings for the two threads, as well as the reading of the centre thread, is noted in the field book.

For the observation of the line of direction of other stations, the circles are fixed by the clamping screws, and the exact adjustment obtained by means of the micrometer screws.

3089a. Topographical Instrument, called **"Cleps,"** constructed by Messrs. Salmorraghi, Rizzi, and Co., Milan.

M. Antoine d'Abbadie, Member of the Institute, 120, Rue du Bac, Paris.

This altazimuth has three peculiarities :—

    1st. The telescope is very powerful for its size.

    2nd. Through the small transverse eyepiece the observer may read at the same time the horizontal and the vertical angle; the divided circles, completely covered, being lighted from above. To save time, there is neither microscope nor vernier, the angles being estimated in 10ths of divisions by three apparent wires, the means of these three estimations only being used.

3rd. The division is decimal; each hundredth of the quadrant is numbered, and divided into five parts.

The error of collimation is ascertained by reversing the ring that supports the wires. This instrument, constructed for topography, is of small dimensions In the larger model the division is carried to the thousandth of the quadrant.

3089b. Aba, or new **Altazimuth.** Designed by the exhibitor.

Antoine d'Abbadie, Membre de l'Institute.

The sextant, so useful at sea, is not a traveller's instrument, fo · it requires on land an artificial horizon, and cannot give, without much trouble, the small angular heights and true bearings of distant terrestrial objects, so useful in mapping. With common altazimuths time is lost in adjusting the levels and collimation error. These two adjustments are avoided in the " aba ;" by construction the errors are brought down to narrow limits, are determined by reversals, and remain practically constant. The telescope is very powerful for the size of the instrument in order to ensure easy vision. The observer can read his verniers without shifting his position. There is no inferior or fiducial teleseope, as the observer, after going round his horizon, takes care to repeat his first bearing, so as to provide against any motion in azimuth. A pinion and toothrack supply the place of clamp and screw, thus saving time and trouble.

The verniers read to the ten thousandth part of the quadrant or right angle (equal to 32″·4), this decimal division being preferred, on trial, by persons already accustomed to sexagesimal divisions. It is likewise more easily learnt by beginners. The figures indicate hundredths of the quadrant, and are marked so as to be very easily read. From the jutting position of the object glass it is possible to take the bearing of a signal at the very foot of a tower or wall. As screws get loose in travelling, the maker has employed fewer of them than it is usual in altazimuths. A powerful screw is attached to the stand and fastens the box on its top, before sliding off its superior part, which can thus be quickly put on again in case of a sudden shower of rain. The small box is destined to contain a thermometer for calculating atmospheric refraction, sun glasses, &c.

By taking corresponding azimuths of the sun on the same day or on two ensuing ones, the latitude is obtained together with true bearings of all visible objects, while the compass gives the needle's variation.

"A person familiar with the use of the sextant only, on observing circummeridian altitudes for his first attempt with the " aba," obtained his latitude to within 4″

3101. Geodetical Tachygraph, a representation of an instrument invented and constructed by the exhibitor, for conducting measuring operations in their minutest details.

Prof. Joseph Schlesinger, Vienna.

This instrument consists of a circle with divisions, mounted in a movable frame, that can be adjusted so as to place the circle in any given direction. It is also provided with a rule to plot angles.

3102. Eckhold's Omnimeter, for measuring linear distances by one and the same operation. *Elliott Brothers*

This instrument is a transit theodolite, with an apparatus for measuring by one operation the distance of an object or staff of a determined length, and its height over or under the level line of the instrument ; the former with the accuracy of a good chaining, the latter with the precision of a perfect levelling, and accomplishes the work of theodolite, level, and chain with a great economy of time.

3109. Tacheometer of Gentilli, a telescopic instrument for measuring heights and distances, in surveying difficult country. Without calculations it measures accurately distances up to 400 meters, or over 1,300 feet, with an error of less than $\frac{1}{3000}$. *Dr. Karl von Scherzer.*

This instrument was invented by M. Amadeo Gentilli, an eminent Austrian engineer at Vienna. Its use is for the measurement of heights and distances in the survey of difficult ground, and it has proved especially useful in surveying the contour-lines of mountainous districts. The means by which it measures the distance is an apparatus which obliges the telescope to traverse a precise and unvarying angle. The test of value of this instrument is the fact that, with a magnifying power of 40, it measures distances up to 400 meters with such exactness that the maximum error is less than $\frac{1}{2000}$ of the distance.

The horizontal axis of the instrument revolves in the head of a vertical arm, which can be clamped to it when the measurement of a distance is to be made. The motion of the lower end of this arm is effected by a slow-motion screw, and limited in each direction by an adjustable stop.

3109a. Laslett's Metroscope. For measuring inaccessible heights and distances, and for levelling. *Thomas N. Laslett.*

For the use of architects, engineers, and the military. Heights are read off as decimal parts of the horizontal distance. Distances are measured by readings from the two ends of a base line, which base line may be vertical, or lie in the direction of the object, or at right angles to that direction. *Lineal dimensions only are used;* no degrees angles, or tables. With the index at zero the instrument becomes simply a spirit level.

3109b. Drawing of Metroscope. For the determination of dimensions of distant objects.
Dr. Snellen, Physiological Laboratory and Ophthalmological School, Utrecht.

Before the objective of a telescope are placed two mirrors, one above the other, each occupying one half of the field of view, and inclined to the plane of the objective at an angle of 45°, the angle between them being 90°. On a cross bar, at each side, a mirror parallel to the first is movable. Looking through the telescope, by means of these two sets of mirrors, two objects, seen straight above each other, must be, at whatever distance they are from the observer, at a mutual distance equal to the distance of the two outer mirrors. (Handbuch der Ophthalmologie, von Graefe und Saemisch, III. p. 203.)

3124. Universal Instrument, by Breithaupt and Son, Cassel. *Royal High School of Industry, Cassel (W. Narten).*

Universal Instrument, with distance-measurer for levelling and measuring horizontal and vertical angles.—This instrument is intended for large surveys and levellings, which require to be performed with great accuracy and as little expenditure of time as possible. It corresponds with the tacheometer, but exceeds this in the amount of work that can be got through with it. The telescope rests with prisms and screw-heads upon the steel plates of a bearer, and, like the spirit-level, may be reversed ; the bearer, furnished with a double altitude arc, lies with its axis in closed rests, and has a vertical movement and a fine adjustment by means of micrometer screws ; a spirit-level on the head of the axis serves for its accurate adjustment in a perpendicular position. Horizontal circle, with glass coverings to the verniers. Constructed in 1854 by F. W. Breithaupt and Son.

4561. Photograph of the Universal Instrument of Luigi Pelli, for levelling operations, trigonometrical and graphic, by being adapted to the Prætorius table. *Luigi Pelli, Florence.*

The universal instrument of Luigi Pelli serves for levelling operations, being mounted like the most exact telescope-levels. It serves for the same trigonometrical operations as the theodolite, having an azimuth circle similar to that of the repeating theodolite, but with only two verniers. The alidade is placed so as to coincide with the plane of the tablet by means of a central screw situated in the column which sustains the telescope. A point fastened with springs within the screw marks on the paper the centre of the azimuthal circle. In this manner angles may either be measured or plotted at once.

3130. Photograph of an Altitude Circle, objective of 2·1″ aperture, diameter of circle 10½″. Latest construction of the instruments of a similar kind extensively in use in Russia for determining geographical positions.
A. Repsold and Sons, Hamburg.

b. REFLECTING CIRCLES.

3061. Ten-inch reflecting Prismatic Circle, similar to that of Pistor-Martin. *Julius Wanschaff, Berlin.*

3087. Reflecting Prism Circle, 10 in. diameter.
Meissner, Berlin (H. Müller and F. Reinecke).

3088. Reflecting Prism Circle, 6 in. diameter.
Meissner, Berlin (H. Müller and F. Reinecke).

The reflecting prism-circles can be recommended on account of their excellent execution, in optical as well as in mechanical respects, and also on account of their cheapness.

3115. Reflecting Circle, according to Pistor and Martins.
F. W. Breithaupt and Son, Cassel.

3115a. Improved Reflecting Circle of Pister and Martins. *The Pulkowa Observatory.*

A full description of the patent reflecting circle of Pister and Martins in its original construction is given by the inventors in the Astronomische Nachrichter, vol. xxiii. In the form exhibited general improvements are introduced by the inventors on Professor Dallen's suggestion ; for instance, all correction screws of the prism are removed, the stiffness of the main body is considerably increased, the different parts of the instrument are more symmetrically disposed, the telescope and the dark glasses are separated from the radius, on which mirror and prism are fixed, means are given for 'cross observation to the extent of 288°, &c. Also, the sfand for observations on shore is in parts changed and endered more practical.

2993a. Allen's Reflecting Circle, for repeating the observations and reading the angle on any portions of the circle, so as to eliminate errors of graduation or workmanship.

W. Watson and Son.

1722b. Double Reflecting Circle, by Capt. Owen, R.E. The property of the Royal Astronomical Society.

Robert J. Lecky, F.R.A.S.

This invention of the late Capt. Owen, R.E., is peculiarly valuable for the readiness with which two observations or "sights" can be taken with it, without removing the instrument from the eye, *e.g.*, in a lunar observation at sea the altitude of the star and the lunar distance may be taken with great rapidity, as well as in other observations.

The two circles each six inches diameter are divided on silver and read by three verniers on each to 20 seconds of arc, and the glasses being placed between the circles are well defended from accident at sea. It has other adaptations of much ingenuity.

3038a. Repeating Circle, of 1 foot diameter, by Troughton.

Royal Society.

3111. Portable Repeating Circle (14 centimetres).

J. & A. Molteni, Paris.

3088a. Prismatic Compass (german silver).

G. W. Strawson.

3088b. 3½-inch Prismatic Compass (bronze).

G. W. Strawson.

3088c. Singer's Gilt Compass.   *G. W. Strawson.*

c. LEVELS.

3037. Large Levelling Instrument by Cary.

Royal Society.

3038. Large Levelling Instrument by Troughton and Simms.   *Royal Society.*

3063. Levelling Instrument with telescope of 15″ focal distance, 1·6″ aperture ; a so-called compensation plane with rotating level.   *Ott and Coradi, Kempten, Baviera.*

This "compensation level" is a levelling instrument with telescope, which can be turned round its optical axis and reversed in its sockets. The water level is firmly fixed to the telescope, and can be turned in sockets round its own axis in such a manner that on rotating the telescope through 180° the scale of the water-level can be always turned upwards.

The object of this arrangement is to eliminate the errors occurring by damage of the telescope rings, and to give to the engineer at all times the possibility of correcting the instrument as easily from any position as in a levelling instrument with reversible telescope, with exactly equal ring diameters and adjustable water-level. By this arrangement the tedious reversing of the telescope and the adjusting of the water-level, which easily give rise to injury, will be rendered superfluous. Other constructors have endeavoured to obtain his result by the employment of double-cut water-levels ; a correct construction, however, of such instruments is very difficult, and any errors incurred cannot be corrected.

3046. Patent Level. A combination of the and Dumpy patterns. *Patrick Adie.*

3045. Theodolite-Level. A combination of the two instruments. *Patrick Adie.*

3046a. Dumpy Level, of improved construction.
Joseph Casartelli.

With graduated circle, for taking horizontal angles, and ball-and socket motion greatly facilitating its adjustment, especially on hilly or uneven ground, and saving the wear of the adjusting screws. The instrument being made of hard gun-metal, and the centre being long and accurately ground, renders it little liable to derangement.

3086. Levelling Instrument.
Meissner, Berlin (H. Müller and F. Reinecke).

This levelling instrument is distinguished by its simple and unchangeable arrangements for correction.

3089. New Levelling Instrument, constructed by A. Geppert, completed by F. Miller. *F. Miller, Innsbruck.*

This instrument is adapted especially for surveyors and engineers in mountainous countries. The arrangement is as follows :—The vertical axis is placed perpendicularly by means of 4 screws. It is provided with a horizontal graduated circle reading by means of a vernier, to two minutes. The astronomical telescope magnifies 15 times, and is capable of 14 degrees of depression or elevation. The inclination of the telescope is measured by a micrometer screw. Its uses are : Direct and indirect levelling, measurement of height and distance, horizontal angles, &c., and plotting the same.

3104. Abney's Level. The instrument was designed by Capt. Abney, R.E., F.R.S., for military reconnaissances. It measures vertical angles by means of a spirit level attached to a graduated are, which is capable of moving round an axis. The object whose angle of depression or elevation is to be ascertained, is brought into the field of view, and the spirit level is then moved till the bubble is in the centre. When in this position the bubble is seen in a reflector placed in the body of the tube; an opening being left for the purpose. *Elliott Brothers.*

3105. Elliott's 14 inch improved Dumpy Level, with compass for taking levels and bearings. *Elliott Brothers.*

This instrument, being provided with an object glass of large aperture and short focal length, and sufficient light being thus obtained to admit of a higher magnifying power in the eyepiece, the advantages of a much larger instrument are obtained. A mirror is placed over the bubble, so that the operator can see, while reading the levelling staff, if the instrument keeps its proper position; this being necessary on soft and spongy ground.

3105a. 14-in. Dumpy Level, with improved rack adjustment to cross webs. *Henry Porter.*

3105b. Twelve-inch Gun Metal Engineer's Level, with channelled bottom, for setting flat work, shafting, &c. Very delicate. *Joseph Casartelli.*

3105c. Small Improved Builder's Level, working parts all made of gun-metal. *Joseph Casartelli.*

3105d. Ten-inch Dumpy Level, with long centre improved construction, and the working parts all made of gun-metal. *Joseph Casartelli.*

3105e. 12-inch Level. *G. W. Strawson.*

3105f. 10-inch Level. *G. W. Strawson.*

3110. Level, with independent bubble, Gravet's system, with stand. *M. Tavernier Gravet, Paris.*

3112. Patent Surveying Level. *Francis Pastorelli.*

This instrument combines several important improvements, including increased facility in use, greater steadiness and freedom from vibration and possibility of derangement, accurate adjustments, and great durability.

Tripod and Staff Head.—Stability is of the utmost importance. This is secured in the staff head (which is the adoption of a plan by Wm. Froude, C.E., F.R.S.), which has the cheeks set wider apart, the leg joints being similar to an inverted mortar with strong trunnions, which can be tightened in their bisected cylindrical bearings by means of capstan headed screws. As wear goes on they can always be kept perfect; this cannot be done with the ordinary staff head.

Ball Joint and Clamp.—The instrument is free to move with 20° of inclination. This is most important, as much valuable time is saved, more especially upon hilly ground, as it can be almost set instantaneously.

Adjustments.—The parallel screws work in movable hemispherical nuts, which are held in seats in the parallel plates, those in the lower being held in brackets. Their action permits the upper plate to be worked at an inclination of about 15° without their being strained or twisted. By their means the upper parallel plate is made to clamp or set free the inverted cup on the vertical axis of the instrument; in addition, they cause it to heel over to perfect the adjustment, so as to bring the main bubble in the centre of its run whenever a force is applied to them greater than is necessary to clamp the inverted cup.

Suspension of Telescope.—The telescope is solidly fixed to its base, which is parallel to the axis of it; it has a female screw with a true surface which fits on to the gun metal centre, so that when the instrument is reversed in any direction the main and circular bubble is retained in the centre of their run. Around the telescope are two gun-metal collars, accurately turned and ground to a perfect circumference; by their means are adjusted the mechanical and optical axis of the telescope and the line of collimation; the main bubble is dead fitted; it neither admits nor requires adjustment, as it is done in construction.

Diaphragm.—The diaphragm with its collimating screws cannot be disturbed. A grooved channel is formed in the eye end of the telescope, into which they are sunk, and a cylindrical ring conceals them from sight, thus keeping them intact. This is one of the important adjustments of the instrument.

3112a. 12-inch Level mounted on the Ball movement.
H. Husbands.

The 12-inch level is mounted on the ball arrangement, the combined invention of Mr. Metford and Mr. Froude.

3113. Levelling Instrument with Telescope, 40 cm. long, arranged for rotation and reversal in its sockets, with stand.
A. Bonsack, Berlin.

The bearers are closed by springs. The horizontal circle is divided on silver to ½°, and adapted for reading to one minute.

3114. Compensation Level.
F. W. Breithaupt and Son, Cassel.

A portable instrument. It possesses the important advantage that, without previous correction, even if the telescope-cylinders are unequal, correct levels may be obtained, provided that the spirit-level is correctly adjusted, which may be easily ascertained by turning it round in its points of suspension. By means of the tangent screw the spirit-level is brought to equilibrium in both positions of the telescope, and the mean of the two readings is taken. By this operation all errors are compensated. The divided head of the tangent screw serves at the same time for distance-measurement; and, besides, a distance measure on glass is inserted in the eyepiece of the telescope for direct measurement.

3116. Levelling Instrument of Precision, with telescope of 15″ focal distance, 1·6″ aperture, a so-called **Compensation Level** with rotating water-level, according to new patent construction. *Ott and Coradi, Kempten, Baviera.*

3117. Levelling Instrument with horizontal circle.
Imperial German Navy (August Lingke and Co.).

3122. Large Reversible Level, for precision levelling. Length of the telescope 0·50 m., magnifying power 36, 65 diameters, in steel sockets, with reversing level of exactly equal tangents, indicating five seconds, latest construction. *Ed. Sprenger, Berlin.*

Length of telescope 0·44 meter, magnifying power 40 times, with a new construction of double spirit-level in one individual of exactly equal tangents, by which the employment of the second position of the telescope is greatly facilitated, and the instrument works with great precision, as any inequality in the bearing rings of the telescope immediately becomes visible. The support with the bearers is of cast steel in one piece, and to protect it from rust is coated with chloride of platinum. From this construction the instrument has great stability. The object glass, eyepiece, and field glass are removable, so that the telescope is free from optical errors. The bearing rings of the telescope are of aluminium-bronze, and there are removable distance-threads.

3123. Large Reversible Level, for precision levelling. The telescope 0·36 m. in length, magnifying ·power 28 times, in all other respects like the previous one. *Ed. Sprenger, Berlin.*

3125. Levelling Instrument, with rotating and reversing telescope, by Breithaupt and Son, Cassel.
Royal High School of Industry, Cassel (W. Narten).

The telescope and the spirit-level, as in No. 3124 arranged upon prisms, screws, and steel plates for reversal and distance-measuring. This arrangement, fulfilling the same purpose as the cylinder on the telescope, has the further advantage that any wearing that may be produced by the screws may be at once corrected without the aid of the mechanician. The instrument is particularly adapted for levellings of precision ; that now exhibited was employed in this way in the European measurement of a degree by Prof. Spangenberg.

d. QUADRANTS AND SEXTANTS.

3083. Quadrant, of boxwood. One side engraved as usual, on the other are numerous figures and a movable index.
William Sykes Ward.

3084. Gunter's Quadrant, of boxwood.
William Sykes Ward.

3085. Quadrant, of brass ; one side divided as a geometrical square, the other with movable circle and index, signs of the zodiac, and various figures. *William Sykes Ward.*

3067. Sextant of Aluminium, constructed by A. Petri, Rostock. *Prof. H. Karsten, Rostock.*

3085a. Jacob's Staff, made to take astronomical observations, with double arcs ; 17th century.
Ministry of Marine, Madrid.

3085b. Surveying Instruments of the 18th Century.
Ministry of Marine, Madrid.

3085c. Two Double Sextants, made in the last century by Davis. *Ministry of Marine, Madrid.*

3031. Double Pocket Sextant, an instrument for travellers. Made by H. Porter. *Capt. C. George, R.N.*

It can be used on shore with artificial horizon, in obtaining altitudes near the zenith ; also as two single sextants, one of which can be used in case of the other being damaged ; or one can be used by an assistant, and the other retained by the observer.

It can measure angles of nearly double the arc which can be measured by the ordinary sextant.

It can be used for the simultaneous measurement of two angles in the same place.

For laying out curves for railways and harbours, it is invaluable to the civil engineer and marine surveyor.

3033. Large Double Sextant (6 in. radius) for taking two lunars simultaneously ; an instrument for travellers.   Made by H. Porter.   *Capt. C. George, R.N.*

3033a. Capt. George's Improved Double box Sextant. Made by H. Porter.   *Capt. C. George, R.N.*

This instrument enables the observer to take right and left angles simultaneously, thus combining rapidity of use with great portability.

3033b. Box Sextant with improvement of wheel head to telescope, and case.   Made by H. Porter.   *Capt. C. George, R.N.*

3107a. Improved Box or Pocket Sextant.   To enable the observer, when sounding from a moving boat or vessel, to take both angles in fixing a position, without delaying to read off one angle before taking the other.   *Capt. J. E. Davis, R.N.*

This is effected by means of a supplementary arc and vernier, not affected by moving the proper arm unless connected by means of the hook and pin.

When used, the arms are connected, and the larger angle is taken first, the hook is then pushed back, and the supplemental arm remains to record that angle, which is read off after the small angle is taken.

Particularly useful when fixing positions in sounding and the boat or vessel is moving quickly.

3107b. Improved Sounding Sextant.   To enable the observer, when sounding from a moving boat or vessel, to take both angles in fixing a position, without delaying to read off one angle before taking the other.   *Capt. J. E. Davis, R.N.*

When used, the pawl of the indicator is attached to the movable arm and moves with it.   When the larger angle has been taken, the finger is applied to the capstan-headed screw, which at once clamps the indicator and frees the pawl from the arm, the smaller angle is then taken and read off ; the arm is then moved up to the indicator and the larger angle read off ; the capstan-headed screw is then loosened and the pawl drops, connecting the indicator again with the arm.

3107c. Improved Sextant.   For observing and recording a number of observations without the necessity of reading off at the time of observation, or removing the eye from the telescope, effected by means of a micrometer movement affixed to the tangent screw, and indicators applied to the arc.

Capt. J. E. Davis, R.N.

It is particularly adapted for observing lunar distances, circum-meridian altitudes, equal altitudes for time.   For position in bad weather, &c.

Its advantages are :—

1. It enables the inexperienced observer to take observations with as much facility as the more practised.
2. By greatly multiplying the number of observations, the instrumental and personal errors are reduced to a minimum.
3. At sea, it enables a number of observations to be made in a short time without being dependent (as is often the case) on one.

4. The differences of the altitudes being all equal, the check on the time-taker is apparent.

5. For night observations it is peculiarly adapted.

6. The micrometer movement can be thrown out of gear at pleasure.

The sextant has been submitted to the Astronomer Royal, who states that *" the arrangement is simple, very little liable to get out of order, and I should " think very effective."*

The many sources of error to which the astronomical sextant is liable, either from the different expansive qualities of the materials of which it is composed or the mechanical difficulties attending its construction, render it a less perfect instrument than it is generally supposed to be.

There is also a personal error, incidental to every observer with the sextant or any other instrument, and this is frequently augmented in night observations by the necessity of dilating the pupil to the utmost when observing, and suddenly contracting it in reading off by the aid of the strong rays of a bull's-eye lantern.

With these sources of error, any improvement that can be devised to bring either the instrument to a greater state of perfection, or to deduce a greater degree of accuracy from the observations taken with it, is a step in the right direction, and it is anticipated that by the use of the micrometer movement combined with the indicator, the number of observations can be so multiplied as to reduce the error to a minimum.

3107h. Sextant by Syeds, to be used in a fog.
Capt. C. George, R.N.

The use and design of this instrument is to give a correct altitude when the horizon is obscured by fog, &c., which desirable end is obtained by this quadrant—the addition of a projecting bar, containing a glass level and the bubble floating therein, which, when seen in the horizon vane in connexion with the sun, gives the true altitude.

3107d. Double Telescope Sextant, with reduced arc and natural angles.   *Patrick Adie.*

This instrument was invented and patented by Mr. Adie some years ago. but owing to difficulties in simplifying the adjustments, not yet surmounted for lack of time, has not been publicly introduced. It is an arrangement of the object ends of two telescopes, united by a common hollow axis; the rays from each are received on total or other reflecting surfaces placed at an angle of 45° at the centre of either end of this axis, along which each of the reflectors sends the rays at right angles; these rays are received by two other reflecting surfaces, each occupying half the field of the eye-piece. Thus both objects are seen at the same time. A motion—half that of the movable telescope—is given to the eye-piece, to correct the parallax of reflection from prisms placed at an angle of 45°.

3107e. Cary's Improved Edge Bar Sextant. Special construction to prevent expansion or contraction in different temperatures.   *Henry Porter.*

3107f. Portable Sextant.   *John Browning.*

3107g. Six-inch Sextant.   *John Browning.*

3107i. Davis's Quadrant.   *H. Porter.*

3107j. Pocket Sextant, by Cary, used by Captain Henry Bristow, of the Quartermaster-General's Department, in making the military maps of the north of Spain, for the use of the British army, under the Duke of Wellington, during the Peninsular war.
H. W. Bristow, F.R.S.

The sextant is a convenient instrument used for measuring the actual angle between any two well-defined objects, in whatever direction they may be placed, so that the angle does not exceed 140°; and without requiring more steadiness than is necessary for seeing the objects distinctly.

The pocket sextant was formerly used in military and maritime surveying, for fixing points and for filling in the details of maps. It has been superseded by the form known as the *box-sextant* (the principle of which is the same as in the larger sextant), which will measure the angle between any two objects to a single minute.

2996a. Rowland's Patent Sextant. Consists of two sextants on the pillar frame principle, mounted parallel, with their faces towards each other, in such a way that one telescope answers for both. One sextant measures angles in a forward direction, the other in a backward, so that if an angle between two points be measured on the larger sextant, and a third point taken between the other two, the angle between this third point and the other two can be measured at one observation, and the sum of these two angles should equal the first. *W. Watson and Son.*

e. MISCELLANEOUS.

3030. Portable Artificial Horizon. Made by H. Porter.
Capt. C. George, R.N.

Its improvement consists, not only in its reduced size and weight, but in its mechanical arrangements, form, and moderate price.

It secures altitudes near the horizon, as low as 3°; saves time, and prevents waste of mercury.

	lbs.	oz.	in.	in.
Large size, weighs in the case -	4	10	measures 9 by 5	
Small „ „ „ -	1	11	measures 6 by 3	

3035. Inclined Reflecting Horizon, an instrument for travellers. Made by H. Porter. *Capt. C. George, R.N.*

This instrument is used with the usual artificial horizon, to which it may be said to form an appendage or adjunct. By its aid such increased power is given to the sextant and artificial horizon that altitudes of the heavenly bodies can be measured from the zenith to 30° below it, and also altitudes from the horizon to 30° *above* it and 30° below it; this the sextant and artificial horizon have hitherto failed to do. It consists of a glass reflector, supported by a framework, which has its underside ground mathematically level, and this side floats on the surface of the mercury, carrying the reflector at an angle of about 30° with the natural horizon; and when properly made will always float on the mercury at the same angle; it has been tested by numerous observations, with satisfactory results. *See* observations and drawings.

3035a. Capt. George's Improved Portable Artificial Horizon. Made by H. Porter.   *Capt. C. George, R.N.*

This combines the reservoir for holding the mercury with the trough for observation, the two being cast in one piece of iron, with a stopcock to let the mercury run from the reservoir into the trough and back again when not in use. This instrument has a perfectly worked parallel glass floating upon the surface of the mercury, thus giving a perfectly brilliant surface and protected from the action of the wind ruffling the surface of the mercury during an observation. This instrument combines the most perfect horizontal surface, with exceeding portability (being less than one sixth the size of the ordinary mercurial artificial horizon) and the greatest facility in use.

3030a. Artificial Horizon, constructed by W. Herbst.
M. W. Herbst, Pulkowa.

M. Herbst's artificial horizon differs from that commonly used in the following points :—It is a box of rectangular shape, the mercury is enclosed in the lower part of the box, and brought up on the silvered copper-plate by means of a screw; it admits of easy cleansing of the reflecting surface without any loss of mercury; its folding rectangular roof is covered with mica.

3030b. Plumb Level, or Artificial Horizon.
Louis Brocher, Geneva.

The instrument consists of a vertical rod, a horizontal disc, and a small suspension chain. In operating, the instrument should be held suspended by the chain, like a plumb-line, the eye being on a level with the disc; the projection of the plane of its upper and lower surface determines the horizon. As the rod is graduated, and as the distance from the eye to the rod can be ascertained by means of a line, the observer can determine approximately the *sine* of the angle of elevation of the object—building, tree, or mountain—and ascertain its height, if not absolutely, at all events relatively. This instrument, which was originally invented with a view of supplying a means (in sketching from nature) of ascertaining in any place the level of the horizon (a matter indispensable for correct perspective), may be, and has already been, useful to architects, engineers, geographers, tourists, and also to agriculturists, for drainage purposes. Its small size renders it easy to carry, and the rapidity with which observations can be taken enables them to be repeated almost without intermission.

3029. Mercurial Barometer, an instrument for travellers. Made by H. Porter.   *Capt. C. George, R.N.*

Its peculiarity is, that the tube and cistern are carried empty, the mercury being secured separately in an iron bottle.

The cistern is used as a funnel.

It is filled, when required, by the traveller, using the spiral cord, which is kept in the tube while being filled.

Circular motion is given to the spiral cord, which, acting on the dense body of the mercury, forces the cord upwards, and out of the mercury, and with it the air bubbles, leaving a superior vacuum, as shown by the mercury always having a convex surface whether rising or falling.

A spare tube is in same box.

The instrument has been tested by R. H. Scott, F.R.S., Director of the Meteorological Office, London, and also at the Kew Observatory, with good results. It is now being used by travellers on the African lakes, and various parts of the world.

3055. Compass Theodolite, a representation of an instrument invented by the exhibitor, for observing, with a microscope, the exact position of the magnetic needle.
Prof. Joseph Schlesinger, Vienna.

3064. Cross-Staff, with box-level.
Zimmer Brothers, Stuttgart.

This instrument, cross-staff, with iron rod, angular mirror, and box-level, serves for setting out right angles, made entirely of metal ; on stony ground or in frosty weather it is preferable to the cross-staff with wooden pole.

3065. Reflecting Hypsometer.
Zimmer Brothers, Stuttgart.

The cross staff and mirror hypsometer.
The cross staff serves likewise for marking right angles and drawing perpendiculars.
The mirror-hypsometer is used for measuring the height of trees, &c.
These instruments are particularly useful for foresters and agriculturists, on account of their being easily carried about.
(See " Holzmessekunst," by Prof. Dr. Baur, Hohenheim).

3068. Hypsometer (thermo-barometer), graduated in $\frac{1}{200}$ths of a degree centigrade, and in millimetres representing the corresponding tensions of aqueous vapour.
W. Haak, Neuhaus am Rennweg, Thüringen.

3068a. Improved Hypsometer, large, with thermometer for the exact measurement of heights by the vapour of boiling water.
L. Casella.

3068b. Hypsometer, small, for travellers, with thermometer, for the exact measurement of heights by the vapour of boiling water.
L. Casella.

3069. Hypsometer for levelling. To ascertain elevations up to 500 metres by observing the boiling point of water.
W. Haak, Neuhaus am Rennweg, Thüringen.

3070. Hypsometer for levelling. To ascertain elevations up to 4,550 metres by observing the boiling point of water.
W. Haak, Neuhaus am Rennweg, Thüringen.

3071. Apparatus constructed for observing the boiling point of water, in the use of hypsometers.
W. Haak, Neuhaus am Rennweg, Thüringen.

3090. Land Surveying Apparatus, of simple construction, by A. Geppert, completed by F. Miller.
F. Miller, Innsbruck.

This apparatus, chiefly designed for work requiring no particular exactitude, and which a beginner with little experience in surveying may be trusted to execute, consists of a " Nativ," in which a round plank is screwed, and serves as a leaf. On the edge of this leaf are put 4 brass plates, with a deviation scale at distances of 90°, for forming right angles. The distance of the two sights is 30 cm.; one bears two horizontal cords, in the other a little slide with movable sight, which can be moved in a vertical direction. If the index of this slide stands at zero on the division fastened to the sights, the line of sight is horizontal when the tablet of the instrument is put in action. By moving the slide, and measuring the way which has been passed, the instrument can be used for indirect levelling distance and mensuration of heights and tracing. A chord drawn over both sights allows the formation of perpendicular lines.

3090a. Surveying Table, with plummet rule as used by the topographical surveys in Russia.

The Topographical Department of the Imperial Russian General Staff, St. Petersburg.

The plummet rule is provided with an altitude circle, and, besides, arranged as an instrument for measuring distances by adjusting the horizontal strings of the two water-levels which are attached to the plummet rule; the lower one is for regulating the horizontal position of the surveying table, the upper one for regulating the horizontal position of the altitude alhidada, by means of a micrometer screw acting against the beam of the alhidada.

3090b. A. C. Bagot's Patent Clinometer.   *A. Apps.*

The object of this instrument is to enable those persons who are engaged in engineering, contractor's work, land agencies, mines, &c. to get sufficient preliminary data with respect to the configuration, &c. of the land as shall warrant a more complete survey later, without having to carry about a theodolite, or Y level, with their various cumbrous appendages.

Foremen platelayers can accurately get the gradients of the railway; the arrow on the brass scale on the right indicates the angle subtended; the arrow on the left simultaneously indicates the number of inches rise or fall per yard.

Thus, if the limbs are opened until the right-hand arrow marks 45°, on referring to the left-hand arrow the number 36 is found; this means a rise of 36 inches per yard, or 1 in 1.

The telescope is reversible and of short focus. Such tables as are frequently necessary when in the field are calculated out, and engraved on the side of the instrument for reference.

The sight of the prismatic compass enables the operator to read off number of degrees, and still keep his eye on the object.

As a draining level for estate work this instrument will be found unrivalled, since an approximate idea of the amount of fall can be got before going to the expense of having the land surveyed, and being told that the fall is inadequate to the purpose.

Mining engineers can solve many problems with this instrument where the Y level could not be set up.

The design and construction, &c. of the instrument is strictly protected by Royal Letters Patent, granted to Alan Charles Bagot, Churchdale, Rugeley.

3091. Jaehn's Polymeter, with plane-table.

Schmidt and Haensch, Berlin.

Jaehn's "polymeter" enables the measurement of the distance of two points, or their relative position in the horizontal direction, and their differences of altitude to be obtained in a mode quite different from previous methods of operation. Both may be effected—1, simultaneously; 2, by one observation;

3, from a single point; 4, without any calculation; 5, by a mechanical setting out of the measured distances and heights in proportion upon the plane of the measuring table; 6, in any scale.

3094. Distance and Altitude Measuring Instrument (telescope ruler), executed, according to the directions of Prof. Helmert, by F. W. Breithaupt and Son, Cassel.

Prof. Helmert, Aix-la-Chapelle.

This instrument is designed for topographical surveys. It has a distance-measure on glass; the altitude-circle is furnished with two verniers, although, as a rule, only one of them need be read off;'moreover, the position of the index may be ascertained with the lens without the employment of the vernier. A spirit level is movable together with the vernier arm, a mirror enabling it to be seen from the eyepiece. The instrument also has levels and adjusting screws for the horizontal axis of rotation of the telescope, and parallel ruler.

3095. Triangular Prism.
Geodetic Institute of the Royal Polytechnic School, Munich, Prof. Dr. von Bauernfeind.

The *Three-sided Angle Prism*, invented by the exhibtor in 1851, serves for the measurement and setting out of right angles, and depends upon the deviation of light by refraction and total reflection. See Bauernfeind's "Elemente der Vermessungskunde," 5th edition, 1876, Stuttgart, F. G. Cotta, Vol. I., pp. 37–39 and 164, 165.

3096. Distance-measuring Prism.
Geodetic Institute of the Royal Polytechnic School, Munich, Prof. Dr. von Bauernfeind.

The *Distance-measuring Prism*, also invented by the exhibitor in 1851 but not applied to measuring distances until afterwards, serves for marking off isosceles triangles in which the equal sizes are definite multiples of the base. It is used in the same way as the three-sided angle prism. See Bauernfeind's "Elemente der Vermessungskunde," 5th edition, Vol. I., pp. 39–40 and 167 (No. 4), and Vol. II., pp. 90, 91.

3097. Pentagonal Prism.
Geodetic Institute of the Royal Polytechnic School, Munich, Prof. Dr. von Bauernfeind.

The *Five-sided Angle Prism*, invented by the exhibitor in 1869, serves not only for the measurement and marking off of right angles and half right angles, but especially for the laying down in position of two inaccessible points, or points which cannot be directly observed. It may replace the prismatic cross No. 3098. See Bauernfeind's "Elemente der Vermessungs-kunde," 5th edition, Vol. I., pp. 44–46 and 166–168.

3098. Prism Cross.
Geodetic Institute of the Royal Polytechnic School, Munich, Prof. Dr. von Bauernfeind.

The *Prism Cross* was invented by the exhibitor in the year 1851, and was then described in a separate memoir ("Das Prismenkreuz," published by I. Palm, Munich). It is described with later improvements in Bauernfeind's "Elemente der Vermessungskunde," 5th edition, Vol. I., pp. 168–175, especially 173–175. Its object is identical with that of the subsequently invented five-sided angle prism (No. 3097), which, as already stated, may be used instead of it.

3041a. Compass Binocular.   *Robert E. Barker.*

A small mariner's compass is fixed between the two tubes of the field-glass. Inside of the cover of the compass case (which is hinged tightly at the edge farthest from the eye, so as to remain set at any angle between 0° and 90°) is a mirror. When the mirror is set at about 45°, the direction in which objects seen through the glass lie, is shown by merely raising the eye and looking at the reflected image of the compass in the mirror.

This combination will be of service to officers of the army or navy, also to exploring parties, travellers, or tourists. With it the compass bearings of any object in the field of the glass can be seen directly in the mirror attached. It adds very little to the weight of the glass, can be fitted to any binocular, and will readily go into the original sling case with it. Rough surveys may be made with it in positions where larger and more costly instruments would not be available.

3042. Clinometer, by General Naeser.
Survey Office, Christiania, Norway.

This apparatus consists of a thin circular brass box, vertically fixed on three screws, by means of which the exact position can be maintained. Through the centre of the box moves an axis, supporting on one side a small telescope, and on the other a needle, following the movements of the telescope, and giving the readings of the tangential scale engraved on the box. By a small change-wheel, the angle between the optical axis of the telescope and the horizon is multiplied 10 times, thus enabling the operator to read off the tangent with sufficient accuracy. The horizontal distance between two objects being known, the difference in height can easily be found. The instrument can be used with advantage for distances up to 15 miles, and generally for all levelling purposes.

3043. Clinometer, by G. Olsen.
Survey Office, Christiania, Norway.

This apparatus consists of a square box, in which a pendulum moves on a horizontal axis. At the lower end of the pendulum a tangential scale is fixed, which doubly reflects in two small mirrors placed over the top of the pendulum. On the outer side of the box a small telescope is fixed, movable in the vertical plane. By directing this telescope to an object whose horizontal distance is known, and by reading off the division on the scale that coincides with the object, the difference in height can be readily found by referring to a table. The instrument needs no corrections, is very handy, and easily transported, but cannot be used at such long distances as the clinometer constructed by General Naeser.

3043a. Casella's Improved Ship's Clinometer, for measuring the inclination of ships or yachts fore and aft or athwart ship.   *L. Casella.*

3044. Pocket " Mensor." An improved arrangement or combination of various instruments for measuring and other purposes.   *Ridley Henderson.*

This instrument, when folded up, measures 3 inches square by $2\frac{1}{4}$ inches deep, and weighs 2 lbs. 2 ounces.

It contains within a box, hinged together in three parts, twenty instruments, as follows :—Anemometer, aneroid barometer, clinometer, goniometer, thermometer, circumferentor, protractor dial, prismatic compass, hypo-thonite, quadrant, spirit level, limb and sights for taking altitudes, sun-dial, callipers, plummet, magnifying lens, Nicol prism, scale of inches, scale of chains; added to which are arranged two tables of constants and useful formulæ, and an easy method of ascertaining the variation of the magnetic needle.

It is a measurer of time, heat, velocity, and pressure, also of height, depth, length, and breadth; and of horizontal, vertical, acute, and oblique angles.

It is made so portable in form and weight as to enable it to be carried in the pocket, and yet it possesses sufficient size and strength to render it a trustworthy and useful companion to military, civil, and mining engineers, geologists, mineralogists, railway and land surveyors, and travellers.

3108. Drawing of a Horizontal Goniometer for determining geographical longitude without a chronometer.

H. Haedicke, Demmin, Pomerania.

The instrument of which this is a drawing serves, in the first place, to determine by direct reading the angle formed by the line joining two stars (a—b) with the horizon. The handling of the instrument is for this purpose similar to that of the sextant; that is to say, the moment must be noted when the star line is covered by the hair line of the instrument. That an observation may be carried out on board ship, the instrument is provided with an arrangement which enables the position of the scale with respect to the artificial horizon to be fixed at the moment of the observation, so that the angle can afterwards be read by means of a vernier.

When in this manner the angle of a second star line (c—d) to the horizon has been determined, a simple subtraction will give the angle between the star lines a—b and c—d. Should there be a planet among the stars a, b, c, d, it becomes possible, by means of a proper astronomical table, to calculate the astronomical time as well as (if the local time be known) the geographical longitude of the place of observation.

3127. Instrument, called **Metrostroph, for the Construction of Altitude Curves** (contours).

Bau Deputation, Hamburg.

This instrument is a variable scale, invented by F. H. Reitz, which facilitates plotting a point on a contour line from the two nearest points, the levels of which are determined. *See* 3217a.

3151. Two Levels, sensitive to show one second of inclination by $\frac{1}{20}$ of an inch in any part of their length. *Adam Hilger.*

These levels were made at the desire of the late Colonel Strange, F.R.S., to stand a hot climate like India, and can be easily filled with ether.

The glass tubes are $9\frac{1}{2}$ ins. long by $\frac{3}{4}$ ins. in diameter and ground to a radius of 1,000 feet. The ends of the tube have glass plates, which are spherically ground and very highly polished, and fit so accurately that no ether can escape, the ends being kept in their places by a spring which has three arms. The ends of the brass mounts have bayonet joints, so that the level can easily be refilled.

3039a. **Collection of Drawings** in illustration of the progress in the construction of instruments employed in the science of geodesy, selected from various works published by the contributor. *Dr. Wilhelm Tinter, Professor of Practical Geometry at the I. R. Polytechnic Institute, Vienna.*

Illustrations of the Improvements in the construction of Surveyors' or Engineers' Tables.

1. Drawing and description of Praetorius's table (1573–1616), after Daniel Schwenter (1623).
2. Drawing and description of Marinoni's table (1676–1755).
3.   „   „   Kraft's table (1827), A., page 334.
4.   „   „   Starke's table (1859 and 1873), A., page 336.

Illustrations of the Improvements in the construction of the appliances of sight required in the use of Surveyors' and Engineers' Tables.

5. Drawing of a simple alidade for taking the level towards one side, A., p. 340.
6. Drawing of an alidade for taking the level on both sides, A., p. 340.
7. Drawing of an alidade with mountain side-vanes, A., p. 340.
8. Simple perspective index (or ruler), after Sadtler (1816), A., p. 346.
9. Essentially improved construction of the perspective index, after Kraft (1854), A., p. 352.
10. Improved construction of the perspective index, after G. Starke (1867), A. and C., p. 356.
11. Improved construction of the perspective index, with Stampfer's surveying screw, after G. Starke (1832–1869), E., pages 53 and 55.
11a. Perspective index with turning water-poise (spirit level), after G. Starke (1874), M.

Illustrations of the Improvements in the Appliances for Reading the Graduated Divisions.

12. Nunnez, Pedro-Nonius (1497–1577), proposed the employment of auxiliary quadrants, variously divided (1542).
13. Hommel, John (1518–1562), proposed using transversals for dividing a circle.
14. Vernier, Pierre (1580-1637), proposed the employment of Vernier's (Nonius's) graduated scale (1631).
14a. Ramsden introduced the micrometer (or reading microscope, 1777). A., page 135, and D.

Theodolites and Universal Instruments.

15. Simple theodolite for land surveying, after G. Starke. A., page 259.
16. Repeating theodolite, after G. Starke. A., page 273.
17. Astronomical universal instrument, after G. Starke. A., page 133.
18. Astronomical universal instrument, after G. Starke. E., page 39.
19. Astronomical universal instrument, with telescope in the axis for travelling purposes, after G. Starke. E., page 41.
20. Transit instrument, after G. Starke. A., page 212.

Illustrations of the Improvements in the Construction of the Hydrometrical Vane.

21. Oldest construction, after Woltmann (1790).
22. Improved construction, after Sadtler and Kraft.

23. Improved construction, with differential wheels.
24. Improved construction, after G. Starke. E., p. 73 (1870).
25. Improved construction, after Amsler-Saffon (1870).
26. Construction according to Amsler-Saffon with electro-magnetical numbering apparatus (1873). E., p. 73.

Illustrations of the Improvements in the Construction of Planimeters.

27. Polar planimeter, after Amsler (1856).
28. Momentum planimeter (Integrator), after Amsler (1856). A., 4 part, p. 36, E., page 81.
29. Polar planimeter, after Miller and Starke (1856). A., 4 part, page 24.
30. Polar planimeter, after Coradi and Ott (1874).
31. Planimeter Wetli, Starke (1850, 1871). A., 4 part, page 8, E., page 82.

Illustrations of the Improvements in the Construction of Levelling Instruments.

32. Vitruv's water-level (63 before Christ, 14 after Christ).
33. Water-level, after Picard (1620-1682).
34. Water-level, after Huyghens (1629-1695).
35. Water-level, after de la Hire (1704).
36. Water-level, after Rômer (1644-1710).
37. Channel level (Picard).
38. Mercury level, after Keith (1790).
39. Vega's levelling diopter (1754-1802).
40. Liesganig's water level (1719-1799).
41. Sisson's water-level (18th century).
42. Stampfer's levelling diopter (1839). F., page 24.
43^1. Hogrewe's levelling instrument (1800).
43^2. Pocket levelling instrument, after G. Starke. F., page 33.
43^3. English levelling instrument, by Cooke and Son. E., page 65.
44. English universal levelling instrument, by Cooke & Son.
44^2. Levelling instrument, with Stampfer's levelling screw (1836) A., page 197.
45. Universal levelling instrument, with Stampfer's levelling screw, after G. Starke (1867), F., page 146, and G.
46. Universal levelling instrument with telescope, after G. Starke (1867). F., page 174, and H.
47. Tachymeter, after G. Starke (1872). F., page 103, and I.
48. Polymeter, after Jähns (1873). K.
49. Tachygraphometer, after Wagner (1869-1871), constructed by Fennel. L.
50. Aneroid, after Naudet. F., 2nd part, page 35.
51. Aneroid, after Goldschmidt. F., 2nd part, page 49.

The foregoing illustrations are selected from the following publications by Dr. Tinter:—

A. Lectures on elementary geodesy (autobiographies).
A^1. Lectures on the theory and practice of geodetical and astronomical instruments.
B. Mathematical, geodetical, and astronomical instruments (History of Trades and Manufactures and Inventions, 1873).
C. The perspective index, especially with regard to the construction given to this instrument by G. Starke.
D. The micrometer.
E. Universal Exhibition Report on astronomical and geodetical instruments.
F. Lectures on elementary geodesy, Part II.

G. G. Starke's universal levelling instrument, with telescope and Stampfer's surveying screw.

H. G. Starke's Universal levelling instrument with telescope.

I. G. Starke's Tachymeter.

K. Polymeter, by Jähns.

M. G. Starke's perspective index with turning water level.

A. Description of Praetorius's table, according to Schwenter.

3134a. Folding Alhidade or Sight-Vane, small size, for rapid military surveys, with plane table with scales from $\frac{1}{2000}$ to $\frac{1}{20000}$. *M. Georges Sarasin, Geneva.*

This instrument is so constructed as to give, with an approximate exactitude which in no way hinders the rapidity of the work (notwithstanding the reduced size of the instrument for portability), the measure of distances by means of a micrometer, and with the optional assistance of an improvised sight, the measure of inclinations by means of a clinometer furnished with a vernier reading to five minutes, and consequently the measure of the differences of level, as well as a sketch of the horizontal lines on the ground, with the aid of a calculating rule. It is possible to check any inaccuracies of centering by verniers diametrically opposite to one another, and by turning the telescope end for end. There are also means of adjustment and correction.

3134b. Folding Alhidade or Sight-Vane, large size, for topographical surveys, with plane table to scales from $\frac{1}{200}$ to $\frac{1}{2000}$. *M. Georges Sarasin, Geneva.*

The plane table, constructed to give with considerable accuracy, notwithstanding its reduced size for portability, measures of distances by the aid of a sight *stadia*, inclinations by means of a clinometer furnished with a vernier reading to three minutes, and consequently the measure of the differences of level, as well as a sketch of the horizontal lines on the ground, to scales of from $\frac{1}{200}$ to $\frac{1}{2000}$. It is possible to *check* any inaccuracies of centering by verniers diametrically opposite to one another, and by turning the telescope end for end. There are also means of adjustment and correction.

3135. Signalling Apparatus (Aëroclinoscope), by Major Kromhaut. *Prof. Buys-Ballot, Utrecht.*

This apparatus consists of four movable discs and two fixed hollow cylinders. Two of the four movable discs *never* enter the space between the cylinders; the other pair remain *constantly* between the fixed cylinders, but may very easily be placed in six different positions clearly to be distinguished from afar.

The two discs f i may be brought very easily close to one another and to the first cylinder or the second cylinder; or they may be more or less separated, one close to one of the cylinders, or both or neither of them. We have then six combinations. Each of the two other movable discs may be hidden, also by another chord, by the cylinder next to it, or be placed close to it or at a double distance. Three positions for each give nine combinations ($6 \times 9 = 54$).

On the whole, when we make use of two pairs of discs we have 54 different signals, visible from far in a very distinct way.

Now, in Holland, the upper outer discs are to represent the barometer-height at Helder, Groningen, and the under outer discs the barometer-height at Flushing and Maestricht, and these form at the same time the direction of the expected wind. They show the azimuth of the gradient (the strike). The two inner discs denote the magnitude of the gradient (the fall). All these positions are to be seen in the joined diagrams.

3135a. Gauss's Heliotrope, ancient construction.
Geodetic Institute of the Observatory, Göttingen, Prof. Dr. Schering, Director.

3135b. Gauss's Heliotrope, modern construction.
Both instruments were constructed in the years 1821 and 1822.
Geodetic Institute of the Observatory, Göttingen, Prof. Dr. Schering, Director.

3135c. Heliotrope, an instrument for throwing the reflected light of the sun in any required direction.
Prof. W. H. Miller, M.A., F.R.S.

The Heliotrope consists of a plane glass mirror, two adjacent edges of which are ground and polished in planes making right angles with one another and with the large planes of the mirror, and then covered with asphalte varnish. A portion of the silvering, not larger than the pupil of the observer's eye, is removed from the angle where the two small polished surfaces meet the hinder plane of the mirror.

If held in such a position that the sun's light falls in the solid angle between the face of the mirror and the two small polished surfaces, a portion of the sun's light that falls upon the face of the mirror is refracted at the first surface, reflected internally at each of the small surfaces, and finally emerges through the space from which the silver has been removed, in a direction parallel to, but opposite to, that in which the reflected light travels from the large plane of the mirror.

Hence, any point with which the faint image of the sun appears to coincide will receive the light of the sun reflected from the mirror.

3136. Sun Signals, for the use of travellers.
Francis Galton, F.R.S.

The difficulty in sun-signalling is to direct the flash aright. The rays of the sun are reflected from a mirror, in a cone of light precisely similar to that which reaches it, the mirror itself (whose size may be disregarded) being the apex of the latter cone, and the sun's disc its base. It follows, that to the signaller, whose eye is near the mirror, the place where the cone of reflected rays falls on the distant landscape would always appear to him as a disc of precisely the same shape and size as the sun itself. In other words, his accuracy of aim must be within 30 minutes of a degree. In the author's heliostat an image of the sun is produced, which overlies the area on which the flash of the mirror falls. A lens is fixed in the instrument at right angles to the line of sight; half of the lens lies within the tube through which the observer looks, and occupies a portion of his field of view, the other half is external to his field of view; it projects beyond the side of the eye tube, and receives the flash of the mirror. The mirror turns on an axis attached to the tube which allows it movement in one direction, while the rotation of the entire instrument in the hand gives movement in the other. When the mirror is so adjusted that the reflected (parallel) rays from any one point of the sun's disc impinge on the lens, they are brought to a focus on the screen, and form a minute speck of light upon it. Rays radiate from this speck in all directions, and those that strike the part of the lens inside the eye tube, are reduced by its means back again to parallelism with the rays that originally left the mirror. Consequently the eye, looking down the tube, sees a bright speck through the lens, which it refers to the same distant point in the landscape seen to the side of it, as that to which the unobstructed rays from the mirror are being flashed. If a telescope be fitted to the tube the speck would overlie the spot on the landscape. Now what is true for any one point in the sun's disc is true for every point, therefore the signaller sees a luminous disc in his field of view, and this exactly overlies the *locus* of the flash. By gently rotating the hand the image can be made to cover or to forsake any point in the landscape that may be desired, and when that is done an observer stationed at that point will see a succession of flashes. Morse's alphabet can be adopted. A flash passing through a square hole of only one-third of an inch in the side, is visible to the naked eye at a distance of 10 miles, if the background be dull and the air perfectly clear. The principle of this heliostat was described in a memoir read before the British Association in 1858.

3137. Optical Telegraph, by Colonel Laussedat, composed of a transmutor and of a receiver, with their stands.

Colonel Laussedat, Paris.

The principle of this optical telegraph is due to Professor Mauras. Several other scientific men, during the siege of Paris in 1870-71, and since have assisted, under the direction of Colonel Laussedat, in modifying and perfecting the construction of the original apparatus. The model exhibited is the one which Colonel Laussedat proposed for country telegraphs and for the application of luminous signals to geodesy. When the sky is cloudless, recourse can be had to the rays of the sun; when the case is otherwise, one must be contented with the flame of the petroleum lamp, of which the range is far more limited. By night its range increases considerably, so much so that when the two small lanterns are exposed it is possible to send and receive signals at a distance of more than 40 kilometres; the communication takes place by means of luminous flashes and eclipses effected by a small screen, worked by the sender of the message. The Morse alphabet can be used by using short flashes for the points, and longer flashes for the lines. The working of the instrument can quickly be learnt, and this very simple apparatus is most useful in cases where the electric telegraph is wanting.

3138. Models illustrating two methods of **verifying Sextants** employed at the Kew Observatory:—

1st. By flashing the sun's rays to distant mirrors, whereby stars of light were visible to the operator at the testing table. This was ceased to be employed on account of the rarity of sunshine. Designed by F. Galton, F.R.S.

2nd. A system of five collimators, fixed firmly to a wall on a circular arc, arranged so as to send parallel rays across the testing table at known angles. Designed and constructed by J. Cooke, and described in the Proceedings of the Royal Society, vol. XVI., page 2.

Kew Committee of the Royal Society.

3099f. Prismatic Compass, with improvement of ring to hold more safely in the hand.　　　*Henry Porter.*

3034. Pocket Compass, for bearings of objects, &c., an instrument for travellers. Made by O. S. Bishop.

Capt. C. George, R.N.

A combination of the ordinary compass and the dipping needle. The advantage aimed at is, that it will act at or near the magnetic pole when the ordinary compass ceases to be of any use.

3032. Universal Tripod Stand, an instrument for travellers. Made by H. Porter. *Capt. C. George, R.N.*

May be used for five instruments, viz.:—
Sextant (for Lunars).
Telescope.
Barometer.
Prismatic compass.
Artificial horizon.

3139. Spectacles for Divers, for use in water.

Francis Galton, F.R.S.

When we look down into still clear water we see all objects in it with perfect distinctness, but the moment that the open eyes touch the water all distinctness of vision ceases. The convex surface of the eyeball has indented the plane surface of the water with a plano-concave lens, and, if we desire to restore distinctness of vision, we must use convex glasses of sufficient power, when immersed in water, to neutralise this effect. A double convex flint glass, each of whose surfaces has a radius of about half an inch, is therefore required. By means of the glasses exhibited it is possible to read the smallest type under water, with perfect ease. The principle of these glasses was described in a memoir read before the British Association in 1865.

3140. Ground Tongs for Sea Soundings; invented by Francis Hopfgartner, Austrian Imperial and Royal Naval Officer of the line. *Lieut. Hopfgartner.*

A hole is bored lengthways through the centre of an ordinary plummet. In this hole is inserted a movable metal rail, at the lower end of which there are attached two scoops or spoons, opening and closing by means of a hinge or link. At the upper end there are two movable bows which are joined to the ladles by small chains. If the plumb-lead is suspended to these hook-like bows by means of two short auxiliary lines, the scoops are opened and the apparatus is then ready to be let down into the water. On reaching the ground the bows will fall back, dropping the auxiliary lines, the weight of the lead presses the scoops into the ground, and by pulling up the main plumbline, which is now acting directly on the metal rail, the scoops are closed and drawn into the hole of the lead so far as to be securely closed.

3141. Ground Tongs for Sea Soundings, with disengaging weight; invented by Francis Hopfgartner, Austrian Imperial and Royal Naval Officer of the Line. *Lieut. Hopfgartner.*

Two scoops or spoons, intended for securing specimens of the sea bottom, are opened and closed like a pair of tongs by means of two levers. A peculiar metal cover, in which the ladles will fit, secures the closure of the same. At the upper ends of the two limbs of the tongs there are two hook-like movable bows, on which the lead is placed, which, on the apparatus being let down into the water, keeps the scoops in an open position. On reaching the ground the bows will drop back, the weight (stone or a ball) will fall off, and the metal cover will encase the scoops, and keep them closed while the apparatus is being pulled up.

3142. New Lead for **Deep-sea Soundings;** invented by Francis Hopfgartner, Austrian Imperial and Royal Naval Officer of the Line, and Moritz Arzberger, Civil Engineer at Vienna.

Lieut. Hopfgartner and·Moritz Arzberger.

The exterior form of this apparatus is that of a tube, of which the lower part forms the contrivance for throwing off the weight, while the centre part contains the indicating apparatus, being a system of metal cases which, through the pressure produced on them by the water, indicate the depth on a scale. The upper part is an arrangement for propelling the apparatus upwards in case the lead is to be used without a line.

3142a. Sinker and Tube with Detaching Appliance for Sir William Thomson's Pianoforte Wire Sounding Apparatus. *Sir William Thomson.*

The pianoforte wire (No. 22 B. W. G. Webster and Horsfall's) weighs about 14½ lbs. per nautical mile (6,086 feet) in air, or 12⅝ lbs. in water.

The strength of this wire to resist pull is such that it bears about 230 lbs. (104 kilogrammes) weight in air, or 29·4 kilometres (or 15·9 nautical miles) of its own length.

The splice, a specimen of which is shown, is made as follows .—The two pieces of wire to be spliced are first prepared by warming them slightly, and melting on a coating of marine glue to promote surface friction. About three feet of the ends so prepared are laid together and held between the finger and thumb at the middle of the portions thus overlapping. Then the free foot and a half of wire on one side is bent close along the other in a long spiral, with a lay of about one turn per inch, and the same is done for the free foot and a half of wire on the other side. The ends are then served firmly round with twine, and the splice is complete.

To the lower end of the wire a ring weighing about ½ lb. is attached, and a chain of two fathoms long, weighing 3 lbs., is shackled to this ring. The lower end of this chain bears an elastic wire double claw. The weight is spherical, with a perforation for the tube and two indentations for the two claws, by which it is hung till it reaches the bottom. The moment its weight is taken off by the bottom and removed from the claws, they spring out and leave it free; but before this is done it presses the tube into the ground by means of a bolt attached to the tube. This bolt is kept by a slight spring in its place until the wire is hauled up. During the time the wire is going down, the tube is kept from falling by this bolt, which after the tube touches the bottom, presses it downwards into the mud. When the wire is hauled up, a piece of small cord two or three inches long, connecting the lower end of the chain with a lever arm belonging to the bolt, which was loose so long as the weight hung by the claws, becomes tightened and draws out the bolt. This leaves the tube free to come away from the weight, and it is drawn up by the cord, chain, and wire.

Two specimens of the apparatus are exhibited, one showing the weight hanging on the claw with the tube resting on the bolt just before touching the ground; the other showing the weight supported partly by the bottom and partly by the tube pressed into the bottom (supposed to be mud) and the cord scarcely yet tightened enough to draw out the bolt.

3142a. Sea Sounding Apparatus (Bucknill and Casella's Patent), for the exact measurement of depths in the sea, unaffected by the influence of currents however strong. *L. Casella.*

3142b. Massey's Hand Lead and Deep-sea Sounding Machine. *E. Massey.*

3144. Deep-sea Water-raising Apparatus, according to
Dr. Meyer.
*Ministerial Commission for the Scientific Examination of
the German Seas, Kiel.*

? 3145. Deep-sea Water-raising Apparatus, according to
Prof. Dr. Jacobsen.
*Ministerial Commission for the Scientific Examination of
the German Seas, Kiel.*
Intended to bring up with certainty a sample of water from any depth in
the sea

3146. Deep-sea Water-raising Apparatus.
L. Steger, Kiel.

3147. Map Drawing Instrument. *A. Bonsack, Berlin.*
Instruments of this kind have come into use in the province of Schleswig-
Holstein.

3148. Coast Station Areometer Case.
*Ministerial Commission for the Scientific Examination of
the German Seas, Kiel.*

The *Station Areometer* gives the specific gravity of sea water compared
with distilled water of + 14° R. within 0·0001.

3149. Two Sets of Ship Areometer Cases, for oceanic
voyages.
*Ministerial Commission for the Scientific Examination of
the German Seas, Kiel.*

The *Ship Areometer* gives but half as accurate results as the preceding;
it is characterised by its shortness, which renders manipulation more convenient on board ship.

3150. Two Sets of Ship Areometer Cases, for voyages
in the east part of the Baltic.
*Ministerial Commission for the Scientific Examination of
the German Seas, Kiel.*

3145. Slow-Thermometer, in ebonite, for measuring the
temperature of the sea.
*Ministerial Commission for the Scientific Examination of
the German Seas, Kiel.*

The *Slow Thermometer* furnishes very accurate readings when it can remain in the water for an hour. For moderate depths it is more certain,
cheaper, and more durable than registering instruments.

**3145a. Apparatus for ascertaining the Temperature of
the Sea at various depths.**
J. L. W. Dietrichson, Christiania, Norway.
The temperature of the sea at different depths cannot always be ascertained
with a maximum and minimum thermometer, because it does not constantly
increase or decrease with increase of depth.
Supposing the temperature to be: at the surface, 15°; at 200 metres, 6°;
at 400 metres, 8°, it would not be possible to gauge this last temperature
with a maximum and minimum thermometer.
An apparatus must then be used which shall, in the water and out of the
water, preserve the observation taken at the depth.

The apparatus now submitted supplies this want, and also enables the temperature at different depths to be ascertained very easily by a single immersion
The principle is this :—
Two points, 0° and 20°, are accurately marked upon a small thermometer tt (Fig. 1) c 8^{cm} long.
If its tube be filed at a given point (x), it will easily break there. A column of mercury, length = xy, will therefore remain in the broken part of the tube. By the outer pressure of the atmosphere, this column will be slightly forced into the tube. If the tube is afterwards exposed to another temperature, the variation resulting therefrom, as affecting the length of the mercury column, will be almost imperceptible. To ascertain the temperature at the time of rupture, it is calculated upon a scale where the distance CD is divided into 20 equal parts. Each part, therefore, will indicate the length of a degree. For instance, if it be seen that the point x is placed at + 6°, and that the length of the column is = 14 parts of the scale, it follows that the temperature at the moment of rupture was = 14 + 6 = 8°.

To effect the rupture of the tube at great depths without the registration of the thermometer being affected by the outer pressure, often enormous, the thermometer is encased within a metal wrapper, as proof as need be, and which can easily be fastened to the lead line by metal wires M and N, shaped corkscrew-wise.

After fixing the thermometer with a screw (z), the ball upwards, between the two cheeks (VH), (Fig. 2), the plan of which is seen only from the back, the filing is done at point x, and the top part AB of the wrapper is tightly screwed up. The bottom part D forms a kind of hinge, covered over with an india-rubber pipe to prevent the water penetrating into it.

The apparatus being sunk at the required depth, sufficient time is allowed for the thermometer to attain the temperature of the surrounding water. A heavy ring is then dropped down the lead line. This ring, following the apparatus through its entire length, strikes the lever CF, which swings, and pressing upon the hinge occasions the rupture of the tube by the sharp end RS, which runs up within the apparatus.

To prevent the hinge being unduly displaced, it is barred by the screw K, which is removed just as the apparatus is plunged into the water. Likewise, to prevent the tube being broken by any unforeseen shock to the apparatus, it is fixed between two india-rubber strings (V), drawn between the two cheeks which support the thermometer.

2909a. Seismochronograph, designed by the exhibitor.
Prof. Dr. von Lasaulx, Royal University, Breslau.

1. The instrument is screwed on, inside, to the back of the station regulators * in such a manner that the pendulum, when the ball is resting on the plate intended for that purpose, and, consequently the lower arm of the lever is directed downwards, can swing past this latter, but is arrested immediately without making any further vibratory motion when the ball is thrown off through concussion, or by shaking, and the lever shifted thereby before the pendulum.

2. The instrument must be screwed on very firmly to the back partition of the regulator, in order to prevent the possibility of spontaneous action.

3. By putting up the instrument the eight round recesses designed for receiving the ball when thrown off must be marked according to their actual position towards the regions of the heavens. The best contrivance will be to engrave the letters N., N.E., E., S.E., S., S.W., W., N.W. on the border, so that the section marked N. actually points towards the north.

---

* In places where there is a station clock without a case, the instrument must be fastened to the wall.

4. Each time the ball is thrown off (provided there is no other circumstance, such as winding up or regulating the clock, to cause it) is noted on a separate sheet (see the annexed form) with the exact date, the time and hour of the day, and the direction in which the ball was thrown off by naming the recess with its corresponding letter in which the ejected ball was found. In places where there are clocks indicating seconds the time must be noted to seconds, otherwise by noticing as accurately as possible the fractions of the minutes.

5. After each reading the ball is placed again directly on the plate, the clock set in motion, and immediately correctly adjusted, if possible.

6. In case of greater commotions of the earth, or shocks of an earthquake, which can be recognised as such also without apparatus, a report as to the time and direction must *immediately* be submitted to the general administration of the telegraph service by the telegraph branch office concerned.

7. All other movements or disturbances which are being indicated by the instrument are registered, as above, and sent in monthly, or, if nothing have occurred, a report to that effect.

8. Particular attention must be paid to the *daily* careful adjustment and regulation of the clocks according to the directions given, in order to render the instruments useful.

OBSERVATION of an EARTHQUAKE by means of the SEISMOCHRONOGRAPH designed by Von Lasaulx, at the Imperial Telegraph Station at

1.	2.				3.	4.	5.	6.	7.
Date of Observation.	Time and Hour of the Day.				Direction of the Commotion or Shock, according to the Position of the ejected Ball.	What kind of Shocks or Commotion.	Duration of the Phenomenon (whether actually observed or estimated).	Simultaneous, Preceding, or Succeeding Noise.	Special Remarks (as to Damages, Injuries, &c.).
	Time of the Day.	Hour.	Minute.	Seconds.					
1875, Aug. 29.	a.m.	9	40	30	The ball was found lying in the recess directed towards N.E. External observation seemed to confirm that the shock came from that direction.	Not so much undulating as abrupt movement; one shock apparently, with wavering after vibration.	From about 2 to 3 seconds (as per estimation).	Dull, hollow, thunderlike noise following immediately after the shock.	A high chimney of a manufactory at fell in the direction towards N.E. Some houses were damaged, receiving cracks, not in the middle of the walls, but in the vicinity of the window panes. The phenomenon was undoubtedly recognised as being an earthquake, and was very violent, &c.

2909b. Mallet's original Seismometer.

R. Mallet, C.E., F.R.S.

IV.—COMPASSES.

3151a. COMPASSES, ADMIRALTY PATTERN, IN USE IN HER MAJESTY'S NAVY.

EXHIBITED BY THE ADMIRALTY—HYDROGRAPHIC DEPARTMENT.

Admiralty Standard Compass. Tripod; azimuth circle; light and heavy cards; spare cap and pivots. Used for steering and taking bearings.

Plain glass cover when the compass is used for steering. Cover and azimuth circle, for taking bearings, with the card read with a prism, and graduated for measuring horizontal angles. Light card, four compound needles of two laminæ each; ruby cap; pivots two of hard steel, two pointed with native alloy. Heavy card, four compound needles, two of four laminæ, two of two laminæ; cap of speculum metal; pivots pointed with ruby. Tripod and sprang, for use of compass on shore.

This compass can be allowed to rest on india-rubber suspension, by turning the screws marked F and A; it has the effect of steadying the card when the compass is exposed to vibration.

Prismatic Azimuth Compass, with tripod, card, spare cap, and pivots. Used for taking bearings on shore, and as a standard compass for small vessels. Card with two compound needles, of two laminæ each; agate cap; pivots of hard steel. Tripod for use on shore.

Steering Compass, large size; with card and spare pivot. For large vessels, card with four compound needles of two laminæ each; cap of speculum metal; pivot pointed with ruby.

Steering Compass, small size, with card, spare cap and pivots. For small vessels. Card with two single needles; cap of agate; pivots of hard steel.

Dent's Liquid Steering Compass. For use when the card of the ordinary compass becomes unsteady.

Liquid, two thirds water, one-third alcohol. Air chamber round upper rim, to allow for the expansion of the fluid from temperature; aperture for filling when fluid has wasted. Lifter for raising the card off its pivot when required.

West's Liquid Steering Compass.

Liquid, of same composition as Dent's. Bowl hermetically sealed, the expansion of the liquid being provided for by having the bottom of the bowl made of flexible corrugated metal.

Boat's Liquid Compass, in binnacle, with lamp. For use in boats.

Construction, the same as Dent's Liquid Steering Compass.

Life-boat Compass, in binnacle, with lamp. For use in life-boats.

Construction, the same as Dent's Liquid Steering Compass

Sledge Compass. Used in Franklin's search-expedition (1850–53).

Card with single needle capable of being moved round under the card, according to the magnetic variation, so that the North on the card coincides with the true North. A light needle in case attached to shoulder strap, for use without card when the horizontal force is weak. Spare card in cover, and directions for use.

Sledge Compass, as furnished to Arctic expedition (1875).

Card with single needle capable of being adjusted to the meridian. Aperture in leather, so that line of direction may be seen without opening cover. Ivory lifter.

Small Azimuth Compass, as furnished to Arctic expedition (1875). For determining the variation of the compass.

Single needle; alloy pivot; ruby cap; graduated circle of alumina metal; sight-vane, and prism.

3151b. MARINERS' COMPASSES OF VARIOUS DATES AND PATTERNS : GENERALLY OBSOLETE.

Exhibited by the Admiralty—Hydrographic Department.

Chinese Compass. Very small needle resting on steel pivot; cover of talc.

Walker's Meridional Compass; with apparatus for determining the latitude when the sun is on the meridian, also the sun's altitude, and hour angle.

Crow's Liquid Compass (Patent 1813).

Card hollow, convex lens shaped, buoyant and pressing upwards against the pivot. Expansion of liquid provided for by air chamber round upper rim of bowl; also by a spring valve allowing the escape of expended air and refilling when required. In the original patent the liquid was entirely alcohol.

Pope's Dipping Needle Compass (Patent 1820); with arc showing dip of the needle.

Graydon's Celestial Compass (Patent 1824); with apparatus for determining the latitude, the angular distance between celestial objects, and their true azimuth.

Danish Azimuth Compass; with telescope for observing distant bearings.

Steering Compass, by Sir William Snow Harris (about 1831); with stout copper ring, to calm vibrations of needle.

Preston's Liquid Compass. A steering compass; same construction as Dent's without air chamber.

Dent's Axis Compass (Patent 1844). Card with four heavy and deep needles, attached to axis working in socket above and below, preserving card parallel to surface.

Steering Compass, by Captain Walker, R.N. Card with single needle, centre of gravity below point of suspension; pivot long and on a brass bell-shaped cap, the latter cap working also on a pivot.

Grey's Vertical Compass (Patent 1854). Liquid between outer and inner bowls; card with two dipping needles; pivot inverted.

Gowland's Liquid Compass (Patent 1854). Rim of card vertical; pair of needles; pivot inverted.

Compass by Mr. Keen (Patent 1854). Porcelain bowl; card two single needles; pivot on springs inverted; cap centred in india rubber ring.

Magnets used in correcting ship's compasses. Hardsteel; 6 inch; 10 inch; and 12 inch.

––––––––––

SPECIMENS OF THREE DIFFERENT SIZES (4-INCH, 6-INCH, 8-INCH, 10-INCH) OF MARINER'S COMPASS, WITH SUN AND STAR AZIMUTH MIRROR, AND BINNACLE CONTAINING CORRECTORS FOR QUADRANTAL SEMICIRCULAR AND HEELING ERRORS.

Sir William Thomson.

3145b. I.—Binnacle, with correctors for quadrantal semi-circular and heeling errors, with mirror azimuth instrument on bowl, for taking bearings of sun, stars, or terrestrial objects. Small size compass (4-inch card), suitable for armour-clads or other ships, with quadrantal error exceeding 11°.

3145c. II.—6-inch Compass Card, suitable when quadrantal error is from 7° to 11°.

3145d. III.—Medium size (8-inch card) suitable for standard or steering compass in any ship, iron or wooden, steam or sailing, having quadrantal error less than 7°.

3145e. IV.—Large size (10-inch card) suitable for standard compass when quadrantal error does not exceed 5°.

In the improvements here illustrated the object primarily aimed at was to obtain a compass to which the Astronomer Royal's correctors could be applied with safety and convenience. The quadrantal correctors must not be so near the needles of the compass as to sensibly affect its direction through magnetization of the soft iron by the influence of the needles, otherwise the quadrantal error will, if truly corrected in middle latitudes, be over corrected in high magnetic latitudes, and under corrected in low magnetic latitudes. Thus when the 12-inch iron cylinders of the Liverpool Compass Committee are applied with their ends at a distance of 7 inches from the centre of an Admiralty compass card in this country, they correct a quadrantal error of 12½°, but of this 7½° is due to magnetization of the iron by the compass needles, and only the remainder or 5° is genuine, that is to say, dependant on the magnetization of the correctors by the terrestrial magnetic force. Blocks of iron weighing many tons would be necessary to safely correct for all latitudes a quadrantal error of even so moderate an amount as 5° or 6°, when the compass needles are of so great magnetic moment as those of the Admiralty standard compass. But if, as in the several sizes of compass now exhibited, the needles are of thin steel wire from an inch and three quarters to three inches long, a quadrantal error of any amount not exceeding 21° may be corrected perfectly in all latitudes by a couple of globes of iron of not more than 6 inches diameter fixed on two sides of the compass.

To correct a quadrantal error of 21½° a couple of globes, each 6 inches in diameter, fixed on the two sides of the binnacle at a distance of 6 inches asunder may be used, with the bearing point of the compass midway between them. Hence, to allow room for the case containing the compass, and for the gimbals supporting it, the diameter of the compass card must not be more than 4 inches (I.). This is the smallest of the four sizes now exhibited.

When the quadrantal error is 11° the globes, if of 6 inches diameter, must be placed 9 inches asunder and the 6 inches diameter compass card (II.) may be used. If with the same size of globes the distance asunder is 12 inches the quadrantal error corrected is 6⅓°, and, therefore, when the quadrantal error is of this amount or anything less the 8-inch card (III.) may be used.

The binnacle exhibited is suitable for 4-inch or 6-inch compass cards. It contains two adjustable magnetic correctors for the semicircular error, one for neutralizing the athwart ship component, the other the fore and aft component of the ship's magnetic force. The athwart ship corrector suffices to correct an error of about 23°, whether to port or to starboard, when the ship's head is north or south; the fore and aft corrector corrects an error of like amount when the ship's head is east or west. The binnacle also contains, in the four edges of its square top, provision for placing securely a pair of bar magnets athwart ship and another pair fore and aft in convenient positions a little below the level of the compass card; the former pair to be used when the athwart ship component of the ship's magnetic force, the latter when its fore and aft component exceeds the amount neutralizable by one or other of the adjustable correctors. Thus the binnacle with its several appliances now exhibited supplies convenient means for thoroughly carrying out the complete system of compass correction set forth by the Astronomer Royal in his paper on the correction of the compass, published in the Transactions of the Royal Society for 1839, according to the following very simple rule in three parts, as follows :—

I. Place the ship's head north or south magnetic and bring the compass to point correctly by the athwart ship correcting magnets.

II. Place the ship's head east or west and bring the compass to point correctly by the fore and aft correcting magnets.

III. Place the ship's head N.E., or S.E., or N.W., or S.W., and bring the compass to point correctly by the quadrantal correctors (a pair of 6-inch globes now recommended).

The whole process may be thoroughly performed with all needful accuracy for a new ship in a quarter of an hour; though of course it will be desirable to take an hour or two for verifying or perfecting the correction by testing it on other points than the three on which it was first made. When the quadrantal correctors have been once accurately placed they have never again to be changed for the same ship, and the same place in it (except of course in the case of taking on board a cargo of iron or introducing or shifting masses of iron so near the compass as to sensibly modify the quadrantal error). At any time afterwards the semicircular error (which is always liable to change through changes of the ship's sub-permanent magnetism, and also through change of magnetic latitude in the course of a voyage) is readily annulled by placing her head north or south and using the athwart ship corrector, and again east or west, and using the fore and aft corrector. In a steamer this may be done at sea on any clear enough night to allow stars to be seen, or day when the sun's altitude does not exceed 50°. When the weather is moderate enough to allow her to be steered steadily for two or three minutes first on one and then on the other of the two cardinal points nearest to her course, the detention at worst (that is, when the course is on one of the cardinal points) need not exceed five minutes.

The binnacle also contains an appliance for an adjustable magnet below the compass in a line through its centre perpendicular to the deck, for correcting the heeling error in iron sailing ships.

Each magnet supplied for this purpose is in two parts, joined together by a hinge or chain, so that when out of use they cannot be placed in the box provided for containing them without folding them together with unlike poles close one to the other, so that wherever they may be placed in the ship, they cannot disturb any of the compasses. A stout bar magnet brought carelessly on board a ship without this precaution may be as dangerous as dynamite.

An important objection had weighed with the Compass Department of the British Admiralty against the use of quadrantal correctors in the navy. It was, that they would obstruct the taking of bearings of celestial or terrestrial objects for the purpose of correcting the compass or of terrestrial objects for the navigational use of it. On this account the mirror azimuth instrument now exhibited was designed. It not only does away with that objection to the application of quadrantal correctors, but it is much more convenient for ordinary use at sea than the prismatic arrangement hitherto in use. It facilitates very much the taking of star azimuths by throwing (as in the *camera lucida*) an image of the star upon the divided circle of the compass card, illuminated by the ordinary binnacle lamp), or more properly speaking, on a virtual image of this scale at an infinite distance as seen through a convex lens. It is easy when there is not much motion in the ship to read the positions of the star accurately to a small fraction of the white space between two dark degree divisions on which its image is seen. The focal length of the convex lens is a little greater than the radius of the circle, and thus for objects on the horizon or at any altitude not exceeding 30°, no farther adjustment of the azimuth appliance than just to bring the object fairly into the field of view is necessary.

The compass consists of a light aluminium boss, with a central sapphire cap (by which the compass is supported on an iridium point), and a rim of aluminium of from 4 to 10 inches diameter, according to the size of the compass. There is an even number of holes in the rim, and half that number in the circumference of the boss. The rim and boss are connected together by means of fine silk threads forming, as it were, 32 spokes, and the compass-card is partly supported by these threads and partly by the rim. Two or four small magnets having their corresponding ends tied together by silk threads of equal lengths, so that the magnets may be as nearly parallel as possible, are attached to the rim by means of four silk threads.

The compass thus obtained, being extremely light, and having a large radius of gyration, has very small frictional error, with small enough magnetic movement to give a very long period of free vibration. The smallest compass (I.) has just about the same period of free oscillation as the Admiralty standard compass; and the same quality of steadiness at sea, while the larger sizes have considerably longer periods of from 28 to 43 seconds, and therefore much greater steadiness at sea.

3145aa. Marine Equatorial for correcting Compass by the Sun. *Sir William Thomson.*

A circle corresponding to the earth's equator is set upon gimbals with adjustment, by which the inclination of the circle to the vertical is made equal to the latitude of the place. A lens is mounted on a doubly-pivoted frame, which keeps its centre on the centre of the equatorial circle· while allowing the lens to turn round its own diameters through this point.

Part of the arrangement for effecting this consists of an outer frame, which has a motion round an axis perpendicular to the equatorial circle. This outer frame carries a portion of a spherical surface arranged as a screen to receive the sun's image during an hour. This screen is marked with declination circles, and these declination circles are divided by portions of meridional circles into spaces corresponding to five minutes of time. When the instrument is set with the axis of the equatorial circle truly parallel to the earth's axis, the sun travels along the declinational circle corresponding to his declination at the time. The frame carrying the screen is turned into such a position that the two extreme meridional circles of the screen agree with numbers on the equatorial circle, corresponding to the integral hours of apparent time before and after the time of observation. Then the instrument is turned in azimuth till the sun's image falls on the proper declination circle, and the point of it corresponding to the apparent time at the moment of observation. If there are no instrumental errors and if the adjustment for latitude is perfectly correct, then it is sufficient to set the instrument in azimuth till the sun's image falls with absolute accuracy on the proper declinational circle. Besides showing the true North, the instrument then shows as a sun-dial the apparent time, or if the apparent time is known with perfect accuracy, and the instrument is turned in azimuth till the sun's image is brought to mark correctly the apparent time, then the instrument shows the sun's declination. In practice both indications are looked to. When the instrument is used in equatorial regions, however, it is solely or almost solely by the sun's declination that it is set so as to give the true North.

In high latitudes, North or South, it is almost wholly from independent knowledge of the apparent time that the instrument is set. Generally the directions to the navigator using it are to set it to agree as well as he can both with the apparent time and with the sun's declination.

The instrument now exhibited has been tested at sea on board the Cunard steamers "Russia" and "Scythia" in voyages from Liverpool to New York and back, and has been found to act well for the purpose for which it was designed. It is, however, defective, and inconvenient in several details, which will be improved in instruments of the kind to be made in future.

3131. Photograph of Paugger's Patent Dromoscope.
Dr. F. Paugger, Trieste.

This dromoscope is circular, and has almost the shape of the ordinary chronometer.

It is about 22 centimeters in diameter, and 7 centimeters in height. On the upper part there is a fixed compass card, with two hands, besides a graduation reading to 45 degrees to the right and left, which lies. below these hands (scale of deviation and variation). The inner mechanism of the instrument is a perfect calculating machine; it exhibits for any desired course the exact tangent formula of the deviation of the compass when the contrivance for turning it has been put in motion. On the lower or back part of the instrument there are five scales or graduations marked A, B, C, D, und E; and corresponding with these graduations are five set screws (not visible in the photograph). With these adjusting screws, and the scales belonging to them, the five co-efficients of deviation for any ship can be indicated.

By means of this instrument the deviation of the compass, either of the course or azimuth, is indicated merely by stopping the hand.

A printed description accompanies the instrument.

Compass, made in Venice in 1564 by Antonio Blanchini, watchmaker.
The Royal Institute of " Studii Superiori," Florence.

Nautical Compass, made in 1607 by Baldassare Lancaeuj, of Urbino.   *The Royal Institute of " Studii Superiori," Florence.*

Goniometric Compass, of Paolo Massucsiuj, of Lucca.
The Royal Institute of " Studii Superiori," Florence.

3099b. Normal Compass of the Imperial German Navy,
with stand, sights, two compass-cards, &c. *Carl Bamberg, Berlin.*

The normal *card A.* of the normal compass is arranged for reversal so that at any time a determination of the error of collimation of the zero-line to the axis of the needle may be effected, and we are enabled to make a determination of the collimation of other cards of ordinary construction, but of the same diameter, by comparison. For measurements on land the gimbals may be fixed and the compass itself be brought to a horizontal position by means of screws.

3099c. Boat's Compass, adapted to be employed as a small
azimuth compass.   *Carl Bamberg, Berlin.*

The case of the boat compass serves for its transport, and also (by the removal of two pieces) as a binnacle, thus making it a small, handy, and accurate azimuth compass.

3099d. Boat's Fluid Compass.   *Carl Bamberg, Berlin.*

The card of the fluid compass is furnished with a float, so that pressure, and therefore also the friction on the pivot, are reduced to a *minimum*. The cap and pivot are made of ruby, so that they do not easily wear, and a stop becomes unnecessary.

3099e. Albini's Registering Steering Compass.
Elliott Brothers.

The instrument consists of a steering compass, with clock and apparatus attached, for printing on a slip of paper the direction of the ship's course every five minutes, the clock giving the exact time for eight days without winding. An instrument of this description placed in the captain's cabin would thus enable him to have a record of a whole voyage.

3099a. Portable Compass. *Colonel Degen, Bobruisk.*

3151. Azimuth Compass, by Brauer.

            T. Brauer, St. Petersburg.

3155aa. Bronze Compass taken from a wrecked Japanese junk.   *Capt. Murray, per F. Buckland.*

3151d. Floating Mariner's Compass.

            Dumoulin Froment, Paris.

3151e. Eclimétre for floating compass forming a spherical spirit level for quick elevation. *Dumoulin Froment, Paris.*

3151k. Dipping Needle and Compass used by Captain Cook during his voyage round the world.

          Royal Naval Museum, Greenwich.

3151l. Chinese Compass, from a collection made by Mr. Coryton, Barrister-at-law of the Temple.

          Royal Naval Museum, Greenwich.

3151m. Symon's True North Compasses, with improved indicator arranged to show the True North instead of the Magnetic North as usually shown.   *L. Casella.*

3151n. Coloured Compass Cards. *Max. Raphael, Breslau.*

3151o. Mica-plates for uncoloured **Compass Cards.**

           Max. Raphael, Breslau.

For the manufacture of compass cards, mica is now exclusively used. The advantage to be derived from its use consists in the greater facility with which the magnetic needle can be fixed, the transparency of the card, its non-conductivity for electricity, its lightness and indifference to ordinary changes of temperature, its durability, &c.

3151p. Compass, Hart's.   *M. Bréguet, Paris.*

3151q. Patent True Course Finder. *W. H. Roberts.*

The object of this instrument is to avoid arithmetical calculation in finding the true course from the compass or magnetic course, with corrections for lee-way, deviation, and variation. This instrument will give by three movements of the index the true course to fifteen minutes. The outer rim of the dial is accurately divided into degrees, and the inner rim into points, ½ points, and ¼ points. The index has two arcs, A and B, and moves freely on a pivot, C. The arc A indicates the points, ½ points, and ¼ points. The arc B is so made that its centre line will point to the exact number of degrees which are contained in the number of points shown by the arc A. The arc B has also a vernier, by which the observer can read off to ¼ of a degree. On the extreme inner rim are arrows, half encircling the dial, to remind the observer in which direction to allow easterly and westerly deviation and variation.

3151r. Self-registering Ships' Compass, designed by J. M. Napier.—D. Napier and Sons, Vine Street, Lambeth.

South Kensington Museum.

3151s. Gimbal Compass, for use in small crafts.

South Kensington Museum.

3151x. Chinese Compass.   *Ministry of Marine, Madrid.*

This differs from those employed by other nations, in the inversion of the meridian line.

3151y. Compass or Azimuthal Needle, which was employed in making the plans of the coast, This instrument was made at Carthagena, at the end of the last century.

Ministry of Marine, Madrid.

3151z. Compass made at Carthagena, Spain, according to the plan adopted by Don Antoino Doral, the Commander of Fleet.   *Ministry of Marine, Madrid.*

V.—MAPS, BOOKS, &c.

COLLECTION OF MAPS, &c., LENT BY THE ORD-
NANCE SURVEY. MAJOR-GENERAL CAMERON,
R.E., C.B., F.R.S., DIRECTOR-GENERAL.

MOUNTED MAP of part of the city of WINCHESTER, scale $\frac{1}{500}$, or 10·56 feet to one mile, with corresponding photographic reduction on $\frac{1}{2500}$ scale from MS. plan.

MOUNTED MAP of part of LONDON, scale 5 feet to one mile.

MOUNTED MAP of SOUTHAMPTON and its environs, scale $\frac{1}{2500}$, or 25·344 inches to one mile.

MOUNTED MAP of part of HAMPSHIRE around Southampton, scale 6 inches to one mile.

MOUNTED MAP of part of SCOTLAND (in outline), scale 1 inch to one mile.

MOUNTED MAP of part of SCOTLAND (hill features engraved), scale 1 inch to one mile.

MOUNTED MAP of JERUSALEM, showing the hill features, scale $\frac{1}{2500}$, or 25·344 inches to one mile.

PORTFOLIO, containing specimens of MAPS of TOWNS on the $\frac{1}{500}$ scale and 5 feet scale, viz. :—

$\frac{1}{500}$ scale, zincographed.
Cheshire, Chester, Sheets XXXVIII, 11, 17, 18; XXXVIII.
15, 2.
Cheshire, Hyde, Sheet XI, 1, 19.
,, Crewe, Sheet LVI, 7, 10.
Kent, Canterbury, Sheet XLVI, 3, 14.
Sussex, Chichester, Sheet LXI, 7, 16, 17, 21, 22.
Gloucester, Cirencester, Sheet LI, 10, 10. With shading of houses transferred to zinc from lines engraved on a copper plate.
Gloucester, Cirencester, Sheet LI, 11, 6. With shading of houses transferred to zinc from lines engraved on a copper plate.
5 feet scale, engraved.
London, Sheets VI, 30.
,, ,, VII, 53, 56, 64, 65, 66, 73, 76, 77.
,, ,, VIII, 12, 31, 41, 61, 71.

PORTFOLIO, containing specimens of MAPS of the CADASTRAL SURVEY on the $\frac{1}{2500}$ scale, or 25·344 inches to a mile, approximately 1 square inch to an acre, viz. :—

London, Sheets XLII, LXV, LXXIV, LXXV.
Denbigh, Llangollen, Sheet XXXIV, 15.
Cheshire, Chester, Sheet XXXVIII, 11.
Hants, Winchester, Sheet XLI, 13.
Kent, Canterbury, Sheet XLVI, 3.
„   Dover, Sheet XLVIII, 14.
Sussex, Rye, Sheet XLV, 7.
Forfar, Dundee, Sheet LÍV, 5.
Lanark, Hamilton, Sheet XVII, 4.
Essex, Chigwell parish, Sheets LVII, 15 and 16 ; LVIII, 9, 13, 14 ; LXV, 3, 4, 7, 8 ; LXVI, 1, 2, 5, 6.
One book of areas of Chigwell parish.

Sheets of London reduced by photography from the 5 feet scale to the $\frac{1}{2500}$ scale, or 25·344 inches to one mile.

Sheets VII, 3, 13, 14, 23, 24, 73, 83, 93.

Photographic reduction used in producing the Cadastral Map on the $\frac{1}{2500}$ scale from the Town Map on the $\frac{1}{500}$ scale.

Sheets LXI, 7, 6, 7, 11, 12, 16, 17, city of Chichester.

PORTFOLIO containing specimens of COUNTY MAPS on the scale of 6 inches to one mile, viz. :—

Hants, Sheet XLI, LXV.
Kent, Sheets VII, XLII.
Middlesex, Sheets VII, XI, XVI, XXI.
Surrey, Sheets VII, VIII, XIV, XXIII, XXIV, XXV.
Aberdeen, Sheet LXXV.
Argyll, Sheets XCVIII, CXLIX, CLXXIX, CXC.
Bute, Sheet CCIV.
Dublin, Sheet XVIII.

Surrey, Sheet XVI. Reduced by photography from the $\frac{1}{2500}$ scale to the 6-inch scale and photo-zincographed.

Hampshire, Sheet LXXII. With a reduction to the 3-inch scale by photo-zincography.

Hampshire, Sheet LXXX. With a reduction to the 3-inch scale by photo-zincography.

12 mounted photographs reduced from MSS. plans or the $\frac{1}{2500}$ scale, or 25·354 inches to one mile, to the $\frac{1}{10560}$ scale, or 6 inches to one mile, by photography.

Photographic reductions used in producing the County Map on the scale of 6 inches to a mile ($\frac{1}{10560}$) from the MSS. sheets of the Cadastral Map. Scale, 25·344 inches to a mile ($\frac{1}{2500}$), viz:—

Essex, Sheets XLV, 10, 12 ; LX, 5.
Hants, Sheet VIII, 8.
Shropshire, Sheet XIII, 2.
Sussex, Sheet XXI, 16.

Map of the Turco-Persian frontier; made by Russian and English officers in the years 1849 to 1855, on the scale of $\frac{1}{73050}$, and reduced to the scale of $\frac{1}{253440}$, or four English miles to 1 inch, at the Ordnance Survey Office, Southampton. Drawn on Sir Henry James's Rectangular Tangential Projection of the Sphere; and photo-zincographed at the Ordnance Survey Office, Southampton, 1873.

PORTFOLIO, containing specimens of MAPS on the 1-inch scale, viz. :—

Geological Survey. England, Sheets XCII, S.W.; XCVIII, S.E. ; CV, S.W.

Geological Survey. Scotland, Sheets IX, XXXII, XXXIII.

General Maps :—

England.

Sheets CCLXX, CCLXXI, CCLXXII, CCLXXXV, CCC.

Sheet XCVIII, N.W., outline and hills, two sheets.

   „ XCVIII, S.W.,   do.

   „ XCII, N.W.,   do.

   „ CII, S.E.,   do.

   „ CII, S.W.,   do.

   „ CIV, S.W.,   do.

Scotland.

Sheets IX, XIII, XXI, XXIII, outline and hills, two sheets of each.

   „ LXII, LXIII, (in outline), showing the parallel roads of Glen Roy.

Ireland.

Sheets XXVIII, XXXVI, outline and hills, two sheets of each.

   „ XXIV, showing hill features.

ORDNANCE SURVEY of JERUSALEM, bound in three volumes.

ORDNANCE SURVEY of SINAI, bound in five volumes.

FACSIMILES of NATIONAL MANUSCRIPTS made by photo-zincography at the Ordnance Survey Office, Southampton, viz. :—

Domesday Book (1084–6), two volumes.

Black Letter Prayer Book of. 1636, one volume.

Extracts from Parker's Register, 1559, one volume.

National Manuscripts, England, four volumes.

   Do.   Scotland, two volumes.

   Do.   Ireland, two volumes.

SURVEY OF PALESTINE, EXECUTED FOR THE PALESTINE EXPLORATION FUND.

Country from the Mediterranean to the Jordan Valley, including the greater part of the Plain of Esdraelon. Area, 907 square miles. Scale, 1 inch to the mile, by Lieut. C. R. Conder, R.E.

Part of the Wilderness of Judea on the west of the Dead Sea, and the hill country west of Jericho (Kurn Sartabeh), by Lieut. C. R. Conder, R.E.

Special plans, comprising the Roman Amphitheatre at Beisan (Bethshan), Kaṣr el Hajlah a crusading ruin south of Jericho, Joshua's Tomb at Tibneh, the ruins of Cæsarea, and details of a Temple at Abu Amr, by Lieut. C. R. Conder, R.E.

Reconnaissance survey of the Sea of Galilee and vicinity, by Captain Anderson, R.E.

Details of temples, &c., in the neighbourhood of Mount Hermon, by Captain Warren, R.E.

Reconnaissance survey of the Jordan Valley and neighbourhood, by Captain Warren, R.E.

MAPS EXHIBITED BY THE ORDNANCE SURVEY DEPARTMENT OF THE ROYAL PRUSSIAN GENERAL STAFF, BERLIN.

3152. Map A. Copy of an Original Survey taken by means of Surveyor's Table. (Plane table sheet.)

Map B. 1. Nine surveyor's table sheets, north-west from Berlin, original survey on the scale of $\frac{1}{25000}$ (2·534 inches to 1 mile), lithographed, with mountain features, marked in lines.

Map B. 2. The same district, scale $\frac{1}{50000}$ (2·534 inches to 1 mile).

Map B. 3. The same district, copper plate, scale $\frac{1}{100000}$ (0·634 inch to 1 mile).

Map C. 1. Eastern Prussia composed of the sections of the Ordnance Map of the General Staff, copper plate print, in $\frac{1}{100000}$, northern part.

Map C. 2. The same province, southern part.

Map D. The western part of the province Hesse-Nassau, copper plate, scale $\frac{1}{100000}$ (0·634 inch to 1 mile).

Map E. Environs of Berlin, copper plate print, scale $\frac{1}{100000}$, with watercourses, meadows, and villages, indicated by special colouring.

The maps A, B1, B2, B3, represent the process from the fieldwork to the publication of the maps on three different scales, namely, $\frac{1}{25000}$ (original survey); $\frac{1}{50000}$, and $\frac{1}{100000}$. The last is specially intended for military purposes.

The maps C1, C2, and D, show larger surfaces, the two former in hilly, the latter in mountainous country. On copper on the scale of $\frac{1}{100000}$.

The map E represents a surface of 70 geographical square miles in level country. On copper to the scale of $\frac{1}{100000}$, with the water, meadows, and districts coloured.

SECRETARY OF STATE FOR INDIA.

3153. Selection of Maps to illustrate the progress of **Cartography and Surveys in India.**
The Secretary of State for India.

(1.) Fortelezas de Mombaim e Ilha de Carania.
Old Portuguese plans of Bombay and Karanja, extracted from the " Livro de Antonio Bocarro, Guardamor de Archivo Real da India," &c., 1646. (Bombay was ceded to England in 1661, and became the seat of the presidency in 1683. The Portuguese retained the Island of Karanja in the harbour of Bombay until it was seized by the Mahrattas in 1683. In 1774 Karanja was taken by the English.)

(2.) Dutch chart of the Deccan (properly Concan) and Malabar Coasts, with the Laccadive and Maldive Islands and the Padua Bank, &c., including the western coast of India from Diu to Cape Comorin and part of the Gulf of Manaar, with soundings. Date, 18th century. MS. (No. 227).

Dutch chart of the Malabar Coast and Backwaters, from Coilang (Quilon) to Cranganor, with soundings. By Jan Tim, revised by H. G. Farrant, Engineer. Date, 1697. MS. (No. 229).

Dutch chart of the coast of Coromandel, from Tondy to Point Goddewarre (Godavery). Also a chart of the coast of Orissa. MS. (No. 254).
These are selected from a series of MSS., the originals of which are in the State Archives at the Hague, and catalogued in the " Inventaris der Verzameling Karten berustende in het Rijks Archief," 2 vols., 8vo.

(3.) Collection of Dalrymple's charts in 1 vol.

(4.) Bengal atlas, containing maps of the theatre of war and commerce on that side Hindoostan, compiled from the original surveys, and published by order of the Honourable the Court of Directors for the affairs of the East Indian Company, by James Rennell, F.R.S., late Major of Engineers and Surveyor-General in Bengal, 1781. (Second edition. 21 maps, folio, bound.

(5.) Rennell's map of India.

(6.) Geographical and statistical map of the north-east part of the Mysore dominion or the ceded districts, &c. By Colin Mackenzie, Surveyor-General of India, 1815. Scale, 4 miles to 1 inch; size, 59 inches by 74. MS.

(7.) Topographical survey of Madras and its environs, by the Officers of the senior class of the Military Institution, 1806. Scale, 2 inches to 1 mile; size, 78 inches by 49. MS.

(8.) Mullangoor Circar, surveyed in 1834, on 15 small sections. MS.

(9.) Map of Mullangoor Circar, executed under the superintendence of Lieut. H. Du Vernet in the year 1834. Lithographed by Trel. Saunders.

(10.) Survey of the River Ganges from Hurdwar to Nahgul, taken in the month of February 1800, under the immediate instructions of Major-General Sir James H. Craig, K.B., &c. By Lieut. Thomas Wood, Corps of Engineers. Part 1. Size, 51 inches by 21. MS.
The survey is complete in 7 parts.

(11.) Survey to Gangotri, by Lieut. Webb, 1810. The general map scale, 24 miles to 1 inch. MS.
The survey on the scale of 1 inch to 1 mile is on 12 sheets.

(12.) . . . . Map of India, compiled from all the latest and most authentic materials . . . . By A. Arrowsmith. London, 1816.

(13.) Atlas of Southern India, in 18 sheets, from Cape Comorin to the river Kistnah, delineated on a scale of 4 English miles to 1 inch, principally from original surveys communicated by the Honourable Court of Directors of the East Indian Company. By A. Arrowsmith. London, 1822. Folio, half-bound.

(14.) The Indian atlas. The index map and sheets of the Punjab, mounted on roller.

(15.) Chart of the Red Sea, on three sheets. By Captain Moresby.

(16.) Atlas of the North-Western Provinces. Settlement maps compiled and lithographed under the order of H. M. Elliott, Esq., 1840 to 1843. Folio, half-bound.
Atlas of the district maps of the Lower and North-Western Provinces of Bengal, including the second edition of the district maps of the North-Western Provinces. Folio, 2 vols., half-bound.

(17.) Map of the Eastern Frontier of British India, with the adjacent countries extending to Yunan in China. By Captain R. Boileau Pemberton, 44th Regiment Native Infantry, 1838.

(18.) The districts Jhelum, Rawalpindi, and Shahpoor. By Captain D. G. Robinson, Engineers.
A specimen of the drawings.

(19.) The same lithographed on 28 sheets.

(20.) Map of Kashmir, with part of adjacent mountains. By Captain T. G. Montgomerie, Engineers, &c. Scale, 2 miles to 1 inch. On four sheets.

Jamoo, Kashmir, &c. Scale, 4 miles to 1 inch. On four sheets.
(21.) Madras revenue survey :—
Nellore district map.
Bapatla Taluk map.
Kelampakkam village map.
(22.) Revenue surveys :—
District Peshawur, 10 sheets.
Peshawur Cantonment City, &c., four sheets.
District Seebsaugur, Assam, 11 sheets.
Topographical surveys :— .
Rewah survey, a selection mounted.
(23, 24.) General map of the Punjab, four sheets.
General map of the North-West Provinces, four sheets.
General map of Assam.
(25.) S. G. O. engravings of Indian atlas.
(26.) Himalayan sheets of Indian atlas.
(27.) Kumaon and Gurwhal, on 37 sheets, a selection mounted together.
(28.) Kattywar survey, a selection mounted together.
(29.) Guzerat survey, the three published sheets.
(30.) Colonel Walker's map of Turkistan, four sheets, mounted together.
(31.) Colonel Montgomerie's Trans-frontier maps.
(32.) Sholapore Collectorate, in a cover.
(33.) Laughton's Bombay, 172 sheets, a selection mounted together.
(34.) Ditto, reduced scale.
(35.) Native explorer's maps of—
Arun River.
Tengri Nor.
Western Nepal.
Rudok.
Sanpu River.
(36.) The Mesopotamian surveys.
(37.) St. John's Persia, six sheets. Beluchistan.
(38.) Prinsep. Atlas of Sealkote district.
(39.) Progress Report maps.
(40.) Greenough's geological map of India.
(41.) Geological atlas of India.

3153a. Selection of Maps and Publications to illustrate the progress of **Cartography and Surveys in Spain.**
Istituto Geografico y Estatistico de Espana.
(1.) Red Geodesica de Primer Orden en 1874.

The actual state of geodetic studies in Spain is represented on a scale of $\frac{1}{2000000}$. The thick black lines mark the triangles in the French and Portuguese network, which is joined to the Spanish; thinner black lines mark the direction of the Spanish system; the black lines of the plan point out the quadrilateral directions or capitals of provinces. The localities of which the latitudes are determined astronomically are marked with a blue semicircle; the localities in which the difference of longitudes is determined are marked with a red semicircle; the azimuths known are marked with a blue circumference. The lines of levelling are traced with red ink, those already carried out with a continued line, and those not yet carried out by dots. The localities in which bench marks have been established are marked with a red cross.

(2.) Publications relating to the International Commission on the Metre.

The determination of the metre and international kilogramme is one of the most interesting problems of measurement.

The International Commission assembled to determine this was formed of members of all nations.

In the sessions held at Paris in 1872 this Committee removed all the general doubts relating to the construction and comparison of the international prototypes of weights and measures, and appointed a permanent committee instructed to carry on the work. The Spanish commissioner, Brigadier Ibañer, was named president.

The permanent Commission have spared no efforts to overcome the numerous difficulties presented in the construction of the metre and kilogramme and in the comparison of the metre of the French Archives.

They have completed the construction of two provisional bars for the metre, and they are now proceeding to melt the irridiated platinum for the metres required.

In the third pamphlet a description is given of the latest experiments.

(3.) Numerical Results of the Levels of Precision taken between Alicante and Madrid, and between Madrid and Santander.

The results given are those which have been obtained in the two levelling lines, giving the heights found in all the points of reference and the errors which appear.

(4.) Memoir upon the general compensation of the errors contained in the geodetic network of Spain, by Don Toaquin Barraquer and Don Francisco Cabello, Madrid, 1874.

This memoir is a most important study of geodesy, applied to the network formed by geodetic lines of the first order, which, following the direction of the meridians, parallels, and coasts, comprehends the Spanish territory of the Peninsula.

(5.) Studies of Geodetic Levelling, by Don Carlos Ibañer, 1 vol., 4to., pl., Madrid, 1864.

The object of the author is to examine the accuracy of the different methods employed up to the present time in geodetic studies to determine the third co-ordinate of the zenith of triangulations, and most especially to apply the practice of geodetic levelling to a theory presented 25 years ago by Biot, which had never been tried in Spain or elsewhere.

(6.) Instructions for Geodetic Studies. 1 vol., 8vo., pl., Madrid, 1872.

This book contains instructions for the observation of angles and the calculations necessary in geodetic triangles, and general instructions for the persons appointed to carry out the work.

(7.) Comparison of the Geodetic Standard Bar belonging to the Government of the Viceroy of Egypt with the one which served to measure the central base of the Map of Spain, by Ismail Effendi and Don Carlos Ibañer.

M. Brunner constructed, by order of the Egyptian Government, a base-measuring apparatus, identical in character with the Spanish pattern; he tested it in Paris in order to ascertain its co-efficient of expansion. It was, however, indispensable to determine its length, and to do so the Spanish geodetic bar was chosen.

Ismail Effendi came to Madrid with the Cairo astronomical apparatus; Don Ibañer was appointed for this commission; they both agreed as to the manner of conducting the operation, and they carried out the observations and calculations until they obtained the equation of the Egyptian bar with a probable error of $\pm 0 \cdot 0011^{\text{mm}}$.

A full account of this is given in the book exhibited; the plate represents the comparator mounted at the Observatory of Madrid, to verify the comparison of the Egyptian and Spanish bars.

(8.) Plan of Madrid on a scale of $\frac{1}{2000}$.

This plan consists of several lithographed sheets, and includes all the details of the capital of Spain, the houses, number of streets, the trees, public and private gardens, and principal buildings.

The sheets are divided into squares, with their corresponding letter 'and number.

The levels are given by contours, traced at metre distances with the corresponding enumeration of the heights referred to the Mediterranean.

(9.) Memoirs of the Geographical and Statistical Institute of Madrid.

This important work comprehends the result of the geodetic studies between Salamanca and Madrid, terminated as far as the establishment of the interdepending equations between the calculations of every station and those that are required for the compensation of the whole polygonal system, the observations taken at every station, an account of these, the calculations of the most probable directions relating to every isolated station by the establishment of the equations already alluded to, &c.

The second memoir gives an account of the levelling of the line from Alicante to Madrid.

The third memoir describes the determination of latitudes and azimuths, and comprehends the astronomical geodetic studies carried out by the members of the Madrid Observatory for the Geographical Institute, in order to ascertain with proper exactitude the latitudes, longitudes, and azimuths, of a certain number of trigonometrical points of the first order. These studies were carried out by Don Manuel Morino.

The volume contains other papers of high topographical interest.

(10.) Topographical Map of Spain.

The map of Spain on a scale of $\frac{1}{50000}$ is composed of about 1,078 sheets, engraved in five colours, with numerous details. The sheets comprising Madrid, Colmenar, Viejo, and Yetafe are already engraved.

The general plan of the levellings projected for Spain (plate 20) comprehend, besides the line of Alicante to Santander, passing by Madrid, other lines which, starting from Madrid run one to the N.E. which is to connect itself in Perpignan with the French lines, another towards the south, which will serve to determine the difference of levels between the observatories of Madrid and St. Ferdinand, and two others which starting from Madrid will go to Lisbon and Oporto. The remaining lines will complete this plan, and bring out the levels of the capitals of provinces and towns of importance.

The system of lines is so disposed that it offers numerous polygons in order to establish conditions which may give greater precision to the results.

In conformity with the instructions of the Commission appointed to carry out this map, the levelling of precision will follow in general the lines of communication, such as railroads and bye-roads of every description. The levelling will be duplicate, and carried out from the centre by different observers, and with different instruments. The calculations will be taken from Madrid. The greatest error admissible, deduced from a double levelling, is not to exceed 4 millimeters per kilometer, and in general will be less than $5^{mm}\sqrt{R}$, R being the number of kilometers.

The general lines will be divided into independent sections, and these into portions of a kilometer, with benchmarks at the extremities of the section, important towns, &c.

In August 1871, the levelling of the line between Alicante and Madrid was begun; touching Madridejos at its base, 540 kilometers were levelled in 916 days, of which only 451 were useful. The number of stations verified was 11,782.

The line from Santander to Madrid was levelled in 1873 in the same manner, and this will determine the difference of level between the ocean and Mediterranean.

Without taking into account the datum points of the seaport towns, a standard datum has been established irrespective of mean sea level. This point is situated at the observatory of Madrid, although, considering its geological conditions, the hill upon which it is placed does not offer the securities which might be desired. Four others have been placed in different parts of Madrid.

(11.) Experiments made with the Base-measuring Apparatus belonging to the Commission of the Map of Spain. Madrid, 1859.

The authors, Don Carlos Ibañer and Don Frutos Saavedra Meneses, were appointed in 1853 to design the system of microscopes and bars with which the base of the geodetic network of the Peninsula was to be measured. They went to Paris to superintend the construction of an apparatus of their invention, intrusted to M. Brunner.

This apparatus principally consists of a platinum bar, which forms a metallic thermometer with another of latteen; they both rest on an iron bench placed upon movable supports resting upon wooden tripods. Various micrometer microscopes are placed in the centres of graduated circles fixed in other tripods, and these divide the base which is to be measured into small intervals, the lengths of which are determined by placing the bar between every two microscopes, and observing the divisions on the platinum and latteen. An eyepiece and two lenses are used to range the tripods of the microscopes in the line of the base, and another glass settles the beginning and end of the day's work.

This delicate apparatus cannot be sent out of the country ; it is preserved at the Geographical Institute of Madrid.

After this study was made, the expansion of the bars was investigated and compared with that of Borda, at the Paris Observatory, which has served as the standard of all the geodetic bases measured in France since 1798.

The result of the studies is given in this work, and the manner of using this apparatus and carrying out the calculations.

These plates represent—

(a.) Two of the projections of the bar and level mounted upon its supports and tripods, and the microscopes placed to observe the lines of division at the extremities.

(b.) Details of the construction of the supports of the bar, the slow and quick movements which it is capable of making, the construction of the latteen and platinum bars, the details of the same divided into decimilimeters, and a drawing of the level of the bar.

(c.) Details of the micrometer microscopes, circles, reference glass, lenses, and tracers.

(d.) The comparator. Experiments to verify the studies of expansion in oil bath.

(e.) Same in detail as the former.

(f.) Apparatus used by the observers in the construction and comparison of thermometers.

(g.) Borda's bar with which the Spanish one was compared.

(12.) Geodetical description of the Balearic Islands, by Don Carlos Ibañer. Madrid, 1871.

The author was commissioned in 1864 to study the first of the geodetic Catastralian districts, which are comprehended in the provinces of Castellon, Valencia, Alicante, and Balearic Islands ; he verified it in the campaigns of 1865, 1866, 1867, and 1868, and made the necessary studies to describe and interlace the three groups formed by the islands of Tviza and Formentera with surrounding islets, with those of Mallorca, Cabrera, and Dragonera, and with the island of Minorca and adjacent islets. Each of these groups with its corresponding triangulations is divided by first, second, and third class nets.

Each local triangulation was founded upon a base measured by the Ibañer apparatus, constructed by Messrs. Brunner.

In the first part of this work will be found a description of the instruments and accessories employed in this study, beginning with the new apparatus for measuring bases. This consists principally of a plated iron bar, with four thermometers of mercury and a plane table, placed upon tripods and supports, microscopes, and lineation glasses, and reference to the beginning and end of the daily work. This apparatus offers the greatest advantage over other analogous instruments, for the rapidity and ease with which the measures are taken, without losing the degree of exactitude required for a geodetic base. It is to be employed to measure the three bases still wanting in the Spanish geodetic net projected in the provinces of Cadiz and Lugo.

The probable errors resulting from the measurement of the Balearic Islands were—

Base of Ibiya $\pm 0\cdot401^{mm}$, or $\pm 0\cdot00000240$ of the length measured.

Base of Prat, Mallorca, $\pm 1\cdot681^{mm}$, or $\pm 0\cdot000000794$ of the measured length.

Base of Mahon (Minorca) $\pm 0\cdot770^{mm}$, or $\pm 0\cdot000000326$ of the length measured.

The Southampton Institute has made several metrological studies. The apparatus designed by Ibañer, which was connected with the bar constructed by Borda, produced the most satisfactory results.

The three other parts of this work refer to triangulations of the three groups of the above-mentioned islands. The plates represent—

(*a.*) The bar of the Ibañer apparatus in two projections, mounted for observation.

(*b.*) Details relating to the supports of the bar and its construction.

(*c.*) Microscopes; glasses and other details of this apparatus.

(*d.*) Form, dimensions, and construction of the station marks established at the trigonometrical points of the 1st, 2nd, and 3rd order.

(*e.*) The theodolites employed for the azimuth and zenith observations in 1st class triangles of the Balearic Islands.

(*f.*) Theodolite of 2nd class.

(*g.*) Geodetic local networks interlacing the groups of the Balearic Islands. Printed in three colours. The first class network is in black, the 2nd in red, the 3rd in blue.

(13.) Central base for Geodetic Triangulation in Spain, by Don Carlos Ibañer, Lieut.-Col., &c., Don Frutos Saavidra, Don Fernando Monet, and Don Cesario Quiroga ; Madrid, 1865.

The parting line chosen for the geodetic triangulation of Spain starts from the central base of Madridejos (province of Toledo). The direct measurement gave a result of 14,664^m, $\pm 0 \cdot 0025^m$, with a difference of level between the extremes (Bolos y Carbonera) of $2 \cdot 482 \pm 0 \cdot 010^m$.

These studies were carried out by the observers mentioned on the title-page of the book, by means of the apparatus of Ibañer.

The measurements adopted to fix the extremities of the base, and establish a trigonometrical network by which it might be determined whether it was necessary or not to measure bases of great extent, are all given in detail in this work. The observations and calculations for the measurement and levelling of the base, the triangulating studies and compensation of the network, and the final results of this operation are also mentioned.

The results of these operations, and their precision, may be judged by the probable error of measurement, which has already been mentioned, as $\pm 0 \cdot 0025^m$, or $\frac{1}{5800000}$ of the length of the base.

In order to decide the question of the extent of base line requisite, a comparison was made between the direct measurement of the sections of the base and their length deduced from the calculations of the trigonometrical network, which gave the following results :—

—	Measure.	Triangulation.	Difference.
	m	m	
First section - -	3077·459	3077·462	−0·003
Second „ - -	2216·397	2216·399	−0·002
Third „ - -	2766·604	2216·399	−0·002
Fourth „ - -	2723·425	2723·422	+0·003
Fifth „ - -	3879·000	3879·002	−0·002
Base - - -	14662·885	14662·889	−0·004

These remarkable results proving the direct measurement, at the same time give authority for reducing the length of the geodetic base to two or three kilometres, that is, whenever they are connected with the sides of the net by means of a system of lines suitable for compensation. This is the first experimental result of the kind given in geodesy.

PLATES.

(a.) Measurement of Madridejos wooden gallery formed of nine moveable houses in which the apparatus was sheltered from sun and wind.

(b.) Plan on scale of $\frac{1}{50000}$ of the country surrounding the base Madridejos with division in sections.

(c.) Vertical sections of the central base, on a scale of $\frac{1}{20000}$, for horizontal distances, and $\frac{1}{1000}$ for the heights.

ADMIRALTY—HYDROGRAPHIC DEPARTMENT.

ORIGINAL (MS.) JOURNALS AND LOG BOOKS, &C. KEPT BY CELE-
BRATED ENGLISH NAVIGATORS BETWEEN THE YEARS 1768
AND 1825.

3154. Dampier's original MS. Chart of his voyage to New Guinea, 1699–70.

3155. Cook, Captain James. Journal of the proceedings of H.M. bark the "Endeavour," in a voyage round the world; performed in the years 1768–69–70–71. Folio.

3156. Cook, Captain James. Journal of the proceedings of H.M. sloop "Resolution," in exploring the South Atlantic, Indian, and Pacific Oceans, in the years 1772–73–74–75. Folio, with charts.

3157. Bligh, Captain William. Log of the proceedings of H.M. armed vessel "Bounty," in a voyage to the South Seas, to take the bread-fruit from the Society Islands to the West Indies, in the years 1787–88–89. Folio. Including an account of the mutiny.

3158. Bligh, Captain William. Log of the proceedings of H.M.S. "Providence," on a second voyage to the South Seas, in the years 1791–92–93. Folio.) 2 vols.

3159. Franklin, Sir John. Log of the proceedings of H.M. hired brig " Trent," in a voyage towards the North Pole, in the year 1818. Folio.

3160. Parry, Sir Edward. Log of H.M.S. " Alexander," in a voyage for the discovery of a North-west passage, in the year 1818. Folio.

3161. Parry, Sir Edward. Official journal of H.M.S. " Hecla," in the years 1819–20. Folio. 3 vols.

3162. Parry, Sir Edward. Official journal of H.M.S. " Fury," in the years 1821–22–23. Folio. 2 vols.

3163. Parry, Sir Edward. Log of H.M.S. " Hecla," in the years 1824–25. Folio.

CHARTS AND PLANS.

3164. North Atlantic Ocean (Chart). Compiled in the Hydrographic Office of the Admiralty. Scale, 0·5 inches to a degree of longitude.

3165. British Islands (Chart). Compiled from the latest Admiralty Surveys. Scale, 1·4 inches to a degree of longitude.

3166. Lough Swilly, Ireland (Plan). Surveyed by Captain G. A. Bedford, R.N. Scale, 2·3 inches to a nautical mile.

3167. West Coast of Scotland (Chart). Compiled from the latest Admiralty Surveys. Scale, 5·3 inches to a degree of longitude.

3168. Davis Strait (Chart). Compiled from the latest Danish Government Charts ; and information derived from British and American Exploring Expeditions. Scale, 0·9 inches to a degree of longitude.

3169. South Atlantic Ocean (Chart). Compiled in the Hydrographic Office of the Admiralty. Scale, 0·5 inches to a degree of longitude.

3170. Gulf of Suez (Chart). Surveyed by Captain G. S. Nares, R.N. Scale, 0·23 inches to a nautical mile.

3171. Port Royal and Kingstown Harbours, Jamaica (Plan). Surveyed by Staff-Commander George Stanley, R.N. Scale, 2·5 inches to a nautical mile.

3172. Indian and Western Pacific Oceans (Chart). Compiled in the Hydrographic Office of the Admiralty. Scale, 0·2 inches to a degree of longitude.

3173. Pacific Ocean, between the parallels of 37° N. and 37° S. (Chart in four sheets). Compiled in the Hydrographic Office of the Admiralty. Scale, 0·62 inches to a degree of longitude.

3173a. Port Simpson, British Columbia (Plan). Surveyed by Staff-Commander Daniel Pender, R.N. Scale, 3 in. to a nautical mile.

3174. Candia or Crete (Chart). Surveyed by Captain Thomas Spratt, R.N. Scale, 0·5 inches to a nautical mile.

3175. Spithead (Plan). Surveyed by Captain Sheringham and Staff-Commander Daniel Hall, R.N. Scale, 2·75 inches to a nautical mile.

3176. North Sea, Dover to Scheveningen (Chart). Compiled from the most recent British and Foreign Surveys. Scale, 0·3 inches to a nautical mile.

3177. Dover Strait (Chart). Compiled from the latest British and French Surveys. Scale, 0·7 inches to a nautical mile.

3178. Zanzibar Harbour, east coast of Africa (Plan). Surveyed by Commander W. J. L. Wharton, R.N. Scale, 1·5 inches to a nautical mile.

3178a. Mikindani Bay, East Coast of Africa (Plan). Surveyed by Lieut.-Commanding F. J. Gray, R.N. Scale, 2 inches to a nautical mile.

3179. Plymouth Sound and Hamoaze (Plan). Surveyed by Commander H. L. Cox, R.N. Scale, 6·0 inches to a nautical mile.

3180. Curves of equal Magnetic Variation, 1871. Compiled by Staff Captain F. J. Evans and Navigating Lieutenant E. W. Creak, R.N.

3181. Current Chart of the Pacific, Atlantic, and Indian Oceans. Compiled by Staff Captain F. J. Evans and Staff Commander Thomas A. Hull, R.N.

3182. Ice Chart of the Southern Hemisphere. Compiled by Staff Captain F. J. Evans and Staff Commander G. F. McDougall, R.N.

3183. Antarctic Charts lent by the Hydrographic Office.

S 91. New South Britain (South Shetland Islands), Captain Smith, 1819 (original drawing).

S 92. South Shetland Islands. Mr. Bransfield, 1820.

543. Kerguelen Island. Captain Cook, 1776–9 (original drawing).

445. Do. Captain Rhodes, 1799 (original drawing).

A 4323. Do. Compiled in Hydrographic Office, 1874.

A 3711. Heard and McDonald Islands. Captain Nares, 1874.

A 508. Powell's group (South Orkneys). Captain Powell, 1821-2.

L 1752. Louis Philippe Land. Captain Dumont D'Urville, 1838.
L 2115. Adelie Land and Côté Clarie.   Do.   1840.
L 3125. Antarctic Continent. Lieutenant Wilkes, United States
          Navy, 1840. Auckland and Campbell Islands.
L 4589. Victoria Land. Captain James Clark Ross, 1841.

ATLASES.

3184. Atlantic Pilot Charts.

These charts show in a graphic form the prevailing winds and other phenomena for periods nearly corresponding to the seasons of the year. A general chart of the ocean currents is also given.

3185. Wind and Current Charts for the **Atlantic, Pacific,** and **Indian Oceans.**

To these are added smaller synoptical charts showing lines of equal air temperature, barometric pressure, and sea-surface temperature.

3186. Officers' Atlases, containing a selection from the Charts supplied for the navigation of Her Majesty's Ships.

3187. Index to the Charts and Plans published by the Hydrographic Office of the Admiralty.

3187a. Globe illustrative of the **Earth's Magnetism.**

3187b. Maps illustrative of the **Earth's Magnetism.**

ROYAL GEOGRAPHICAL SOCIETY OF LONDON.

3188. Collection of Arctic Charts.

3188a. Pictorial Diagrams of the Arctic Regions.

3188b. Pictorial Diagrams of the Aurora Borealis.

3189. Arctic Maps.

1. Facsimile of the ancient chart of the Zeni. (R.G.S.)*
2. The map of Hondius, from Pontanus. (H.S.)
3. Petermann's map, showing the track of Barents. (R.G.S.)
4. Gerrit de Veer's map of the voyage of Barents. (R.G.S.)
5. Map of Spitzbergen from Purchas. (G.M.)
6. Hudson's map of 1612. (H.S.)
7. Van Kéulen's chart of Spitzbergen. (G.M.)
8. Dutch chart of Davis Strait after 1721 : translated into English. (A.)
9. Phipps's MS. chart of North Spitzbergen. (A.)
10. Circumpolar chart of 1818. (R.G.S.)
11. Chart of Baffin's Bay, by John Ross, 1818. (A.)
12. Buchan's voyage to Spitzbergen, with tracks. MS. by Beechy. (A.)

13. Chart of Parry's discoveries. First voyage. MS. by Liddon. (A.)
14. Parry's discoveries. First voyage. MS. by Bushnan. (A.)
15. Parry's discoveries. Second voyage. MS. by Bushnan. (A.)
16. North of Spitzbergen and Parry's track in 1827. MS. by Foster. (A.)
17. Graah's chart of Julianshaab, Greenland. (R.G.S.) Danish map of Greenland with Eskimo names. (R.G.S.) Moller's chart of Baal's river and the Godthaab district. (R.G.S.)
18. Circumpolar chart of 1835. (A.)
19. Circumpolar chart of 1838. (A.)
20. Simpson's chart from Point Barrow to Return Reef. (R.G.S.)
21. Facsimile of the chart supplied to Sir John Franklin. (R.G.S.)
22. Circumpolar chart of 1850. (R.G.S.)
23. Parry Islands and Arctic America, 1850. (A.)
24. Track of H.M.S. *Assistance* to the Cary Islands, 1851. (A.)
25. Sutherland's circumpolar physical map, 1851. (R.G.S.)
26. Charts by Arrowsmith of the coast of Arctic America, showing Dr. Rae's journeys. (S.)
27. Herald Island, MS. and views. (A.G.S.)
28. Chart of Baffin's Bay, with Capt. Inglefield's corrections. (A.)
29. Discoveries of M'Clure (4 sheets MS.) (A.)
30. Arrowsmith's chart showing the discoveries of M'Clintock in 1859. (R.G.S.)
31. M'Clintock's discoveries. MS. by Allen Young. (A.)
32. Circumpolar chart of 1859. (R.G.S.)
33. Admiralty chart of Kane's discoveries up Smith's Sound. (R.G.S.)
34. Kane's original MS. chart of his discoveries. (R.G.S.)

35. Petermann's map of Wrangell Land, 1869. (R.G.S.)
36. Captain Long's chart and sketch of Wrangell Land, 1867. (R.G.S.)
37. Swedish chart of Spitzbergen.
38. Petermann's map of Nordenskiold's voyage to Spitzbergen in 1868. (R.G.S.)
39. Petermann's map of the north end of Novaya Zemlya, 1872. (R.G.S.) Petermann's maps of Novaya Zemlya, Waigat Isles, Matotchin Shar, &c. (R.G.S.)
40. Track of Captain Koldewey's voyage in 1868, by Petermann. (R.G.S.)
41. Discoveries of the German expedition on the east coast of Greenland. (R.G.S.)
42. Franz Joseph Land. Discoveries of the Austrian expedition. (R.G.S.)
43. Map showing the "Hypothesis Petermann," & making Smith's Sound a *cul de sac*. (R.G.S.)
44. Polar discoveries of Hall in the *Polaris*, American chart. (R.G.S.)
45. Map showing the drift of the boat of the *Polaris* down Baffin's Bay, by Petermann. (R.G.S.)
46. Track of the *Arctic* in 1873, by Commander A. H. Markham, R.N. (G.M.)
47. Circumpolar chart by Stanford, 1875. (S.)
48. Track of the *Alert* across the Atlantic. (A.)
49. Disco Island. (G.M.)
50. Track of the *Alert* from Upernivik to the Cary Isles. (G.M.)
51. New chart of Baffin's Bay. (A.)
52. New chart of Smith'sSound. (A.)
53. New half circumpolar chart. (A.)
54. Chart showing winter quarters of all expeditions. (A.)
55. Projection for the use of the Arctic Expedition, 1875-6. (A.)

*REFERENCES :—(R.G.S.) From the collection of the Royal Geographical Society. (A.) From the Admiralty. (H.S.) From the Hakluyt Society's volumes. (G.M.) From the *Geographical Magazine*. (S.) From Mr. Stanford.

3189a. Four Views, Antarctic Regions.

3190. Manuscript Maps of Celebrated Explorers.

Year.	Title.	Name of Explorer.
1850	River Zambesi - -	Dr. Livingstone.
1853	River Leeba - - -	Dr. Livingstone.
1854	Victoria Falls to Loand -	Dr. Livingstone.
1854	River Coanza - - -	Dr. Livingstone.
1854	River Quango - -	Dr. Livingstone.
1855	Zanzibar to Uniamesi Sea -	Erhardt and Redmann.
1855	Loand to Victoria Falls -	Dr. Livingstone.
1855	Country South of Zambesi -	Dr. Livingstone.
1856	Victoria Falls to Tétè -	Dr. Livingstone.
1857	River Pangani - -	Burton and Speke.
1858	Khartum to Gondokoro -	J. Petherick.
1858	Zanzibar to Ujiji - -	Burton and Speke.
1858	Kasé to Lake Victoria -	Speke.
1859	Lake Shirwa - - -	Dr. Livingstone.
1860	Sources of the Nile - -	G. G. Miani.
1860	Zanzibar to Gondokoro -	Speke and Grant.
1861	Zanzibar to Gondokoro -	Speke and Grant.
1861	Lake Nyassa - - -	Dr. Livingstone.
1862	Khartum to Albert Nyanza -	Sir S. Baker.
1862	Lake Albert Nyanza - -	Sir S. Baker.
1862	Zanzibar to Gondokoro -	Speke.
1863	Sources of Nile - -	J. Petherick.
1868	Eastern Turkistan - -	G. H. Haywood.
1875	Lake Victoria - -	H. M. Stanley.

3190a. Diagram of Africa, showing Comr. V. L. Cameron's route across that Continent, also the general river basins and drainage.

3190b. Sectional Route, by Comr. V. L. Cameron, R.N., C.B.

3191. Specimens of Printed Maps in the Journal of the Society.

Glaciers of the Mustakh Range, India, by Capt. H. G. Austen.

Canterbury and Otago, New Zealand, Messrs. McKerron, Haast, and Hector.

Vancouver Island, Dr. C. Forbes, R.N.

Small portrait (1 foot square) of Dr. Livingstone, considered the best likeness.

3192. Model of the Victoria Falls, River Zambesi. By T. Baines, Esq., F.R.G.S.

3193. Views. Descriptive Geography.

12 Water-coloured views of glaciers, Middle Island, New Zealand.

10 views of the Victoria Falls, Zambesi River, by T. Baines, Esq.

3 views in Eastern Turkistan, near Kashgar, &c., by G. H. Haywood.

Sketches of the natives, Island of Formosa, with map. Exhibited by Admiral Sir R. Collinson, K.C.B.

3194. Photographs taken in China. Four volumes. By J. Thomson (Paris Medallist).

3195. 21 Views on the River Zambesi (oil paintings), by the late T. Baines, F.R.G.S.

3196. Collection of Publications of the Society.

A complete set of the Journals, 45 vols.
Arctic papers for the Expedition of 1875.
The lands of Cazembe, Lacerda's journey, &c.
Hints for Travellers.
A complete set of the "Proceedings," 19 vols.

3197. Geographical Diagrams (Wall Maps).

1. Smith's Sound.
2. Central America.
3. Vancouver's Island and British Columbia.
4. Abyssinia.
5. Mount Sinai.
6. Korea and Eastern China.
7. Pigafetta's Congo (Ancient Africa).
8. Caucasus.
9. Zeno's Map.
10. Spitzbergen.
11. South-west Pacific, Fiji Islands, &c.
12. Burma.

3198. Pictorial Diagrams (Geographical).

1. View of the volcano of Santorin, large.
2.   ,,   small.
3. Crater of Kilauea, Sandwich Islands.
4. Murchison Falls (River Nile).
5. View in Fiji Islands.
6.   ,,   ,,
7. Ruined Esquimaux hut.
8. Aurora Borealis (five diagrams).
9. Section of an ice-blink or glacier.

3199. Ancient Maps of the Sources of the Nile.

1.	1508	-	-	- By Ptolemy.
2.	1550	-	-	- Sebn. Munster.
3.	15th century	-	-	?
4.	1618	-	-	?

3199a. Map, showing the principal **Caravan Routes** of Eastern commerce, ancient and modern. By Dr. Yeats, F.R.G.S.

3200. Miscellaneous Contributions by the **Geographical Society.**

Three cases of select specimens of Cartography (105 in No.)
Educational models, various.
Engraved Portraits of celebrated explorers.
Barent's relics, 1597 ; water-coloured drawing.
Pizigani's Map of the World, 1367.

MAPS, GLOBES, AND MISCELLANEOUS OBJECTS
BY VARIOUS CONTRIBUTORS.

3201. Models to illustrate the arts of Camp Life.

Francis Galton, F.R.S.

This case of models and specimens was made at the same time as 10 others that were constructed in 1858, by order of the War.Office, after the design of a set which Mr. Galton caused to be made, and had presented to the Royal Institution at Woolwich in the previous year. Their object was not only to interest and instruct individual soldiers, but rather to suggest to instructors the precise subjects on which practical classes in the arts of camp life may most usefully be engaged. A catalogue accompanies the case in which the models are contained, and in this an asterisk is placed opposite to those objects which Mr. Galton's experience leads him to prefer for the purpose. "An old cam-
" paigner's acquirements consist partly in knowledge and partly in handiness.
" Field lectures, illustrated by experiments, may convey the first to an
" intelligent novice, and it was hoped that these models might serve to
" explain what kind of things must be made by his hands, before he can
" acquire the latter."
The examples illustrate the various modes of the production of fire, the procuring, purifying, and carriage of water, cooking, the uses of portable food, substitutes for boats, cattle enclosures, expedients for tools and appliances in various handicrafts, tenting, hutting, and various other needs of camp life.

3202. Stereoscopic Maps, taken photographically, from models in relief. *Francis Galton, F.R.S.*

A much clearer notion of the physical features of a country may be obtained from a model in relief than from an ordinary map, however carefully executed, but the great weight and cumbrousness of models makes them unsuitable for the library or for travel. The chief advantages of both methods of illustration may in great measure be secured by stereoscopic maps taken photographically from models. The accompanying specimens were exhibited in illustration of a memoir read before the Royal Geographical Society in 1865, a copy of which is placed beside the photographs. It is there shown that the proposed plan has two other unexpected advantages. First, owing to causes there explained, we are able to deal with models of considerable dimensions both laterally and longitudinally, for when such a model has been photographed stereoscopically in separate squares, and the prints have been properly united, it becomes possible to view any part of the map with pseudo-stereoscopic if not with stereoscopic effect. Secondly, it is shown, that the insertion of names improves the appearance of relief in models, and consequently in the stereoscopic representations of them, while it spoils the effect of shading in ordinary maps.

3176a. Maps of the Baltic, Western Part, compiled from the surveys of the Imperial Admiralty, and published by the Hydrographic Office of the Admiralty.

Hydrographical Department of the Imperial Admiralty, Berlin.

This is a sheet from a series of charts of the littorals of the Baltic, which it is proposed to publish in the course of the next few years, and may thus serve as an example of the scale on which the work will be done.

3202c. Perspective Map of Africa, according to French, English, and German travellers.

M. Launay, Professor at the Lyceum of Caen.

This map was executed previously to the publication of the accounts of Mr. Stanley, respecting Lake Ukériné, of Comr. Cameron respecting Lake Tanganiyka, and of M. Grandidier respecting Madagascar.

3203. Map of Gaul, showing by different coloured tracings the relative antiquity and importance of the Roman roads.

M. Hayaux du Tilly, Paris.

3203a. Old Spanish Map of the Province of Aragon, &c., with the roads, bridges, hill-shading, &c., inserted in MS. by Captain H. Bristow, by means of the pocket-sextant, No. 3107j.

H. W. Bristow, F.R.S.

During the Peninsular War, great difficulty was experienced in getting accurate maps of Spain for the British army.

The most accurate Spanish maps then procurable were old, very defective, and nearly useless as route maps, owing to the roads, hill-shading, &c. not being shown upon them. These details had to be supplied for the use of the army by the officers attached to the Quartermaster-General's department, the engraved maps serving as the groundwork for the new survey.

The old Spanish map of the province of Aragon and surrounding districts is one of those maps upon which the roads, bridges, the defences of Saragossa, and other details, were so supplied in MS. by Captain Bristow, by means of the pocket-sextant, No. 3107j.

3203b. Collection of Water-colour Sketches of the glaciers of the **Mustak Range,** Little Thibet, &c.

Major Goodwin Austen.

3203c. Map of the Great Glaciers on the North of the Swiss Alps; with index card.

Prof. Alphonse Favre, Professor of Geology at the Academy of Geneva.

3203d. General Map of Ireland, on roller.

General Valuation of Ireland, Dublin.

3203e. Model of Victoria: 70″ × 50″.

The Agent General of Victoria.

3203f. Map of Victoria: 38″ × 26″.

Sketch of new geological map of Victoria : 38″ × 26″.

The Agent General of Victoria.

3204a. Specimen Sheets of the two most important Productions of the Topographical Department at St. Petersburg. Copper plate.

The Topographical Department of the Imperial Russian General Staff, St. Petersburg.

1. Topographical map of the western part of European Russia in 507 sheets. Scale, 1:126,000.

2. Special map of European Russia in 145 sheets. Scale, 1:420,000.

 a. Copies from copper plate in two colours.

 b. Transfer print from lithographic-stone, and chromo-lithography in four colours, black for the skeleton and the writing; brown for the ground of the country; blue for the watercourses, and green for forests and woods.

3204b. Map of Trans-Caspia, in five sheets. Scale 1:840,000. Lithographed in Tiflis.

The Topographical Department of the Imperial Russian General Staff, Tiflis.

The dry bed of the Amu-Darja has been indicated on the map according to the latest survey.

3205. Map of the Hop-growing Districts of Central Europe, by Joh. Carl, ̈editor of the General Hop Gazette, and C. Homann, containing :—

1. Agrarian statistical general map of the European hop-growing districts on the Continent and in England.
2. Special map of Bavaria.
3.   „   Bohemia.
4.   „   Wurtemberg and Baden.
5.   „   Belgium.
6. Tabular and graphical representations of the cultivation of hops, and of the hop consumption of the whole world.
7. Classification of the various sorts of hops.
8. Comparative tables of the agrarian measures and commercial weights.   *J. Carl and C. Homann, Nüremberg.*

The maps represent the cultivation of hops in a manner not to be found elsewhere. The agrarian statistical general map shows the hop cultivation of the Continent and England, and contains special maps of the principal hop-districts of England, Bavaria, Bohemia, Württemburg, Baden, Belgium, &c. The tabular and graphical representations contain very important statistics as to the cultivation, production, and consumption of the whole world, which may be summed up as follows:—

—	Surface under Cultivation.	Production.	Consumption.	Production+ or−Consumption.
	hectares.	cwts.	cwts.	cwts.
Germany - - -	37,910	477,111	321,500	+155,611
England - - -	25,696	384,090	600,000	−215,910
Austria - - -	7,711	92,532	100,000	− 7,468
Belgium - - -	6,500	97,500	15,000	+ 82,500
France - - -	4,000	48,000	48,000	—
The rest of Europe - -	619	8,454	25,500	− 16,546
Total Europe - -	82,346	1,107,687	1,109,500	− 1,813

The produce and surface under cultivation of the hop districts of the different countries are also shown in the same way. The average production of one hectare in an average crop is as follows:—

In Prussia per hectare 12 cwts., in all   59,400 cwts.
   „ Bavaria   „   12   „   „   212,556   „
   „ Württemburg „   15   „   „   73,595   „
   „ Baden   „   15   „   „   26,310   „
   „ Alsace and Lorraine 12   „   „   9,000   „
   „ the rest of Germany 12–15   „   15,150   „

Total produce of an average crop in Germany   396,011   „

The most noted hop districts are all indicated by the colouring; the Principal localities of production are also indicated.

3206. Wetzel's Wall Map of Mathematical Geography.
Dietrich Reimer, Berlin (Reimer and Hoefer).

3207. Kiepert's Physical Wall Maps, &c., viz.:—
Wall map of the Eastern Planiglobe.
Wall map of the Western Planiglobe.
Wall map of Palestine. Scale, 1 : 200,000.
Wall map of Palestine. Scale, 1 : 300,000.
Map of Europe.
Map of Asia.
Map of Africa.
Map of North America.
Map of South America.
Map of Australia.
Small hand atlas.
Small school atlas.
Adami-Kiehert's school atlas. .
Dietrich Reimer, Berlin (Reimer and Hoefer).

3207a. Physical Map of the Colony of Natal, showing the central highlands, the northern one-river basin, and the southern many-river system of watersheds. *Dr. Mann.*

The central highlands are shown to be a spur from the inland mountain frontier, or Drakenberg range, with finger-like subdivisions descending to the coast to constitute the many-river system which is developed there.

3208. Möhl's Orohydrographical Wall Map of Germany. *Th. Fischer, Cassel.*

3209. Map of Hesse-Nassau. *Th. Fischer, Cassel.*

3210. Map of South-western Germany.
Th. Fischer, Cassel.

These maps, Nos. 3208, 3209, 3210, are lithographed in the exhibitor's offices after a method of representing the earth's surface, first published by Professor Möhl. The printing is effected by scraped and chalk stones, besides the ordinary colour stones.

3211. Carta geografica de la Républica de Costa Rica (Central America). Scale, 1 : 500,000. Constructed by L. Friederichsen, 1876. *L. Friederichsen and Co., Hamburg.*

This map has been drawn and produced, on the foundation of important original materials, by L. Friederichsen, under the authority of the Government of Costa Rica. It shows for the first time the geography of the Republic of Costa Rica on a large scale.

3213a. Map of the Central Part of the Chain of the Caucasus ; scale, $\frac{1}{565000}$. *M. Ernest Favre, Geneva.*

Geological map of the central part of the Caucasus, executed by M. Ernest Favre from his own observations and extracted from " Recherches géologiques " dans la partie centrale de la chaine du Caucase " (H. Georg. Geneva and Bâle, 1875) by the same author.

3212. Portfolio, containing **Maps** in **Copper-plate Print.**

3213. Portfolio, containing **Maps** in **Lithographic Print.**
Bau Deputation, Hamburg.

3213b. Portfolio, containing reductions of the survey sheets to scale of 1 : 4000, and trigonometrical chart, scale, 1 : 50,000 lithographed. *Bau Deputation, Hamburg.*

3213c. Portfolio, containing reductions of the survey sheets in scales of 1 : 1000 ; 1 : 4000, and 1 : 20,000 on copper plates.
Bau Deputation, Hamburg.

3215. Portfolio, containing **Maps** and **Drawings.**
Bau Deputation, Hamburg.

3216. Plan, 1 : 4000. Sect. IV. galvanoplastic relief plate.
Bau Deputation, Hamburg.

3128. Plan, 1 : 20,000 ; in frame under glass.
Bau Deputation, Hamburg.

3128a. Hamburg and its Environs. Scale, 1 : 20,000, engraved by H. Petters. *Bau Deputation, Hamburg.*

3129. Plan, 1 : 4000 ; in frame under glass.
Bau Deputation, Hamburg.

3129a. Plan of Hamburg and its Environs. Official edition. According to the land survey, &c., engraved by S. Siebert. Scale, 1 : 4000. *Bau Deputation, Hamburg.*

3216a. Galvanoplastic Relief Plate of Section IV. of the before-mentioned copper plate print. *Bau Deputation, Hamburg.*

3217. Relief Maps of Bergedorf.
Bau Deputation, Hamburg.

3217a. Two Relief Maps, one of the town of Bergedorf and its environs, produced from the lithographed sheets. Longitudinal scale=scale of height=1 : 4000, the other parts of the area of Bergedorf, produced from the transfer-print sheets drawn for that purpose on cartoon paper. Longitudinal scale=scale of height= 1 : 1000. *Bau Deputation, Hamburg.*

As a basis for the survey of Hamburg, the geodetic labours of Councillor Schumacher, Astronomer, are made use of, and in particular the base measured by him in the neighbourhood of Hamburg, in Holstein, in the years 1820 and 1821. At the suggestion of Engineer Lindley, the technical adviser of the public buildings deputation until 1860, a new survey and the construction of a new map on the scale of 1:250 was resolved upon in the year 1845, chiefly for technical purposes, and the necessary triangulation for that object was executed by Dr. A. C. Petersen, under the direction of Schumacher.

In the year 1863, the survey of the entire territory was commenced in connexion with that of the city for the Doomsday Book (land registry), and concluded in the year 1868. The additional triangulation work required for that purpose, which was commenced simultaneously with the divisional survey, was connected with the already determined trigonometrical points of the town, and then with the Holstein base. The number of trigonometrical points amounts to 1663, and the station points among the same, 993 in number, have been marked, with a very few exceptions, by granite pillars sunk into the ground. The angles were repeatedly measured by means of an 8-inch theodolite; the trigonometrical network corrected and adjusted according to the method of least squares, which was followed by the calculation of the horizontal rectangular co-ordinates, from which the geographical position was deduced by a very simple method. As origin of the system of co-ordinates, the spire of the large St. Michael's Church had already been fixed upon, and its geographical position determined as follows:—

Latitude = 53°, 32′, 55·7″

Longitude E from Greenwich = 9°, 58′, 41·75″.

The bailiwick of Bergedorf, formerly belonging to the joint jurisdiction of Hamburg and Lübeck, having been incorporated with the territory of Hamburg, the survey was extended to this area; and for this district a new division of survey sheets was adopted since the introduction of the metre standard.

The scales employed in drawing the maps are:—1:250, and ·1:1000; for the new division of the sheets 1:200, 1:500, and 1:1000. Exact levels are given all over the entire territory, and bench marks are durably established in sufficient number. Connected with these, special levelling will be carried out by means of distance and altitude measurement.

In the adjoining parts of the territory not belonging to Hamburg, so far as it is to be included in the maps to be published, levels will be taken by means of aneroid barometers, on the Reitz system.

The reduction of the survey sheets is effected by means of a specially constructed pantograph.

The multiplied copies will be published in scales of 1:1000, 1:4000, 1:20,000, and 1:50,000 (**3213b** and **c**), by O. Meissner and Behre, Hamburg.

According to the directions given in the plates for the distance and the altitude measurements, the latter will be executed by the tachymeter, **3079a**. An ordinary levelling staff, divided into centimetres, will be employed for the purpose. A survey field book illustrates the work actually executed for a portion of the bailiwick of Bergedorf.

The co-ordinates of the station points required for levels will be calculated from those of known trigonometrical points, or from the polygon points of the detail survey, and entered in the survey sheets. In the principal polygons the distances between the station points will be measured with chains, or steel tapes, and the altitudes determined by levelling. A synoptical table of results obtained by distance and altitude measurements, with the directly executed longitudinal measurements and with the actual levellings, is given. It should be mentioned that this work was executed in November 1874, in unfavourable weather, and that more satisfactory results are obtainable by this method.

When the station points have been plotted on the survey-sheets, and the levels of the altitude points have been calculated from the field book, these latter are plotted on tracing paper, containing the sheet divisions, the station points, and more or less the situation of the respective survey-sheets. For plotting the altitude points a transferrer is used, made of paper; it is ad-

justed with its centre point under the corresponding station point, according
to the horizontal angular measurements, and then all the altitude points
based on this station are sketched on the tracing paper, their levels being
noted. When all the altitude points have been marked down, the position
of the intervening contour lines, on the assumption that the slope of the
the ground is uniform between two altitude points, can be determined
by the construction of similar triangles; this operation can, however, be
avoided by the use of the instrument (**3127a**) devised for this purpose. If,
for example, the altitude heights marked 30.5^m and 32.2^m have been ascertained, the difference of which amounts to 17 decimeter, the two contours for
31 and 32 metres will fall between these two points. The edge of the mirror of
the variable scale is applied to the two points ; the cylinder is turned until
17 parts of the division projected on the paper by the mirror fall between the
points ; the cylinder is then moved in the direction of its axis until the corresponding divisional marks 30.5^m and 32.2^m have been reached, and the points
of section of this line with the contours for 31 and 32 metres are marked off.
A better general view of the surface of the country divisions is afforded by
the models produced from these plans.

3229a. Portfolio, containing various publications; viz.—

(*a.*) " Vermessung der Stadt Hamburg, Verzeichniss der trigo-
" nometrisch bestimmuten Punkte mit Netzkarte, Ham-
" burg, 1848." (Survey of the city of Hamburg, table of
the trigonometrically determined points with reticulated
map of Hamburg, 1848), containing the rectangular
co-ordinates of the points in the town and the nearest
vicinity, with lithographed sketches, and division of
sheets.

(*b.*) " Vermessung der Stadt Hamburg, Verzeichniss der trigo-
" nometrischen Punkte nebst Dreieckskarte, Hamburg,
1872." (Survey of the city of Hamburg, table of the
trigonometrical points with trigonometrical chart. The
latter in a scale of 1 : 50,000.)

(*c.*) " Nivellements und Höhenbestimmungen von Hamburg
" und Ungebung, ausgeführt abseiten des Vermessungs-
" bureaus in den Jahren, 1869–1872. Mit einer Karte,
" Hamburg, 1872." (Levellings and determination of
the heights of Hamburg and its environs, executed by
the Survey Department in the years 1869–1872. With
a map, 1872.)

(*d.*) " Verzeichniss von Höhenpunkten in Hamburg und Umge-
" bung, bestimmt durch geometrische Nivellements
" abseiten des Vermessungs-Bureaus in den Jahren 1869
" bis 1872, Hamburg, 1872." (Table of the heights of
Hamburg and its environs, as determined by geometrical
levellings, executed by the Survey Department in the
years 1869 to 1872, Hamburg, 1872.)

(*e.*) " Distanz- und Höhenmessung, Formeln und Tabellen
" behufs Aufnahme und Höhenbestimmung, u.s.w. Ham-
" burg, 1873." H. Stück. (Distance and altitude sur-

vey. Formulæ and tables for survey and determination of the heights, &c., Hamburg, 1873.) H. Stück.

(*f.*) " Ueber die Ausführung von Höhen-Messungen mit dem " Aneroid-Barometer. System Reitz, aus der Fabrik von " R. Deutschbein in Hamburg, u.s.w. Hamburg, 1874." (Levelling with the aneroid barometer. System of Reitz, from the manufactory of R. Deutschbein in Hamburg, &c., Hamburg, 1874.)

(*g.*) " Vermessung von Hamburg, u.s.w. 1873 ; " Kurze Darstellung der im Hamburgischen Gebiete ausgeführten geodätischen Arbeiten. (Survey of Hamburg, &c., 1873 ; short description of the geodetic works executed within the territory of Hamburg).

Bau Deputation, Hamburg.

3229. Tables for the Measurement of Heights and Distances. *Bau Deputation, Hamburg.*

3230. Field-Book of executed height and distance measurements. *Bau Deputation, Hamburg.*

3230a. Portfolio containing representations of the hypsometrical works, namely :—

(*a.*) Tables for distance and levels.

(*b.*) Tables for the staffs used for distance measurement and levelling.

(*c.*) Field-book, with distances and levels taken in a part of the area of Bergedorf.

(*d.*) Transferrer for plotting heights.

(*e.*) Table containing the stations and the polygon lines for the distances and levels executed in a part of the area of Bergedorf, with a synoptical summary of the results obtained by this method with reference to direct measurement of length, and with regard to actual levelling.

(*f.*) Sheet with annotations of the altitude points determined by distance and height measurement, and the one metre contours constructed from them ; scale 1 : 4000 of part of the adjoining Holstein territory.

(*g.*) Sheet of one metre contours, with the stations and polygon lines ; scale 1 : 1000, of part of the area of Bergedorf.

(*h.*) Plan of the town of Bergedorf, with one metre contours ; scale 1 : 4000, lithographed.

(*i.*) Plan of part of the area of Bergedorf, with one metre contours ; scale 1 : 1000, lithographed print on cartoon paper. *Bau Deputation, Hamburg.*

3218. Map of the Earth, projected on a regular **Dodecahedron.** *George H. Darwin.*

3219. Map of the Earth, projected on a regular **Icosahedron.** *George H. Darwin.*

In these maps, the representation of the earth is effected by supposing the globe to be inscribed to a regular dodecahedon or icosahedron, and then projected on the faces of the solid by lines radiating from its centre. The faces of the solid are then opened out and laid flat, in such a manner as to represent the equator by a straight line. In the case of the icosahedron the polar axis is supposed to pass through two corners of the solid. In the case of the dodecahedron, the polar axis is supposed to pass through the middle points of two opposite faces, and these two faces are cut up into triangles.

3219a. Great Spheroidal and Universal Atlas.
M. Hanicke de St. Senoch, Paris.

This work is divided into seven parts the first two of which are devoted to general geography.

1st part. Maps developed in the shape of a Planisphere, and showing from different points of view the comparative status of geographical knowledge. (9 maps.)

2nd part. Aspects of the terrestrial globe, seen in projection upon plans perpendicular to divers great circles corresponding to the position of the chief countries. (8 maps.)

The five other parts are devoted to detailed maps of the five parts of the world.

3219b. Map of France. *M. Hanicke de St. Senoch, Paris.*

3220. Maps in Relief, embossed in paper, showing rivers, towns, railways, &c.; also, in relief, mountains, valleys, and other physical features of the earth's surface.

Larger series, 23 in. by 21 in.

ENGLAND AND WALES.	EUROPE.
SCOTLAND.	AFRICA.
INDIA.	PALESTINE.

Smaller series, 10 in. by 8 in.

ENGLAND AND WALES.	EUROPE.
SCOTLAND.	PALESTINE.

Henry F. Brion.

3221. Four Tableaux of Original Models for composing a new relief atlas by means of stamp-printing.
Ed. Uhlenhuth, Anclam.

3223. Section of the Brocken Mountain-group of the large relief map of the Hartz Mountain, composed in 1 : 25,000, according to the altitude curves of the plane-table sheets of the Royal Prussian General Staff, showing the layer formation in papier-mâché and wax. *Ed. Uhlenhuth, Anclam.*

3224. Brocken Group, executed in gypsum.
Ed. Uhlenhuth, Anclam.

3225. Terrestrial Globe of 80 cm. diameter.
Dietrich Reimer, Berlin.

3226. Terrestrial Globe of 34 cm. diameter.

Dietrich Reimer, Berlin.

3227. Terrestrial Globe on an iron stand with semi-meridian.

Dietrich Reimer, Berlin.

3228. Terrestrial Globe. *Dietrich Reimer, Berlin.*

3231. Terrestrial Globe, 48 cm. diameter.

Ernst Schotte, Berlin.

3232. Four Maps. *M. Delesse, Paris.*

1. *Agricultural Map of France.*—On this map the various cultures or crops are shown by conventional colours, the shades of which are proportionately darker according as the returns or yields are more important.

If we consider the arable lands, which occupy the largest portion of France, they will of course vary yearly in their produce, but it is possible to form a money estimate of their average annual return per hectare.

By studying the figures for the cantons, taking into account their shape, as well as their elevation, and the mineral composition of the soil, curves have been traced showing the annual revenues of 20, 40, 60, 80, 100, and 120 francs.

For the woods, meadows, and vineyards, where the culture is permanent, there have also been traced curves of equal revenues, showing the average returns per hectare.

Giving to the cultivated lands, the woods, fields, and vineyards, their conventional colours, the depth of shade of these has been graduated according to the various curves to which they correspond.

Although the map is on a scale of 1:4,000,000, it nevertheless enables an estimate to be formed of how the agricultural riches of France is subdivided.

2. *Hydrological Map of the Seine and Marne.*—The hydrological map of the departments of the Seine and Marne is on a scale of 1:100,000, and shows the underground stores of water of the region of La Brie. These water supplies are represented by certain conventional colours, and their form has been determined by the geological investigation of the subsoil, from surveys and borings made over a network of wells. In fact, the surface represents the levels which have been ascertained for each well below the level of the sea, and these are shown by horizontal curves of 20 metres distance, so that it is very easy to trace the course of the flow of the water supplies.

In the region of La Brie, the principal subterranean water beds correspond with the most important argillaceous strata, that is to say, the green clays, the plastic clays, and in certain points, the building limestones of La Beauce.

Moreover the subterranean water-bearing beds known as those of infiltration correspond to the various watercourses which traverse La Brie, and especially the rivers, such as the Seine and Marne.

The water-bearing beds over the green clays is by far the most important in the plateau of La Brie, whilst in the valleys the wells are supplied by the beds of infiltration.

The hydrological map of the Seine and Marne enables a judgment to be formed of the depth it would be necessary to attain in order to obtain a water supply, and an opinion can also be formed of the geological nature of the soil which would be met with in boring. A judgment can also thus be formed of how the water sources and beds act in an absorbent soil like chalk; a question which is of considerable interest at the present time, when a tunnel is proposed between France and England.

3. *Lithological or Submarine Chart of the Seas of Europe.*—This chart, compiled from hydrographical observation, serves to convey an idea of the nature of the rocks which are covered by the seas of Europe.

The bed of the seas is characterised by horizontal curves traced from the depths ascertained by a great number of soundings. The beds distinguished are the stony rocks of various kinds, clays, mud, muddy sand, gravelly mud, sand, grit, and pebbles.

These are represented by conventional colours, as in geological maps, but one colour indicates the same lithological character. The localities most rich in shells are also determined and marked by shadings.

Among the rocks which form the bed of the seas, some are anterior to the present epoch, and are constantly worn or beaten, and are covered by deposits. They may be stony, like the granites or paving stones and the limestones, but they are occasionally tender and brittle like the clay, and sometimes movable like the sand and the pebbles.

The rocks anterior to the present period are seen principally in the parts which jut out along the coasts, in the straits, and in the localities where the tides run strongly. The rocks belonging to the deposits of the present epoch are essentially of the movable character, they fill especially the places having hollows or depressions; they cover the plateaux, and accumulate chiefly in depths where the tides or currents are not rapid.

The geological study of the coasts occasionally enables the prolongation of the rocks which have emerged to be traced beneath the sea, and portions of the geological submarine chart to be constructed.

4. *Ancient and present Seas of France.*—An endeavour has been made to represent the ancient seas of France in the silurian, the trias, the lias, the eocene, and the pliocene periods. Starting with the facts furnished by geology, there have been comprised in one blue tint all the points in which the existence of soils deposited by these various seas have been established. But these soils are not now what they were when originally formed; they have been partially destroyed by the atmosphere and by flowing water; they have been sundered by faults; they have been subjected to numerous and complicated overthrows; and it is chiefly their actual condition that it is proposed to describe. With this view, in order to study the lands in the parts where they are now visible, as well as in those where they are covered up, they have been represented in relief by means of horizontal curves. The curve having the side O is particularly interesting, as it shows the intersection of the actual level of the sea with the surface of the land under consideration, and, if we admit that the sea level remains constant, all the portions of land which are formed above this curve have necessarily been elevated.

The total of horizontal curves shows the orography of the superior surface of land, which has resulted from the wasting and changes since its deposition.

By the aid of these maps we can well see that while a basin has received the superposed soil, and the original elevations and depressions are to a certain extent maintained, they have thinned out successively in the more recent soils. Moreover, where a soil has been raised on the flanks of mountains, it presents generally a strong declivity, which is well shown by the approach of the horizontal curves, but disappears at a short distance from the mountains. The sixth map of the plate represents the actual seas of France, of which it shows the orography. It indicates also the proportion of carbonate of lime in the marine deposits which form on the shores, the upheaval and subsidence of the coasts, the distribution of the rainfall, &c.

3232a. Specimens of Charts published by the Navy Department.
Général Direction of the Depôt of Marine Charts and Plans, Paris.

Cherbourg Roads, 1836.
Directions of currents in the Channel, 1855.
Coasts of the mouths of the Rhône, 1848.
Coasts of Italy, 1863.
Approaches to Brest, 1868.
Island of Ouessant, 1868.
Guadaloupe, 1874.
Japan, 1874.

3235. Collection of Maps and Plans.
M. Biard, Paris.

3236a. Collection of Maps, Plans, Plaster Casts, and Photographs. *M. Delagrave, Paris.*

3236b. Map of Travel, setting off by differently coloured tracings the relative antiquity and importance of the Roman roads. *M. Hayaux du Tilly, Paris.*

3237. Manuscript Map of British Isles, on rollers.
Wichmann, Hamburg.

3238. Collection of Models and Maps.
Frère Alexis, Gochet.

3239. Diagrams showing the amount of the **Misrepresentation of the Sphere** by plane maps in the most usual projections.
C. W. Merrifield, F.R.S.

3240. Lecture Diagrams (4) of Alpine Glaciers, of Switzerland, by the late G. V. du Noyer. Used in illustration of lectures at the Royal College of Science, Dublin.
Edward Hull, F.R.S.

3241. Working Model of a Geyser.
Prof. W. F. Barrett.

This is a lecture table arrangement constructed upon the method first suggested by Prof. Wiedemann, and illustrated on a large scale by Prof. Tyndall. Water is put into the tube and heated by a gas flame below. Under the pressure of the column of water, the boiling point is raised, and the pressure being lessened by the overflow of water into the basin, steam is suddenly generated, which ejects the water periodically to a considerable height. When the tube is cooled, the eruption from this small model can eject the water at least 20 feet high. A large form, on the same pattern, similarly ejects the water 30 or 40 feet high. The size of the larger model is as follows: Tube of galvanised iron, 7 feet long, 4 inches in diameter below, and 2 inches diameter above. Basin, say 4 feet in diameter. A spiral row of gas jets heats this larger tube one-third of the way up, as well as a large gas flame below.

3242. "Cadre-Mobile." Frame for wall maps and pictures used in schools. *Ernest Recordon (Geneva).*

3243. Guthe's School Wall Map of Hanover. Newly drawn by W. Keil.   *Th. Fischer, Cassel.*

3244. G. Rohlfs' Three Months in the Lybian Desert, with photograph and map.   *Th. Fischer, Cassel.*

Rohlf's volume forms the first part of the report on the Expedition for the Investigation of the Libyan Desert, carried out under the auspices of His Highness Ismail Pasha, Khedive of Egypt.

3245. Album of Printing Specimens. Colour-printing specimens for works by Lischke, von der Decken, Heuglin, Rohlfs, Pfeiffer, Dunker, &c.   *Th. Fischer, Cassel.*

The album contains proofs in colour printing of plates from the geographical and natural history works of Heuglin, Von der Decken, &c.

3246. Journal of the Museum Godeffroy. Edited by L. Friederichsen. 3 vols., 4to.   *T. C. Godeffroy, Hamburg.*

This journal is the scientific organ of the "Museum Godeffroy." It is almost exclusively devoted to the publication of the results of the investigations of the travellers of the Museum in the Pacific, and contains geographical, ethnographical, and natural history memoirs. The management lays particular stress upon the production of technically perfect figures of the novelties described, so that the plates accompanying the journal must be regarded as in many respects patterns of the most various modes of pictorial reproduction. Particular care is also devoted to the production of original maps.

3246a. Reproduction of the **Chart** made in 1500 by Tuan de la Cosa, Columbus's pilot, after his second journey with Columbus in 1493.   *Archæological Museum, Madrid.*

This is taken from the original at the Ministry of Marine, at Madrid. It may be considered a Mapa-mundi. The part applying to America is most important, as it was drawn out by this pilot in the second journey which he took with Columbus in 1493. This chart was made for Queen Isabella, and contains the following inscription :—*Tuan de la Cosa la fizo en el Puerto de Santa Maria en Anno de* 1500.

This reproduction is accompanied by a memoir on the same subject, written by Dn. Fernander Duro, and published in the "Museo Español de Antiguedades."

3246b. System of Photo-Lithography as. adopted in the Surveyor-General's Office, Adelaide.

The Royal Commissioners of the Exhibition of 1851.

3246c. Plans, Southern portion of South Australia.

The Royal Commissioners of the Exhibition of 1851.

3153e. Collection of Dried Plants from **Arctic Regions,** presented to Elizabeth Gurney. Collected by Sir Edward Parry.

Samuel Gurney.

Photographs of the **Yellowstone.**

J. Norman Lockyer, F.R.S.

SECTION 16.—GEOLOGY AND MINING.

WEST GALLERY, ROOM L and M.

I.—GEOLOGY.

1. SPECIAL COLLECTIONS. GEOLOGICAL SOCIETY OF LONDON.

a. GEOLOGICAL INSTRUMENTS AND APPARATUS.

3247. Apparatus, &c. employed in **Sir James Hall's** celebrated experiments. *The Geological Society of London.*

A series of specimens, with the instruments employed in their preparation, illustrating the remarkable experiments carried on by Sir James Hall, Bart., between the years 1787 and 1805, to confirm some of the positions taken up by Dr. James Hutton in his celebrated "Theory of the Earth." The series consists of the following:—

(1.) A selection of whinstone and lava rocks which have been fused and allowed to cool under various conditions. These are the products of the earliest experiments, which were carried on in ordinary clay crucibles.

(2.) Porcelain tubes employed in the first attempts to submit carbonate of lime and other substances to a high temperature under pressure.

(4.) "Gun-barrels," employed in the final and successful experiments for the same purpose; with the cylinders of rock operated upon.

(The collection of apparatus and specimens from which the above are selected was presented to the Geological Society of London by Captain Basil Hall, R.N., at his father's death.)

3248. Geological Model of the **South-east of England** and **part of France,** including the **Weald** and the **Bas Boulonnais.** Founded upon the maps of the Ordnance, Admiralty, and Geological Surveys. Scale, 1 inch to 4 miles.

W. Topley and J. B. Jordan.

This model illustrates the close connexion which exists between the geological structure of a district and the "form of the ground." The chief hill ranges and the broad longitudinal valleys correspond with the outcrops of hard or soft rocks; the transverse valleys, in which the main rivers run, cut across all the rocks alike.

3249. Agricultural Map of **Kent.** Scale, 1 inch to 4 miles.

William Topley.

This map is founded upon the published maps of the Geological Survey, additions and modifications of colouring for agricultural purposes being made by the author. Stiff and retentive soils are coloured dark tints—grey, purple, and brown; light and absorbent soils, yellow or light green; calcareous and absorbent soils, blue.

The index notes briefly the quality of the soils; the character of the land, whether chiefly arable, pasture, or woodland; and mentions a few of the more noteworthy crops of Kent.

3250. Diagram illustrating the comparative **Agriculture** of **England and Wales.**   *William Topley.*

The object of this is to show the relative agricultural value of the various counties, which can best be done by exhibiting the per-centage acreage of the different crops. The numbers are obtained from the " Agricultural Returns " for 1869 ; the area of " woodland," and that of permanent pasture as divided between grass and hay, being added from the returns for 1871. The absolute numbers vary slightly in different years, but the variation of the per-centage numbers is very small. [For the results obtained from these maps and diagrams, *see* Journ. Roy. Agric. Soc., ser. 2, vol. vii., p. 268, 1871.]

3251. Maps and **Table** illustrating **William Smith's** first efforts towards producing his **Geological Map** of England.
                              The Geological Society of London.

(1.) Original table of the order of strata and their imbedded organic remains in the vicinity of Bath, as examined and proved by William Smith prior to 1799. The MS. is in the handwriting of the Rev. B. Richardson, of Farley, who wrote it down from the dictation of William Smith, in the year 1799.

(2.) A geological map of the district around Bath, prepared by William Smith about the same period.

(3.) Smith's first small geological map of England ; in this the identification was effected by mineral characters, not by fossils, and the formations were traced by physical features.

(These maps were presented by William Smith to the Geological Society of London in the year 1831, on the occasion of his receiving at the hands of their president, Professor Sedgwick, their award of the Wollaston medal and donation fund.)

3252. Maps illustrating the rise and gradual progress of the art of **Geological Surveying** in the British Islands and the Colonies.   *The Geological Society of London.*

Wm. Smith's first large geological map of England, published in the year 1815.

G. B. Greenough's geological map of England, the first edition, published in the year 1819.

Bain's first geological map of South Africa.

Bain's smaller geological map of South Africa.

John Phillips's geological map of Yorkshire, first edition.

Farey's section across the Weald.

Section across the Weald by John Farey in 1806.

This horizontal section from London to Brighton, upon the scale of 1 inch to a mile, was constructed by John Farey in 1806. It was never published, but several MS. copies were made of it. It is of great historical interest as being the first illustration of the anticlinal structure in the south-east of England, of the truncated chalk escarpments, and of the proofs of enormous denudation which must have taken place in the district.

3252a. Series of Indexes of Colours and Signs employed in the Maps and Sections of the Geological Survey of the United Kingdom.

The gradual advance of English geology (as expressed by maps) is shown to some extent by the successive editions of the Index of Colours used by the Geological Survey.

The earliest index (in MS. only) was drawn up by Sir Henry De la Beche, and shows the classification of rocks in use, and the extent to which geological surveying was carried in the year 1832.

In the first published index the volcanic rocks were indicated by green colours, but Sir R. I. Murchison, on becoming director-general in 1855, introduced the sytem of lettering the various formations, beginning with A for the Cambrian rocks, and altered the colouring of the igneous rocks to various shades of red. The results of these changes are shown on the second sheet.

In the index of colours published in 1856, the system of colouring, lettering, &c. then introduced will be seen to be essentially the same as that now in use. Various modifications have, however, since been made from time to time, for the purpose of representing the more numerous subdivisions of certain formations which it has become necessary to map, and some alterations have also been made in that part of the index which relates to the igneous and metamorphic rocks, so that the nomenclature of the maps of the Survey may approximate more closely to that of modern petrology.

3252b. Geological Map of Cornwall and West Devon.

Ten Sheets on the scale of one inch to one mile, mounted together.

In this map the granitic masses of the S.W. of England are shown partly surrounded by Devonian rocks and partly by rocks of Carboniferous age, the latter overlying the Devonian rocks conformably, and filling up part of a synclinal trough, so that the Devonian rocks occupy superficial areas north and south of that composed of the Carboniferous rocks (culm-series). Dykes of Elvan (quartz-porphyry), which proceed from the underlying granite, are seen to trend in an east and west direction, while interbedded greenstones (rocks of a doleritic type, including beds of volcanic ash) follow the strike of the Devonian and Carboniferous rocks in various places. Masses of serpentine and diallage rock are likewise marked on the map, and also the tin, copper, iron, and lead lodes which occur in different localities, the stream-tin deposits being also indicated.

This map has a historical interest, since it is almost entirely the work of Sir Henry De la Beche, the founder and first Director of the Geological Survey of the United Kingdom. The publication of this map and the accompanying report on the geology of Cornwall, Devon, and West Somerset led to the recognition of the scientific and economic value and importance of a geological map of the country; to the institution of a Government geological survey under Colonel Colby, R.E., who was then Director of the Ordnance Survey; and afterwards to the foundation of the Museum of Economic Geology (now the Museum of Practical Geology), to illustrate the work of the Survey, and its practical application in mining, agriculture, architecture, and the fine arts. The Royal School of Mines was subsequently added, to afford facilities for the teaching of geology and the collateral sciences.

3252c. Geological Map of East Devon, Somerset, Dorset, and part of Wilts. (Scale, one inch to one mile.)

Ten sheets, on the scale of 1 inch to a mile, mounted together, illustrating not less than 53 formations in East Devon, Somerset, Dorset, and part of Wilts. The map includes the various tertiary beds of East Dorset, the chalk of Salisbury Plain and of Dorsetshire, including that of Ballard Down, together with the S.W. range of the Upper Greensand of Shaftesbury and Westbury, and to the west that of Black Down and Great Haldon.

The southern portion of the Jurassic formation in England forms a prominent feature in the map.

The Triassic Strata flanking the Mendips occupying extensive valleys between Watchet and Torquay.

The rocks composing the great anticlinal ridge of the Mendip Hills mark the boundary of the Somersetshire coal-field, which here is generally concealed by Triassic, Liassic, and Oolitic overlying formations; when, however, the coal measures are exposed, the coal seams are indicated by continuous black lines; these lines, when broken, show the extension of the seams beneath the overlying deposits. In the S.W. portion of the map the Devonian rocks of Totnes, Torquay, and Start Bay are shown, and also the Carboniferous and Devonian rocks of Mid Devon, and the easternmost portion of the granite of Dartmoor.

3252d. Geological Map of the Wealden Area.

In addition to the Wealden area, this map comprises the southern and main parts of the London Tertiary Basin and the eastern portion of the Hampshire Basin, which are separated from each other by the intervening parallel range of chalk hills, commonly known as the North and South Downs respectively.

The upper and lower cretaceous rocks, from the Upper Greensand to the Lower Greensand inclusive, appear in succession from beneath the chalk, forming an outer margin surrounding the Wealden area, which is separated from the central portion, consisting of the alternations of sand and clays (called the Hastings Beds), by the Weald clay, a broad belt of low and generally flat ground.

The model constructed by Messrs. Topley and Jordan (No. 3248) will convey a better notion to the eye of the general configuration of the ground than can be afforded by the map.

The latter, however, shows the undulations of the chalk, which forms large synclinal and anticlinal folds or elevations and depressions, and the enormous amount of cretaceous strata which once extended over the space between the North and South Downs, and which must have been removed by denudation.

3252e. Geological Map of North Wales, on the scale of one inch to a mile.

The rocks consist of the Cambrian and Lower Silurian strata, from the bottom of the Menevian beds to the top of the Bala beds, including the Bala limestone, with associated eruptive rocks and contemporaneous lavas and ashes, also of Upper Silurian strata of the Wenlock series, and of Old Red Sandstone, Carboniferous Limestone, Coal Measures, and Permian beds.

In Anglesea, the Cambrian rocks consist of grits and gneiss, and the Silurian strata of the island are also partly gneissic and associated with granite rocks. In Carnarvonshire, the Cambrian rocks largely consist of grits and purple slates, which are extensively quarried for slate, especially at Llanberis and Penrhyn. In Merionethshire, the same formations consist chiefly of massive purple and green grits, with a little purple slate. The contemporaneous volcanic series occurs at the top of the Arenig, and also in the Bala series, the top of Snowdon being a calcareous volcanic ash contemporaneous with the Bala limestone. · Most of this series is fossiliferous.

The Upper Silurian beds consist of the Denbighshire flag-stones and Wenlock shales, also fossiliferous.

The Old Red Sandstone (unfossiliferous) is sparingly developed in Anglesea and Denbighshire. The Carboniferous Limestone and overlying millstone grit and Coal Measures are shown in Anglesea, Denbighshire, and Flintshire, and these are in places succeeded by red Permian strata and by the New Red series.

3252f. Geological Map of Mid Wales, on the scale of one inch to a mile.

This map contains 17 quarter sheets, on the scale of one inch to one mile. It comprises part of the Silurian rocks of Wales, and the Old Red Sandstone of Herefordshire and Shropshire. Rocks of Lower Silurian age occur in the western portion of the district, the north-western area displaying intrusive and interbedded igneous rocks. To the S.E. there is a succession of beds of Upper Silurian and Old Red Sandstone age, while to the N.E. Lower Silurian rocks crop out, together with a small exposure of the Cambrian rocks in the neighbourhood of Church Stretton. A portion of the Shropshire coalfield is included in the N.E. part of the map, and also the overlying Permian and Triassic formations of Shrewsbury. The localities where lead ores occur are indicated by the usual symbols.

3252g. Geological Map of the Midland Counties, N.W. Division.—Six sheets on the scale of one inch to one mile, mounted together.

The Carboniferous Limestone of Derbyshire, coloured pale blue on the map, occupies almost the central portion. The interbedded basalts (Toadstones) are shown, and also the metalliferous lodes (mostly lead ores). This tract is traversed by the Pennine anticlinal axis from which the strata dip away to the east and west, the Yoredale Shales, Sandstones, and Limestones, and the Coal Measures, appearing successively on either side. To the eastward these form the Nottinghamshire and Yorkshire coal-fields, and are overlaid by Permian Sandstone, Magnesian Limestone, and various Triassic strata, but they are cut off to the westward by the boundary fault, which running nearly N.N.W. by Congleton brings the Carboniferous Limestone against the Trias. Near Stockport the Coal Measures reappear as the continuation of the Lancashire coalfield; and further, south, near Stoke-on-Trent and Newcastle-under-Lyne in the N. Staffordshire coalfield, south of the limestone tract, the Carboniferous rocks are concealed by Triassic beds, but they reappear in the Leicestershire coalfield, part of which is seen in the S.E. corner of the map. Towards the west a large area is occupied by the Triassic Marls of Cheshire, in which extensive deposits of rock-salt occur.

3252h. Geological Map of the Midland Counties, S.E. Division.—Six sheets, on the scale of one inch to one mile, mounted together, illustrating the geology of the whole or parts of nine counties in central England.

The Oolitic rocks cross the map diagonally from N.E. to S.W. The strata in which the Northamptonshire iron-ores are worked are indicated on the map. Towards the S.E. these limestones and sandstones are succeeded by the Oxford Clay, which, after forming the flat districts around Huntingdon and Bedford, passes beneath a second escarpment of the Cretaceous rocks, running roughly parallel to the first. A small portion of the Lower Tertiaries of the London basin occurs in the S.E. corner of the map. To the N.W. of the Oolitic feature the Lias and Trias form the broad undulating country of Leicestershire, while in the N.W. corner of the map a great part of the Warwickshire and Leicestershire coalfields are shown.

3252k. Geological Map of the Lancashire Coalfield, scale six inches to one mile.

The area comprised in the sheet includes on the north Preston and Burnley, and on the south St. Helens and Manchester.

The principal coal-seams of the district are shown, and also the great lines of fault which displace the coal-bearing strata. The horizontal sections, on the scale of six inches to one mile (for height and distance), illustrate the sheet, one showing the Dukinfield Colliery and the various seams of coal through which the shafts are sunk ; the other, the structure of the Manchester coalfield and adjoining country. A vertical section, on a scale of 40 feet to an inch, of the Lancashire and Cheshire coalfields, shows the various seams of coal worked at Dukinfield and other collieries, and the relative thickness of the sandstones, shales, and other strata connected with them.

Horizontal and vertical sections descriptive of the Yorkshire coalfield are also shown.

32521. Geological Map of a portion of the Yorkshire Coalfield (on the scale of six inches to the mile) ; the topography by the Ordnance Survey, the geology by the Geological Survey of England and Wales.

The geological formations shown on the map are (top) Magnesian Limestone (blue), Middle Coal Measures (black), Lower Coal Measures (pale black), Miistone Grit (still paler black) ; in the Coal Measures and the Millstone Grit thick beds of sandstone are denoted by yellow and other colours spread over the general black tint ; the positions of faults at the surface are shown by white lines, and where faults have been proved in workings, their corresponding positions in beds of coal are denoted by yellow lines ; black lines are the outcrops of coal seams. The depths of coal pits, and where possible the thickness of the coals passed through, and sundry other geological and mining details, are engraved on the maps. In looking at the map it must be borne in mind that its object is to show what rock it is that forms the surface at each spot, and that the fact that the surface at different spots is formed of different rocks is owing to the inclination of the strata. If the beds lay flat, we should have the same rock extending over nearly all the area. But the beds do not lie flat ; over a great part of the country they slope or dip in a general way from south-west to north-east. In consequence of this dip the lowest beds (Millstone Grit) come up to the surface at the south-west corner ; as we go thence in a north-easterly direction the Millstone Grit passes away underground and is covered up by the Lower Coal Measures ; these in their turn are carried down and hidden beneath the Middle Coal Measures. The dip is steepest on the south-west, and decreases towards the north-east till the beds become flat and at last turn up and begin to dip towards the south-west, so that the measures, as far as we can see them, lie in a basin. How far towards the east this basin extends we cannot say, because, soon after the change of dip occurs, the Magnesian Limestone comes on and hides the Coal Measures from view. Besides the rocks mentioned the flats of alluvial river-mud and gravel are distinguished by a pale straw colour ; old river gravels, deposited when the rivers flowed at a higher level than now, by a reddish brown tint ; and some few patches of Boulder Clay and Gravel by stippling.

3252m. Index Geological Map of Wales, scale four miles to one inch, being a reduction of the one inch Geological Survey map of the same district.

Vertical section of the Purbeck strata of Dorsetshire.
Vertical section of the Tertiary strata of the Isle of Wight.
Horizontal sections of the Isle of Wight.
Horizontal sections of Dorsetshire coast, &c.

3252n. Geological Survey Map of Ireland, scale one inch to one mile.

The Geological Survey map of Ireland shows the state of progress to the present time. The coloured maps are published, those uncoloured are in various stages of progress, while the northern portions of Ireland have yet to be geologically surveyed.

The map shows the central plain of Ireland formed of Carboniferous Limestone (coloured blue), having on the south the Galty, Commeragh, and other mountains, formed of Lower Silurian and Old Red Sandstone, rising to elevations of over 3,000 feet. Along the east are the granitic and Silurian mountains of Wicklow, reaching in Lugnaquilla a height of 3,039 feet. On the west are "the Western Highlands" of Mayo and Galway, consisting of granite, quartzite, schist, and other metamorphic rocks of Lower Silurian age, in the midst of which are Upper Silurian beds on both sides of Killary Bay. To the extreme south-west are the mountains and promontories of Kerry, formed of Old Red Sandstone, with some Upper Silurian beds at Sybil Head and Smerwick Bay. Carrantuohill, near Killarney, reaches an elevation of 3,404 feet, being the highest point in Ireland. Small coal basins (tinted dark), resting on flagstones and shales representing the Millstone Grit series, occur in counties Cork, Tipperary, Kilkenny, Leitrim, and Tyrone.

Edward Hull, F.R.S., Director.

3252o. Drawings illustrative of the Microscopic Structure of various English Eruptive Rocks, by Frank Rutley. *Geological Survey of England and Wales.*

1. Micaceous Felstone, × 37. Potter Fell, Long Steddale, Westmoreland. This rock is one of the numerous dykes which occur in the Upper Silurian rocks of the lake district, and may be regarded as a good typical example of very many of them. It consists of crystals of Orthoclase and magnesian mica embedded in a felsitic matrix. A little Magnetite is also present. Of the different component minerals, only a few of the Felspar crystals are sufficiently large to be apparent to the naked eye. The mineral components of these dykes are the same as those of Minette, but the former rocks differ from the latter in containing less magnesian mica, and in some of the Orthoclase being well crystallized.

1a. Key to Fig. 1.

2. Devitrified Pitchstone Porphyry, × 25. Stoke Lane, Mendips, Somersetshire. In this section crystals of both orthoclastic and plagioclastic felspars are shown, containing some Viridite, and surrounded by the devitrified magma which characterises this rock.

2a. Chromo-lithograph of Fig. 2, by Messrs. Vincent Brooks, Day, and Son.

3. Dolerite, × 75. Charfield Green, 150 yards north of Railway Station, Gloucestershire. The larger drawing shows a much altered crystal of Orthoclase, as seen by polarized light, exhibiting twinning upon the Carlsbad type. Several of the larger crystals which occur in this rock, but which are not shown in the drawing, may also be Orthoclase, while some of the smaller ones appear to be Plagioclase greatly altered. Pseudomorphs after Olivine are also plentiful in the rock, and the smaller drawing represents one of them.

3a. Chromo-lithograph of Fig. 3, by Messrs. Vincent Brooks, Day, and Son.

4. Basalt, × 50. Damery, Gloucestershire. Composed of Plagioclase, altered Augite, and serpentine pseudomorphs, some possibly after Olivine and Magnetite.

822   SEC. 16.—GEOLOGY AND MINING.

4a. Chromo-lithograph of Fig. 4, by Messrs. Vincent Brooks, Day, and Son.

5. Minette, × 37. Near first Railway Bridge from Windermere Station, Westmoreland.

In this section crystals of magnesian mica are shown lying in a felsitic matrix.

6. Diabase, × 25. Knowles Hill, Newton Bushel, Devonshire.

This rock is mainly composed of an altered triclinic Felspar, Augite, Chlorite, Serpentine Quartz, and Pyrites apparently altered in places into Limonite. A little Apatite is also present. In the drawing the upper half of the field represents the section by ordinary illumination, while the lower half shows the remaining portion as seen by polarized light.

7. Minette, × 150. Washfield, three miles N.W. of Tiverton, Devonshire.

A much decomposed Minette, full of small cavities, which cause it to resemble a scoriaceous lava, but when examined microscopically many of these holes are seen to present definite hexahedral forms, from which no doubt crystals of mica have been removed. The black granules and crystals represent Magnetite in some, if not in all, instances. The matrix is felsitic.

8. Microscopic drawings illustrative of structure in some of the eruptive rocks of Somersetshire, Fig. 1, volcanic breccia × 25. Wrington Warren, near Bristol. Section etched with acid and drawn as seen by reflected light. The upper portion of the field represents part of a small fragment of fossiliferous limestone (showing crinoid remains).

Fig. 2. Portions of crinoid stems from one of the above limestone fragments, × 105.

Fig. 3. Doleritic matrix in which the above limestone fragments are imbedded, × 50.

Fig. 4. Crystals of Magnetite in Felstone, × 150. Near Downhead, Mendips.

Fig. 5. Amygdaloid in Dolerite, × 25. Uphill Cutting, Great Western Railway.

Fig. 6. Amygdaloid of Calcspar in Dolerite, × 45. Uphill Cutting, Great Western Railway.

8a. Mechanical autotype of Plate 8, by Messrs. Vincent Brooks, Day, and Son.

9. Microscopic drawings illustrative of structure in some of the eruptive rocks of Somersetshire.

Fig. 1. Pseudomorph after Augite, from Dolerite, × 175. Uphill Cutting, Great Western Railway.

Fig. 2. Outline of an Augite crystal (natural size), given for comparison with preceding figure.

Fig. 3. Plagioclase from Doleritic matrix of Volcanic Breccia. Wrington Warren, near Bristol.

Fig. 4. Ditto, showing banding by polarised light, × 100.

Fig. 5. Pseudomorph after Augite, from Dolerite, × 175. Uphill Cutting, Great Western Railway.

Figs. 6, 7, and 10. Calcareous bodies in Volcanic Breccia. Wrington Warren, near Bristol.

Fig. 8. Crystal of a Felspar in devitrified Pitchstone Porphyry. Stoke Lane, Mendips.

Fig. 9. Calcspar Amygdaloids in Dolerite, × 50. Uphill Cutting, Great Western Railway.

Fig. 11. Crystal of Titaniferous Iron from Felstone, × 350. Near Downhead, Mendips.

Fig. 12. Fragment of Volcanic Breccia (natural size). Wrington Warren, near Bristol.

The matrix is Doleritic, the imbedded fragments are Fossiliferous Limestone, probably of Carboniferous age.

Figs. 13 and 14. Magnetite crystals in Felstone, × 350. Near Downhead, Mendips.

9a. Mechanical Autotype of Plate 9, by Messrs. Vincent Brooks, Day, and Son.

b. MAPS, DIAGRAMS, &c.

COLLECTION OF MAPS AND SECTIONS LENT BY THE GEOLOGICAL SURVEY OF THE UNITED KINGDOM OF GREAT BRITAIN AND IRELAND.

3253a. Geological Map of the Keswick District, on the scale of 6 inches to the mile, with appended one-inch maps, sections, &c. *J. Clifton Ward.*

With this are exhibited three one-inch maps of the same district, one of which serves as a coloured index sheet to the large map, and shows the varied dips of the strata, while the other two serve as cleavage and glacial maps. Also a horizontal section at the foot of the large map on the six inch scale, ten horizontal sections on the one-inch scale, a section through the plumbago mine in Borrowdale, and twenty-four chromo-lithographic figures of the microscopic structure of rocks of the district.

3253aa. William Smith's original Geological Map of Hackness Hill, Yorkshire, being one of the earliest geological maps ever constructed on a large scale. *R. Turnbull.*

The maps of William Smith "the Father of English Geology" of the outlying mass of the Middle Oolite at Hackness, exhibit in a striking manner the great knowledge of their author and the able manner in which he traced and mapped their geological structure, and also pointed out its bearing on the agriculture of the district. The geological survey map of the same area, while it helps to form a comparison of the old and new style of mapping, at the same time shows that William Smith was well acquainted with the general details of this unique and somewhat obscure group of rocks.

Series of Views in the Cam Valley, near Bath, coloured geologically. This was the valley in which William Smith, the father of English geology, made his earliest discoveries. *W. Stephen Mitchell, LL.B.*

3253ab. Geological Survey Map of the same district for comparing the old and new style of geological surveying. *R. Turnbull.*

3253b. One-inch quarter-sheet, 101 S.E. of the Ordnance Survey (Keswick District), converted into a **Model** by Mr. Jordan, and this into a Geological Model by the exhibitor, showing the various geological divisions, dips, faults, and mineral veins. *J. Clifton Ward.*

The work here embodied is the result of the labours of the contributor as an officer of the Geological Survey, but the colouring of the model has been privately undertaken with a view to its presentation to the Keswick Museum of Local Natural History.

3253c. One-inch sheet, 101 S.E. of the Ordnance Survey (Keswick District), converted into a Model by Mr. Jordan, and with glacial information, including ice-scratches, boulder-transportation, and old lake-beds.   *J. Clifton Ward.*

Prepared by the contributor for the Keswick Museum of Local Natural History.

3253d. Diagrams of Microscopic Rock - structure ; photographs, coloured and uncoloured, enlarged from small water-colour drawings 1¼ inches in diameter, made direct from the microscope.   *J. Clifton Ward.*

The photographs have been taken by the firm of Hennah and Kent, of Brighton, from small water-colour drawings made by the contributor, direct from the microscope, and 1¼ inches in diameter.

A. Spherulitic felsite, Carrock Fell, Cumberland.
B. Hypersthenite, Carrock, Cumberland.
C. Fine-grained band in Hypersthenitic rock, Carrock Fell.

Three uncoloured photographs, also taken from small water-colour drawings made by the contributor:

D. Chiastolite slate, Skiddaw, Cumberland.
E. Quartz felsite, St. John's Vale, Cumberland.
F. Skiddaw Slate, passing into quartz felsite, St. John's Vale, Cumberland.

The small coloured drawings from which these photographs have been taken, are (except in the case of D) also exhibited (*see* Plate C. 3253f.).

3252dd. Dunn's Geological Map of South Africa.
Edward Stanford.
Marcon's Geological Map of the World.
Edward Stanford.
Table of British Sedimentary and Fossiliferous Strata, by W. H. Bristow, F.R.S.   *Edward Stanford.*

Diagrams of Natural History.   *Edward Stanford.*

3253e. Coloured Diagrams enlarged from small water-colour drawings, 1¼ inches in diameter, made direct from the microscope by the contributor.   *J. Clifton Ward.*

A.—*Cumberland Volcanic Rocks* (see also Plate A of small water-colour drawings contributed, 3253f.).

1. Lava, Brown Knotts, Keswick.
2. Lava, Wastwater. (Polarized light.)
3. Felsite, amid altered ash, Bleaberry Fell, Keswick.
4. Ash, Steel Fell. (Polarized light.)
5. Altered streaky ash, Base Brown.
6.   „   „   „   (Polarized light.)
7. Altered ash, Hart Side.
8. Highly altered fine bedded ash, Great Gable.

B.— *Welsh Volcanic Rocks.*

9. Felstone, a metamorphosed ash, Rigghause End, Vale of St. John, Cumberland. (Polarized light.)
10. Felstone, Aran Mowddwy. Also represents the structure of Fig. 9. when viewed with plain light.
11. Felstone, Aran Mowddwy. (Polarized light.)
12. Felstone, Aran Benlynn.
13. Felstone, Llanberis Route, Snowdon.
14. Slaggy felstone or altered streaky ash, Llanberis Route, Snowdon.

NOTE.—The Welsh felstone probably represents the modern trachytes or quartz-trachytes, and the Cumberland lavas the modern trachy-dolerites and basalts.

3253f. Three Plates of small water - colour drawings, 1¼ inches in diameter, made direct from the microscope by the contributor, illustrating **Microscopic Rock-structure.**

J. Clifton Ward.

A.—*Modern Volcanic Rocks.*

1. Trachyte, Solfatara, × 40.
2.   ,,   ,,   (Polarized light), × 15.
3. Leucitic lava of 1631, Vesuvius, × 15.
4.   ,,   ,,   Alban Mount, Rome, × 14.
5.   ,,   ,,   ,,   ,,
                       (Polarized light).
6. Lava of 1794, Vesuvius, × 6.
7.   ,,   ,,   × 45.

Ancient Cumberland Volcanic Rock.

8. Lava, Eycott Hill, × 25.
9.   ,,   Brown Knotts, Keswick, × 30.
10.   ,,   near Wastwater. (Polarized light), × 15.
11. Felsite, amid altered ash, Bleaberry Fell, Keswick, × 60.
12. Ash, Steel Fell. (Polarized light), × 15.
13. Altered streaky ash, Base Brown, × 15.
14. Altered ash, Hart Side, × 20.

B. *Illustrating the Metamorphism of Volcanic Rocks around the Granites of Eskdale and Shap.*

1. Altered ash, Yewbarrow, × 50.
2. Altered contemporaneous trap (?) close to granite, × 50.
3. Same field of view, in polarized light, × 50.
4. Highly altered ash, close to granite. (Polarized light), × 50.
5. Bastard granite. (Polarized light), × 50.
6. True granite, × 50.
7. Junction of highly altered ash with Shap granite. (Polarized light), × 25.

C.

1. Altered Skiddaw slate passing into quartz felsite, Clough Head, St. John's Vale, × 30.
2. Quartz felsite, St. John's Vale, × 30.
3. Altered Skiddaw slate, Red Pike, × 70.
4.   ,,   ,,   Buttermere, × 70.

5. Syenitic granite, Scale Force, Buttermere, × 30.
6. Spherulitic felsite, Carrock Fell summit. (Polarized light), × 10.
7. Hypersthenite, White Crags, Carrock. (Polarized light), × 10.
1. Fine-grained band in hypersthenite. (Polarized light), × 10.

3253g. Plates illustrative of **Structures** in **Felspars, Leucite, Obsidian,** and **Perlite,** from drawings made partly in pencil, partly in lamp-black, by Frank Rutley, F.G.S., H.M. Geological Survey. Plates I. and II. illustrate a paper in the Quart. Journ. of the Geological Society of London for Aug. 1875. Plates III. and IV. illustrate a paper in the Royal Microscopical Society's Transactions, March 1876.

Frank Rutley, Geological Survey Office.

PLATE I.

Fig. 1. Orthoclase, Arendal, showing crystals arranged in two directions, other than those of striation. × 50 (polarized light).
2. Felspar crystal in perlite, Schemnitz, Hungary, showing partial cross-hatching. × 55 (polar.).
3. Crystal in trachyte, Berkum, near Bonn, Rhine, showing internal curved divisional markings, with lines crossing at right angles in the lateral areas. × 115 (polar.).
4. Partially formed, or partially disintegrated crystal in obsidian, Mexico. × 175 (dark-ground illumination).
5. Patches of cross-hatched felspar in oligoclase, Twedestrand, Norway. × 115 (polar.).
6. One of the above patches, showing partial cross-hatching. × 350 (polar.).

PLATE II.

Fig. 1. Minute crystal in obsidian, Mexico. × 240.
2. Diagrammatic exaggeration of the above.
3. Crystal from trachyte, Berkum, Rhine. × 90 (polar.).
4. Crystal from basalt, Cleveland, showing rectangular banding and subordinate oblique striation, the latter possibly indicating planes connected with twinning on the Baveno type. × 45 (polar.)
5. Crystal in obsidian, Mexico. × over 100.
6. Crystals of plagioclase in gabbro, Volpersdorf, Silesia, showing bands at right angles. × 90 (polar.).
7. Crystal in basalt, Cleveland, Yorkshire. × 45 (polar.).
8. Fragment of plagioclase in basalt, Cleveland, showing alternately dark and light bands fringing an obliquely cut twin lamella, as in fig. 10. × 45 (polar.).
9. Crystals on edge of section of basalt, Cleveland, showing cross-hatched striation and a rectangular cleavage at *x* on the outer edge of the section. × 45 (polar.).
10. Diagrammatic vertical section through a section of plagioclase, cut obliquely to the twinning planes; the spaces marked *w* would alternately appear light and dark in different positions of the Nicols when seen by polarized light, the overlap of the complementary colours giving rise to white light.
11. Diagrammatic section similar to fig. 10, but cut more obliquely to the twinning planes, so that, instead of coloured bands separated by light or dark ones, an unbroken surface of white light might result, although the section might be that of a felspar many times twinned.

PLATE III.

Figs. 1, 4, 5, 6, 7. Structures seen in a thin section of spherulitic obsidian from the lava stream of Rocche Rosse, in the Isle of Lipari.
8. Portion of one of the spherulitic bands in the above rock.   × 22.
12. Magnetite crystals partially converted into peroxide of iron, occurring in a crystal of leucite from Vesuvius.   × 760.
2, 3, 9, 10, 11, 13, 14. Structures seen in a crystal of leucite, from Vesuvius.

PLATE IV.

Perlite, from Buschbad, near Meissen, showing the spheroidal structures lying between divisional straight lines F. F. F. F.

3253. Map showing the work of the Geological Survey of **Scotland,** on the scale of one inch to the statute mile. All the published sheets of the map are given here, and a few of which the survey is completed, but which are not yet engraved, are inserted in MS. At the foot of the map an example of the horizontal sections is given which are run across the country on the scale of six inches to the statute mile.
Geological Survey of Scotland, Prof. Geikie, F.R.S., Edinburgh, Director.

3254. Geological Map of the Ayrshire coal-field and adjoining districts, on the scale of six inches to the statute mile. This map has been selected as an illustration of the detailed work of the Geological Survey of Scotland. The whole county is surveyed on this scale, though only the mineral districts are published on these maps, the general map of the country (*see* No. 3253) being on the scale of one inch to a mile. At the foot of the map a MS. sheet is inserted to show the stages of progress in the field work of the survey. Two specimens are likewise given of the detailed vertical sections, on the scale of forty feet to the inch, which are published in illustration of the coalfields.
Geological Survey of Scotland, Prof. Geikie, F.R.S., Edinburgh, Director.

3254a. Geological Map of **England and Wales,** by the late Sir Roderick I. Murchison, Bart., K.C.B., &c. Fifth edition, with all the railways. Scale, 28 miles to an inch ; size, 14 inches by 18.   *Edward Stanford.*

3254b. Geological Map of **Ireland,** by the late Joseph Beete Jukes, M.A., F.R.S., Director of H.M. Geological Survey of Ireland. This map is constructed on the basis of the Ordnance Survey, and coloured geologically. It also shows the railways, stations, roads, canals, antiquities, &c., and when mounted in case forms a good and convenient travelling map. Scale, 8 miles to 1 inch ; size, 31 inches by 38.   *Edward Stanford.*

3254c. Geological Map of **London** and its Environs. Scale, 1 inch to a mile; size, 24 inches by 26. Compiled from various authorities by J. B. Jordan, of the Mining Record Office, and printed in colours exhibiting the superficial deposits. It includes Watford on the north, Epsom on the south, Barking on the east, and Southall on the west, and shows the main roads in and around the Metropolis, the railroads completed, and the sanctioned lines.   *Edward Stanford.*

3254d. Geological Map of **Canada** and the adjacent regions, including parts of the other British provinces and of the United States, by the late Sir W. E. Logan, F.R.S., &c., Director of the Geological Survey of Canada. Scale, 25 miles to an inch; size, 102 inches by 45. On eight sheets.   *Edward Stanford.*

3254e. Geological Map of **India.** General sketch of the physical and geological features of British India, by G. B. Greenough, F.R.S. With tables of Indian coalfields, minerals, fossils, &c. Scale, 25 miles to an inch; size, 68 inches by 80. On nine sheets.   *Edward Stanford.*

3254f. Geological Map of the **World,** by Jules Marcou. Constructed by J. M. Ziegler. Size, 72 inches by 50.

J. N. Ziegler.

3254g. Geological Sketch Map of **South Africa,** from personal observations (combined with those of Messrs. A. G. Bain, A. Wylie, T. Bain, jun., Dr. Atherstone, and R. Pinchin, in Cape Colony, together with those of Dr. Sutherland in Natal, and of Mr. E. Button), north of 24° latitude. Scale, about 35 miles to an inch; size, 33 inches by 28.   *E. J. Dunn.*

3254h. Geological Map of the British Isles.

E. Best.

2. MAPS, DIAGRAMS, FOSSILS, &C., EXHIBITED BY VARIOUS
CONTRIBUTORS.

3257. Section of a **Well** at the **Hampstead Road,** showing detail depths of strata, 1840.   *R. W. Mylne, F.R.S.*

3258. Five Outline Sections of the strata under **London,** with a block index plan, 1850.   *R. W. Mylne, F.R.S.*

3259. Topographical and **Geological Map** of **London** and its environs.
100 square miles, scale 1·43 inch to a mile, 1851.

R. W. Mylne, F.R.S.

3256. London and its Environs Topographical and Geological.
131 square miles, scale $\frac{1}{17032}$, English and French geological references, 1855. *R. W. Mylne, F.R.S.*

3255. Contours of London and its **Environs.** Plain and coloured geologically.
176 square miles, scale $\frac{1}{17032}$, 1856. *R. W. Mylne, F.R.S.*

3261. Geological Map of the **London** and **Paris Basins.**
The tertiary and cretaceous districts of England and the north of France, Belgium, Holland, and Denmark.
The coalfield areas and contoured depths of the adjacent seas, 1862. *R. W. Mylne, F.R.S.*

3260. Geological Map of **London** and its **Environs.**
159 square miles, scale 1·43 inch to a mile, with a longitudinal section of 18¼ miles, 1871. *R. W. Mylne, F.R.S.*

3253b. Table of Sedimentary Rocks, with a pamphlet.
Prof. E. Renevier, Lausanne, Switzerland.

3253c. Set of 12 Sopwith's Geological Models, with letter-press description. *Prof. Tennant.*

3253d. Ten Sheets of Geological Maps and Diagrams of Switzerland.
Prof. B. Studer, Geological Survey Commission of Switzerland, Berne.

3253e. Geological Map of the British Isles.
Society for Promoting Christian Knowledge.

3262. Map of the Bristol Coalfields and country adjacent, geologically surveyed by William Sanders, F.R.S.
Bristol Museum and Library.

The area of the map is 720 square miles. The topographical basis consists of a reduction to the scale of 4 inches to the mile of 218 parish maps. The geological lines are entirely the result of the author's personal surveys; about 20 years were devoted to the work. The map was published, in 1864, at the author's cost. In nineteen sheets. Folio. J. Lavars, Bristol.

A reduced copy, on the scale of one inch to the mile, in one sheet, was published in 1873.

3264. Geological Map of the British Isles, by Professor John Phillips, F.R.S.
Society for promoting Christian Knowledge.

3265. Geological Map of the Arctic Regions.
C. E. De Rance.

The topography is taken from the Chart accompanying the Admiralty correspondence connected with the British Arctic Expedition of 1875. The geological boundaries of Parry Islands, and the north coast of America, from the determination by Conybeare, Murchison, Salter, and Dr. Haughton—of the specimens brought back by the expeditions of Franklin, Parry, Back,

John and James Ross, Sabine, Buchan, Beechey, Sherard Osborn, and M'Clintock. Those of West Greenland from the observations of Giesecké, Nordenskiöld, O. Heer, and Dr. Brown. Those of East Greenland and Spitzbergen from the results of the various Austrian, Swedish, and North German expeditions. Those of Hall Basin, and the channels lying north of Smith's Sound, from the labours of Drs. Kane and Bessels, which prove that the upper Silurian rocks, noticeable along the southern fringe of the Arctic Archipelago, reappear in this tract. The Lower Carboniferous Coal-bearing Sandstones, and overlying Carboniferous Limestones, lying in a basin.

3267. Model of the Cleveland Hills and district showing the outcrop of the main seam of Ironstone. *John Bell.*

3268. Geological Maps and Model of New Zealand.

1. Copy of the first geological map of the whole of New Zealand, prepared by Dr. Hector, and exhibited at the New Zealand Exhibition in 1865. This map was-engraved by the Geological Department and published in 1869.

2. Geological sketch-map of New Zealand constructed from the official surveys of the Geological Department by James Hector, C.M.G., M.D., F.R.S., &c., director of the Governmental Geological Surveys, Wellington, 1873.

3. Relief-model of New Zealand on same scale as the Geological map (2), and with a vertical scale four times as great as that of the horizontal. This model is placed beside the geological map to illustrate the forms of the surface of the parts of country occupied by the different geological formations. Modelled by Dr. Hector, 31st March 1876. *James Hector, M.D., F.R.S.*

3268a. Geological Map and sections of a part of the Province of Koutais, published by the "Administration des Mines," 1873. *Mining Institute of St. Petersburg.*

3268b. Geological Map of a part of the Province of Bakou, published by the "Administration des Mines," 1872.
Mining Institute of St. Petersburg.

3268c. Geological Map of Russia in Europe, begun in 1868 by the Members of the Mineralogical Society of St. Petersburg.
Mining Institute of St. Petersburg.

3268d. Geological Map of the mining district of Outka, in the Oural, by M. Valérien de Moeller, 1875.
Mining Institute of St. Petersburg.

3268e. Geological Map of the mining district of Ilim, in the Oural, by M. Valérien de Moeller, 1875.
Mining Institute of St. Petersburg.

3268f. Sections of the mining districts of Ilim and Outka, in the Oural, by M. Valérien de Moeller, 1875.
Mining Institute of St. Petersburg.

3268g. Geological Map of the southern part'of the Province of Nijnii Novgorod, by M. Valérien de Moeller, 1875.
Mining Institute of St. Petersburg.

3268h. Geological Map of the Province of Wladimir, by M. Dittmar. *Mining Institute of St. Petersburg.*

3268i. Geological Map of Russia in Europe and the Oural mountains, executed by Murchison de Verneuil, and Keyserling, and completed up to 1870 by M. Helmersen.
Mining Institute of St. Petersburg.

3268j. Geological and Industrial Map of the western portion of the carboniferous district of Donetz, by Messrs. An. and Al. Nossow, 1873. *Mining Institute of St. Petersburg.*

3268k. Geological Map of the carboniferous district of the country inhabited by the Cossacks of the Don, executed by Messrs. Antipon Teltonochkine and Wassiliew, 1866-69.
Mining Institute of St. Petersburg.

3268l. General Geological Map of the carboniferous district of Donetz, executed under the direction of M. Helmersen, 1871. *Mining Institute of St. Petersburg.*

3268m. Geological Map of the southern portion of the Oural chain, by Messrs. Meglitsky and Antipon, 1855.
Mining Institute of St. Petersburg.

3268n. Geological Map of the western slope of the Ourals, by M. Valérien de Moeller. *Mining Institute of St. Petersburg.*

3268o. Geological Map of. the Province of Kherson; by M. Barbot de Marny. *Mining Institute.of St. Petersburg.*

3268p. Geological Map of the Province of Kiew, by M. Theophilaktow, 1872. *Mining Institute of St. Petersburg.*

3268q. Geological Map of the Town of Kiew, by M. Theophilaktow, 1874. *Mining Institute of St. Petersburg.*

3268r. Geological Map of the Province of Twer, executed by Messrs. Lahousen and Dittmar.
Mining Institute of St. Petersburg.

3268s. Geological Map of a part of. the Province of Erivan, published by the "Administration des Mines."
Mining Institute of St. Petersburg.

3268t. Geological Map of the Province of Elisavetpol, published by the "Administration des Mines," 1869.
Mining Institute of St. Petersburg.

3268u. Geological Map of a part of the Province of Koutais, published by the "Administration des Mines," 1874.
Mining Institute of St. Petersburg.

3268v. Geological Map of the Province of Nowgorod, by M. Lahousen.   *Mining Institute of St. Petersburg.*

4571. Drawing of **Two Sections** of the **Carnic** and **Julian Alps.** (One plate, m. 0·80 by m. 0·40.)
Technical School at Udine; Director, Torquato Taramelli, Professor of Physical and Natural Science.

3153b. Selection of Maps and Publications to illustrate the progress of the **Geological Survey of Spain.**
Comision del Mapa Geologico de España, Cath. de Isabel la Catolica, 23, Madrid.

(1.) Junta general de Estadistica.

Descripcion fisica y geologica de la provincia de Santander, por Don Amalio Marstre, second class inspector of the Society of Mining Engineers. 1 vol. Madrid, 1864.

(2.) Junta general de Estadistica.

Ensayo de Descripcion geognostica de la provincia de Teruel, by Don Juan Villanova y Pura. 1 vol. Madrid.

(3.) Boletin de la Comision del Mapa geologico de España. Vols. 1 and 2.   Madrid, 1874–75.

(4.) Memorias de la Comision del Mapa Geologico de España.

Bosgnejo de una descripcion fisica y geologica de la provincia de Zaragoza, by Don Filipe Martin Donayre, first class engineer of the School of Mines. Madrid, 1874.
Descripcion fisica geologica y agrologica de la provincia de Cuenca, by Don Daniel de Cortarjar, mining engineer, and member of the French Geological Society. 1 vol. Madrid, 1871.
Trabajos geognosticos y typograficos practicedos por la Comision de Estudio de la Cuenca Carbonifera de Asturias. Madrid, 1874.

(5.) Mapa geologico de la provincia de Madrid, published by the Council of Statistics of the kingdom in 1864, and formed by Don Casiano de Prado, engineer of mines, member of the Commission of the Spanish Geological Map.   1854.

(6.) Mapa geologico de la provincia de Palencia, traced by Don Casiano de Prado, member of the Commission of the Spanish Geological Map, 1856.

(7.) "Cuadro grafico de altitudes de la parte septentrional de la provincia de Palencia," traced by the section under the direction of Don Casiano de Prado, member of the Commission appointed to form the Spanish Geological Map, 1856.

(8.) Plano geologico de la Cuenca carbonifera de San Juan de las Abadesas, province of Gerona, by Amalio Maestre, Madrid, August 15, 1855.

(9.) Comision del Mapa geologico. Secion geografica metero-
logica. Directed by the member and first class engineer, Dr. Jose
Subercase.

(10.) Mapa geologico estratigrafico de las montanas de la provincia de Palencia, by Don Casiano de Prado, 1857.

(11.) Sections of the N. and N.E. of the province of Madrid,
which include the Sierra of Truda, and cerros called de Concha
within the limits of La Puebla de la Muger muerta.

3153c. Illustrations of the progress of **Geology, Mining,**
and **Metallurgy** in **Spain,** selected from works published by the
contributor. *Don B. Federico de Botella y de Hornos, Madrid.*

(1.) "Descripcion Geologica Minera de las Provincias de
Murcia y Albacete (antigno reino de Murcia)," by B. Federico
Botella y de Hornos, first class engineer of the National School of
Mines, 1 vol., grand folio, 22 plates, and numerous drawings in the
text.

(2.) Geological and agronomical plan of Murcia and the surrounding country, giving the system of irrigating canals employed
at the Huerta of Murcia.

(3.) Sierra of Carthagena. Lead foundry. Spanish slag
hearths. Atmospheric furnaces. Blast furnaces. Cementing
furnaces. Compressed air furnaces used at the Escombrera
Works.

(4.) Sierra of Carthagena. Desilverization. Condensing furnaces. Refining furnaces. Pattison's crystallizing boilers. Implements and tools employed in working these boilers.

(5.) Mazarron. General plan of alum works. Serrata of
Lorca. Advantages of sulphur. Furnaces of horizontal cylinders.
Furnaces of vertical cylinders.

(6.) Sierra of Carthagena. Metallurgy. Moulds used for
mixing the bricks used at the furnace. Atmospherical furnaces.
Moulds employed for pressing the lead out of the furnace. Silver
in cupella. Blast furnaces, with ventilators and bellows. Moulds
for casting leaden bars. Boilers and other implements.

(7.) Sierra of Carthagena. Mechanical preparations employed
at the works. Works at Escombreras. Round buddles worked
by steam. M. Victor Simon's classifier. Water clarifier.

(8.) Topographical and geological plan of the Sierra of Carthagena.

(9.) Sierra of Carthagena. Geological section from Cabera
Gordo to Cerro de Santo Espiritu.

40075. 3 G

(10.) Sierra of Carthagena. Mechanical preparations used in metal works. American crushing machine. Classifying trommels. Dividing trommels. Refining trommel. Sieves. Classifying cones. Classifying cases. Concave rolling table. Lifter used for minerals.

(11.) Sierra of Carthagena. Cupellation. Cupellating furnace. General plan of desilverizing works.

(12.) Drawing representing a statuette of Hercules 0·152 met. II. found at the Esperanza mine in 1840.

Lamps, jars, and vessels of different kinds found at the mines of the Sierra of Carthagena.

Bronze lamp found at Lorca. Bas-reliefs. Fragment of a statue found in a mine of the Sierra of Carthagena. Lead bar of the Roman time, with inscription, found at the Sierra of Carthagena.

(13.) Sierra of Carthagena. Mechanical preparation used in lead works. Works at Escombreras. Crushing cylinders worked by steam. Vertical grinders. Classifying trommels. Sieves. Tables for shaking minerals.

(14.) Topographical and geological plan of the mines and works of San Juan de Riopar (province of Albacete).

(15.) Sierra of Carthagena. Geological sections.

(16.) A geological map of the provinces of Murcia and Albacete. (Spain).

(17.) Sierra of Lorca. Fossil fish found in the sulphur mines.

(18.) Cerro of Mazarron. View of the vein of mineral of San Juan. Sierra of Orihuela. Peña del Gato. Sulphur mines at Hellin. View of the volcanic cerro of Monagrillo. Cerro of Mazarron. Section of an ancient excavation at Pedreras Viejas. View of the sulphur mines at Hellin.

(19.) Sierra of Carthagena. Mechanical preparation of minerals. Drawer sieves. Round buddles employed at the sierra. Cylinders used for crushing minerals worked by horse power. Sifters.

3331. Portions of the Geological Special Map of Prussia and the Thuringian States.
District of ILFELD, HARTZ ; 6 sections.
District of the OHM MOUNTAIN; 6 sections.
District of JENA, on the SAALE ; 12 sections.
District of RIECHELSDORF; 6 sections.
District of HALLE ; 4 sections.
COAL BASIN of SAARBRÜCK ; 13 sections.

TOPOGRAPHICAL BASIS to a part of the above-named sections, representing the OUTCROP of the FLÖTZE; 6 sections.
The same sheets representing the FLÖTZE in PROJECTION; 6 sections.
COAL MEASURES (CARBONIFEROUS) and PERMIAN formation of the vicinity of HALLE, with the diluvium removed; 4 sections. The same in PROFILE.
GEOLOGICAL AGRICULTURAL MAP of a part of the environs of BERLIN.
MAP, containing eight parts of the SPECIAL GEOLOGICAL MAP OF PRUSSIA.
SEVEN SMALLER MAPS containing the explanations to the above.
Royal Geological Institution and Mining Academy of Berlin (Prof. Hauchecorne, Director).

3332. Geological and Agricultural Map of the manor of Friedrichsfelde, near Berlin, an example of a new cartographic method.

1. Map, scale $\frac{1}{5000}$.
2. Map, scale $\frac{1}{25000}$.
3. Length profiles, scale $\frac{1}{10000}$.
4. Map of ground rent, scale $\frac{1}{10000}$.
5. Die geognostisch-agronomische Kartirung, Text, Berlin, Ernst und Korn, 1875.
Prof. Dr. Orth, Berlin.

3333. Six large " Pedological " Plans.
I. & II. Profiles of soils which filter water.
III. & IV. Profiles of soils which do not filter water.
V. & VI. Profiles of rich soil suitable for the cultivation of the sugar beet.
VII. Text, Berlin, Wiegand, Hempel and Parey, 1876.
Explanation to the above.   *Prof. Dr. Orth, Berlin.*

The learned English physician, Dr. Lister, in the year 1683, proposed to the Royal Society of London, to prepare a map which should represent English soils and minerals, their distribution to be shown by special colours, &c. The above maps represent a new mode of geological " cartography," and exhibit the soil in its geographical position, and with its economical uses in a more precise manner than has hitherto been possible.

3339. Geological Map of the Bavarian Alps in a single plate.   Scale $\frac{1}{100000}$.
Geological Survey of Bavaria (Dr. Gümbel).

3340. Geological Map of the Boundary Mountains, Eastern Bavaria. Scale $\frac{1}{100000}$.   Sheet I.

,,   ,,   ,, II.
,,   ,,   ,, III.
,,   ,,   ,, IV.
,,   ,,   ,, V.
Geological Survey of Bavaria (Dr. Gümbel).

3341. Geological Sheet. Views of the Bavarian Alps.
,,   Fichtelgebirge I.
,,   Views from the eastern Bavaria boundary mountains.
,,   Fichtelgebirge II. (in progress).
,,   Views of the Fichtelgebirge.
Geological Survey of Bavaria (Dr. Gümbel).

3342. Four Single Sheets of the Original Geological Survey, showing the different stages of progress, scale $\frac{1}{5000}$.
Geological Survey of Bavaria (Dr. Gümbel).

3343. Geological Detail Sheet reduced from the single sheets, scale $\frac{1}{25000}$.
Geological Survey of Bavaria (Dr. Gümbel).

3344. Geological Sheet of a part of the Upper Bavarian Lower Miocene (Oligocene) beds.
Geological Survey of Bavaria (Dr. Gümbel).

These maps, while exhibiting the geological work done in Bavaria, also illustrate the mode in which the maps published on the scale of 1:100,000 have been produced. The land tax maps, scale 1:5,000, have been used for this purpose, and the observations made were entered on the spot (sheets 13–15), and afterwards geologically coloured. These observations have been then transferred to maps of a different scale 1:25,000 (Sheet No. 16), then to the military atlas 1:50,000 (No. 18), and lastly to the scale of those published 1:100,000 (1–11). Each original sheet may be obtained for the actual cost of production.

3269. Agricultural Map of Belgium.
(*a.*) Geology and mining. Geological models, horizontal and vertical sections.
(*b.* and *c.*) This map shows, by means of colours and signs, the mineral constitution of the soil of Belgium, based on the geological data, and the different kinds of cultivation. The agricultural land is divided into regions, and the latter into zones.
(*d.*) Height, $0\cdot45^{m}$; width, $0\cdot50^{m}$.
C. *Malaise, Professor at the Government Agricultural Institute, Gembloux (Province of Namur, Belgium).*

3270. Illustrations of the **Sub-Wealden Boring,** at **Ne-therfield,** near **Battle, Sussex.**
Contributed, on behalf of the Sub-Wealden Exploration Committee, by Henry Willett and W. Topley.

The Sub-Wealden boring was commenced in 1872, with the view of ascertaining the order and thickness of the secondary rocks beneath the south-east of England, and of determining, if possible, the depth and age of the Palæozoic rocks which are believed to underlie them. Geologists have for some years believed, chiefly in consequence of a paper published by Mr. R. Godwin-Austen in 1856, that beneath the secondary strata of the south-east of England there exists a floor of Palæozoic rocks, prolonged from South Wales, Gloucestershire, and Somersetshire on the west, and from Belgium and the north of France on the east. Amongst these Palæozoic rocks there is a possibility that coal-measures may occur; coal, in fact, is now worked beneath the Oolites in Somersetshire, and beneath the Cretaceous rocks in the north of France. It is only in this subordinate sense that the Sub-Wealden boring can be described, as it sometimes has been, as a " search for coal." The primary object is to learn what rocks underlie the Weald, this being a point of high scientific interest.

Such were the problems presented for solution. The methods employed and the results obtained are as follows :—A committee of reference was formed in London, with Prof. A. C. Ramsay as chairman ; Mr. H. Willett, of Brighton, has throughout acted as hon. sec. and treasurer. The money has been mainly raised by private subscription, aided by grants from the Government, the Royal Society, and the British Association. Two borings have been made ; the first was abandoned at a depth of 1,030 feet, owing to an accident to the rods. The second boring was commenced in Feb. 1875, and is now (in March 1876) 1,903 feet from the surface.

The specimens exhibited are arranged in two series ; those on the top and second shelves are examples of fossils from the Kimeridge Clay, named by Mr. Etheridge. Those on the lower shelves are arranged in order of depth, the ends nearest the surface being placed on the left hand side of the bottom shelf ; the examples in this series are mainly from the second boring, but a few specimens are added from the first boring—these are marked as such.

The work is performed by the Diamond Rock-boring Company. The rock is bored by a rapidly revolving " crown " set with diamonds ; the débris are carried up to the surface by a stream of water which is forced down the hollow boring-rod. The " core " of rock rises within the enlarged space, or " core-tube, at the bottom of the rods ; it is thus preserved, and is afterwards drawn to the surface. The machinery employed is shown in the photo-

graphs; an example of the boring " crown " is placed on the bottom
shelf of the case; specimens of long " cores " extricated in solid
masses as here shown are placed by the side of the case.

The second boring commenced with a hole 8 inches in diameter,
yielding a 7-inch core; this was gradually reduced, owing to the
necessity of lining the hole, until, at 1,670 feet, it was reduced to
a 2-inch hole, yielding a 1-inch core.

The long diagram gives a detailed section of the strata passed
through in the second boring, which may be grouped as follows :—

			Thickness.	Depth from Surface.
Purbeck Beds -	-	{ Shales, limestones, cement-stones, and gypsum - - - }	200 -	—
Portland Beds -	-	Sandstone - - - -	57 -	257
Kimeridge Beds	-	{ A variable set of strata; chiefly shales, and cement stones in the upper part, with many beds of sandstone and limestone in the lower part - - }	1,512 -	1,769
Coralline Oolite	--	Oolitic limestone - - -	17 -	1,786
? ?	-	Shales, sandy shales, and limestone	117 -	1,903
			1,903	—

There is still some uncertainty as to the relations of the lowest
strata; they may belong to the Oxford Clay, or they may represent
the lower Coralline series of Dorsetshire. This point will be
cleared up as the boring progresses. The greater part of the
Kimeridge Clay is very fossiliferous, some of the fossils being new
to England; the least productive parts are the highly calcareous
shales or cement stones; some of the limestones are mainly composed of small oysters; an example is shown in the small core at
the side of the case.

The smaller diagram gives a section of the Cretaceous and Oolitic
rocks of the south-east of England; distinguishing those previously
known in that area from those discovered by the boring. The
same point is brought out in the horizontal section, which also
serves to illustrate the structure of the Wealden area. A simple
inspection of these diagrams will show the amount of information
already obtained by the boring. If the Palæozoic rocks should not
be reached, the boring will still have yielded most valuable results,
for we shall have acquired a knowledge of the Oolitic strata of this
area, such as could not possibly have been obtained in any other way.

Some valuable beds of Gypsum were discovered in the Purbeck
Beds; these are now being worked, giving rise to a new branch of
industry in Sussex.

The position of the boring is shown in the Geological Model of the Weald which is exhibited in an adjoining case. Further details of the structure of the districts can be seen in the maps and sections of the Geological Survey.

3271. Geological Maps in relief, embossed in paper, of ENGLAND AND WALES, after Sir R. I. Murchison, and THE ISLE OF WIGHT, after H. W. Bristow, F.R.S. *Henry F. Brion.*

3272. Model of Etna. *Th. Dickert, Poppelsdorf, near Bonn.*
The model exhibited is a scientific work, much valued and used by all the larger institutes of Germany for illustration in their science classes.

3273. The Great Crater of Vesuvius. Photograph, showing the interior after the eruption of April 1872.
Robert James Mann, M.D.
This photograph was taken by Mr. J. M. Black from the gap in the broken edge of the rim. The top of the great rent, extending north and south through one side of the cone into the Atreo del Cavallo, is shown on the further side of the crater between the rounded and pointed eminences.

3274. Four Sketches representing a **Volcanic Eruption,** to illustrate the form.
Mineralogical and Geological Cabinet of the School of Industry, Cassel (Dr. H. Möhl).

3275. Four Views of Crater Eruptions.
Mineralogical and Geological Cabinet of the School of Industry, Cassel (Dr. H. Möhl).
The drawings are used as wall maps to illustrate geological lectures. For more information, v. Papers XI., XIII.

3276. Photographic Views (2) of Mount Sorrel granite quarries, Charnwood Forest, Leicestershire, by Messrs. T. and J. Spencer. *W. J. Harrison, Town Museum, Leicester.*
This rock is a tough hornblendic granite, largely used for revatt setts, kerbs, &c.

3277. Photographs of the slates and syenites of Charnwood Forest, and of a column of the hard rocks of Leicestershire erected in the grounds of the Leicester Town Museum.
W. J. Harrison, Town Museum, Leicester.
These rocks are coloured as Cambrian on the maps of the Geological Survey, but they have yielded no fossils, and the evidence of superposition is not clear, as the oldest rocks near are of Mountain Limestone. The views are by Messrs. J. Burton and Sons.

3278. Rock Sections. Series prepared for microscopical examination. *James How & Co.*

3281. Relief Map of the Habichtswald, near Cassel, province Hesse-Nassau ; petrographically coloured.
Friedrich Sievers, Wehlheiden, near Cassel.

Constructed from the maps of the former electorate of Hesse, the level lines in which were taken at a distance of 60 Prussian feet apart, so that the surface of the North Sea is represented by 0.

The distribution of the rocks is shown in colours according to the observations of Dr. Möhl and the exhibitor, as follows :—

Carmine -   -   - the lower group of the " Bunter " sandstone, or New Red Sandstone.

Reddish yellow -   - the upper group of the same.

Blue -   -   - " Muschelkalk."

Light-greenish grey - clay and sand with lignite.

Dark-greenish grey - marine sand rich in fossils, and septarium clay (Upper Oligocene).

White -   -   - alluvium.

Black -   -   - basalt.

Brown -   -   - basalt conglomerate.

Villages are shown in vermilion, railways and high roads in black, standing and running waters in blue, with corresponding figures referring to index, which is attached to the relief.

3285. Geological Sections.   *University Museum, Oxford.*

Coast of Dorset, from Lyme Regis to Isle of Portland.
Slapton Sands, between Dartmouth and Plymouth.
Dunolly Castle, near Oban.
Brent Tor, near Tavistock.
View from Exmouth.
Parallel Roads of Glen Roy.
Section from Beerhead to Axmouth.
Country between Malverns and Cotswolds.
Coast of Dorset between Charmouth and Abbotsbury.
Submarine Forest of Stolford.
Views of the Coast of Devon E. and W. of Sidmouth (2).
Sections of Landslip.
Bird's Eye View of part of Devonshire.
View of Cliffs, Lyme Regis.

3287a. Microscopical Preparations of Spanish Rocks.
Francisco Quiroga, Madrid.

1. Spanish silicious diatoms.
2. Rutile from Horcaguelo (Madrid).
3. Molibdate of lead from Quentar (Granada).
4. Tourmaline from Buitrago (Madrid).
5. Tremolite from the province of Guadalayera, composed of tremolite and glass.
6. Meteorolite from Molina (Murcia), 24th December 1858 ; composed of pyroxene, peridot, and other metallic substances.
7. Meteorolite from Molina (Murcia), 24th December 1858 ; insoluble silicate produced by the action of acids ; fòrmed of pyroxene.
8. Meteorolite from Cangas de Onis (Asturias), 6th December 1866 ; it is formed of pyroxene, peridot, and metallic substances.

9. Dimyte from the Serrania de Ronda, formed of peridot picotite.*
10. Dimyte from the Serrania de Ronda, formed of peridot, picotite.*
11. Chrysolite from the Serrania de Ronda, formed of peridot, diopside.*
12. Serpentine from Sierra Parda (Serrania de Ronda), formed of peridot, serpentine, hematite.*
13. Serpentine from Sierra Parda (Serrania de Ronda), formed of serpentine, peridot, magnetite, hematite.*
14. Serpentine from Istau (Serrania de Ronda), formed of serpentine and magnetite.
15. Serpentine from Ystau (Serrania de Ronda), formed of serpentine and magnetite.
16, 17. Serpentine from Real del Dugue (Serrania de Ronda), formed of serpentine and magnetite.
18. Serpentine from Ronda, formed of serpentine and magnetite.
19. Serpentine from Torrevieja (Alicante), formed of serpentine and magnetite.
20-22. Serpentine from Barranco de San Juan (Granada), formed of serpentine and magnetite.
23. Serpentine from the Sierra of Guejar (Granada), formed of serpentine and magnetite.
24. Chlorite state of Cogulludo, formed of chlorite and magnetite.
25. Anthrakonite from Galicia, formed of fetid calcite.
26. Nummulitic limestone from San Vicento de la Barguera (Santander).
27. Cipolin from Robledo de Chavela (Madrid), formed of calcite and muscovite.
28. Chloritic limestone from Almaden, formed of calcite chlorite.
29. Obdurate clay of Segovia (cretaceous period), formed of clay, quartz.†
30. Clay slate of Peña de Penilla (silurian age), Segovia, formed of clay, feldspar, and quartz.
31. Talcose slate of the Escorial, formed of talc, quartz, and muscovite.
32. Felsyte from the Barranco del Cebollon, Granada, formed of felsyte, quartz, amphibole.
33, 34. Felsyte from the Escorial, formed of felsyte alone.
35. Felsyte from Peguerinos (Avila) formed of felsyte, quartz.‡
36. Granite from Robledo de Chavila (Madrid), formed of orthoclase, quartz, muscovite, and hornblende.
37. Gneiss from Chapas de Marbella (Serrania de Ronda), formed of orthoclase, quartz, muscovite, apatite.§
38. Gneiss from Mijas (Serrania de Ronda), formed of orthoclase, quartz, and muscovite.
39. Garnetiferous gneiss from Huertal (Granada), formed of orthoclase, quartz, muscovite, garnet, and picotite.
40. Gneiss from Robledo de Chavila (Madrid), formed of orthoclase, quartz, muscovite, and hornblende.
41. Gneiss from the Escorial, formed of muscovite and quartz and orthoclase.
42. Micaceous gneiss from the Escorial, formed of muscovite and quartz.

* Brevis apuntis auria del origen peridotico de la serpentina de la Serrania de Ronda, by Don T. Macpherson. Anales de la Sociedad Española de Historia Natural, J. iv.
† Excursion geologica por la provincia de Segovia, por Don Alfonso de Aristo y Sarrinaga y Don Francisco Quiroga y Rodriguer. Anales de la Socd. Española de Hist. Natural, J. iii.
‡ Anales de la Soc. Esp. de Hist. Nat. J. iv. Actas, p. 73, sesion del 1 de Setz. de 1875. Paper read by F. Quiroga.
§ See (*), Memoria 10bre la estructura de la Serrania de Ronda, by T. Macpherson. Cadiz, 1874.

43. Coccolite with garnetite from the Escorial, formed of diopside and garnet.
44. Coccolite from Riara (Segovia), formed of diopside, garnet, hematite, quartz with fluid cavities, and apatite.*
45. Coccolite from Riarja (Segovia), formed of diopside and hematite.
46. Coccolite with garnetite from Riarja, Segovia, formed of diopside and garnet.
47. Garnetite from Riarja (Segovia), formed of garnet and diopside.
48. Hyalomicte from Riarja (Segovia), formed of quartz and muscovite.
49. Ophite from Puerto Real, Cadiz, formed of plagioclase, augite, diallage, epidote, hornblende, magnetite.†
50. Ophite from Puerto Real (Cadiz) formed of felspar, hornblende, magnetite.
51. Diabase from the Sierra de Cordova, formed of plagioclase, chlorite, augite, magnetite.
52. Garnetiferous amphibolite from Pedraya de la Sierra (Segovia), formed of amphibole and garnet.
53. Garnetiferous amphibolite from Pedraya de la Sierra, composed of amphibole and quartz with fluid cavities.
54. Quartziferous amphibolite from the escorial, formed of amphibole and quartz.
55. Quartziferous amphibolite from Dilar (Granada), formed of amphibole and quartz.
56. Garnetic quartziferous amphibolite from Toril de Dilay (Granada), formed of amphibole, quartz, garnet, and picotite.
57. Quartziferous amphibolite from Laguna de Bacares (Granada), formed of amphibole, quartz, and picotite.
58. Quartziferous amphibolite from Barranco de les Araligos (Granada), formed of hornblende and quartz.
59. Quartziferous amphibolite from Barranco de los Araligos (Granada), formed of hornblende and quartz.
60. Quartziferous amphibolite from Barranco de los Araligos (Granada), formed of hornblende, quartz, garnet, and hematite.
61. Quartzdiorite from Almaden, formed of plagioclase, hornblende, magnetite, quartz, with apatite.
62. Quartzdiorite from Almaden, insoluble from the action of acids.
63. Diorite from Peguerinos (Avila), formed of plagioclase, hornblende.‡
64. Idem from Peguerinos (Avila), formed of plagioclase, hornblende.
65. Idem from Peguerinos (Avila), formed of plagioclase, hornblende, magnetite, augite.
66. Idem from Peguerinos (Avila), formed of plagioclase, hornblende, magnetite, viridite, calcite, and quartz.
67. Diorite from Peguerinos. (Avila), formed of plagioclase, hornblende, magnetite, viridite, and augite.
68. Trachyte from Cartagena, formed of sanidin, hornblende, magnetite.
69. Trachyte from Telde (Grau Canaria), formed of sanidin, magnetite.
70. Trachyte from Guia (Grau´Canaria), vide No. 8 of the collection of Grau Canaria Rocks, by M. S. Calderon, formed of sanidin, magnetite, hornblende.§

---

* Observaciones sobre algunas rocas de Riarja (Segovia), by Quiroga. An. de la Soc. Esp. de Hist. Natural, J. v.
† Bosquijo geologico de la provincia de Cadiz, por T. Macpherson. Cadiz, 1872, and An. de la Soc. Esp. de Hist. Nat., J. v.
‡ Anales de la Soc. Esp. de Hist. Nat., J. iv., Actas, p. 78, sesion del 1 de Setz. de 1875. Paper read by F. Quiroga.
§ Reseña de las rocas de la Isla Volcanica Grau Canaria, por Salvador Calderon. An. Soc. Esp. de Hist. Natural, J. iv.

71. Trachyte from Palmas (Grau Canaria), vide No. 5 of the collection of Grau Canaria rocks of M. S. Calderon, formed of sanidin, magnetite.
70. Trachyte from Grau Canaria, vide No. 7 of collection of Grau Canaria rocks of M. S. Calderon, formed of sanidin, magnetite, augite, hornblende.
73. Trachyte from Grau Canaria (No. 9 of collections of Grau Canaria rocks, by M. S. Calderon, formed of sanidin, magnetite, hornblende.
74. Trachyte from La Cumbro (Grau Canaria), No. 3 of the collection of Grau Canaria, by M. S. Calderon, formed of sanidin, magnetite, augite, aragonite.
75. Tufa from Artenaza (Grau Canaria), No. 53 of the collection of Grau Canaria of M. S. Calderon, formed of sanidin, clay.
76. Plagioclase dolerite from Salto del Castellano (Grau Canaria) No. 45 of the collection of Grau Canaria rocks, by M. S. Calderon, formed of plagioclase, augite, and magnetite.
77. Basalt from Las Palmas, Grau Canaria, No. 40 of collection of Grau Canaria rocks of M. S. Calderon, formed of augite, magnetite, olivine, plagioclase.
78. Basalt from Las Palmas, Grau Canaria, No. 39 collection Grau Canaria rocks of M. S. Calderon, formed of augite, magnetite, olivine, plagioclase.
79. Plagioclasbasalt from La Cumbre (Grau Canaria), vide No. 58 of collection of Grau Canaria rocks of M. S. Calderon, formed of plagioclase, augite, magnetite, hematite.
80. Nephelinebasalt, from La Cumbre (Grau Canaria), No. 36 of the collection of Grau Canaria of M. S. Calderon), formed of nepheline, augite, magnetite, plagioclase, hematite.
81. Rosed felsyte from the Escorial, formed of felsyte and hornblende.
82. Serpentine from Burgos, formed of serpentine, diallage, magnetite.
83. Phonolyte from Agacte, Grau Canaria (No. 23 of collection of Grau Canaria rocks of M. S. Calderon), formed of sanidin, nepheline, hornblende, sphene.
84. Leptynyte from Sierra de Guadarrama, formed of orthoclase, quartz, muscovite, apatite.
85. Diorite from Peguerinos (Avila), formed of plagioclase, hornblende, magnetite, augite.
86. Diorite from Buitrago (Madrid), formed of plagioclase, hornblende, magnetite.
87. Ophite from Pando (Santander), formed of plagioclase, augite, diallage, hornblende, viridite epidote, magnetite.*
88. Ophite from Pando (Santander), treated by acids, formed of plagioclase, augite, diallage, epidote.
89. Altered ophite from Pando (Santander), formed of plagioclase, augite, diallage, magnetite, and viritite.
90. Garnet fils of Riaza (Segovia), formed of garnet, hornblende, plagioclase.†

3288. Geological Sections of five proposed lines of railway through the counties of Surrey and Sussex, from Brighton to London, made and surveyed in 1837, by Joseph Gibbs, C.E., and Arthur Dean, C.E. ☐*Free Library and Museum, Brighton.*

South-Eastern, Brighton line, from Carlton Hill, Brighton, to junction with the Greenwich Railway.) Direct line, from Church Street, Brighton, to junc-

---

* Opita de Pando (Santander), por F. Quirago au de la Soc. Esp. de Hist. Nat., J. V.
† Vide Elemente de Pelographic von Du. A. von Lasaulx. Bonn, 1875.

tion with the Greenwich Railway. Gibbs' line, from Western Road, Brighton, to junction with the Croydon Railway. Stephenson's line, from North Lane, Brighton, to Nine Elms. Line without a tunnel, from Brighton to Kennington Oval.

3289. Sections showing the positions of the **Palæozoic Rocks** beneath the Tertiary or Secondary Strata, with some specimens of the former.   *Prof. Prestwich, Oxford.*

1. Section of the Kentish Town Well, London.
2. Specimens of some of the strata passed through in the above well.
3. Section of the Artesian Well boring at Ostend.

3290. Geological Section from **Paris** to **Brest.**
Delesse, Paris.

This outline section has been executed, under the direction of M. Mille, Chief Engineer, Ponts et Chaussées by Messrs. Triger, Delesse, and Guillier. It is on a scale of 1:40,000 horizontal, and 1:2,000 vertical. It follows the line of the railway which, leaving Paris, passes by Bonneval, Chateaudun, Vendome, Tours, Angers, Nantes, Vannes, and Quimper, and comes out at Brest. Many classic regions are traversed in this line, as the Paris basin, Beauce, Touraine, the Valley of the Loire, and a part of Brittany.

On this section can be followed the succession of the different geological strata, which are marked the whole length of the railway ; their position is determined by their height above the level of the sea, so that, notwithstanding the exaggeration of the scale of elevation, it is easy to ascertain their relative position.

In the regions of La Beauce the subterranean sources of water supply have been specially studied.

This section also affords information as to the materials of construction furnished by each geological stratum, and the vegetable soil and the nature of the crops.

Geological studies of this kind are eminently useful if they precede the formation of railways, because they supply a knowledge of the difficulties likely to be encountered during the course of construction, as well as the resources which may be reckoned on in each region traversed ; and even when made after the completion of a railway, they supply facts highly useful to science and industry.

In pursuance of the orders of M. de Franqueville, Director-General, Ponts et Chaussées, these geological surveys have been carried on by M. Mille over a great part of France.

3291. Original Sketches, illustrative of **Geological Scenery** and sections, taken by Dr. Buckland between 1815 and 1840.   *The Oxford University Museum, Geological Section.*

1. Landslip, Lyme Regis.
2. View of Coast near Lyme Regis.
3. View of Coast near Sidmouth.
4. View of Coast near Sidmouth.
5. Submarine Forest of Stolford near Bridgewater.
6. View of Coast between Charmouth and Abbotsbury.
7. Bird's eye view of Dartmoor and south coast of Devon.
8. View of Vale of Severn between Malverns and the Cotswolds.
9. Coast view between Beerhead and Axmouth.

10. Parallel roads of Glen Roy.
11. View of Exmouth.
12. Brent Tor near Tavistock.
13. Dunolly Castle near Oban.
14. View of Shapton Sands.
15. View of south coast of Lyme Regis with Portland in the distance.

3292. Geology. Collection of specimens of felspars and amphibolitic rocks from Belgium and the French Ardennes. This collection comprises about 20 specimens, average dimensions 10 centimètres by 15.

Six chromolithographs of drawings of minerals and rocks as seen through the microscope. *A. Renard, Louvain.*

3293. Specimens illustrating the production by **Compression** of **Natural** and **Artificial Slaty Cleavage.** *H. C. Sorby.*

Specimens of slate rocks, showing, by various facts, that they have been greatly compressed in a line perpendicular to the cleavage.

Pipe-clay mixed with portions of blue paper, and also with iron scales, being the results of the first experiments made to show that a structure like that which causes the cleavage in slates can be artificially produced by pressure.

Artificial cleavage in compressed flaky graphite, being as perfect as that in any slate rock.

3294. Specimens illustrating the **Metamorphic Origin of Mica Schist,** and the difference between stratification-foliation and cleavage-foliation. *H. C. Sorby.*

" Ripple drift " in slate rocks in which the cleavage cuts the stratification at a considerable angle.

" Ripple drift " in contorted and highly metamorphosed mica schist, thus proving the original stratified nature of the rock.

Mica schist with foliation in the plane of stratification, being rock metamorphosed before being compressed.

Mica schist with foliation in the plane of cleavage, developed by compression, before the work was metamorphosed.

3295. Microscopical Photographs of sections of iron and steel. *H. C. Sorby.*

The above were photographed by means of strong surface illumination, and show structures due to the arrangement of crystals of iron combined with a varying amount of carbon, of portions of slag, and of crystalline plates of graphite. Note the contrast between the structure of cast iron, cast steel, and meteoric iron, although all have solidified from fusion.

3296. Microscopical Sections of iron and steel. *H. C. Sorby.*

The above were prepared by very carefully grinding down and polishing the surface so as to avoid all such burnishing action as would alter the form or structure of the ultimate crystalline particles. The whole section is then placed in very dilute nitric acid, and carefully examined in water under the microscope, time after time, until the irregular action of the acid on the different constituents has advanced so far as to show the general structure to the greatest advantage. The surface is then well washed, dried, and gently wiped, and finally protected by a thin glass cover cemented down by Canada balsam.

3297. Working Model illustrating the **Formation** of **" False Bedding "** in **Stratified Rocks.**   *H. C. Sorby.*

The drifting action of the current of water is represented by the screw, which carries forward the sand until it falls down and accumulates on the slope at the angle of rest. The larger fragments roll to the bottom, and the fine particles are sorted or not into thin or thicker bands according as the screw is moved in an irregular or regular manner, just as occurs when a current moves with varying velocity. The arrangement of the materials thus produced is in all respects similar to what is so often seen in certain beds of stratified rocks.

3298. Microscopical Sections of **Shells** and **Rocks.**
    H. C. Sorby.

These are prepared by grinding down one side to a perfectly flat and smooth surface, which is then fixed to a glass with Canada balsam. After this, the other side is ground down, first with emery and at last on a very hard piece of water of Ayr stone, so as to leave a thin portion of the rock in a perfectly undisturbed condition attached to the glass, with a smooth and almost polished surface. According to the nature of the rock, the thickness of such sections should vary from about $\frac{1}{100}$ to $\frac{1}{1000}$th of an inch. Thin glass is then mounted over the whole with Canada balsam. As examples of rocks presenting special difficulties, attention may be drawn to the sections of soft chalk, and of slate and mica schist, cut perpendicularly to the cleavage or foliation.

3299. Lithographed Plates, illustrating the **Microscopical Structure** of **Limestones.**   *H. C. Sorby.*

These show structure due to larger or smaller fragments of organic bodies or grains of sand, and to the more or less complete development of crystals formed either during or after the deposition of rock.

3300. Working Model, illustrating the **Movement** of **Waves** in forming ripple marks.   *H. C. Sorby.*

The model shows the movement of the particles at the surface and at the bottom of a wave which has advanced from deep to shallow water, so as to give rise to ripple marks. At the surface the water moves nearly in circles, which in the model is represented by the white discs attached to arms connected with each alternate wheel. At the bottom the water moves forwards and backwards, drifting the sand in the line of the movement of the wave when its crest is passing, and backwards when the trough is passing. In the model this movement is produced by the action of small eccentric wheels on pieces of brass fixed at one end and carrying white discs at the other.

3300a. Specimens relating to the exhibitor's researches on synthetical and experimental **Geology.**

1. Minerals formed at about 400° by superheated water (1860).
2. Product of the reactions of vapours.

3. Contemporaneous minerals, produced by the action of hot springs (1858 to 1876).

4. Transformation of terrestrial rock into meteoric rock (1866).

5. Imitation, by experiment made with native magneto-polar platina (1875).

6. Experiments upon the possibility of capillary infiltration through the pores of rocks, notwithstanding the resistance of a high steam pressure (1861).

7. Experiments upon the schistose properties of rocks, geological deductions (1860 to 1876).

M. Daubrée, Director, School of Mines, Paris.

3252p. Stand and Glass Cover, with eight specimens of minerals of contemporaneous formation.

M. Daubrée, Director, School of Mines, Paris.

3252q. Stand and Glass Cover, with four ispecimens of minerals formed in water heated to 400°.

M. Daubrée, Director, School of Mines, Paris.

3252t. Stand, with three tubes of iron burst by expansion of steam in the above experiments.

M. Daubrée, Director, School of Mines, Paris.

3252r. Apparatus to demonstrate the possibility of capillary infiltration in the pores of rocks.

M. Daubrée, Director, School of Mines, Paris.

3252s. Specimen, illustrating the transformation of terrestrial rocks into meteoric stones.

M. Daubrée, Director, School of Mines, Paris.

3252u. Three Specimens of Earth, illustrating the schistosity of rocks. *M. Daubrée, Director, School of Mines, Paris.*

3252v. Small Stand and Glass Shade, with three specimens of magneto-polar platinum (contained in glass covered cardboard case). *M. Daubrée, Director, School of Mines, Paris.*

3252w. Bound Volume, " Programme des Cours de l'Ecole des Mines."

M. Daubrée, Director, School of Mines, Paris.

3252x. Bound Volume, " Institution et But de l'Ecole Programme de l'Admission," etc.

M. Daubrée, Director, School of Mines, Paris.

3307. Sections of Typical Rocks (30 specimens), a selection of Prof. J. Roth, Berlin; with a commentary by Prof. H. Rosenbusch, Strâsburg. *R. Fuess, Berlin.*

3308. Sections of Typical Rocks (30 specimens), with a commentary; a selection by Prof. F. Zirkel, Leipsic.

R. Fuess, Berlin.

3309. Sections of Typical Basalts (30 specimens), a selection by Prof. H. Möhl, Cassel; with a commentary.

R. Fuess, Berlin.

3310. Sections of Rock-forming Minerals (30 specimens), a selection by Prof. H. Rosenbusch, Strasburg; with a commentary.

R. Fuess, Berlin.

3311. Sections of the Rocks of the Kaiserstuhl (30 specimens), a selection by Prof. H. Rosenbusch, Strasburg.

R. Fuess, Berlin.

3312. Sections of the Eruptive Rocks of Hungary and Servia (30 specimens), a selection by Prof. Szabo, Budapest; with a commentary.

R. Fuess, Berlin.

3313. Sections of Typical Rocks (30 specimens), a selection by Prof. F. Zirkel, Leipsic ; with a commentary.

R. Fuess, Berlin.

3314. Sections of the most Characteristic Rocks of the Huronian iron region, Lake Superior, U.S.A. (30 specimens), a selection by Major T. B. Brooks, Wisconsin Geological Survey, and microscopically examined by Dr. Wichmann, Leipsic.

R. Fuess, Berlin.

3315. Collection of Thin Sections of all the rocks occurring in the Fichtelgebirge.

Geological Survey of Bavaria (Dr. Gümbel).

The collection of thin sections illustrates the method of collecting and examining, large numbers of slices from the same rock and locality being prepared. The accompanying plate shows sections of a few of the most important of these rocks magnified and printed in colours.

3316. Pictures of the most important of these **Rocks,** as drawn from sections under the microscope.

Geological Survey of Bavaria (Dr. Gümbel).

3317. Collection of Hand Specimens of Typical Tertiary Basalts and altered enclosures (68 specimens), illustrated by one hundred thin sections.

Mineralogical and Geological Cabinet of the School of Industry, Cassel (Dr. H Möhl).

The classification of the Basalts is that begun by Prof. F. Zirkel, and extended by the exhibitor, as illustrated in Paper No. I. Special descriptions of these rocks and their structure will be found in the Papers II.-VII., the figures being contained in Paper X.

3318. Collection of Hand Specimens of Typical Phonolites and nearly related Andesites (38 specimens), illustrated by seventy thin sections.

Mineralogical and Geological Cabinet of the School of Industry, Cassel (Dr. H. Möhl).

The classification of the Phonolites by the exhibitor is contained in Paper No. IX., the felspar phonolites of other authors being omitted as approaching more to the trachytes. Drawings and descriptions of normal types are found in Paper No. X., with the original plates G and H.

3319. Collection of Hand Specimens of Typical Basaltites (Melaphyre, Palatinite, Minette, Kersanton and Kersantite), also Porphyrite (48 specimens), illustrated by ninety thin sections. *Mineralogical and Geological Cabinet of the School of Industry, Cassel (Dr. H. Möhl).*

For reasons in favour of adopting the name Basaltite, *see* Paper No. 8, 1st and 2nd part. The descriptions of two so-called minettes will be found in Paper No. VII., 3rd part.

3320. Thirteen Papers relating to the above rocks and sections, with drawings, also ten coloured plates of drawings of thin sections (amongst them six original plates).
Original water-colour drawings.
Mineralogical and Geological Cabinet of the School of Industry, Cassel (Dr. H. Möhl).

3319a. Collection of Specimens of Rocks of the Canary Islands. *Don Salvador Calderon, Madrid.*

These specimens are accompanied by a memoir by Señor Calderon, on the rocks of the Canary Islands, entitled "Resina de las rocas de la isla volcanica Grau Canaria." 4° Madrid, 1876.

3321. Thin Sections of typical rocks in a box (90 specimens). *Voigt and Hochgesang, Göttingen (Gust. Voigt).*

Contains rocks selected according to F. Zirkel's "Microscopical Characters of Minerals and Rocks."

3322. Thin Sections of typical rocks in a box (20 specimens). *Voigt and Hochgesang, Göttingen (Gust. Voigt).*

Collected by Dr. v. Seebach, and described by Dr. F. Zirkel, Leipzic.

3323. Series of the most characteristic Porphyries of Silesia, with thin sections of the same, in a case.
University of Breslau (Prof. A. von Lasaulx).

3324. Collection of Thin Sections of the rocks of Saxony.
Royal Saxon Mining Academy, Freiberg.

3266. Maps, illustrative of theories of relative directions of lodes, joints, mountain chains, coast lines, limits of geological formations and rivers. *Jos. P. O'Reilly.*

Map of Europe (geographical); two base lines, viz., Algerian coast and Syrian coast lines.
Map of Spain (Francisco Cuillo's geographical); base line taken from the map of Europe. (Compare with de Verneuil's map for correspondence of geographical and geological lines of direction.)

Alpine Club Map of Switzerland; two base lines from globe, east coast of Sardinia and south-east coast of Red Sea.

Jukes's geological map of Ireland; base line, eastern coast of Adriatic.

The base lines are portions of great circles having the directions of, and passing along, the coast lines indicated, and have been transferred from the globe.

3326. Globe mounted for the study and demonstration of the angular relations of directions presented by lodes, dykes, lines of dislocation, mountain chains, coast lines, and river valleys.

Jos. P. O'Reilly.

This globe allows of the tracing of great circles through given points, with great facility and rapidity. These great circles can be transferred to maps, and serve as bases of comparison for the lode systems of mining districts, and in general for all geological lines of direction.

The above was designed by the exhibitor in 1874, and used for public lectures. A similarly mounted globe was shown in Paris at the Geographical Exhibition of 1875, by M. Chancourtois.

3327. Portable Apparatus, for preparing Sections of rocks, minerals, fossils, &c., for microscopic examination.

F. G. Cuttell.

3328. Specimens of Sections prepared with the above machine.   *F. G. Cuttell.*

3328. Lapidary Apparatus for slitting, grinding, and polishing rocks, pebbles, fossils, &c., especially adapted for the

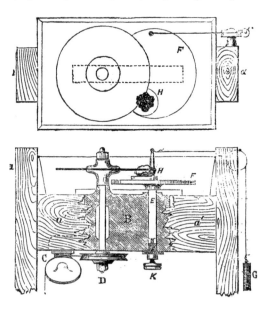

preparation of thin sections of rock or other hard material for microscopical examination. Designed by the exhibitor, and manufactured by Messrs. Cotton & Johnson, Soho. **James B. Jordan.**

The machine consists of a mahogany, oak, or beech framework (*a a*), similar to that of an ordinary turning lathe, and is supplied with crank, fly-wheel, and treadle, occupying a floor space of 2 ft. 2 in. by 20 in. To the bed of this frame is fitted an iron casting (B), bored to receive two spindles, viz. :—a vertical spindle (D), which carries the slitting disc, grinding laps, &c., and another spindle (E), to which is fixed a horizontal plate (F), carrying on its surface a cup containing cement, in which the material to be cut is imbedded, and which is retained and uniformly pressed against the slitting disc by means of a cord attached to the plate, which, passing over a pulley having a suspended weight (G), gives a constant and regular pressure to the plate containing the specimen. There is also a convenient screw arrangement (K) by which this plate can be accurately adjusted for the purpose of regulating the thickness of the slice. The slitting disc and laps are made to revolve, at the rate of about 500 revolutions per minute, by means of a treadle used in the ordinary way.

3329. Apparatus for the preparation of rock sections, and plates of crystals for examination in polarised light.

R. Fuess, Berlin.

3330. Apparatus for the preparation of **Thin Microscopic Sections** of rocks and minerals :—

1. Cutting apparatus.
2. Grinding and polishing apparatus.

Christ. Weber, Eisenach.

3353a. Ceteosaurus Oxoniensis.

Left femur.
Dorsal vertebra, 2 views.
Left humerus.
Ilium, 2 views.

Dr. Acland, F.R.S., on behalf of the Radcliffe Trustees.

3353b. Megalosaurus Bucklandi.

Upper jaw.
Lower jaw.
Scapula.
Pubis.
Right tibia.
Ilium.
Caracoid.
Rib.
Ischium, 2 views.
Sacrum.

Dr. Acland, F.R.S., on behalf of the Radcliffe Trustees.

3354. Plates of British Fossils, from the annual volumes of the Palæontographical Society.   *The Rev. Thomas Wiltshire.*

Illustrating :—
Pleistocene Mammalia (skull of Felis spelæa).
Upper Greensand Urchins (Caratomus rostratus).
Lower Greensand Urchins (Trematopygus faringdonensis).
Wealden Reptilia (Radius and Ulna of Iguanodon Mantelli).
Purbeck Mammalia (Teeth and Jaws of Stylodon, Leptocladus, Bolodon, &c.).
Oxford Clay Belemnites (Belemnites Puzosianus ; B. porrectus).
Inferior Oolite Corals (Thecosmilia Wrighti ; T. Flemingi).
    ,,   ,,   Trigoniæ (Trigonia signata ; T. Moretoni).
Liassic Corals (Astrocænia gibbosa, Helistræa Moorei, &c.).
  ,, Starfish (Ophioderma Gaveyi ; O. Egertoni).
Carboniferous Flora (Halonia regularis).
Devonian Crustacea (Stylonurus Powriei ; Cheirurus articulatus).
    ,,   Brachiopoda (Spirifera disjuncta, Spiriferina cristata).
    ,,   Fish (Cephalaspis Lyelli ; C. Murchisonii).
Silurian Trilobites (Phacops, Amphion, Asaphus).
    ,,   Eurypterida (Slimonia acuminata, Pterygotus bilobus).

3355. Collection of Fossil Plants, proving the existence of the elements of recent flora in that of the Tertiary period, comprising examples of the elementary forms of salt and fresh water flora, and of the various continents, including Australia.

Fossil plants from various localities in the Tertiary formation are also exhibited, exemplifying the origin of our recent local flora from ancient types.

This is shown by coloured tickets attached to the specimens.

The change of the fossil *Castanea atavia* into the recent *Castanea vesca* is also illustrated.

Prof. Dr. Constantin Baron von Ettingshausen, Graz.

I.—The Tertiary Floras, Elements in General.

A.—Elements of the Flora of Salt Water.

Caulerpa pyramidalis, *Sternb.* Wiener Sandstein.
Caulerpa, *sp. n.* Wiener Sandstein.
Sphaerococcites, *sp. n.* Podsused.
Chondrites furcatus, *Brongn.* Wiener Sandstein.

Chondrites intricatus, *Brongn.* Wiener Sandstein.
Cystoseira communis, *Ung.* Radoboj.
Zostera Ungeri, *Ett.* Radoboj.

B.—Elements of the Flora of Fresh Water.

Confervites bilinicus, *Ung.* Schoenegg.
Chara Meriani, *A. Braun.* Savine.
Typha latissima, *A. Braun.* Schoenegg, Savine, Leoben.
Sparganium, *sp. n.* Schoenegg.

Najadopsis lucens, *Ett.* Schoenegg.
Ceratophyllum, *sp. n.* Schoenegg.
Anoectomeria Brongniarti, *Sap.* Leoben, Podsused, Haering.
Nelumbium Buchii, *Ett.* Leoben, Kutschlin.

C.—Elements of the Flora of Australia.

Casuarina Haidingeri, *Ett.* Leoben, Haering, Schoenegg.
Leptomeria Benthami, *Ett.* (Styriaca, *Ett. MS.*). Schoenegg.
Leptomeria gracilis, *Ett.* Haering.
Santalum, *sp. n.* Schoenegg.
Santalum osyrinum, *Ett.* Haering.
Hakea, *sp. n.* Parschlug.
Hakea macroptera, *Ett.* Sagor, Schoenegg.

Banksia longifolia, *Ett.* Savine.
Banksia haeringiana, *Ett.* Haering.
Banksia Ungeri, *Ett.* Monte Promina.
Banksia Haidingeri, *Ett.* Trifail.
Dryandra acutiloba, *Ett.* Bilin.
Eucalyptus oceanica, *Ung.* Sotzka.
Eucalyptus haeringiana, *Ett.* Haering.

D.—Elements of the Asiatic Floras.

Glyptostrobus europaeus, *Heer.* Schoenegg.
Planera Ungeri, *Ett.* Parschlug.
Ficus Deschmanni, *Ett.* Savine.
Ficus tenuinervis, *Ett.* Islaak.
Iuglans costata, *Ung.* Putschirn.
Iuglans parschlugiana, *Ung.* Parschlug.

Engelhardtia Brongniarti, *Sap.* Radoboj, Savine, Leoben.
Cinnamomum polymorphum, *A. Braun.* Savine, Promina, Schoenegg.
Nerium, *sp. n.* Fohnsdorf.
Ailanthus, *sp. n.* Schoenegg.

E.—Elements of the American Floras.

Taxodium distichum miocenicum. Bilin.
Libocedrus, *sp. n.* Leoben.
SequoiaSternbergii, *Goepp.* Haering.
Sequoia Couttsiæ, *Heer.* Savine, Bilin.
Sequoia Tournalii, *Brongn.* Haering.
Sequoia Langsdorfii, *Brongn.* Leoben.
Pinus Palaeostrobus, *Ett.* Leoben.
Myrica lignitum, *Ung.* Parschlug, Schoenegg.
Myrica salicina, *Ung.* Radoboj.
Quercus daphnes, *Ung.* Parschlug.
Fagus Feroniæ, *Ung.* Leoben, Bilin.
Ulmus Braunii, *Heer.* Parschlug, Schoenegg.

Populus latior, *A. Braun.* Parschlug, Leoben.
Liquidambar europaeum, *A. Braun.* Parschlug.
Ficus arcinervis, *Heer.* Savine.
Andromeda protogaea, *Ung.* Savine.
Acer trilobatum, *A. Braun.* Bilin, Schoenegg, Leoben.
Acer megalopteryx, *Ung.* Radoboj.
Tetrapteris, *sp. n.* Savine.
Ternstroemia radobojana, *Ett.* Radoboj.
Carya bilinica, *Ung.* Bilin.
Robinia Hesperidum, *Ung.* Parschlug, Fohnsdorf.

F.—Elements of the African Floras.

Widdringtonia Ungeri, *Endl.* Parschlug.
Callitris Brongniarti, *Endl.* Radoboj, Leoben, Podsused.
Myrica subæthiopica, *Ett.* Parschlug.
Pterocelastrus, *sp. n.* Parschlug.

Celastrus oxyphyllus, *Ung.* Parschlug.
Celastrus Acoli, *Ett.* Schoenegg.
Celastrus, *sp. n.* Parschlug.
Celastrus deperdidus, *Ett.* Parschlug.
Rhus cuneolata, *Ung.* Parschlug.
Rhus, *sp. n.* Parschlug.

G.—*Elements of the European Flora.*

Pinus Laricio, *Poir.* Leoben Schoen-egg, *Podsused.*

Pinus Palæo-abies, *Ett:* Sagor.

Alnus Kefersteinii, *Goepp.* Schoen-egg, Leoben.

Alnus gracilis, *Ung.* Leoben.

Quercus mediterranea, *Ung.* Par-lug.

Fagus deucalionis, *Ung.* Parschlug, Putschirn.

Ulmus Bronnii, *Ung.* Leoben, Bilin.

Acer decipiens, *A. Braun.* Parschlug, Leoben.

II.—THE MIXTURE OF THE TERTIARY ELEMENTS OF THE FLORAS.

1. *Haering in Tyrol.*

Xylomites Zizyphi, *Ett.*

Puccinites lanceolatus, *Ett.*

Casuarina Haidingeri, *Ett.* (Australian form.)

Elaeodendron degener, *Ett.* (Asiatic form.)

Callitris Brongniarti, *Endl.* (African form.)

Sequoia Sternbergii, *Goepp.* (American form.)

Sequoia Sternbergii, *Goepp.*

Sabal major, *Ung.* (American form.)

Persoonia daphnes, *Ett.* (Australian form.)

Sequoia Sternbergii, *Goepp.* (American form.)

Santalum salicinum, *Ett.* (Australian form.)

Sequoia Tournalii, *Brongn.* (American form.)

Santalum salicinum, *Ett.* (Australian form.)

Sequoia Tournalii, *Brongn.* (American form.)

Casuarina Sotzkiana, *Ett.* (Australian form.)

Sequoia Couttsiae, *Heer.* (American form.)

Sabal major, *Ung.* (American form.)

Persoonia daphnes, *Ett.* (Australian form.)

Celastrus protogaeus, *Ett.* (African form.)

Banksia Ungeri, *Ett.* (Australian form.)

Banksia haeringiana, *Ett.* (Australian form.)

Rhus prisca, *Ett.* (African form.)

Dryandra Brongniarti, *Ett.* (Australian form.)

Zizyphus Ungeri, *Heer.* (Asiatic form.)

Ceratopetalum haeringianum, *Ett.* (Australian form.)

Cassia ambigua, *Ung.* (American form.)

Persoonia daphnes, *Ett.* (Australian form.)

Dodonaea salicites, *Ett.* (Australian form.)

Zizyphus Ungeri, *Heer.* (Asiatic form.)

Embothrites leptospermos, *Ett.* (Australian form.)

Zizyphus Ungeri, *Heer.* (Asiatic form.)

Caesalpinia Haidingeri, *Ett.* (Asiatic form.)

Eucalyptus haeringiana, *Ett.* (Australian form.)

Cassia pseudoglandulosa, *Ett.* (Asiatic form.)

2. *Sotzka in Styria.*

Davallia Haidingeri, *Ett.* (Canarian form.)

Sequoia Sternbergii, *Goepp.* (American form.)

Betula eocenica, *Ett.* (European form.)

Ficus apocynoides, *Ett.* (American form.)

Ficus Laurogene, *Ett.* (American form.)
Banksia Ungeri, *Ett.* (Australian form.)
Notelaea eocenica, *Ett.* (Australian form.)
Apocynophyllum ochrosioides, *Ett.*
Sapotacites minor, *Ett.*
Ceratopetalum haeringianum, *Ett.* (Australian form.)
Hiraea Ungeri, *Ett.* (American form.)
Cupania juglandina, *Ett.* (Australian form.)
Acer sotzkianum, *Ung.* (European form.)
Zizyphus Druidum, *Ung.* (Asiatic form.)

Engelhardtia sotzkiana, *Ett.* (Asiatic form.)
Andromeda protogaea, *Ung.* (American form.)
Callistemophyllum diosmoides, *Ett.* (Australian form.)
Getonia macroptera, *Ung.* (Asiatic form.)
Callistemon eocenicum, *Ett.* (Australian form.)
Cassia Phaseolites, *Ung.* (American form.)
Davallia Haidingeri, *Ett.* (Canarian form.)
Acacia caesalpiniaefolia, *Ett.* (Australian form.)

3. Monte Promina in Dalmatia.

Sphenopteris eocenica, *Ett.*
Cinnamomum lanceolatum, *Ung.* (Asiatic form.)
Cassia Phaseolites, *Ung.* (American form.)
Cinnamomum polymorphum, *Ung.* Asiatic form.)

Sequoia Sternbergii, *Goepp.* (American form.)
Bambusium sepultum, *Ung.* (Asiatic form.)
Apocynophyllum plumeriaefolium, *Ett.* (Tropical form.)
Rhamnus Eridani, *Ung.* (American form.)

4. Kutschlin in Bohemia.

Libocedrus salicornioides, *Endl.* (American form.)
Sequoia Sternbergii, *Goepp.* (American form.)
Cinnamomum lanceolatum, *Ung.* (Asiatic form.)

Celastrus elaenus, *Ung.* (American form.)
Engelhardtia Brongniarti, *Sap.* (Asiatic form.)

5. Sagor in Carniola.

Sequoia Sternbergii, *Goepp.* (American form.)
Zostera Ungeri, *Ett.* (Salt water form.)
Quercus daphnes, *Ung.* (American form.)
Ficus bumeliaefolia, *Ett.* (American form.)
Carpinus grandis, *Ung.* (European form.)
Banksia longifolia, *Ett.* (Australian form.)

Dryandra sagoriana, *Ett.* (Australian form.)
Cinnamomum polymorphum, *A. Braun.* (Asiatic form.)
Cinnamomum lanceolatum, *Ung.* (Asiatic form.)
Andromeda protogaea, *Ung.* (American form.)
Engelhardtia Brongniarti, *Sap.* (Asiatic form.)

6. Savine in Carniola.

Sphaeria Suessii, *Ett.*

Chara Langeri, *Ett.*

Hypnum sagorianum, *Ett.*

Glyptostrobus europaeus, *Heer.* (Asiatic form.)

Sequoia Langsdorfii, *Brongn.* (American form.)

Sequoia Couttsiae, *Heer.* (American form.)

Pinus Palaeo-taeda, *Ett.* (American form.)

Ostrya Atlantidis, *Ung.* (American form.)

Ficus bumeliaefolia, *Ett.* (American form.)

Ficus lanceolato - acuminata, *Ett.* (Tropical form.)

Pisonia eocenica, *Ett.* (Asiatic form.)

Cinnamomum lanceolatum, *Ung.* (Asiatic form.)

Laurus stenophylla, *Ett.* (American form.)

Banksia longifolia, *Ett.* (Australian form.)

Neritinium majus, *Ung.* (Tropical form.)

Apocynophyllum Reussii, *Ett.* (Tropical form.)

Andromeda protogaea, *Ung.* (American form.)

Tetrapteris sagoriana, *Ett.* (American form.)

Celastrus deperditus, *Ett.* (African form.)

Celastrus Hippolyti, *Ett.* (African form.)

Berthemia, *sp. n.* (American form.)

Engelhardtia Brongniarti, *Sap.* (Asiatic form.)

Eucalyptus oceanica, *Ung.* (Asiatic form.)

Sophora europaea, *Ung.* (Asiatic form.)

Cassia Phaseolites, *Ung.* (American form.)

7. Fohnsdorf in Styria.

Proteoides lanceolatus, *Ett.* (African form.)

Pinus Palaeo-taeda, *Ett.* (American form.)

Banksia, *sp.* (Australian form.)

Enteromorpha stagnalis, *Heer.* (Fresh-water form.)

Glyptostrobus europaeus, *Heer.* (Asiatic form.)

Sequoia Langsdorfii, *Brongn.* (American form.)

Pinus rigios, *Ung.* (American form.)

Alnus Kefersteinii, *Goepp.* (European form.)

Quercus Drymeja, *Ung.* (American form.)

Fagus Feroniae, *Ung.* (American form.)

Lomatia angustifolia, *Ett.* (Australian form.)

Banksia longifolia, *Ett.* (Australian form.)

Banksia Ungeri, *Ett.* (Australian form.)

Cinnamomum Scheuchzeri, *Heer.* (Asiatic form.)

Carya bilinica, *Ung.* (American form.)

Cinnamomum polymorphum, *A. Braun.* (Asiatic form.)

Nerium, *sp. n.* (Asiatic form.)

Acer trilobatum, *A. Braun.* (American form.)

Celastrus deperdidus, *Ett.* (African form.)

Cassia ambigua, *Ung.* (American form.)

Sophora europaea, *Ung.* (Asiatic form.)

Acacia parschlugiana, *Ung.* (American form.)

8. Eibiswald in Styria.

Glyptostrobus europaeus. (Asiatic form.)

Pinus Palaeo-taeda, *Ett.* (American form.)

Phoenicites, *sp. n.* (Trop. form.)

Ficus tiliaefolia, *Ung.* (American form.)

Embothrites borealis, *Ung.* (Australian form.)

Lomatia angustifolia, *Ett.* (Australian form.)

Callistemophyllum, *sp. n.* (Australian form.)

9. Radoboj in Croatia.

Cystoseira communis, *Ung.* (Salt-water form.)
Cystoseira affinis, *Ung.* (Salt-water form.)
Woodwardia Roessneriana, *Ung.* (European form.)
Pteris firma, *Ett.* (African form.)
Libocedrus salicornioides, *Endl.* (American form.)
Callitris Brongniarti, *Endl.* (African form.)
Pinus Palaeo-taeda, *Ett.* (American form.)
Zostera Ungeri, *Ett.* (Salt-water form.)
Smilax grandifolia, *Ung.*
Myrica deperdita, *Ung.* (American form.)
Ulmus prisca, *Ung.* (European form.)
Laurelia rediviva, *Ung.* (American form.)
Pinus Palaeo-taeda, *Ett.* (American form.)
Zostera Ungeri, *Ett.* (Salt-water form.)
Pisonia radobojana, *Ett.* (American form.)
Embothrites sotzkianus, *Ung.* (Australian form.)
Cinnamomum polymorphum, *A. Braun.* (Asiatic form.)
Fraxinus primigenia, *Ung.* (European form.)

Diospyros bilinica, *Ett.* (Asiatic form.)
Olea Osiris, *Ung.* (African form.)
Weinmannia europaea, *Ett.* (American form.)
Ceratopetalum radobojanum, *Ett.* (Australian form.)
Cissus radobojensis, *Ett.* (American form.)
Acer trilobatum, *A. Braun.* (American form.)
Acer campylopteryx, *Ung.* (American form.)
Engelhardtia Brongniarti, *Sap.* (Asiatic form.)
Embothrites borealis, *Ung.* (Australian form.)
Zostera Ungeri, *Ett.* (Salt-water form.)
Engelhardtia Brongniarti, *Sap.* (Asiatic form.)
Pinus Palaeo-taeda, *Ett.* (American form.)
Getonia petraeaeformis, *Ung.* (Asiatic form.)
Dalbergia sotzkiana, *Ung.* (Asiatic form.)
Rhynchosia Isidis, *Ung.* (African form.)
Rhus Pyrrhae, *Ung.* (American form.)
Rhus obovata, *Ett.* (African form.)

10. Podsused in Croatia.

Cystoseira communis, *Ung.* (Salt-water form.)
Pinus Palaeo-taeda, *Ett.* (American form.)
Pinus Laricio, *Poir.* (European form.)
Myrica salicina, *Ung.* (American form.)

Quercus Lonchitis, *Ung.* (American form.)
Pisonia eocenica, *Ett.* (Asiatic form.)
Laurus tetrantheroides, *Ett.* (American form.)
Sapindus Pythii, *Ung.* (American form.)

11. Bilin in Bohemia.

Glyptostrobus europaeus, *Heer.* (Asiatic form.)
Taxodium distichum, *Mioc.* (American form.)
Sequoia Couttsiae, *Heer.* (American form.)
Alnus Kersteinii, *Goepp.* (European form.)
Fagus Feroniae, *Ung.* (American form.)

Ulmus plurinervia, *Ung.* (European form.)
Cinnamomum Rossmaessleri, *Heer.* (Asiatic form.)
Acer trilobatum, *A. Braun.* (American form.)
Carya bilinica, *Ung.* (American form.)

12. Leoben in Styria.

Sphaeria Dryadum, *Ett.*
Dothidea, *sp. n.*
Xylomites Lonchitidis, *Ett.*
Xylomites, *sp. n.*
Rhytisma Geinitzii, *Ett.*
Glyptostrobus europaeus, *Heer.* (Asiatic form.)
Sequoia Couttsiae, *Heer.* (American form.)
Pinus Laricio, *Poir.* (European form.)
Alnus Kefersteinii, *Goepp.* (European form.)
Castanea atavia, *Ung.* (European form.)
Alnus gracilis, *Ung.* (European form.)
Fagus Feroniae, *Ung.* (American form.)
Quercus Lonchitis. (American form.)
Populus Geinitzii, *Ett.* (American form.)
Urtica miocenica, *Ett.* (American form.)
Ficus Deschmanni, *Ett.* (Asiatic form.)

Ficus Iynx, *Ung.* (Asiatic form.)
Cinnamomum polymorphum. A. Br. (Asiatic form.)
Cinnamomum Scheuchzeri. (Asiatic form.)
Cinnamomum lanceolatum, *Ung.* (Asiatic form.)
Nectandra arcinervia, *Ett.* (American form.)
Santalum microphyllum. *Ett.* (Australian form.)
Banksia, *sp. n.* (Australian form.) ⎫
Glyptostrobus europaeus. (Asiatic form.) ⎬
Apocynophyllum haeringianum, *Ett.* (Tropical form.)
Evonymus, *sp. n.* (Asiatic form.)
Juglans Reussii, *Ett.* (Asiatic form.)
Cassia Zephyri, *Ett.* (American form.)
Cassia Feroniae, *Ett.* (Asiatic form.)
Cassia leptodictyon, *Ett.* (American form.)

13. Schoenegg in Styria.

Blechnum Goepperti, *Ett.* (American form.)
Phegopteris stiriaca, *Ung.* (American form.)
Glyptostrobus europaeus. (Asiatic form.)
Fagus Feroniae, *Ung.* (American form.)
Ulmus Braunii, *Heer.* (American form.)
Juglans parschlugiana, *Ung.* (Asiatic form.)
Alnus Kefersteinii, *Goepp.* (European form.)
Glyptostrobus europaeus. (Asiatic form.)
Santalum, *sp. n.* (Australian form.)
Myrica lignitum, *Ung.* (American form.)
Pinus Palaeo-taeda, *Ett.* (American form.)
Sequoia Langsdorfii, *Brongn.* (American form.)

Typha latissima, *A. Br.* (European form.)
Ceratophyllum, *sp. n.* (European form.)
Celastrus pachyphyllus, *Ett.* (African form.)
Alnus Kefersteinii, *Goepp.* (European form.)
Betula grandifolia, *Ett.* (European form.)
Ficus daphnogenes, *Ett.* (Asiatic form.)
Laurus primigenia, *Ung.* (European form.)
Glyptostrobus europaeus. (Asiatic form.)
Quercus mediterranea, *Ung.* (European form.)
Ulmus Bronnii, *Ung.* (European form.)
Myrica lignitum, *Ung.* (American form.)
Callistemophyllum melaleucaeforme, *Ett.* (Australian form.)

Celastrus pachyphyllus, *Ett.* (African form.)
Glyptostrobus europaeus. (Asiatic form.)
Cinnamomum polymorphum, *A Braun.* (Asiatic form.)
Cinnamomum Scheuchzeri, *Heer.* (Asiatic form.)
Santalum, *sp. n.* (Australian form.) ⎫
Myrica, *sp. n.* (American form.) ⎬
Leptomeria Benthami, *Ett.* (Australian form.)
Celastrus, *sp. n.* (African form.)
Banksia, *sp.* (Australian form.) ⎫
Pomaderris, *sp. n.* (Australian form.) ⎪
Glyptostrobus europaeus. (Asiatic form.) ⎬
Fagus Feroniæ, *Ung.* (American form.) ⎪
Myrica lignicum, *Ung.* (American form.) ⎭
Banksia Ungeri, *Ett.* (Australian form.) ⎫
Laurus Swoszowiciana, *Ung.* (European form.) ⎬

Fraxinus, *sp. n.* (European form.)
Sapotacites minor, *Ett.* (Tropical form.)
Acer trilobatum, *A. Braun.* (American form.)
Sapindus Pythii, *Ung.* (American form.) ⎫
Pinus Laricio, *Poir.* (European form.) ⎪
Quercus Lonchitis, *Ung.* (American form.) ⎬
Glyptostrobus europaeus. (Asiatic form.) ⎭
Celastrus Acherontis, *Ett.* (African form.)
Eugelhardtia Brongniarti, *Sap.* (Asiatic form.) ⎫
Pterocelastrus Oreonis, *Ett.* (African form.) ⎬
Eucalyptus oceanica, *Ung.* (Australian form.) ⎭
Podogonium Knorrii, *Heer.*
Cassia Phaseolites, *Ung.* (American form.)
Santalum salicinum, *Ett.* (Australian form.)

14. *Parschlug in Styria.*

Sphaeria Braunii, *Heer.*
Xylomites granulifer, *Ett.*
Xylomites liquidambaris, *Ett.*
Hypnum Schimperi, *Ung.* (European form.)
Glyptostrobus europaeus. (Asiatic form.) ⎫
Podogonium Knorrii, *Heer.* ⎬
Pinus Goethana, *Ung.* (American form.) ⎫
Widdringtonia Ungeri. *Endl.* (African form.) ⎬
Sapindus falcifolius, *A. Braun.* (American form.) ⎭
Myrica lignitum, *Ung.* (American form.) ⎫
Widdringtonia Ungeri, (African form.) ⎪
Quercus daphnes, *Ung.* (American form.) ⎬
Planera Ungeri, *Ett.* (Asiatic form.) ⎭
Planera Ungeri, *Ett.* (Asiatic form.) ⎫
Macreightia germanica, *Heer.* (American form.) ⎬

Ulmus Braunii, *Heer.* (American form.)
Populus latior. *A. Braun.* (American form.)
Liquidambar europaeum, *A. Braun.* (American form.)
Banksia, *sp.* (Australian form.) ⎫
Ilex aspera, *Ett.* (American form.) ⎬
Fraxinus primigenia, *Ung.* (European form.)
Sapotacites minor, *Ett.* (Tropical form.)
Ledum limnophilum, *Ung.* (American form.)
Vaccinium acheronticum, *Ung.* (American form.)
Acer trilobatum, *A. Braun.* (American form.)
Sapindus Pythii, *Ung.* (American form.) ⎫
Celastrus, *sp. n.* (African form.) ⎬
Sapindus, *sp. n.* (American form.)
Pterocelastrus, *sp. n.* (African form.)

Elaeodendron, *sp. n.* (African form.)

Ramnus aizoon, *Ung.* (American form.)

Myrica lignitum, *Ung.* (American form.)
Vaccinium vitis Iapeti, *Ung.* (European form.)
Juglans parschlugiana, *Ung.* (Asiatic form.)

Myrica lignitum. (American form.)
Ilex stenophylla, *Ung.* (American form.)
Quercus mediterranea, *Ung.* (European form.)

Myrica deperdita, *Ung.* (American form.)

Quercus serra, *Ung.* (American form.)

Planera Ungeri, *Ett.* (Asiatic form.)
Myrica lignitum, *Ung.* (American form.)
Vaccinium vitis Iapeti, *Ung.* (American form.)
Juglans parschlugiana. (Asiatic form.)

Rhus, *sp. n.* (American form.)
Rhus elaeodendroides, *Ung.* (African form.)
Glyptostrobus europaeus. (Asiatic form.)

Rhus retine, *Ung.* (American form.)
Rhus zanthoxyloides, *Ung.* (American form.)
Populus latior, *A. Braun.* (American form.)
Quercus Lonchitis, *Ung.* (American form.)
Quercus mediterranea, *Ung.* (European form.)
Myrica lignitum, *Ung.* (American form.)

Rosa Penelopes, *Ung.*

Physolobium orbiculare, *Ung.* (Australian form.)

Physolobium microphyllum, *Ett.* (Australian form.)

Glycyrrhiza Blandusiae, *Ung.* (European form.)

Cassia Phaseolites, *Ung.* (American form.)

Acacia parschlugiana, *Ung.* (American form.)

The above are original specimens collected in Styria, Carniola, Croatia, Dalmatia, the Tyrol, Hungary, and Bohemia.

The recent plants bearing closest analogy to the fossil plants are shown for comparison.

In the ancient Tertiary strata the *Castanea atavia* has slightly toothed, or nearly toothless leaves, devoid of thorny points. From the primary vein spring, at a distance from one another, curved secondary veins.

In the Middle Tertiary strata are found leaves of the *Castanea atavia* closely approaching the *Castanea vesca* by approximate secondary veins, and by more numerous and protruding teeth, which, however, are still without thorny points.

In still more recent strata of the Tertiary formation there are found here and there thorny points on the teeth.

In the latest Tertiary strata the leaves of the fossil species are nearly identical with those of the *Castanea vesca.* The secondary veins are more approximate and almost rectilinear, while the teeth are set with thorns.

In all strata where leaves of the *Castanea atavia* have been found, the male catkins of that species have also been collected, which differ from those of the *Castanea vesca* by having somewhat smaller flowers. Examples of these are also added to the collection.

3356. Eocene Fossils. Typical collection, comprising 100 species. *R. Damon.*

3357. Rocks, sedimentary, volcanic, and plutonic. Typical collection, comprising 100 specimens. *R. Damon.*

3357a. Table of British Strata, showing the order of their superposition, and the relative thickness of the Formations.

Henry William Bristow, F.R.S., Director of the Geological Survey of England and Wales.

In the construction of this table, the object has been to produce a cheap diagram for educational purposes, and for illustrating lectures on geology.

When thus used, the four columns into which the table is divided may be printed in separate slips, and then coloured and mounted side by side. Each slip being complete in itself, any one, two, or more may be employed to suit the special purposes for which they are required. Thus, for elementary classes the first three columns may be all that are needed, while for more advanced classes, and in systematic lectures on geology, all may be made use of.

The mode here adopted of showing the grouping and order of succession of the various geological formations in parallel columns in juxtaposition with each other, places the facts before the eye in the most striking and comprehensive way. It has also the recommendation of cheapness, an important consideration in the case of schools.

For reference in the study as well as *on an enlarged scale* as a diagram, the table is susceptible of extension, as additional parallel columns may be filled up with (*a*), the names of the fossils most common in the several formations in different districts; (*b*), the lithological and local characters of the various strata; (*c*), the minerals usually found in or associated with them; (*d*), the useful purposes to which the strata and the mineral substances they contain are applied in the arts, manufactures, &c.; and (*e*), the prevailing agricultural characters and peculiarities of the different formations.

3358. British Rocks. A typical collection of aqueous, metamorphic, plutonic, and igneous rocks. *J. R. Gregory.*

3359. Collection of Fossils.
Dr. Acland, F.R.S., on behalf of the Radcliffe Trustees.

3360. British Typical Fossils. An elementary collection arranged stratigraphically. *J. R. Gregory.*

3361. Diagrams for Lecture Purposes, Two. The drawings enlarged from Davidson's Monograph of the British Fossil Brachiopoda.

George Sharman, Museum of Practical Geology.

Illustrations of characteristic British Carboniferous Brachiopoda.
Illustrations of characteristic British Brachiopoda, from various formations.

3361a. Collection of Fossils most characteristic of the different formations in Russia. 214 specimens.

Mining School, St. Petersburg.

1. Eurypterus Fischeri, Eichw. -	Upper Silu-rian.	Rotzeküll, Oesel.
2. Asaphus expansus, Lin. -	Lower Silu-rian.	Duboviki, St. Petersburg.
3. Illænus crassicauda, Schlth. -	,,	Popovka, ,,
4. Amphion Fischeri, Eichw. -	,,	Pulkova, ,,
5. Lituites convolvens, Schlth. -	,,	Reval, Esthonia.
6. Orthoceras vaginatus, Schlth. -	,,	,, ,,
7. ,, duplex, Wahlb. -	,,	Narva, ,,
8. Euomphalus qualteriatus, Schlth.	,,	Reval, ,,
9. Spirifer superbus, Eichw. -	Upper Silu-rian.	Petropavlovsk, Ural.
10. Atrypa prunum, Dalm. -	,,	Lode, Esthonia.
11. ,, arimaspus, Eichw. -	,,	Petropavlovsk, Ural.
12. Rhynchonella Wilsoni, Sow. -	,,	Johannis, Esthonia.
13. ,, nucella, Dalm. -	Lower Silu-rian.	Popovka, St. Petersburg.
14. Pentamerus esthonus, Eichw. -	Upper Silu-rian.	Kattentak, Esthonia.
15. ,, borealis, Eichw.	,,	Hapsal, ,,
16. ,, acutolobatus, Sandb.	,,	Petropavlovsk, Ural.
17. ,, Knightii, Vern. -	,,	Petuhovsk, ,,
18. ,, baschkiricus, Vern.	,,	Satkinski Pristan, Ural.
19. ,, vogulicus, Vern. -	,,	Bogoslovsk, Ural.
20. ,, ,, Vern. -		,, ,,
21. Porambonites æquirostris, Schlth.	Lower Silu-rian.	Sommerhausen, Estho-nia.
22. Porambonites Tscheffkini, Vern.	,,	Pulkovo, St. Petersburg.
23. ,, reticulatus, Pand.	,,	Gostinopolskaia Pristan, Novgorod.
24. Platystrophia lynx, Eichw. -	,,	Pulkova, St. Petersburg.
25. Orthis obtusa, Pand. - -	,,	Popovka, ,,
26. ,, extensa, Pand. -	,,	,, ,,
27. ,, calligramma, Dalm. -	,,	,, ,,
28. ,, parva, Pand. -	,,	Pulkovo, ,,
29. Orthisina plana, Pand. -	Lower Silu-rian.	River Ishara, ,,
30. ,, anomala, Schlth. -	,,	Reval, Esthonia.
31. ,, hemipromites, Buch.	,,	Pulkovo, St. Petersburg.
32. ,, adscendens, Pand. -	,,	Popovka, ,,
33. ,, Verneuilli, Eichw. -	,,	Reval, Esthonia.
34. Leptæna Humboldtii, Vern. -	,,	Erras, ,,
35. ,, imbrex, Pand. -	,,	Teve, ,,
36. ,, transversa, Pand. -	,,	Pulkovo, St. Petersburg.
37. ,, oblonga, Pand. -		,, ,,
38. ,, uralensis, Vern. -	Upper Silu-rian.	River Tsvestka, Ural.
39. Siphonotreta unguiculata, Eichw.	Lower Silu-rian.	Popovka, St. Petersburg.
40. Siphonotreta verrucosa, Eichw.	,,	Pulkovo, ,,
41. Obolus Apollinis, Eichw. -	,,	Narva, Esthonia.
42. Crania antiquissima, Eichw. -	,,	Popovka, St. Petersburg.
43. Echino-encrinus angulosus, Pand.	,,	,, ,,

44. Cryptocrinus lævis, Pand. -	Lower Silurian.	Popovka, St. Petersburg.
45. Hemicosmites pyriformis, Buch	„	Pulkovo, „
46. Heliocrinus balticus, Eichw. -	„	Reval, Esthonia.
47. Glyptosphærites Leuchtenbergii, Volb.	„	Popovka, St. Petersburg.
48. Echinosphærites aurantium, Wahlb.	„	„ „
49. Chætetes petropolitanus, Pand.	„	Pulkovo, „
50. Astylospongia globosa, Eichw.	„	Popovka, „
51. Pseudosiphonia cylindrica, Eichw.	„	„ „
52. Pterichthys (Asterolepis) Milleri, Ow.	Devonian -	River Aa, Livonia.
53. Holoptychius nobilissimus, Ag.	„ -	River Priksha, Novgorod.
54. Dendrodus biporcatus, Ow. -	„ -	River Aa, Livonia.
55. Chelyophorus, sp. - -	„ -	Orel.
56. Gomphoceras rex, Pacht. -	„ -	Zadonsk, Voroneje.
57. Gomphoceras sulcatulum, Vern.	„ -	River Siass, St. Petersburg.
58. Arca oreliana, Vern. - -	„ -	Orel.
59. Avicula Woerthii, Vern. -	„ -	Gostinopolskaia Pristan, Novgorod.
60. Spirifer Archiacii, Sow. -	„ -	Teletz, Orel.
61. „ Anosofii, Vern. -	„ -	Voroneje.
62. „ curvatus, Schnur. -	„ -	Lechki, Voroneje.
63. „ aculeatus, Schnur. -	„ -	Muraievna, Riazan.
64. Athyris concentrica, Buch. -	„ -	Teletz, Orel.
65. „ subpyriformis, Sem. Möll.	„ -	Muraievna, Riazan.
66. Athyris Puschiana, Sem. Möll.	„ -	„ „
67. „ pectinata, Sem. Möll.-	„ -	Ulabie, Tula.
68. Retzia tulensis, Pand.- -	„ -	Malevka, Tula.
69. Atrypa reticularis, Lin. -	„ -	Gostinopolskaia Pristan, Novgorod.
70. Atrypa reticularis, Lin. -	„ -	Kadinskaia Pristan, Ural.
71. „ Duboisii, Vern. -	„ -	„ „
72. „ latilinguis, Schnur. -	„ -	„ „
73. Rhynchonella Panderi, Sem. Möll.	„ -	Malevka, Tula.
74. Rhynchonella livonica, Buch. -	„ -	Zadonsk, Voroneje.
75. „ Meyendorfii, Vern.	„ -	Pskof.
76. Rhynchonella cuboides, Sow. -	„ -	Sulem, Ural.
77. Orthis striatula, Schlth. -	„ -	Gostinopolskaia Pristan, Novgorod.
78. Leptæna, Asella. - -	„ -	Voroneje.
79. Chonetes nana, Vern.- -	„ -	Malevka, Tula.
80. Stropholosia subaculeata, Murch.	„ -	Zadonsk, Voroneje.
81. Productus Panderi Sem., Möll.	„ -	Malevka, Tula.
82. „ fallax, Sem., Möll.-	„ . -	„
83. Phillipsia mucronata, M'Coy -	Mountain Limestone.	Borovitshi, Novgorod.

84. Phillipsia pustulata, Schlth. -	Mountain Limestone.	Tshernishino, Kaluga.
85. Allorisma regularis, King. -	„	Borovitshi, Novgorod.
86. Spirifer mosquensis, Fisch. -	„	Verei, Moscow.
87. „ „ „ -	„	Deviatino, Olonetz.
88. „ striatus, Mart. -	„	River Indiga, Timan.
89. „ trigonalis, Mart.	„	Steshovo, Tver.
90. Spiriferina Saranæ, Vern. -	„	River Indiga, Timan.
91. Athyris ambigua, Sow. -	„	Steshovo, Tver.
92. Camarophoria plicata, Kut. -	„	Taroslavka, Ural.
93. Streptorhynchus crenistria, Phill.	„	Staritza, Tver.
94. Productus scabriculus, Mart.-	„	River Nara, Kaluga.
95. „ costatus, Sow. -	„	Sloboda, Tula.
96. „ tubarius, Keys. -	„	Kremenskvie, Kaluga.
97. „ undatus, Vern. -	„	Tarussa, Kaluga.
98. „ punctatus, Mart. -	„	Sloboda, Tula.
99. „ Humboldtii, D'Orb.	„	River Indiga, Timan.
100. „ cora, D'Orb. -	„	Kassimof, Riazan.
101. „ longispinus, Flem.	„	Steshovo, Tver.
102. „ semireticulatus, Mart.	„	Staritza, Tver.
103. „ striatus, Fisch. -	„	Borovitshi, Novgorod.
104. „ giganteus, Mart. -	„	Voronovo, Kaluga.
105. „ „ „ -	„	Borovitshi, Novgorod.
106. „ mammatus, Keys. -	„	River Zilma, Timan.
107. „ semireticulatus var. Boliviensis, D'Orb.	„	Saraninsk, Ural.
108. Coscinium cyclops, Keys. -	„	River Indiga, Timan.
109. Polypora orbicribrata, Keys.	„	„ „
110. Chaetetes radians, Fisch.	„	River Tagashma, Olonetz.
111. Syringopora conferta, Keys. -	„	River Sopliussa, Timan.
112. Lonsdaleia floriformis, Flem.	„	Gurieva, Tula.
113. Amplexus arietinum, Fisch. -	„	River Kumish, Ural.
114. Fusulina cylindrica, Fisch. -	„	Zarel Kurgan, Samara.
115. Clydophorus Pallasi, Vern. -	Permian -	Kazan.
116. Pecten Kokscharofii, Vern. -	„ -	Kirilof, Novgorod.
117. Avicula speluncaria, Schlth.-	„ -	Kazan.
118. Terebratula elongata, Schlth.	„ -	River Dioma, Ural.
119. Spirifer Schrenkii, Keys. -	„ -	Kirilof, Novgorod.
120. „ rugulatus, Kut. -	„ -	River Termak, Samara.
121. Athyris pectinifer, Sow. -	„ -	Kirilof, Novgorod.
122. „ Royssiana, Keys. -	„ -	„ „
123. Camarophoria superstes, Vern.	„ -	„ „
124. Productus hemisphaerium, Kyt.	„ -	River Termak, Samara.
125. Productus tenuituberculatus, Barbt.	„ -	Kirilof, Novgorod.
126. Productus Cancrini, Vern. -	„ -	Makarief, Kazan.
127. Stropholosia horrescens, Vern.	„ -	Kirilof, Novgorod.
128. Stropholosia Wangenheimii, Vern.	„ -	Grebeni, Orenburg.

129. Avicula Dalialama, Vern.	-	Trias	-	Bogdo, Steppe of Kirghises.
130. Ichthyosaurus sp.	-	Jura	-	River Volga, Simbirsk.
131. Ammonites Jason, Ziet.	-	,,	-	Koroshovo, Moscow.
132. ,, alternans, Buch.		,,	-	River Petshora, Timan.
133. ,, Brightii, D'Orb.	-	,,	-	Skopin, Riazan.
134. ,, sublaevis, Sow.	-	,,	-	Telatma, Tambof.
135. ,, Koenigii, Sow.	-	,,	-	Koroshovo, Moscow.
136. ,, virgatus, Fisch.	-	,,	-	,, ,,
137. ,, catenulatus, Fisch.		,,	-	,, ,,
138. ,, Kaschpuricus, Trautsch.		,,	-	Syzran, Simbirsk.
139. ,, Ischmae, Keys.	-	,,	-	River Petshora, Timan.
140. ,, uralensis, D'Orb.		,,	-	,, Sosva, Ural.
141. ,, Kirghisensis, D'Orb.		,,	-	Saragul, Orenburg.
142. Belemnites absolutus, Fisch.	-	,,	-	Koroshovo, Moscow.
143. ,, curtus, Eichw.	-	,,	-	River Petshora, Timan.
144. ,, Panderianus, D'Orb.		,,	-	Gorodishte, Simbirsk.
145. ,, magnificus, D'Orb.		,,	-	,, ,,
146. ,, russiensis, D'Orb.		,,	-	,, ,,
147. Pleurotomaria Bloedeana, D'Orb.		,,	-	Koroshovo, Moscow.
148. ,, Buchiana, D'Orb.		,,	-	,, ,,
149. Astarte Duboisiana, D'Orb.	-	,,	-	,, ,,
150. Lucina Phillipsii, D'Orb.	-	,,	-	Gorodishte, Simbirsk.
151. Lyonsia Alduinii, D'Orb.	-	,,	-	Koroshovo, Moscow.
152. Panopæa peregrina, D'Orb.	-	,,	-	,, ,,
153. Pholadomya mediana, Eichw.		,,	-	River Sosva, Ural.
154. Thracia Frearsiana, D'Orb.	-	,,	-	,, ,, ,,
155. Mytilus vicinalis, D'Orb.	-	,,	-	,, ,, ,,
156. ,, Strajevskianus, D'Orb.		,,	-	,, ,, ,,
157. Lima consabrina, D'Orb.	-	,,	-	Koroshovo, Moscow.
158. Pinna sublanceolata, Eichw.	-	,,	-	River Sosva, Ural.
159. Avicula cuneiformis, D'Orb.	-	,,	-	Koroshovo, Moscow.
160. Ancella mosquensis, Fisch.	-	,,	-	,, ,,
161. Gryphaea dilatata, Sow.	-	,,	-	Telatma, Tambof.
162. Terebratula Rogeriana, D'Orb.		,,	-	Koroshovo, Moscow.
163. ,, Fischeriana, D'Orb.		,,	-	,, ,,
164. ,, Strogonofii, D'Orb.		,,	-	River Sosva, Ural.
165. Rhynchonella varians, Buch.		,,	-	Telatma, Tambof.
166. ,, loxiae, Fisch.	-	,,	-	Koroshovo, Moscow.
167. ,, oxyoptycha; Fisch.		,,	-	,, ,,
168. ,, grosse-sulcata, Eichw.		,,	-	River Sosva, Ural.
169. Ammonites versicolor, Trautsch.		Neocomian	-	Simbirsk.
170. ,, discofalcatus, Lahus.		,,	-	,,
171. ,, fasciato-falcatus, Lahus.		,,	-	,,

172. Ammonites elatus, Trautsch.	Neocomian	-	Simbirsk.
173. Astarte porrecta, Buch.	,,	-	,,
174. Avicula Cornueiliana, D'Orb.	,,	-	,,
175. Pecten crassitesta, Roem	,,	-	,,
176. Inoceramus ancella, Trautsch.	,,		,,
177. Rhynchonella obliterata, Lahus.	,,	-	,,
178. Ancyloceras simbirskensis, Tazyk.	Gault	-	,,
179. Ammonites Deshayesii, Leym	,,	-	,,
180. ,, bicurvatus, Mich.	,,	-	,,
181. Belemnitella mucronata, D'Orb.	Upper Chalk		Shilovka, Simbirsk.
182. Janira simbirskensis, D'Orb.	,,	-	Tazykovo, ,,
183. Lima bistriata, Lahus.	,,	-	,, ,,
184. ,, semisulcata, Desh.	,,	-	,, ,,
185. Pecten undulatus, Nils.	,,	-	,, ,,
186. Avicula lineata, Roem	,,	-	Simbirsk.
187. Inoceramus Cripsii, D'Orb.	,,	-	,,
188. Ostrea semiplana, Sow.	,,	-	Maza, Simbirsk.
189. ,, vesicularis, Lamk.	,,	-	,, ,,
190. Terebratula obesa, Sow.	,,	-	Tazykovo, Simbirsk.
191. ,, carnea, Sow.	,,	-	,, ,,
192. Magas pumilus, Sow.	,,	-	,, ,,
193. Rhynchonella octoplicata, Sow.	,,	-	,, ,,
194. Crania parisiensis, Defr.	,,	-	,, ,,
195. Ananchytes ovatus, Lamk.	,,	-	,, ,,
196. Nummulites distans, Desh.	Eocene	-	Simferopol, Crimea.
197. Buccinum baccatum, Bast.	Miocene	-	Stavrofka, Cherson.
198. Trochus podolicus, Dub.	,,	-	Kamenka, Podolia.
199. ,, patulus, Brocc.	,,	-	Vishnevetz, Volhynia.
200. Turbo chersonensis, Barbt.	,,	-	Stavrofka, Cherson.
201. Cerithium disjunctum, Sow.	,,	-	,, ,,
202. ,, scabrum, Oliv.	,,	-	Potshaievo, Volhynia.
203. Natica millepunctata, Lamk.	,,	-	Vishnevetz, ,,
204. Maetra podolica, Eichw.	,,	-	Stavrofka, Cherson.
205. Ervilia podolica, Eichw.	,,	-	Kremenetz, Volhynia.
206. Tapes gregorea, Partsch.	,,	-	,, ,,
207. Lucina borealis, Lin.	,,	-	Potshaievo, ,,
208. Pectunculus pilosus, Lin.	,,	-	Vishnevetz, ,,
209. Cardium obsoletum, Eichw.	,,	-	Stavrofka, Cherson.
210. ,, protractum, Eichw.	,,	-	Kremenetz, Volhynia.
211. ,, Fittani, D'Orb.	,,	-	Kalfa, Bessarabia.
212. ,, littorale, Eichw.	,,	-	Vosnessensk, Cherson.
213. Congeria simplex, Barbt.	,,	-	Gnilolofskaia, ,,
214. Ostrea digitalina, Eichw.	,,	-	Vishnevetz, Volhynia.

3325. Series of Leaf Remains, from the Lower Bagshot beds (Middle Eocene), collected on the coast between Poole Harbour and Bournemouth. *J. Starkie Gardner.*

These leaves occur in isolated lenticular patches, usually of small extent, and were deposited in fresh water, which probably flowed from the north-west.

Their horizon is slightly higher than the well-known leaf beds of Alum Bay, the flora of which presents considerable differences in character. Collections from Alum Bay are in the British and Jermyn Street Museums.

3362. Collection of Fossils from the Gault, &c. at Folkestone. *J. Starkie Gardner.*

Crustacea, including about 115 specimens of brachyura, 10 anomura, 115 macrura, mostly with limbs perfect. Several of the species are still undescribed.
Gault coniferæ. Original specimens described by W. Carruthers, F.R.S.
Gault corals.
Grey chalk and gault echinoderms of the genera Cidaris, Pseudodiadema, Salenia, and allied genera.
Gasteropoda of the gault, illustrating the families aporrhaïdæ, scalidæ, and rissoidæ, described in the Geological Magazine for 1875-76.

3363. Collection of Typical Trachytes, consisting of some fifty specimens, arranged according to the system of the contributor. *Dr. J. Szabo, Buda-Pesth.*

3364. Thin polished plates of Trachytes, prepared at the University Institute for microscopic observation.
Dr. J. Szabo, Buda-Pesth.

3365. Illustrations of a New Method of determining Felspar by the character of the blow-pipe flame, exhibited in diagrams according to the system of the exhibitor.
Dr. J. Szabo, Buda-Pesth.

3366. Collection of Nummulites, systematically arranged with specimens prepared by Hantke and Madarász.
Maximilian Hantken, Buda-Pesth.

3367. General Collection of Foraminifera.
Maximilian Hantken, Buda-Pesth.

3368. Collection of Bryozoa, consisting of finished specimens. *Maximilian Hantken, Buda-Pesth.*

3369. Hammer Holder and Waist-strap; one with hammer in the position when worn. Specially adapted for geologists in the field. *Thomas J. Downing.*

3370. Card-board Trays, showing various sizes suitable for holding and arranging specimens in the cabinet.
Thomas J. Downing.

3371. Tablet-wood, covered with coloured papers, for mounting geological specimens. *Thomas J. Downing.*

3372. Improved Cement for mending and mounting fossils.
Thomas J. Downing.

3373. Collection of Volcanic Rocks of the Prussian provinces of Nassau and the Rhine, and especially from the district round the Laacher See and the Sieben Gebirge, River Nahe, and Nassau. *B. Stürtz, Bonn.*

3374. Local collection of Trachytes from the neighbourhood of Kremnitz in Hungary. *B. Stürtz, Bonn.*

3375. Collection of Rock Specimens from the Island of Elba. *B. Stürtz, Bonn.*

3376. Collection of Volcanic Rocks from the neighbourhood of Naples. *B. Stürtz, Bonn.*

3377. Collection, for illustrating geological lectures, of 150 **Fossils,** in good condition, arranged according to the age of the formations in which they occur, and containing in systematic order the characteristic fossils of all the formations and most of their subdivisions. *B. Stürtz, Bonn.*

3379. Impressions on large plates of **Extinct Species** of **Animals** (saurians and fish), suitable for illustrating the study of palæontology. *B. Stürtz, Bonn.*

3379a. Collection of Minerals.
Royal School of Mines, Madrid.

3379. Collection of various kinds of Rocks, from different parts of Russia. 95 specimens. Value, 4*l.*
Mining School, St. Petersburg.

1. Granite. Serdobol, Finland.
2. ,, (Rappa-Kivi). Wyborg, Finland.
3. Granite. Katharinenburg, Ural.
4. Green granite. Ilmen mountains, Ural.
5. Granite. Ilmen mountains, Ural.
6. Corundum-granite. Kyshtyn, Ural.
7. Gneiss. Katharinenburg, Ural.
8. ,, Zlatoust, Ural.
9. ,, River Wytim, Siberia.
10. Syenite. Bogoslovsk, Ural.
11. Uralite syenite. Turgoyak, Ural.
12, 13. Miascite. Ilmen mountains, Ural.
14. Diorite. Goroblagodatsk, Ural.
15. ,, Miask, Ural.
16. Epidote-diorite. Olonetz.
17. Porphyrytic-uralite-diabasite. Katharinenburg, Ural.
18. Aphanitic-diabasite. Katharinenburg, Ural.
19. Green slate. Katharinenburg, Ural.
20. Oligoclase-hypersthenite. Valamo, Lake Ladoga.
21. Hornblendite. Goroblagodatsk, Ural.
22. Hornblendite, with zirkon. Miask, Ural.
23. Actinolite. Ufaleisk, Ural.
24, 25. Labradorite. Kamennoi Brod, Kiev.
26. Epidosite. Goroblagodatsk, Ural.
27. Garnet-rock. Turinsk, Ural.
28. Serpentine. Miask, Ural.
29. ,, with diallage. Katharinenburg, Ural.
30. Schistose-serpentine. Katharinenburg, Ural.
31, 32. Beresite. Beresovsk, Ural.
33. Borsowite-rock. Kyshtym, Ural.
34. Mica-schist, with staurolite. Taganai, Ural.

35. Mica-schist, with garnet and staurolite. Taganai, Ural.
36. Mica-schist, with garnet and staurolite. Zlatoust, Ural.
37. Chlorite-slate. Werchneivinsk, Ural.
38. Chlorite slate, with tourmaline. Roshkino, Ural.
39. Talcoze-slate. Regewsk, Ural.
40. Listwenite. Ufoleisk, Ural.
41. Stacolumite. Voitzk, Archangel.
42. Quartz-porphyry. Isle Hoghland.
43. Felsite. Lake Aouschkul, Ural.
44. Variolitic-felsite. Altai, Siberia.
45. Orthoclase-labradorite porphyry. Isle Hoghland.
46. Hornblendic porphyry. Bogoslovsk, Ural.
47. Porphyrite. Goroblagodatsk, Ural.
48. Plagioclase-porphyry. Altai, Siberia.
49. Augite-porphyry. Blagodat, Ural.
50. Uralite-porphyry. Katharinenburg, Ural.
51. Oligoclase-diabasite porphyry. Aiatsk, Ural.
52. Felsitic-retinite (Kulibinite). Nertchinsk, Siberia.
53. Amorphic-trapp (Sordowalite). Serdobol, Finland.
54. Trachyte. Kasbek, Caucasus.
55. Hornblendic-andesite. Koktebel, Crimea.
56. Augite-andesite. Kasbek, Caucasus.
57. Andesitic-lava. Kasbek, Caucasus.
58. Obsidian. Goktcha, Caucasus.
59. Anamesite. Rowno, Volhynia.
60. Basalt. Goktcha, Caucasus.
61. Rock salt. Iletzk, Orenburg.
62. Marble. Gornoshitsk, Ural.
63. Limestone. Sanarka, Ural.
64. Oolite. Kamenka, Podolia.
65. Ophiocalcite. Kussinsk, Ural.
66, 67, 68. Dolomitic-marble. Tivdia, Olonetz.
69. Dolomitic-marble. Kapselga, Olonetz.
70. Magnesite. Poliakovsk, Ural.
71. Porous quartz. Beresovsk, Ural.
72. Lydite with anthracite. Isle Volk, Lake Onega.
73. Hornstone with silver. Altai, Siberia.
74. Jasper. Lake Kalgan, Ural.
75. Jasper. Poliakovsk, Ural.
76. Magnetite. Blagodatsk, Ural.
77. Specular-iron-schist. Krivoi-Rog, Cherson.
78. Spathose-iron. Satkinsk, Ural.
79. Sphaerosiderite. Kromy, Orel.
80. Chromite. Ural.
81. Graphite. Tunkinsk, Siberia.
82. Graphite. Tunguska, Siberia.
83. Anthracite. Grushevka, Chain of the Donetz.
84. Coal. Chain of the Donetz.
85. Coal. Regewsk, Ural.
86. Coal. Tovarkova, Tula.
87. Boghead-coal. Muraievna, Riazan.
88. Iet. Tkvibul, Caucasus.
89, 90, 91. Sandstone. Shoksha, Olonetz.
92. Sandstone. Vytegra, Olonetz.
93. Apatite-sandstone. Kursk.
94. Apatite-sandstone. Tambof.
95. Alumstone. Zaglic, Caucasus.

3379b. Specimens of Minerals of the Porphyry Class (23). *Prof. von Lasaulx, Breslau University.*

3379c. Microscopic Slides (33).
Prof. von Lasaulx, Breslau University.

II.—MINING.

3383. Ramsay's Water Gauge, for measuring the friction of ventilating currents in mines or other places. *D. P. Morison.*

The india-rubber tube is connected with the return airway of the mine, the other compartment of the apparatus being in communication with the external atmosphere; the friction, or difference of density of the two atmospheres, is indicated by the difference of level of the two columns of water. Advantages claimed are especially the facility of observation and the steadiness of the columns.

3384. Model of a Ventilator (Guibal).

Royal Saxon Mining Academy, Freiberg.

3384a. Indicator for Mining Purposes for the prevention of Explosions, and for demonstrating the presence of choke-damp in disused pits or old shafts.

Dr. F. Schöpfleuthner, Vienna.

This is a kind of balance to be set up in coal mines. It announces immediately any evolution of gas, enabling the miners to protect themselves at once against this dangerous substance. The existence of the gas is signalled by an electric telegraph. A very light sphere of glass hangs from the one end of the beam of a balance, and is supplied with pure air from the downcast shaft in such a way that between the admission tube and the sphere some air always escapes to prevent the gas from penetrating the sphere. When, however, by the presence of fire-damp the atmosphere surrounding the apparatus is specifically lighter than pure air, the sphere, in consequence of its weight, will sink and make contact with the electrical conductor of the signalling apparatus. By this means both the miners and the men at the head of the mine receive early notice of the presence of gas in the mine. When several instruments are employed an ordinary numbered table is necessary.

3384a. Drawing of Guibal Ventilator for Mines.

David Pemberton Morison.

Drawing of Patent self-contained Guibal Ventilator for Mines, Ships, &c.   *David Pemberton Morison.*

3384b. Model of Cooke's Displacement Ventilator.

John Cooke.

This is a rotary engine, as distinguished from centrifugal exhausters or blowers, and is particularly applicable to the ventilation of mines.

The model exhibits how in place of having the drum cranks, levers, and connecting rods the full length joining the centres of the arcs and drums to their working axes respectively their dimensions, weights, and speed of reciprocation are reduced to a minimum, while at the same time the deflection of the shutter is almost eliminated, by getting firm hold of it at each end, as far down as is desirable.

The principles involved are two : —

1. Maintaining the same relative proportions of the crank, lever, and connecting rod ; and
2. Adjusting the angle of the cranks and of the radius rods or intermediate shafts to harmonise with the oscillation of the shutter.

3385. Mining Barometer.   *Elliott Brothers.*

3386. Mining Thermometer.   *Elliott Brothers.*

3386a. Davis's Mining and Surveying Aneroid Barometer, reading altitudes to 1 foot.   *John Davis and Son.*

This instrument is specially for the use of mining engineers and surveyors, for the purpose of readily ascertaining slight variations in gradients, levels, &c., and, from its extreme sensitiveness, will be found of considerable utility in mining and surveying work generally. Besides its extreme sensitiveness, the speciality claimed for the instrument is an arrangement of the scale of altitudes which admits of subdivision by a vernier, hitherto impracticable, owing to the altitude scale in ordinary use being a gradually diminishing one, to which a vernier cannot be applied. In the present instrument the action has been so adjusted as to give accurate readings upon a regular scale of altitudes, the barometrical scale of inches being made progressive in length, so as to afford the correct relative readings with the scale of altitudes. For mining operations the entire circle of the dial is graduated to represent 6 in. of the mercurial column—that is, from 27″ to 33″. This scale affords observations from about 2,000 ft. below sea level to 4,000 ft. above. The finest division of the altitude scale (1-100th) represents 10 feet measurement, which can be again divided by the vernier scale to single feet. The vernier scale is moved by rack-work adjustment, and a lens, which rotates on the outer circumference of the instrument, facilitates the reading of minute quantities. For surface surveying purposes, where it is not required to be used below sea level, the instrument is made with the scale divided from 25 to 31 in., thus giving an altitude scale of 6,000 ft. above the sea level only ; and with this open scale, and the assistance of the vernier, the same minute readings to single feet may easily be taken. The instruments are also constructed for measuring much greater altitudes, that is 10,000, 15,000, or 20,000 ft., but with these scales the measurement cannot be made quite so minute as in the more open scales. The instrument is 4½ inches in diameter, and is provided with a leather sling case, thus making it sufficiently portable for all practicable purposes.

3386b. Davis's Improved Colliery Barometer, specially adapted for moist climates. *John Davis and Son.*

The necessity of a travelling screw is dispensed with by choking partially the neck of the tube, thus preventing the possibility of the mercury breaking the end of the tube by violent concussion in transit. The travelling screw (or the screw at bottom for driving the mercury to the top of the tube) being dispensed with, the cistern can be made entirely of boxwood. All the parts of the case are well screwed together, thus making the instrument suitable for moist atmospheres, such as pit banks.

3387. Air Meter, used for the ventilation of mines and other large buildings. *Francis Pastorelli.*

This consists of a horizontal box, with a dial, upon which are divided circles and indexes ; this is mounted upon three vertical pillars, fixed into a brass base ; solidly attached is a vertical ring ; within this are eight vanes fixed to one end of a horizontal axis, the other terminating with an endless screw, which works a series of wheels in the box, and their revolutions are recorded by indexes on the dial ; some of the working parts are jewelled to obtain the minimum amount of friction.

The large circle is divided into 100 parts which represent feet ; one revolution of its hand is equal to 100 feet ; now as the hand of the small circle goes 10 times as slow, it is evident, that, while the former makes one complete revolution, the latter will only have made one-tenth of a revolution ; therefore, when it has made one complete revolution, it will indicate 1,000 feet. The small circles are lettered hundreds, thousands, &c., up to 10 millions, and numbered 1, 2, 3, 4, 5, 6, 7, 8, 9, 0.

For measuring an air current of a low velocity the readings on the large circle are taken, for a high velocity on both.

3887a. Improved form of **Air-meter** for use in coal mines, hospitals, &c.   *E. Cetti & Co.*

3388. Patent Electric Velocimeter, for ascertaining the velocity of air currents in any part of the workings of a coal mine at a distance of two or more miles from it, in a chosen station above ground.   *Francis Pastorelli.*

This consists of three parts :—

1. Four hollow hemispherical cups are fixed to the ends of four strong metal arms (at right angles) radiating from a central boss. At right angles to it is attached the horizontal axis, which is mounted in a rectangular metal box ; each revolution of the cups causes a contact to be made.

2. The receiving instrument has a circular box with a dial mounted upon a metal base ; it is worked by an electro-magnet ; by its means each revolution of the cups is indicated, and motion given to a series of wheels. On the face of the instrument are divided circles and indexes, which register from 10, 100, 1,000 and so on up to 10,000,000 feet.

3. Attached to its side is a commutator, so that the current can be opened or closed for timing or other purposes ; it is an ivory handle carrying an index, which can be moved to point to off or on at pleasure ; it will also serve as a means to prevent unnecessary exhaustion of the battery.

4. A Leclanché battery of six No. 2 cells is connected with the above, so that each revolution of the cups may be electrically transmitted and indicated on the large circle of the receiving instrument which represents 10 feet of velocity. The receiving instrument can be placed in any convenient position above ground. The cup arrangement (Dr. Robinson's with mechanical modifications) is intended to be used in the mine ; in the air-ways or workings where the velocity of the air current is to be ascertained, its velocity is registered on the dials of the receiving instrument.

The inventor does not introduce this instrument with the idea that explosions will be prevented if it were generally adopted, but he has a strong impression that, by its use under proper regulations, it will be the means of diminishing them.

3388b. Air Meter for testing ventilation in mines, hospitals, and public buildings.   *L. Casella.*

THEODOLITES, COMPASSES, &c.

3390. Mining Compass, with independent vernier readings.
Patrick Adie.

3390a. Casartelli's Improved Miners' Dial.
Joseph Casartelli.

The first improvement (1861) consisted in mounting the sight-plate on axes cast on the compass-box, and attaching the arc for giving the angles of inclination to one of the axes, with the index so fixed as to be moved by the sight-plate when inclined to sight up or down the roads, and so give the angle of inclination. The second improvement (1874) consists in substituting for the arc a semi-circular limb fixed to the compass-box by pivots in the line N and S in such manner as not to obstruct the view through the sights. The degrees of angle are graduated on the face, and read off by indexes attached to the sight-plate, which ride over the face of the semi-circle when the plate is inclined for the purpose of taking a sight in steep mines. When the semi-circle is not required, it is simply folded down on the outside of the compass-box. In practice this is found to be an exceedingly convenient arrangement.

3390b. 12-inch Circular Protractor, divided to 15 minutes for plotting in connexion with the dial. *Joseph Casartelli.*

3390c. German Compass, which has been used for a great number of years in the different operations, made at the mines of Almaden. *Royal School of Mines, Madrid.*

3390d. 5-inch Miners' Compass. *G. W. Strawson.*

3390e. 2¼-inch Miners' Compass with folding sights. *G. W. Strawson.*

3391. Mining Compass, with arrangement for suspending needle 7 cm. long, circle graduated in degrees. *A. and R. Hahn, Cassel.*

3392. W. König's Telescope Mining Compass, with level and graduated arc, together with Hörold's centre foot-plate in box with lock.
Royal Prussian Upper Mining Court for the Provinces of Silesia, Posen, and Prussia (Breslau).

This instrument is used by mine surveyors, and at official revisions of mine surveys.

3394. Steel Measuring Tape for mining surveyors, in case. *C. Osterland, Freiberg.*

To be used principally in conjunction with a mine theodolite constructed by the exhibitor.

3395. Head with screw for the steel tape.
C. Osterland, Freiberg.

3396. Travelling Box Compass, with brass support adapted for levelling with a small suspending apparatus, graduated arc, and appurtenances. *C. Osterland, Freiberg.*

Constructed for travellers in thinly populated countries, and is adapted for attaching to the saddle.

3397. Mine Surveying Instrument, consisting of compass, suspending apparatus, and protracting plate clinometer, and two cases. *A. Lingke and Co., Freiberg, Saxony.*

3397a. Henderson's Hypo-thonite, with stand, is an improved form of the same kind of instrument generally in use for rough surveying, especially in underground workings of collieries and other mines. *Ridley Henderson.*

The improvements are additions to the ordinary Miner's Dial or Circumferentor, viz. :—1st, a radiating limb and sights worked by rack and pinion attached to a quadrant, for taking vertical angles. 2nd, a similar arrangement, working a circular plate with sights and vernier, for taking horizontal angles. 3rd, a circular level, ground to a radius of about 15 feet, fitted *inside* the dial on the base plate, in. lieu of the usual cross levels *outside*. 4th, fitted into the same case, with sockets screwed to receive three legs either permanent or temporary, is a ball and socket stand, with adjusting rod which enables the instrument to be correctly adjusted almost instantaneously, while the parallel plates are in any position within 25 degrees of the perpendicular. Where the ground to be surveyed is slippery, or the vein thin, with a low roof, and its inclination steep, this arrangement of stand is of great advantage to correct and rapid surveying.

3397b. Davis's Improved Clinometer with tripod, which folds up and forms a walking-stick.   *John Davis and Son.*

This clinometer is much used in geological surveys ; the arc being of a large size, gives by aid of the levels the gradient of stratum very accurately.

The special arrangement in this instrument is in the stand, also in the large size of the compass, which turns over, rendering the clinometer capable of doing the work of a level and dial approximately. Where great accuracy is not required, it will save time and a more expensive instrument, and on account of its extreme portability may be used when a level or dial could not. Its outside dimensions are, when folded up, $6\frac{1}{2}''$ by $\frac{1}{2}''$ by $3''$.

3397c. Davis's Improved Hedley, Dial, or Circumferentor, for mine surveying.   *John Davis and Son.*

This dial combines all the latest improvements of the best Hedley, with the outside vernier of the theodolite.

The figuring is so arranged that the readings of the needle and the vernier tally, thus keeping one another in check. If the needle be used, the vernier will detect the slightest local attraction.

The Hedley sights and the vernier plate may be clamped, and the instrument can then be used as an ordinary rigid dial. One great advantage of this construction is that if the dial be out of adjustment, it is at once detected by comparing the vernier and needle readings.

The weight is not more than that of the ordinary Hedley, and it is equally compact.

3398. Prof. Junge's Levelling Staff for underground use. The divisions are cut out of a brass plate and illuminated by a sliding transparent glass screen placed in front of a candle, so that the divisions appear as alternate light and dark bands of about an inch in breadth.

Royal Saxon Mining Academy, Freiberg.

3399. Collection of various Mine Signals for subterranean measurement of angles with the theodolite.

Royal Saxon Mining Academy, Freiberg.

3400. Pocket Compass apparatus with stand.

F. W. Breithaupt and Son, Cassel (G. Breithaupt).

3418. Photograph of a small **Box Compass Apparatus.**

F. W. Breithaupt and Son, Cassel (G. Breithaupt).

3401. Six-inch **Theodolite,** with two **Telescopes,** by Messrs. Troughton & Simms, with adaptations for underground surveying, suggested by the contributor.   *Walter Rowley, C.E.*

3393. Complete Mine Surveying Instruments, consisting of compass, suspender, protracting clinometer plate, pocket and case for the additional plate.   *Otto Fennel, Cassel.*

3402. Mine Theodolite (No. 159), with stand and leather case.   *Otto Fennel, Cassel.*

Differs from the preceding principally in the apparatus for adjusting the telescope.

3403. Photographs of mine-surveying instruments.
          C. Osterland, Freiberg, Saxony.

3404. Small Repeating Theodolite for mine surveyors.
    A. Lingke and Co., Freiberg, Saxony (M. Hildebrand and E. Schramm).

3405. Compass Attachment for the Theodolite.
    A. Lingke and Co., Freiberg, Saxony (M. Hildebrand and E. Schramm).

3406. Mining Repeating Theodolite, with—
    a. Box compass.
    b. Two separate tripods.
    c. Two signals.
    d. Two signal lamps.
    e. Suspender and appurtenances to the compass.
    F. W. Breithaupt and Son, Cassel (G. Breithaupt).

With horizontal circle 12 cm. in diameter, and silver limb graduated into half degrees, the vernier indicating minutes and is covered with glass; the vertical circle and vernier are similarly divided, provided with transparent signals for use either below or above ground. The suspensory apparatus attached to the setting up compass was constructed by F. W. Breithaupt for the survey of the Hartz.

3407. Mining Theodolite, by Breithaupt and Son.
    Royal High School of Industry, Cassel (W. Narten).

Mining Theodolite, with lateral telescope, constructed by F. W. Breithaupt and Son in 1864.—The two circles are divided upon silver and are protected by the covering invented by Breithaupt, the verniers also are enclosed in glass. The vertical circle is placed opposite to the telescope as a counterpoise. On the cylinder of the telescope axis a compass and a spirit-level can be placed. The spirit-level of the telescope for levelling lies with cylindrical pivots in bearers in front of the latter, and when the telescope is turned over it may be again turned upwards and observed. This theodolite also forms a convenient small universal instrument for astronomical observations. It has already been described in the fourth part of the "Magazin Mathematischer Instrumente," published by Breithaupt's establishment in the year 1860. A similarly constructed eccentric mining theodolite, referred to in 1876 by Prof. Bauernfeind, in his " Vermessuhgskunde," has been made by Ertel, of Munich.

SAFETY LAMPS.

3407a. Davy's Original Safety Lamp.   *Royal Society.*

3408. Collection of Miners' Lamps of various designs.

Catalogue No.	Name of Lamp.	Description of Lamp.	Mainly used.	Approximate Date of Manufacture.	Velocity of Explosive Current requisite to render Lamp unsafe in Feet per Second (per Experiments of Lamp Committee of Mining Institute).
1	Common Davy -	Single gauze, without shield.	By Hewers and all classes of underground men in Northumberland, land, and generally used by Putters and Drivers.	Recent (Mills and Son).	8 feet.
2	Do. - -	Do. - - -	Do. - -	Recent (H. Watson).	„
3	Do. - -	Single gauze, with shield.	Do. - -	Do. - -	8·3 feet.
4	Modified type of Common Davy.	Single gauze, with shield and glass.	- - - -	About 1850 (Abbot).	—
5	Do. - -	Single gauze, with internal tube.	- - - -	No date (Errington).	—
6	Do. - -	Single gauze, self-extinguishing.	- - - -	Recent (Waring)	—
7	Do. - -	Do. - - -	- - - -	No date.	—
8	Do. - -	Do. - - -	- - - -	„ (Jones)	—
9	Common Davy -	Single gauze, ordinary type.	Generally - -	About 1840 - -	8 feet.
10	Do. - -	Single gauze, original type.	- - - -	„ 1820 - -	„
11	Modified Davy -	Single gauze, with modified lock.	- - - -	„ 1852 (Watson).	„
12	Ordinary Davy -	Exposed to fire for two days.	- - - -	About 1860 - -	,
13	Copper Davy -	Surveying lamp with shield and bull's-eye.	Mining Surveyors.	Recent (Watson)	9 feet.
14	Clanny - -	Glass round the flame and gauze above.	Stonemen, Waggonway-men, &c., in Northumberland and Durham, and generally.	„ (Mills) -	
15	Do. - -	Do. - - -	Do. - -	„ (Watson)	„
16	Do. - -	Glass round, with patent lock.	- - - -	„ (Hall).	—
17	Do. - -	Glass round, with double glass.	- - - -	„ (Henderson).	—
18	Stephenson (Geordie).	Glass surrounded by gauze.	Yorkshire and generally.	„ (Mills) -	9 feet.
19	Do. - -	Do. - - -	Do. - -	„ (Unknown)	—
20	Do. - -	Do. - - -	Do. - -	„ (Watson)	9 feet.
21	Do. - -	Glass, with close top over glass.	- - - -	About 1865 (Hann).	Safe.
22	Do. - -	Glass, ordinary -	Generally - -	1816 (original type).	—
22a	Do. - -	Glass, with perforated shield.	- - - -	First experimental.	—

Catalogue No.	Name of Lamp.	Description of Lamp.	Mainly used.	Approximate Date of Manufacture.	Velocity of Explosive Current requisite to render Lamp unsafe in Feet per Second (per Experiments of Lamp Committee of Mining Institute).
23	Mueseler	Glass, with internal cone and gauze over top of glass.	Generally in France and Belgium.	Recent type, authorised by Belgian Government.	25 feet.
24	Do.	Do.	Do.	Ordinary type	11 feet.
25	Do.	Do.		Modified (Abbot)	„
26	Wood's	Gauze cylinder surrounding flame inside chimney.		1865 (experimental).	Safe.
27	Cail and Glover's	Double glass and gauze above.		About 1850	9 feet.
28	Morrison's, No. 2	Double glass, with internal chimney.		„ 1866	Safe.
29	Do.	Do.		„ „ (common type).	„
30	Do.	Double glass, with J. A. R. Morison's lock.		1868	„
31	Upton and Roberts.	Glass and close brass funnel.		About 1840	11 feet.
32	Morrison's, No. 1.	Glass with internal gauze.		„ 1866	Safe.
33	Watkin and Evans.	Double glass		„ 1867	Unsafe.
34	Watson and Lambert.	Mica plates, gauze, and cone.		„ 1850	„
35	Bainbridge	Glass and brass funnel.		„ 1867	Safe.
36	Unknown	Do.		Antique	Unsafe.
37	Hall's	Double glass and Geordie top.		Recent.	—
38	Eloin	Brass with gauze and mica front.		About 1850	Unsafe.
39	Dubrule's	Glass and gauze top.		Antique	„
40	Stable Lamp	Davy with paraffin chimney and reflectors.		Do.	„
41	Steel mill	Flint and steel striker.		Antique	„
—	Owen's improved Davy.	Single gauze, improved body, and patent levered lock.	By Hewers and general use.	Owen's, 1875.	—
—	Owen's improved glass Davy.	Strong glass tube, with perforated copper air caps, top and bottom, with patent levered locks.	By Officials and Waggonway-men, Hookerson, and in places where a good light is required.	„ 1876.	—
—	Owen's improved Clanny.	Glass round flame, and gauze above and patent levered lock.	Stonemen, Waggonway-men, and general use.	„ 1875.	—

Catalogue No.	Name of Lamp.	Description of Lamp.	Mainly used.	Approximate Date of Manufacture.	Velocity of Explosive Current required to render Lamp unsafe in Feet per Second (per Experiments of Lamp Committee of Mining Institute).
—	Owen's improved patent Stephenson's.	Body and gallery for glass combined, and improved shape, with patent levered lock.	Yorkshire and generally.	Owen's; 1876.	—

North of England Institute of Mining and Mechanical Engineers.

3409. Rudiments of Apparatus constructed by **Sir Humphry Davy** during his researches on the **Safety Lamp.**
The Royal Institution of Great Britain.
Phil. Trans. 1815 to 1817.

3409a. Mining Lamp.   *M. Bréguet, Paris.*

3410. Specimen of the General Map of the Belgian Mines (a chromo-lithographic sheet, 0·46ᵐ by 0·60ᵐ, with a MS. or printed note).
This map, to a scale of 1 : 5000, represents a section of the coal basin comprised between two horizontal planes at the levels of 140 and 190 below the sea. It was executed by order of the Public Works Department, and the General Map will be completed in six or seven years.   *J. van Scherpenzeel Thim, Liége.*

3411. Bidder's Patent Magnetic Lock for Miners' Safety Lamps.   *S. P. Bidder.*
All previous inventions have failed to prevent the men from opening their lamps with forged keys or other implements, and it is well known that a large proportion of explosions have occurred from this dangerous practice. It is evident that safety can only be secured by rendering the opening of the lamp impossible by any means, short of its destruction, which are available to the colliers.
This important desideratum is supplied by these lamps, which are secured by a self-locking catch, which comes into action when the bottom of the lamp is screwed home, and it can only be opened by means of a powerful magnet, which is kept in a strong box, under lock and key, in charge of the lamp cleaner.
Another magnet may be put under the care of a fireman in the pit, whose duty it will be to re-light lamps which have been put out by accident

and thus the inconvenience and loss of time involved in conveying lamps to the lamp-room on bank will be entirely obviated.

The locking arrangement, as will be seen, is extremely simple, and completely enclosed within the lamp, and protected from injury.

A number of these lamps can be opened in the time required to open one of the common locks, and a consideration of the accompanying engraving will at once convince the mining engineer of their superiority over every other lamp in use.

Every further information, and list of collieries where the lamps may be seen in use, can be obtained from S. P. Bidder, 24, Great George Street, Westminster, London, S.W., or from the makers.

The unlocking of these lamps by the apparatus for the purpose is very simple, and much more easily performed than with the ordinary key. The lock consists only of a piece of flat iron $2\frac{3}{4}$ inches by $\frac{3}{4}$ inch wide and $\frac{1}{8}$ inch thick, to the end of which is attached the pin; this is kept in position by a strong spring underneath it, and projects into the upper rim of the lamp. It is impossible to unscrew the top until this pin is drawn by a powerful eight-bar magnet, the magnetism passing through the brass bottom, and drawing the pin down on the inside. If a large number of these lamps are required to be opened in a short time for cleaning, a set of electro-magnets, worked by a galvanic battery, placed in the lamp cabin is the most convenient arrangement.

3412. Apparatus for showing externally the mixture of gases existing in underground explorings.　*M. Lemaire Douchy, Paris.*

2. Mining Models and Plans.

3413. Working Model of a proposed new system of Hand Drill for mining purposes.　*Jos. P. O'Reilly.*

The drill consists of the following parts:—

(1.) The bit, made in the model of those employed for artesian borings; it adapts itself by a screw to the

(2.) Barrel.　This is of cast steel and tubular, thus allowing the passage of a stream of water through the bit into the hole.　It is fitted with a bearing on its upper extremity on which plays the

(3.) Water box, the water passing through holes in the barrel from the box.　The water box is kept from turning with the drill by means of

(4.) An armlet or ring passed on the arm, and connected with the box by a short connecting rod.　The end of the barrel is terminated by

(5.) A solid head, changeable, when worn.

Each blow of the hammer drives the water into the hole with a pressure proportional to the force of the blow, thus clearing away the débris from the bottom of the hole, and keeping the cutting edge cool.　The cutting edge, or bit, may be renewed frequently and easily without involving a change of drill, and the carriage of an equal number of drills, as at present.

3414. Two Plans, illustrating the principal modes of **working Coal** in the Yorkshire mining district.　*Walter Rowley, C.E.*

Showing the advantages for economy of working and ventilation of the "Long Wall" system, and the disadvantages resulting from the "Pillar and Stall" mode of working.

3305. Glass Model of the **Anthracite Coal Beds** of the Wurm basin, near Aix-la-Chapelle.

United Coal Mining Co., Aix-la-Chapelle (Director Hilt).

In this model the excessive contortion of the coal measures is shown by 15 vertical sections, drawn upon glass plates, to a true scale of $\frac{1}{4000}$ of the natural size, horizontal and vertical.

These sections are taken at intervals of 100 lachteus, about 226 yards, except Nos. 10 and 11, which are at double that distance, as between them there is a great fault which cannot be shown on the scale of the model.

The 18 workable seams are represented by special colours. The principal seams and the position of the faults are also shown upon a horizontal plane about 230 yards below the surface, which represents the present main level of the deep workings. The depth to the bottom of the basin is about 3,900 feet, and the total amount of coal contained in it is estimated at about 114,000,000 tons, or about 285 years' consumption at the present rate of working.

3306. Transverse Section through the **Anthracite Coal Beds** of the Wurm basin. Scale, 1 : 1000, 5·28 inches to 1 mile, vertical and horizontal. This is an enlargement of one of the sections in the Model No. 3305.

United Coal Mining Co., Aix-la-Chapelle (Director Hilt).

3282. Model of the Altendorf Trough, a double synclinal in the " Sonnenschein " seam.

" Berggewerkschaftskasse," Bochum, Westphalia.

Shows the stratification of the Flötz, Sonnenschein, and of the Flötz of the fat coal below it in the Altendorf Trough, as well as the (*darauf getriebenen Strecken und Abbaue*). By means of Profils made of glass, the (*überlagernden*) parts of the mountains are indicated.

The ground is removed down to the surface of the seam, upon which are indicated contour lines of equal depths, and the area of the seam worked out. The transverse section of the measures is shown upon the glass plates at the ends of the models.

3283. Fourteen Models of Faults in the Carboniferous rocks.   *" Berggewerkschaftskasse," Bochum, Westphalia.*

Typical models for teaching purposes. The blue stripes represent (*Strecken*) in the first (*Sohle*) ; the red ones those in the second ; the green ones those in the third (*Sohle*).

These illustrate the effects produced by dislocations upon the highly inclined strata of the Westphalian coalfield in producing repetition, overlap of seams, &c.

3415. Typical Plans, for showing the works in coal mines; scale, one thousandth (three chromolithographic sheets, 0·60^m by 0·80, with a printed note).

These plans, which are intended as a model for surveyors, may be used for teaching subterranean topography in technical schools.   *J. van Scherpenzeel Thini, Liége.*

3416. General Vertical Sections, showing the order of the various **Seams** of **Coal** in the Yorkshire coalfield.

Walter Rowley, C.E.

3416a. General Plan of the **Royal Mines** of **Almaden,** presented in 1796 by Dr. Diego de Larrañaga. Copied by Dr. Enrique Bermejo, at Almaden, in April 1827.
Royal School of Mines, Madrid.

3416b. Plan, Outline, and **Section** of two furnaces for melting quicksilver, copied from those used formerly at the mines of Almaden, by Alesandro Sierra.
Royal School of Mines, Madrid.

3416c. Photograph of the **General Plan** of the **Quicksilver Mines** of **Almaden, Spain,** made in 1796, by Dr. Diego Larrañaga. *Royal School of Mines, Madrid.*

3416d. Photograph of a **Plan** made in **1830, by Dr. V. Romero,** of **Watt's Steam Engine,** which was used at the mines for extracting water. *Royal School of Mines, Madrid.*

3416e. Photograph of a **Plan** of **Furnaces** for melting **Quicksilver.** *Royal School of Mines, Madrid.*

3419. Collection of forty-three **Thin Sections** of **Hungarian Trachytes** from specimens of the Imp. geologische Reichsanstalt at Vienna. *Voigt and Hochgesang, Göttingen.*

3420. Collection of fourteen **Thin Sections** of **Rocks** of **Monzoni,** according to the researches of Prof. G. von Rath.
Voigt and Hochgesang, Göttingen.

3421. Three Thin Sections of fossil vertebræ of **Ichthyosaurus, Notosaurus,** and **Squalidarium,** according to the researches of Prof. C. Hasse, Breslau.
Voigt and Hochgesang, Göttingen.

3334. Plans of the **Royal Heinitz-Dechen Coal Mine,** consisting of—
   a. General plan showing position of mine (6 sheets).
   b. Ground plan of the whole mine (6 sheets).
   c. 25 special ground plans (25 sheets).
   d. Profile plan (10 sheets).
      Royal Mining Directory, Saarbrück.

3335. Transverse Section through the **Heinitz** and **Reden** mines, respectively through the "Fat Coal" and the "Flame Coal" part of the Saarbrück Coal Basin (7 sheets).
Royal Mining Directory, Saarbrück.

3336. Special Geological Maps (2 sheets), section Reden and section Dudweiler. *Royal Mining Directory, Saarbrück.*

3337. Sheets to illustrate **Position and Mining Drawing**
in the Saarbrück district (3 sheets with cover).
> *Royal Mining Directory, Saarbrück.*

3338. Explanation of the Exhibited Objects.
> *Royal Mining Directory, Saarbrück.*

3345. Muthung's Survey Map of the mountain district of
Upper Silesia, original drawing, scale $\frac{1}{8000}$ in nine special sections,
from *a* to *i* of the chief section, No. 6, together with two network
maps for the north-eastern and south-western part of the drawing
lithographed on scale of $\frac{1}{100000}$.
> *Royal Prussian Upper Mining Court for the Provinces of*
> *Silesia, Posen, and Prussia (Breslau).*

This map, executed to the scale of 1:8000, is based upon a special triangulation by Prof. Sadebeck, and Messrs. Sartor and Hörold, in which are
inserted reduced copies of the official survey maps and other special plans of
mines, railways, towns, &c.   The rectangular network of the map is determined by parallels to the meridian of the chief trigonometrical point of the
Trockenberg near Tarnowitz.

3346. Photographic reduction of the nine sections on
a scale of $\frac{1}{25000}$, not retouched; carried out by Ed. v. Delden,
Breslau, viz.,

a. Nine unmounted small sheets bound in card.
b. Nine small sheets mounted, like chief section No. 6 (3347) of
the map drawing.
> *Royal Prussian Upper Mining Court for the Provinces of*
> *Silesia, Posen, and Prussia (Breslau).*

a. shows the photographic reduction of the nine original sections above
mentioned, as prepared by Ed. v. Delden of Breslau.   On plate *b.* these
nine small sheets appear mounted and united to a chief section of the network
of the survey map.   The maps 4–6 show the use of the photographic map
on the scale of 1:25,000 as a topographical basis for various mining and
other industrial purposes.   The map shows the important mining districts of
the towns Beuthen, Königshütte, and Kattowitz, being the first sheet of a new
map of Upper Silesia, which was not only destined to exhibit the coal beds,
but also the important deposits of zinc and lead ores of the Muschelkalk basin.
A special wall map is exhibited in which the mineral boundaries are shown
in different colours, but, as in those before mentioned, faint black lines are used
to show positions of places and boundaries of districts and mining rights.   The
map illustrating the projected water-supply represents a larger district on the
same topographical basis, and shows the relation intended to be observed
between the supply and the consumption.

3347. Principal Section, No. 6, of a map of a coal deposit and
ore deposit in Upper Silesia, scale $\frac{1}{25000}$.   To complete and
elucidate this, Hörold's map of the ore deposits of the Polish
Upper Silesian Muschelkalk is added, scale $\frac{1}{50000}$, together with a
number of mountain profiles from Dr. Ferd. Römer's Geology of
Upper Silesia.
> *Royal Prussian Upper Mining Court for the Provinces of*
> *Silesia, Posen, and Prussia (Breslau).*

3348. Muthung's Survey Map of the mountain district of Beuthen, with coloured boundaries of the mining fields and districts, scale $\frac{1}{25000}$.

Royal Prussian Upper Mining Court for the Provinces of Silesia, Posen, and Prussia (Breslau).

3349. Survey Map of the preliminary works for the projected water supply of the industrial districts of Upper Silesia, scale $\frac{1}{8000}$, autographically printed on stone (M. Spiegel, Breslau).

Royal Prussian Upper Mining Court for the Provinces of Silesia, Posen, and Prussia (Breslau).

Example of a special map for industrial use for marking projected railroads, waterways, &c. A reprint from the original survey map scale 1:8000.

3350. Map of the Industrial District of Upper Silesia, scale 1: 8000; autographic lithography.

Royal Prussian Upper Mining Court for the Provinces of Silesia, Posen, and Prussia (M. Spiegel, Breslau).

3351. Attempt towards a Petrographic and " Flötz " Map of the Waldenburg mountain district, drawn by the mining surveyor, Lange, in the year 1807.

Royal Prussian Upper Mining Court for the Provinces of Silesia, Posen, and Prussia (Breslau).

This map is of historical interest, as the first attempt to form a geological map representing the relations between important mining bands in the Waldenburg coal district. The scientific and technical observations appear, even at the present time, to be quite correct.

3352. " Flötz " Map of the Coalfields at Kohlau (now the Consolidated Abendröthe Mine), drawn by the mining surveyor, Schultze, in the year 1804.

Royal Prussian Upper Mining Court for the Provinces of Silesia, Posen, and Prussia (Breslau).

A map some years older than the preceding.

3353. Five Interesting Sections of large **Basalt Quarries.**

Mineralogical and Geological Cabinet of the School of Industry, Cassel (Dr. H. Möhl).

SECTION 17.—MINERALOGY, CRYSTALLO-GRAPHY, &c.

WEST GALLERY, UPPER FLOOR, ROOM (M).

I.—APPARATUS, INSTRUMENTS, &c.

a. BLOWPIPE APPARATUS, &c.

3422. Blowpipe Apparatus for determining minerals. Case as supplied to the Arctic Expedition. *J. R. Gregory.*

3423. Plattner's original Blowpipe.
Royal Mining Academy, Freiberg, Saxony.

3424. Complete Blowpipe Apparatus, according to Plattner, consisting of a large case, with balance, instrument case, &c., reagent case, charcoal box, balance case, and leather cover.
A. Lingke & Co. (Max Hildebrand and Ernst Schramm), Freiberg, Saxony.

3425. Cabinet of Apparatus and reagents, for blowpipe analysis. (Mineralogy.) *James How & Co.*

3426. "The Student's Pocket Blowpipe Case."
Thomas J. Downing.

This comprises the following necessaries, carefully selected for the student's use in the analysis of minerals : Blowpipe, candle, matches, charcoal, fluxes, acids, bone ash, mould for making cupel, anvil, hammer (handle loose), file, forceps, glasses for acids, streak plate, lens, platinum wire and foil, magnet, and glass tubes. Fitted in japanned tin case, with fastening, and adapted for the pocket.

3427. Collecting Bag for the geological and mineralogical collector. *Thomas J. Downing.*

3428. Mohs' Scale of Hardness. Nine minerals, with file, in strong case. *Thomas J. Downing.*

3429. Von Kobell's Scale of Fusibility. Six minerals, for determining the degrees of fusibility, in strong case.
Thomas J. Downing.

3430. Scale of Hardness.
Royal Mining Academy, Freiberg, Saxony.

3431. Collection of Minerals for Blowpipe experiments.   *Royal Mining Academy, Freiberg, Saxony.*

3432. Illustrations of Metallic Incrustations produced by means of the blowpipe on charcoal, by Dr. Theodor Richter.
Royal Mining Academy, Freiberg, Saxony.

3433. Harkort's self-made Scale for measuring small Globules of Silver.
Royal Mining Academy, Freiberg, Saxony.

b. GONIOMETERS, &C.

3433a. Double-refraction Eyepiece Goniometer, for measuring angles of crystals in the microscope.
University of Oxford.

On the construction of the late Dr. Leeson, and given by Mrs. Leeson to the University of Oxford.

3434. Retort and **Geissler Tube,** used by Prof. Vogelsang in his experiments, proving the liquid contents of certain minerals to be carbonic acid.
Royal Polytechnic School, Delft, Prof. J. Bosscha.

3435. Wollaston's Reflecting Goniometer, with an arrangement for readily centering the crystal ; also a portable pair of adjustable bright signals, the spot being reflected from the crystal, and the line from a black mirror placed underneath it.
Charles Brooke, M.A., F.R.S.

3435a. Reflecting Goniometer, invented and used by Dr. W. H. Wollaston. Accompanying it are notes of some of his measurements of crystals.   *G. H. Wollaston.*

3436. Wollaston's Reflecting Goniometer, belonging to and used by the late Rev. Dr. Whewell, D.D., F.R.S., Master of Trinity College, Cambridge, when Professor of Mineralogy.
Rev. Nicholas Brady, M.A.

3437. Babinet's Goniometer, for measuring angles of crystals and refractive indices of liquids. For the latter purpose the liquid is placed in the hollow prism.   *Frederick Guthrie.*

3438. Working Model of a new and simple form of **Reflection Goniometer,** very compact, and allowing the application of a vernier for exact measurement.   *Prof. Jos. P. O'Reilly.*

In this model a spiral traced on a cylinder is substituted for the divided circle, as in the ordinary Wollaston. The measurements are given by the intersections of the spiral with a divided straight edge parallel to the axis of the cylinder and in contact with it.
See Plate accompanying notice ; Proc. Roy. Irish Acad., ser. II., vol. I., pl. XX. (The model represents the first form, since modified as in the plate.)

3439a. Simple Substitute for a **Goniometer.**

Prof. W. H. Miller, M.A., F.R.S.

By means of this a crystal can be measured by Wollaston's method, and the angle of an edge determined by measuring the angle between two lines drawn on paper along the wooden part of the instrument in its two different positions, by comparing the arc, intercepted by the two lines on a circle drawn through the point of intersection of the lines, with the whole circumference, by means of compasses.

3440. Goniometer, with additional pieces of apparatus, by which it may be converted into an instrument for determining refractive indices, and an instrument for measuring the angle between the optic axes of biaxial crystals.

University of Oxford.

3441. Contact Goniometer, being the instrument used by the Abbé Haüy.   *University of Oxford.*

It was formerly in the collection of Haüy's minerals, acquired by the late Duke of Buckingham. His Grace presented it to Dr. Buckland, by whom it was placed with the collection of minerals belonging to the University of Oxford, in which Dean Buckland was professor of mineralogy and geology. The Dean's autograph is on the morocco case.

3442. Goniometer, with horizontal circles, and adjusting level.

Prof. Baron von Feilitzsch, Greifswald.

This reflective goniometer has a dividing circle of 7·5 inch diameter. The division runs direct as far as $\frac{1}{2}°$, and with the application of the vernier 0·5 minutes may be read. The instrument can likewise be used for spectrum experiments.

3443. Goniometer.   *R. Fuess, Berlin.*

3443a. Goniometer.   *M. Lutz, Paris.*

3443b. Goniometer, Wollaston's.   *M. Lutz, Paris.*

3443c. Goniometer, Babinet's.   *M. Lutz, Paris.*

3444. Reflecting Goniometer, on marble slab, in mahogany case.   *F. W. Breithaupt and Son (G. Breithaupt), Cassel.*

A reflecting goniometer of the most perfect and improved construction. The circle is 21 centimeters in diameter, divided on silver to 10′, the vernier giving readings to 10″, with lens, telescope, microscope, illuminating tube with prism, and arrangement for centering the crystals; also a lens for the objective of the telescope, &c. On a marble slab. Property of the K. K. Bergakademie of Leoben, in Styria.

3445. E. Mitscherlich's Goniometer.

Prof. A. Mitscherlich, Münden, Hanover.

A description of the goniometer is given in the memoir in " Berichte der königl. preussischen Akademie der Wissenschaften, 1843." E. Mitscherlich used this goniometer in much of his crystallographic work.

3446. Goniometer (made by Messrs. Powell and Lealand), with adjustments for mounting a crystal (as described in the introduction); and with fittings adapting it for the purposes, 1, of an instrument for determining refractive indices, and 2, for measuring the angle between the optic axes of biaxal crystals.

Prof. N. S. Maskelyne.

3446a. Goniometer, according to Börsch's principles; executed by Breithaupt and Son.

High School of Industry, Cassel (Dr. E. Gerland).

The goniometer—a deflecting goniometer, according to Babinet—can be altered into a spectrometer by affixing a prism. It has the advantage that all examinations and corrections can be ascertained without special auxiliary means, as well as all angles; refraction coefficients can be found by two entirely different methods. (See Börsch, Poggend. Ann., vol. CXXIX., p. 384.)

3447. Polarising Apparatus, with telescope tube and goniometer. *Wilhelm Steeg, Homburg, Prussia.*

3448. Brezina's Stauroscope.

Wilhelm Steeg, Homburg vor der Höhe.

3449. P. Groth's Universal Apparatus for Crystallographic Optical research. *R. Fuess, Berlin.*

Vide "Physikalische Krystallographie," by Professor P. Groth, Leipzig, Engelmann, 1876.

3450. "Microgoniometer," for measuring with the microscope. *Prof. Friedrich Pfaff, Erlangen.*

The arc which stands horizontally can also be placed so that it can be used for vertical measurements. A small pamphlet on the instrument accompanies it.

3451. Charles' original Goniometer.

Conservatoire des Arts et Métiers, Paris.

3453. Apparatus for studying and exhibiting the Optical Characters of Crystals. *M. Werlein, Paris.*

3454. Instruments of Observation.

Three prisms for demonstrating the different dispersions of axes in crystals.
Tourmalin pincers (Tourmalin or Lyncurium).
Andalusite pincers.
Prism of dense flint; Rossette substance.
Dichroïsmal lens.

M. Werlein, Paris.

II.—COLLECTIONS OF MINERALS, DIAGRAMS, MODELS OF CRYSTALS, &c.

a. ROCKS AND MINERALS.

3455. Microscopic Sections of rocks and minerals.
J. R. Gregory.

3456. "Explorer's Comparison Mineralogical Cabinet."
J. R. Gregory.

3457. Rough Gems and Stones, 15 varieties, for comparison and testing. *Thomas J. Downing.*

3458. Specimens of **transparent** slices of various **Minerals** for the **Microscope.**
Royal Polytechnic School, Delft, Prof. J. Bosscha.

a. Specimens of minerals described in Prof. Vogelsang's work, " Die Philosophie der Geologie," Bonn, 1867, and depicted on Tables I. and III.
b. Specimens of crystallites described in " Archives Neerlandaises des Sciences Physiques et Naturelles," and depicted in Table VII. of Vol. V., and Table III. of Vol. VI., and in Prof. Vogelsang's posthumous work, " Die Kristalliten," edited by Prof. F. Zirkel, Bonn, 1875.
c. Quartz slices with cavities containing fluid carbonic acid.

3459. Thin Sections of Minerals.
Ludwig Möller, Giessen.

3460. Collection of rare and remarkable specimens of **Zinc Ores** from the north of Spain, and from the county Tipperary, Ireland, to illustrate the modes of occurrence and formation of hydrated silicates and carbonates, particularly the oolitic and imitative forms. *Prof. Jos. P. O'Reilly.*

3461. Photographs (4) of imitative shapes presented by certain hydrated silicates and carbonates of **Zinc** in the collection from the north of Spain. *Prof. Jos. P. O'Reilly.*

3462. Sections or Diagrams illustrative of **Zinc Ore** deposits of silver mines, county Tipperary, Ireland.
Prof. Jos. P. O'Reilly.

3464. Specimens of Minerals prepared for Observation.
Case containing minerals and chemical products.
Case containing specimen of micrographic rock.
Four cubes of different chroïtes crystals.
Rhomboïd in Iceland spar.
Section of a diorite.
M. Werlein, Paris.

3465. Collections of Crystals and Minerals.
B. Stürtz, Bonn.

3466. Collection of 200 separate Crystals, for mineralogical and crystallographical study; their faces have been crystallographically determined; the system of notation used is that of Naumann. *Bernh. Stürtz, Bonn.*

3467. Specimens from Mitscherlich's Collection of Artificially formed Minerals, Felspar, Magnetic Iron Ore, &c. *Prof. A. Mitscherlich, Münden, Hanover.*

3468. Specimens of Mica containing tourmalin, garnet, and quartz. *Max. Raphael, Breslau.*

3469. Collection of Plates of Crystals, Selenite, and Mica combinations, **Dichroiscope,** cubes and plates for showing dichroism; about 210 pieces in mahogany case.
Wilhelm Steeg, Homburg vor der Höhe.

In the collection of plates of crystals are a large number of beautiful, rare, and valuable minerals and chemical preparations, which in part show, besides polarisation figures, also dichroism, asterism, fluorescence, &c. Also preparations exhibiting interesting twin growths in closed crystals, and crystals containing liquids in cavities.

3469a. Russian Minerals taken from the educational collection in the Mining School in St. Petersburg. 417 specimens, value, 180*l.* *Mining School in St. Petersburg.*

1, 2. Gold. Zmeinogorsk, Altai.
3. Silver. Zmeinogorsk, Altai.
4. „ Salair, Altai.
5. „ Nikolaievsk, Altai.
6. „ Urjum, Siberia.
7, 8, 9. Copper. Nishni-Turünsk, Ural.
9*a.* Platinum. Nishni-Tagilsk, Ural.
9*b.* Iridosmine (Newjanskite). Zlatoust, Ural.
10, 11. Meteoric iron (Pallassite). Krasnoiarsk, Siberia.
11 *bis.* Tin. Yenisseisk, Siberia.
12. Aerolite. Pultusk, Sedletz.
13. Native sulphur. Kazan.
14, 15. Graphite. Serdobol, Finland.
16. Graphite. Ilmen mountains, Ural.
17, 18. Graphite. Tunkinsk mountains, Irkutsk.
19. Stibnite. Serentujevsk, Nertchinsk.
20. Molybdenite. Pitzaranta, Finland.
21. Molybdenite. Ilmen mountains, Ural.

22. Galenite. Zadonsk, Caucasus.
23. Tellursilver. Zavodinsk, Altai.
24. Bornite. Nishni-Tagilsk, Ural.
25. Sphalerite or Blende. Klitchkinsk, Nertchinsk.
26. Chalcosite. Olonetz.
27, 28. Chalcosite. Nishni-Turünsk, Ural.
29. Cinnabar. Ildikansk, Nertchinsk.
30. Pyrrhotite. Ersby, Finland.
31, 32, 33, 34. Pyrite. Pitkaranta, Finland.
35. Pyrite. Riazan.
36. Chalcopyrite. Beresovsk, Ural.
37. Jamesonite. Algatchinsk, Nertchinsk.
38. Pyrargyrite. Zmeinogorsk, Altai.
39. Tetrahedrite. Beresovsk, Ural.
40. Patrinite. Beresovsk, Ural.
41. Rock salt. Astrakhan.
42. Cerargyrite. Zmeinogorsk, Altai.
43. Embolite. Michailovsk, Orenburg, Ural.
44, 45, 46. Fluorite. Kadainsk, Nertchinsk.
47. Fluorite (Ratoᵢkite). Iver.
48. Chiolite. Ilmen-mountains, Ural.
49. Cuprite. Nishni-Tagilsk, Ural.

50. Chalcotrichite.   Gumeshevsk, Ural.
51. Hepatinery (Liver-ore). Nishni-Turünsk, Ural.
52. Minium.   Michailovsk, Nertchinsk.
53, 54, 55. Corundum.   Ilmen mountains, Ural.
56, 57. Corundum   (Soimonite). Soimonovsk, Ural.
58, 59. Hematite Gornoshitsk, Ural.
60. Hematite.   Polekovsk, Ural.
61, 62.   Ilmenite.   Ilmen mountains, Ural.
63, 64.   Perofskite.   Achmatovsk, Ural.
65. Perofskite.   Nicolai Maximilianovsk, Ural.
66, 67, 68, 69. Chlorospinel.   Shishimsk mountains, Ural.
70. Magnetite.   Achmatovsk, Ural.
71.   „   Shishimsk mountains, Ural.
72. Magnetite.   Lupiko, Finland.
73.   „   Zlatoust, Ural.
74. Chromite.   Ufaleisk, Ural.
75.   „   Bilimbaievsk, Ural.
76.   Chrysoberyl   (Alexandrite). Katharinenburg, Ural.
77. Cassiterite.   Pitkaranta, Finland.
78, 79, 80. Cassiterite.   Onon, Siberia.
81. Rutile.   Kossoi-Brod, Ural.
82.   „   Sissertsk, Ural.
83.   „   Sanarka, Miask, Ural.
84. Brookite.   Kamenka, Miask, Ural.
85. Diaspore.   Kossoi-Brod, Ural.
86. Göthite.   Volk Ostrov, Olonety.
87, 88, 89. Limonite.   Beresovsk. Ural.
90. Limonite.   Orel.
91. Hydrargillite.   Shishimsk mountains, Ural.
92. Bismuthite.   Beresovsk, Ural.
93. Quartz (Rhombohedron).   Volk-Ostrov, Olonetz.
94, 95. Quartz.   Neviansk, Ural.
96, 97.   „   Serapulka, Ural.
98. Quartz, with green Turmalin. Beresovsk, Ural.
99. Quartz (Yellow).   Neviansk, Ural.
100. Quartz (Smoky).   Neviansk. Ural.

101, 102. Quartz (Smoky).   Mursinsk, Ural.
103, 104, 105. Amethyst.   Lipovaia, Ural.
106. Avanturine.   Tesma, Zlatoust, Ural.
107. Chalcedony.   Kamtchatka.
108.   „   Tshikoi, Nertchinsk.
109. Carnelian.   Shilka, Nertchinsk.
110. Chrysoprase.   Kishtym, Ural.
111. Heliotrope.   Orsk, Orenburg.
112. Jasper.   Korgon, Altai.
113.   Jasper.   Revnevaia-gora, Altai.
114. Jasper.   Dutsharsk, Nertchinsk.
115. Opal.   Kiev.
116. Tripolite.   Simbirsk.
117. Wollastonite.   Pargass (Ersby), Finland.
118, 119, 120.   Diopside.   Achmatovsk, Ural.
121. Baikalite.   Sludianka, Baikal.
121 bis. Vanadoaugite (Lavrowite). Sludianka, Baikal.
122, 123.   Rhodonite.   Schabrovaia, Ural.
124, 125, 126. Pargasite.   Pargass (Ersby), Finland.
127. Tremolite.   Verchneivinsk, Ural.
128. Hokscharowite.   Sludianka, Beikal.
129. Asbestos.   Miask, Ural.
130. Sordawalite.   Serdobol, Finland.
131. Pitkarantite.   Pitkaranta, Finland.
132, 133, 134. Beryl.   Mursinsk, Ural.
135. Beryl. Tigiretsk, Altai.
136.   „   Adun-Tchilon, Nertchinsk.
137. Beryl.   Ubrulga, Nertchinsk.
138.   „   Onon, Siberia.
139.   „   Tammela, Finland.
140. Emerald.   Katharinenburg, Ural.
141. Chrysolite   (Glinkite).   Itkul, Ural.
142, 143.   Phenacite.   Katharinenburg, Ural.
144, 145. Phenacite.   Ilmen mountains, Ural.
146. Helvine.   Ilmen mountains, Ural.

147. Achtaragdite. Achtaragda (Wilui), Siberia.

148, 149. Grossularite. Wilui, Siberia.

150. Almandine. Taganai, Zlatoust, Ural.

151, 152. Garnet. Achmatovsk, Ural.

153. Garnet. Talaia, Siberia.

154, 155, 156. Melanite. Pitkaranta, Finland.

157. Melanite. Achmatovsk, Ural.

158, 159, 160, 161, 162. Zircon. Ilmen mountains, Ural.

163. Auerbachite. Masurenko, Ekatherinoslav.

164, 165. Malacone. Ilmen Mountains, Ural.

166. Vesuvianite. Lupiko, Finland.

167. Frugardite. Frugard, Finland.

168. Vesuvianite. Achmatovsk, Ural.

169. Vesuvianite (Perimorphose). Achmatovsk, Ural.

170. Vesuvianite. Poliakovsk, Zlatoust, Ural.

171. Vesuvianite. Medvedjeva, Zlatoust, Ural.

172. Heteromerite. Shishimsk mountains, Ural.

173. Wiluite. Wilui, Siberia.

174, 175. Epidote. Kumbakssa, Olonetz.

176. Epidote. Heposelga, Finland.

177. „ Sillböhle, Helsinge, Finland.

178, 179. Epidote. Achmatovsk, Ural.

180. Puschkinite. Verchneivinsk, Ural.

180 bis. Bucklandite. Achmatovsk, Ural.

181. Orthite. Werchoturie, Ural.

182, 183, 184. Uralorthite. Ilmen mountains, Ural.

185. Nephrite. Sludianka, Baikal.

186. Axinite. Kumbakssa, Olonetz.

187, 188. „ Berkutovaia-gora, Ural.

189, 190, 191. Steinheilite. Kisko, Oriervi, Finland.

192. Gigantolite. Tammela, Finland.

193. Pyrargillite. Helsingfors, Finland.

194. Mica (biaxal). Kimito, Finland.

195. Mica (biaxal). Alabashka, Ural.

196. Lepidolite. Schaitanka, Ural.

197. „ Serapulka, Ural.

198, 199, 200. Mica (biaxal). Ilmen mountains, Ural.

201. Fuchsite. Kossoi-Broel, Ural.

202. Diphanite. Katharinenburg, Ural.

203, 204, 205. Biotite. Sludianka, Baikal.

206, 207, 208. Wernerite. Laurinkari (Åbo), Finland.

209. Wernerite. Pargass (Ersby), Finland.

210, 211. Stroganowite. Sludianka, Baikal.

212, 213. Glaukolite. Sludianka, Baikal.

214. Elacolite. Ilmen mountains, Ural.

215, 216. Cancrinite. Ilmen mountains, Ural.

217. Sodalite. Ilmen mountains,Ural.

218, 219, 220, 221. Lapis Lazuli. Sludianka, Baikal.

222. Lepolite (Anorthite). Karis-Lojo, Finland.

223. Lepolite. Karis-Lojo, Finland.

224. Lindsayite. Kisko, Oriervi, Finland.

225. Labradorite. Oiamo, Lojo, Finland.

226. Labradorite. Urpala, Sakkiervi, Finland.

227. Labradorite. Rodomisl, Kiev.

228. Oligoclase. Turholm, Helsinge, Finland.

229. Oligoclase. Kyrkslatt, Esbo, Finland.

230. Oligoclase. Stansvik, Helsinge, Finland.

231. Oligoclase. Katharinenburg, Ural.

232. Albite. Nishni Tagilsk, Ural.

233. „ Mursinsk, Ural.

234. „ Alabashka, Ural.

235, 236, 237. Albite. Kiriabinsk, Zlatoust, Ural.

238, 239. Orthoclase. Alabashka, Ural.

240, 241. Orthoclase. Mursinka, Ural.

242, 243. Amazonstone. Ilmen mountains, Ural.

244. Orthoclase. Miakoticha, Altai.
245.   „   (Sunstone). Sludianka, Baikal.
246. Adularia. Mursinsk, Ural.
247. Marekanite.   Marekanka, Ochotsk.
248. Nefedjewite. Klitshkinsk, Nertchinsk.
249. Chonebrodite. Pargass (Ersby), Finland.
250. Tourmaline. Beresovsk, Ural.
251, 252, 253. Tourmaline. Mursinsk, Ural.
254. Tourmaline. Schaitansk, Ural.
255, 256. Tourmaline. Borstchevatchnoi, Nertchinsk.
257, 258. Tourmaline. Ubrulga, Nertchinsk.
259, 260. Tourmaline. Mursinsk, Ural.
261. Tourmaline. Gornoshitsk, Ural.
262. Tourmaline. Adun Tethilon, Nertchinsk.
263. Andalusite. Schaitansk, Ural.
264.   „   Tut Chaltovi, Nertchinsk.
264 bis. Chiastolite. Argan, Nertchinsk.
265. Xenolite. Peterhoff, St. Petersburg.
266. Wörthite. Peterhoff, St. Petersburg.
267. Cyanite. Taganai, Zlatoust, Ural.
268. Cyanite. Kishtym, Ural.
269.   „   Sanarka, Ural.
270.   „   Kamenka, Ural.
271, 272, 273. Topaz. Mursinsk, Ural.
274, 275, 276. Topaz. Ilmen mountains, Ural.
277, 278. Topaz. Adun Tchilon. Nertchinsk.
279. Topaz. Bortstchevotchnoi, Nertchinsk.
280. Sphene. Pargass (Ersby), Finland.
281. Sphene. Achmatovsk, Ural.
282, 283. Sphene. Ilmen mountains, Ural.
284. Sphene. Kazatchia Datcha. Ural.
285. Staurolite. Taganai Zlatoust, Ural.
286. Laumontite. Petropavlovsk, Ural.

287, 288, 289. Dioptase. Altyn Tübeh, Kirghistan Steppe.
290. Chrysocolla. Nishni Turünsk, Ural.
291. Chrysocolla. Nishni-Tagilsk, Ural.
292. Asperolite. Nishni-Tagilsk, Ural.
293. Picrosmine. Miask, Ural.
294. Calamine. Taininsk, Nertchinsk.
295. Calamine. Klitshkinsk, Nertchinsk.
296. Prehnite. Schaitansk, Ural.
297. Apophyllite. Piterlaks, Wiborg.
298. Natrolite. Tchikoi, Nertchinsk.
299. Analcime. Kulinda, Nertchinsk.
300. Chabazite.   „   „
301. Stilbite. Bielaia, Nertchinsk.
302. Talc. Itkul, Perm.
303, 304. Pyrophyllite, Beresovsk, Ural.
305. Cimolite. Alexandrovsk, Eka therinoslav.
306. Pelikanite. Kiev.
307.   „   Lupiko, Finland.
308. Serpentine. Nurali, Miask, Ural.
309. Pinite with chlorite. Shishimsk mountains, Ural.
310, 311, 312. Leuchtenbergite. Shishimsk mountains, Ural.
313. Kämmererite. Bissersk, Ural.
314, 315. Rhodochrome. Itkul, Perm.
316, 317, 318, 319, 320. Clinochlore. Achmatovsk, Ural.
321. Clinochlore (pseud. Ceylanite). Nicolai Maximilianovsk, Ural.
322. Kotschubeite. Ufaleisk, Ural.
323. Kotschubeste. Bilimbaievik, Ural.
324. Chloritoid. Kossvi-Brod, Ural.
325. Xanthophyllite. Shishimsk mountains, Ural.
326. Xanthophyllite. Nicolai, Maximilianovsk, Ural.
327, 328. Pyrochlore. Ilmen mountains, Ural.
329. Tantalite. Kimito, Finland.
330. Columbite. Pvio, Finland.
331, 332, 333. Samarskite. Ilmen mountains, Ural.
334. Mengite. Ilmen mountains, Ural.
335. Apatite. Pargass (Ersby), Finland.

336, 337. Apatite. Katharinenburg, Ural.
338. Talcapatite. Shishimsk mountains, Ural.
339. Pseudoapatite. Shishimsk mountains, Ural.
340. Apatite. Ilmen mountains, Ural.
341, 342. Moroxite. Sludianka, Baikal.
343. Lazur-apatite. Sludianka, Baiknl.
344, 345. Pyromorphite. Beresovsk, Ural.
346. Mimetesite. Beresovsk, Ural.
347, 348. Monazite. Ilmen mountains, Ural.
349. Vivianite. Kertsch, Crimea.
350. „ Bargusin, Nertchinsk.
351. Libethenite. Nishni, Tagilsk, Ural.
352, 353. Pseudomalachite. Nishni, Tagilsk, Ural.
354. Dihydrite. Nishni, Tagilsk, Ural.
355. Planerite. Sissertsk, Tschernoia, Ural.
356. Hydroboracite. Caucasus.
357. Borax. Baikal.
358, 359. Wolframite. Adun, Tehilon, Nertchinsk.
360. Hübnerite. Baiiovka (Bogoriak), Ural.
361, 362. Scheelite. Pitkaranta, Finland.
363, 364. Vanadinite. Beresovsk, Ural.
365. Volborthite. Preobrashensk, Perm.
366, 367. Thenardite. Marmishansk, Altai.
368, 369, 370. Barite. Kusinsk, Ural.
371. Barite. Gasimurovsk, Nertchinsk.

372. Celestine. Archangel.
373, 374, 375. Crocoite. Beresovsk, Ural.
376. Vauquelinite. Beresovsk, Ural.
377, 378. Gypsum. Astrakhan.
379. Gypsum. Bugulma, Samara.
380. „ Tetushi, Kazan.
381. Alunite. Zaglik. Caucasus.
382. Brochantite. Gumeshevsk, Ural.
383. Brochantite. Zyrianovsk, Altai.
384, 385. Calcite. Ural.
386. Calcite. Mulina, Nertchinsk.
387. „ Volk-Ostrov, Olonetz.
388. Calcite. Pavlovsk, St. Petersburg.
389. Calcite. Tmatra, Finland.
390. Dolomite. Pargass (Ersby), Finland.
391. Dolomite. Katharinenburg, Ural.
392. Gurhofian. Orenburg.
393. Siderite. Ural.
394. Sperosiderite. Ural.
395. Smithsonite. Tchagirsk, Altai.
396, 397. Smithsonite. Kadainsk, Nertchinsk.
398. Aragonite. Nishni, Turünsk, Ural.
399. Cerussite. Riddersk, Altai.
400. „ Taininsk, Nertchinsk.
401. „ Kadainsk, Nertchinsk.
402. „ Ildikensk, Nertchinsk.
403. Witherite. Zmeinogorsk, Altai.
404. Malachite. Gumeshevsk, Ural.
405. „ Nishni, Tagilsk, Ural.
406. „ Beloussovsk, Altai.
407. Aurichalcite. Zavodinsk, Altai.
408. Azurite. Nikolaievsk, Altai.
409. „ Semenovsk, „
410. „ Zyrianovsk, „
411, 412. Mellite. Malevka, Tula.

3471. Collection of Crystals of Minerals. *J. R. Gregory.*

3472. Nicol's Prism, whose side is about 60 millimeters.
Wilhelm Steeg, Homburg vor der Höhe.

3473. Prism of Rock Salt, 50 × 60 mm.
W. Steeg, Homburg vor der Höhe.

3474. Lens of Rock Salt, 75 mm. in diameter, and 300 mm. in radius. *W. Steeg, Homburg vor der Höhe.*

3475. Plate of Rock Salt, 60 × 60 mm.
W. Steeg, Homburg vor der Höhe.

3476. Calcspar Rhombohedron, with glasses ground parallel and perpendicular to the axis.   *Wilhelm Steeg, Homburg.*

3477. Calcspar Prism, with plane parallel to the axis.
Wilhelm Steeg, Homburg.

3478. Piece of Sulphur with large Crystals.
Prof. A. Mitscherlich, Münden, Hanover.

E. Mitscherlich.by these crystals of sulphur determined the crystalline form of the sulphur from the molten condition, and discovered the dimorphism of sulphur.

b. MODELS AND DIAGRAMS OF CRYSTALS, ROCKS, MINERALS, &c.

3479. Diagrams and Microscopic Slides, demonstrating the structure of Bohemian Basalt.

1. Diagram representing the most important types of Bohemian Basalt. On six plates.

2. Diagram showing the most important types of Phonolite, Trachybasalt, Trachylitbasalt, and Melaphyr rocks. On four plates.

3. Two boxes containing 10 microscopical preparations of the above-mentioned rocks.

4. Professor Boricky's works on the Basaltic and Phonolithic rocks of Bohemia, with short explanation of the Melaphyr illustrations.   *Dr. Emanuel Borricky, Prague.*

3480. Photograph of the Rittersgrün Meteorite.
Royal Mining Academy, Freiberg, Saxony.

3481. Wall Map of the Natural History of the **Mineral Kingdom,** by Dr. G. Seelhorst.   *P. C. Geissler, Nürnberg.*

The wall map consists of coloured plates, which are mounted on linen, with a portfolio in which to keep the map when not hanging on a wall. The text is printed on both sides of the map. The map is constructed to fold up in the portfolio, as well as to hang on a wall, and consists of one sheet.

3482. Diagrams (four), or "épures," illustrative of **Applications of Descriptive Geometry to Crystallography,** viz. : two illustrating derivation of holohedral and hemihedral tesseral forms ; and two showing construction of crystal projections from angular measurements and from crystallographic formulæ.
Prof. Jos. P. O'Reilly.

3483. Model, mounted with movable arcs and core, for demonstration of the fundamental forms in each crystalline system, and of the geometric method of derivation of the different holohedral and hemihedral forms from the fundamental one in each system.
Prof. Jos. P. O'Reilly.

The model possesses a movable core adapting itself to the six systems and rise of elastic cords to represent the intersections and edges of forms.

3484. Seven Wire Models representing the symmetry of
five of the crystallographic systems (made by Mr. Sparrow, of the
British Museum, for Prof. Maskelyne). *Prof. Maskelyne, F.R.S.*

These models represent the great circles in which the sphere of projection
is intersected by the planes of symmetry characteristic of the different crystal.
line systems ; each plane of symmetry being subdivided by a network of wires
intersecting at distances proportionate to the parametral ratios of some crystal
belonging to the particular system. They are—
  I. The cubic system.
  II. The hexagonal system represented—
     1. By quartz.
     2. By calcite.
     3. By tourmaline.
  III. The tetragonal system represented by apophyllite.
  IV. The prismatic system represented by barytes.
  V. The oblique system.

3486. Series of Twelve Water-colour Drawings, illustrative of the **Optical Phenomena** seen in sections of **Minerals**
cut perpendicular to the optic axis or axes of the crystal under
the influence of polarised light. Drawn by contributor.
Rev. Nicholas Brady, M.A.

PYRAMIDAL AND RHOMBOHEDRAL SYSTEMS.—UNIAXAL.

1. Calcite, negative, plane polarised, analyser and polariser, parallel.
2. „ „ „ „ 45° apart.
3. „ „ „ „ 90° „
4. „ elliptically polarised „ 45° „
  Axis of ¼ wave film inclined about 22½° to plane of polariser.
5. Quartz rotary polarisation. Analyser and polariser. 90° apart.
6. Two similar plates of quartz superposed, the lower showing left-handed
  rotary polarisation, the upper right-handed analyser and polariser, 90°
  apart. This phenomenon is known as *Airy's spirals.*
7. Quartz, circularly polarised, analyser and polariser, 90° apart, and axis of
  ¼ wave film 45° from plane of polariser.

PRISMATIC OBLIQUE AND ANORTHIC SYSTEMS.—BIAXAL.

8. Nitre, plane polarised, analyser and polariser, parallel line joining optic
  axes = to polariser.
9. Nitre plane polarised analyser and polariser, 45° apart, line joining optic
  axes 45° from polariser.
10. Nitre plane polarised analyser and polariser, 90° apart, line joining optic
  axes 45° from polariser.
11. Nitre plane polarised analyser and polariser, 90° apart, line joining optic
  axes parallel to polariser.
12. Aragonite optic axes, widely separated, only one is seen, analyser and
  polariser, 90° apart, line joining optic axes parallel to polariser.

3487. Large Wire Model, containing one example of each
species of simple form in the cubic system of crystallography, with
its corresponding hemihedral forms inscribed within a sphere of
the chief zone circles of the system. *Rev. Nicholas Brady, M.A.*

The coloured wires indicating the junction of the faces are painted as follows:
Cube, vermilion ; octahedron and tetrahedron, French blue ; dodecahedron,

emerald green ; three-faced octahedron and 12-faced scalenohedrons, deep cadmium yellow ; 24-faced trapezohedron and three-faced tetrahedrons, burnt sienna ; four-faced cube and pentagonal dodecahedrons, violet carmine ; six-faced octahedron and six-faced tetrahedrons and irregular 24-faced trapezohedrons, Schele's green.   On the several zone circles are marked the position of all the poles of the chief forms of the cubic system, except those of most of the forms of h.k.l., or six-faced octahedron, which lie in the triangles made by the zone circles.   This model contains more than 700 wires soldered together.   Made by contributor.

All these above forms are shown in solid in their corresponding colours in the accompanying large series of models of the cubic system, and are drawn in plan and perspective on a series of diagrams similarly coloured.

The particular forms contained in this model are 100 ; 111 ; + and − κ111 ; 110 ; 122 ; + and − κ 122 ; 112 ; + and − κ 112 ; 210 ; + and −π 210 ; 531 ; + and −π 351 ; + and − κ 351 ; Miller's Terminology.

3488. Collection of Models of the Cubic System of Crystallography, to a scale of three inches for the circumscribing cube, made in cardboard, and painted according to their simple forms.   *Rev. Nicholas Brady, M.A.*

These comprise all the chief varieties of the several species, both holohedral and hemihedral, together with the chief combinations of the principal examples of each simple form, with the cube, octahedron, tetrahedron, and rhombic dodecahedron, fully illustrating the passage of one combination to another. In this series the positive and negative hemihedral forms and their orientation are denoted by the dark and light tints of the same colour.

3485. Diagrams illustrative of the **Cubic System** of **Crystallography,** painted to match the models..
                                   Rev. Nicholas Brady, M.A.

These give the general position on the sphere of projection of all the chief varieties of the species in the system.   Examples of the different species, both holohedral and hemihedral, in orthographic perspective, in projection on the plane of the paper, and giving the position of their poles on the sphere of projection, with similar drawings for twin crystals, drawn by contributor.

3489. Collection of three-inch **Models** of the regular **Platonic Solids,** the only forms possible with equilateral sides, and equi-angular with the simple forms from which they are respectively derived. Made by contributor. *Rev. Nicholas Brady, M.A.*

3490. Set of Regular Octahedral Models, showing that crystalline form depends upon the number and parallelism of parts, and not upon the mere shape of the face.   Made by contributor.
                                   Rev. Nicholas Brady, M.A.

3491. Set of three-inch **Models** of the **Pyramidal System** of **Crystallography,** giving the chief simple forms of the minerals of the system, with the latitude of their poles.   Made by contributor.   *Rev. Nicholas Brady, M.A.*

3492. Collection of Models of the Rhombohedral System of Crystallography, with its holohedral, hemihedral, and tetrahedral forms, showing their orientation and combination with the form 111 written. Made by contributor.

Rev. Nicholas Brady, M.A.

3493. Large Models of Crystals, for lecture illustration.

Prof. Crum Brown, University of Edinburgh.

The models are made of pasteboard, and painted. Six models are sent— (1) Rhombic Dodecahedron, and (2) Pentagonal Dodecahedron, illustrating the relation of these forms to the cube; (3) and (4), Dextro and Laevo-tartaric acid; (5) Double laevo-tartrate of soda and ammonia; (6) Asparagine.

3494. Collection of 114 **Models of Crystals,** in wood.

Heinrich Piel, Bonn.

3495. Collection of 100 **Models of Crystals,** in wood, to illustrate the most important chemical compounds.

Heinrich Piel, Bonn.

3496. Complete collection of **Models of Crystals,** 300, in three boxes. *Heinrich Piel, Bonn.*

The whole of the collections of the models of crystals have been carefully prepared by hand, without the aid of machinery.

The models of these collections have an average diameter of section of 5 centimeters; if required, the models may also be procured of a section of 10 centimeters, by which the price would be proportionately increased. Besides complete collections of any desired size and number, single models may also be obtained, the price of which will depend on the number of faces and the difficulty of their manufacture. If desired, stands may be obtained. Stands may be procured for exhibiting the models in a position parallel with their axes. Each collection is accompanied by a catalogue containing, besides the name of the model, its form and its crystallographic symbol according to Naumann's—or also, if desired, according to Miller's—system; and, in addition to these, the most important minerals and chemical compounds which the models represent.

3498. Eighteen large **Diagrams of Crystals.**

Prof: G. vom Rath, Bonn.

These diagrams are manuscript drawings, and are used for the illustration of mineralogical and crystallographical lectures.

3499. Models of Crystals, made from glass plates.

W. Apel, Göttingen.

The large size of the models renders them especially useful for demonstration. The angles agree with those of the natural crystals, as exactly as the difficulty of manufacture will permit. The threads showing the axes have the same colour for similar axes, and a different colour for dissimilar axes. In the hemihedral forms the corresponding holohedral forms are included, and from the colours marked on them it can be seen whether they are disappearing or remain in the hemihedral forms. The hexagonal system is, according to Miller, referred to three axes cutting each other at oblique but equal angles.

3500. Models of Crystals, in wood and wire.

Royal Mining Academy, Freiberg, Saxony.

3501. Tables for Instruction in Crystallography.

Dr. F. Pfaff & Th. Bläsing's Library, Erlangen.

3503. Forms of the Isometric Systems, represented in all possible combinations.   *Prof. Dr. Prestel, Emdem.*

3504. Skeleton of a Rhombohedron, with a divided side for crystallographic demonstration.   *Albrecht, Tübingen.*

By the string placed round the dividing rods all the faces of the rhombohedral system, of which the indices of h, k, & l are represented by three of the number 0 to 5, can be illustrated. Compare Miller's Treatise on Crystallography, p. 55.

3505. Book, "Nephrit und Jadeit nach ihren mineralogischen Eigenschaften sowie nach ihrer urgeschichtlichen und ethnographischen Bedeutung." By Heinrich Fischer, Stuttgart, 1875. With 131 woodcuts, and two chromolithographs.

Prof. Leopold Heinrich Fischer, Freiburg, Baden.

In this book is originated a new branch of mineralogy, viz., the archæological and ethnographical. Hitherto the sculptures, amulets, and ornaments of stone in the museums have been looked upon as show objects, but as quite devoid of scientific interest.

The volume also serves as a guide to the series of models (in wax, gypsum, &c.) exhibited by Dr. Adolf Ziegler, of Freiburg, in Baden, of such scientifically important and prominent original sculptures from Asia, New Zealand, Marquesas Island, Otaheite, Central America, Mexico, Peru, &c. as are partly exhibited in the Freiburg Mineralogical Museum and the Freiburg Ethnographical Museum, and partly were lent by other museums to the Freiburg Museum for the preparation of models.

There can be no doubt that the presence of such models in the Archæological and Ethnographical Museums will be a great help to this branch of study, which has until now been quite neglected, but which promises to become of the highest importance in the study of the most ancient races of men; for the stone remains are indestructible proofs of past periods of civilisation.

The remains of sculpture, the archæological records of America and Oceania, so remarkable in their form and in the astonishing hardness of the minerals out of which they are made, and which have hitherto been neglected in Europe, will by means of these models become better known.

3506. Imitations of large and small **Amulets, Idols, Implements,** &c. of archæological and ethnological importance. Compare the work of Prof. H. Fischer, "Nephrit und Jadeit."

Dr. Adolph Ziegler, Freiburg, Baden.

These imitations (of different materials according to the characters of the minerals) were prepared, on the suggestion of Prof. H. Fischer, Director of the Ethnographical Museum of the University of Freiburg, and were used partly to preserve for the ethnographical collection of the University copies of the originals sent to Freiburg for exhibition, and partly for purposes of exchange. Professor Fischer says, in his new work on Nephrite and Jade, "I " had in Freiburg, Baden, the much-desired opportunity for the manufacture " of casts, as Dr. Ziegler, whose preparations in wax are deservedly valued " in the studies of anatomy and the history of evolution, has succeeded so " happily in the imitation in wax of the originals which I handed to him, " even to the extent of copying the original colours so exactly that it is difficult " to say, on looking at the two together, which is the original and which " the imitation."

A list of the different pieces is given with the imitations, containing references to Professor Fischer's work.

3507. Sohncke's Universal Model of the Raumgitter, made by Heckmann of Carlsruhe, for explaining the author's theory of the structure of crystals.

Appendages to the model :—
Key.
12 elastic spiral threads.
6 tin caps.   *Prof. Sohncke, Carlsruhe.*

3508. Model of Octahedral "Raumgitter."
Prof. Sohncke, Carlsruhe.

According to Bravais the molecules of all crystals are so arranged that their centres of gravity lie in the intersection of three co-ordinates originating in parallel equidistant planes forming a grating in space, the so-called Raumgitter of the author, who has further shown that any unlimited regular system of points in space may be reduced to one or more parallel systems of this kind.

The model consists of a series of parallel tubular brass rods arranged as a cube, but which can be lengthened by sliding joints and their angles of intersection may be varied by hinged joints. The intermediate parallel gratings are represented by wires carrying wooden balls which can be shifted to represent the required position of the molecule. With it the 14 possible kinds of positions as defined by the author may be demonstrated.

For a complete description see the author's paper in Carl's Repertorium für Experimental Physik, &c., Munich, Vol. XII.

The universal model of the raumgitter has the object of bringing to view the 14 possible different kinds of Raumgitter or point systems of the parallelopiped form.

Its importance consists in the following points :—

(1.) According to Bravais the molecules of all crystals are arranged in a Raumgitter.

(2.) According to a theory of crystal structure, developed a short time ago by the exhibitor, the molecules are arranged according to regular systems of points, which latter are always composed of many Raumgitters placed in each other.

From this one model all Raumgitters can be derived; it is made so as to be movable. The dimensions can be altered by drawing out eight edges, after the manner of a telescope. Also, the globe must be divided in similar proportion to the edges.

For the purpose of altering the angles, each angle is provided with two joints, by means of one of which it is possible to incline the four unlengthenable edges, which are now vertical. The four horizontal edges must previously be placed parallel to each other.

Elastic threads are sent which can be stretched between different globes to show their lines of combination. The tin caps are used for indicating particular spheres.

The fixed model shows the octahedral Raumgitter; of course the movable model can be brought to this form.

3509. Case, containing **15 Models of the most important Diamonds,** made with Bohemian crystal glass, Koh-i-noor in former and present state, Regent, Pole Star, Orlow, Empress Eugenie, South Star, Nassak, Great Mogul, Tosseau, Piggot, Saucy, Shah of Persia, Hope, Pasha of Egypt.

Dr. Th. Schuchardt, Görlitz.

SECTION XVIII.—BIOLOGY.

SOUTH GALLERY, UPPER FLOOR, ROOM ⓃN

I.—MICROSCOPES AND ACCESSORY APPARATUS.

(a.) MICROSCOPES OF HISTORICAL INTEREST.

3510. Compound Microscope, invented and constructed about 1590 by Zacharias Janssen, spectacle-maker, at Middleburgh, Netherlands.

The Scientific Society of Zeeland, Middleburgh.

Galileo's Microscope, then called **Occhialino.** Only the little tube remains, the lenses are wanting.

The Royal Institute of " Studii Superiori," Florence.

Viviani, in his life of Galileo, and in the inscriptions which he placed upon his house in the Via dell' Amore in Florence, states that the first microscope invented and constructed by Galileo was presented by him to the King of Poland in the year 1612.

Prince Cesi, the founder of the Accademia de' Lincei in Rome, writing to Galileo in 1624, says—"that he had received the instrument lately made by " him (Galileo) for the observation of very small things, &c." And Imperiali, in a letter to Galileo, dated 5th September 1624, writes as follows :— " I have not words enough to thank you for the occhialino which you were " so kind as to send me. It is quite perfect, and has many most admirable " points, as indeed have all your inventions."

3511. Two Microscopes, by Jan van Musschenbroek, a Dutch mechanician (b. 1687, d. 1748).

Prof. Dr. P. L. Rijke, Leyden.

3512. Microscope, of silver, **used by Anthony van Leeuwenhoek,** the Dutch Philosopher (b. 1632, d. 1723), in his investigations, and probably constructed by him.

Prof. Dr. J. A. Boogard, Director of the Anatomical Museum, Academy of Leyden.

3527. Old Microscope, 1705, made by J. Marshall in London, with seven objectives, objective table, object stand of ebony ivory plates, two forceps, needle, and leaden channel. Property of His Highness Prince Pless, Castle Fürstenstein.

Prof. Poleck, The Breslau Committee.

3528. Microscope, by Musschenbroek, in leather case.

Royal Museum, Cassel (Director, Dr. Pinder).

The microscopes exhibited are the oldest in the collection of the Cassel Museum.

The microscope of Jan van Musschenbroek dates from the 17th century. It has been frequently described and represented in drawings. The University of Leyden possesses a similar one. This specimen was seen in 1709 by *Uffenbach* in the collection of Professor *Wolfarth* in Cassel.

3529. Two Microscopes, by Leutmann, in leather case.
Royal Museum, Cassel (Director, Dr. Pinder).

These are microscopes by (or after) *Leutmann*, as is shown by the drawing in *Wolff's* "Allerhand Nützliche Versuche," III. Bd., p. 291. *Harting* gives no drawing of these microscopes.

3530. Microscope, by Hartsoeker, in leather case.
Royal Museum, Cassel (Director, Dr. Pinder).

This microscope by *Hartsoeker* was invented before 1694. *Wolff*, who gives a description and drawing of it, makes no mention of *Hartsoeker* as being its inventor. It was found by *Uffenbach* in Professor *Wolfarth's* collection.

3531. Microscope, by Campani, in leather case.
Royal Museum, Cassel (Director, Dr. Pinder).

One of the microscopes by *Campani*, which Landgrave *Charles* bought in Rome.

Catadioptric Microscope of Amici.
The Royal Institute of " Studii Superiori," Florence.

Dioptric Microscope of Amici.
The Royal Institute of " Studii Superiori," Florence.

One of the two microscopes has, instead of an object-glass, a concave mirror exquisitely worked in metal; the greater enlargements are obtained by varying the ocular; the other has its stage (*porta oggetti*) rendered movable by means of two very delicate micrometrical screws, so that, in this way, it can also be used as a micrometer. Both are for horizontal vision ; in the first instrument, the image of the object being reflected by a plane metallic reflector; in the second one, by a rectangular prism. Particular attention is due to the camera lucida which accompanies them, and to the various systems of illumination. Among the instruments of Amici existing in Florence is his great refractor, more than five meters in length, with an objective 29 centimeters in diameter, an excellent instrument. It could not be sent for exhibition on account of its size.

3526a. Microscope of Nobili.
The Royal Institute of " Studii Superiori," Florence.

3530a. Microscope of Amici.
The Royal Institute of " Studii Superiori," Florence.

3517. Grand Microscope, with solar reflector and appliances. Made by Benjamin Martin about 1740.
The Committee, Royal Museum, Peel Park, Salford.

The instrument stands on a massive brass tripod foot from the centre of which rises a triangular bar of gun metal with a cradle joint at the lower end in which is a tangent movement for the inclination of the instrument to any required angle. The body is attached to the bar by an arm which moves on its centre by a wheel and pinion, and is further capable of a backward and forward motion. The arm can be turned on its centre over the stage for the reception of single lenses, and the body is adapted for movement parallel to the bar by means of a rack and pinion giving motion to a tube sliding within another. The fine adjustment is effected by means of a fine rack and pinion at the object glass end of the body. The stage is capable of movement up and down the bar by means of a rack cut in the back of the latter and a pinion. It is a traversing stage with two pair of dovetail plates at the back at right angles to each other; it is moved by two screws with divided heads for the measurement of objects. The stage can also be made to rotate and there is a divided head and vernier for reading small angles. The instrument is provided with a mirror which is movable up and down the bar by a pinion, a condenser and forceps on stands, stage condensers and forceps, animalcule cage, and other ingenious apparatus. Quekett had probably not seen this instrument of Benjamin Martin's when he asserted that James Smith was the first in this country to make a microscope with traversing stage and quick and slow motion in 1826.

3525. Microscope used by P. Lyonet in his observations with respect to the Cossus ligniperda.   *H. Ottmans, Amsterdam.*

These observations are described in his Treatise on the Anatomy of the Caterpillar, 1760.

3519a. Culpepper Microscope, made about 1790.

E. Russell Budden.

A compound microscope of the form contrived by Mr. Culpepper. It consists of a large external brass body, supported upon three scrolls fixed to the stage, which is supported on three larger scrolls screwed to a brass pedestal. A concave mirror is fitted to a socket in the centre of the pedestal.

3529a. Large Amici Microscope, with apparatus complete, by Chevalier.   *The Royal Microscopical Society.*

3529b. Double Microscope, by Culpepper.

The Royal Microscopical Society.

3529c. Single Microscope, by Dolland.

The Royal Microscopical Society.

3529d. Mechanical Finger, for picking up minute objects, by Bailey, after Professor Smith's pattern.

The Royal Microscopical Society.

3529e. Double Microscope, by Marshall.

The Royal Microscopical Society.

3529f. Single Microscope, by Tulley and Sons.

The Royal Microscopical Society.

3529g. Spectacles (4 pair), used by the late Robert Brown in his botanical researches.   *The Royal Microscopical Society.*

3529h. Single Microscope, by Lindsay, 1742.
The Royal Microscopical Society.

3529i. Large Solar Microscope, complete.
The Royal Microscopical Society.

3529j. Single Microscope, in silver, by Culpepper.
The Royal Microscopical Society.

3531a. The Martin Microscope.
The Royal Microscopical Society.

Said to have been made by the celebrated Benjamin Martin for Geo. III. It bears no date, but was probably constructed about 1770. For full description *see* " Williams, On the Martin Microscope," " Transactions of the Microscopical Society," vol. x., 2nd series. It has a triple eyepiece, the middle glass being almost a bull's eye, and insuring a very large field ; the distance between the two upper glasses and the lower one can be adjusted by a rack and pinion varying the power. The original objectives range from 4 inches to $\frac{1}{10}$th focal length. Higher powers appear to have been added up to $\frac{1}{80}$th ; and there are four high powers up to $\frac{1}{40}$ to be used without the compound body on a small arm which is provided.

It was supplied with an extensive collection of ancillary apparatus, only part of which is in the possession of the Royal Microscopical Society, by whom it was purchased at a sale of Professor Quekett's instruments.

It is remarkable for the variety of its adjustments, general beauty of workmanship, and for special contrivances for various objects of unusually large dimensions. There are, however, obvious faults of construction, interfering both with convenience and steadiness. Its highest optical powers do not produce a good definition of such an object as the scales of Podura easily displayed by very much lower modern achromatics.

3531b. Amician Reflecting Engiscope.
The Royal Microscopical Society.

This is fully described by Dr. Goring in " Micrographia," 1837. The principle is due to Amici ; this particular form to Dr. Goring and Mr. Cuthbert, by whom it was made. The body of the instrument acts as an eyepiece. The narrow tube resembles a Gregorian reflector. A small inclined mirror receives the image of the object and transmits it to a larger one, which forms a magnified image viewed by the eyepiece. The various arrangements for different adjustments are well contrived, and the workmanship of considerable merit. Had not achromatic objectives been constructed, the reflecting engiscope would probably have maintained its ground.

3518. Old-fashioned Simple, Compound, and Solar Microscope combined ; date about 1800.
Essex and Chelmsford Museum, Chelmsford.

Very perfect specimen of old-fashioned simple, compound, and solar microscope combined ; presented to the Essex and Chelmsford Museum by the late J. Disney, Esq., F.R.S., about 50 years ago, believed to have been then old, and to have cost originally 40*l*.

This specimen affords a good example of the progress of mechanical construction. The movement for shading the mirror is so completely superseded as to be a novelty.

Though effecting one object well it has given way to more complicated motions adapted to many uses.

3519. Old-fashioned Solar Microscope ; date about 1800.   *Essex and Chelmsford Museum.*
Presented to the Museum by the late J. Disney, Esq., F.R.S., about 50 years ago.

3526. Achromatic Microscope, made in 1807 by Hermann van Deyl, in Holland, the first maker of true achromatic objective lenses for microscopes.
Prof. Buys-Ballot, Utrecht, Holland.
(See Harting, Das Mikroskop, iii., page 132, sqq., where this very microscope is described.)

3513. Wilson's Pocket Microscope, with three single powers, about $\frac{1}{4}$, $\frac{1}{6}$, and $\frac{1}{8}$ inch, and box of objects of an early period.   *William Sykes Ward.*

3514. Pocket Microscope, for opaque objects, four powers in Lieberkuhns 1 in., $\frac{1}{2}$ in., $\frac{1}{4}$ in., and $\frac{1}{8}$ in.   *William Sykes Ward.*

3515. Achromatic Microscope, by Oberhäuser, of Paris, with powers of an early period; fine adjustment, achromatic condenser, and a curious double movement stage.
William Sykes Ward.

3516. Compound Microscope, by Dollond, formerly belonging to Josiah Wedgwood, potter. It is not inclinable; provided with two movements; the stage is rude in mechanism. There are six object-glasses, frog-plate, and other appliances.
Robert Garner, F.R.C.S.

3520. Microscope used by Sir W. Hooker, and employed by him in describing the British Jungermanniæ, Musci exotici, &c.   *J. D. Hooker, M.D., P.R.S.*

3521. Microscope used by Dawson Turner, and employed by him in describing the "Algæ" for the "Historia Fucorum."   *J. D. Hooker, M.D., P.R.S.*

3522. Microscope which belonged to **Robert Brown,** and was employed by him in his various botanical researches.
J. D. Hooker, M.D., P.R.S.

3523. Copy of Microscope used by Louis Claude Marie Richard in the Analyse du fruit, made for Robert Brown.
J. D. Hooker, M.D., P.R.S.

3524. Large Microscope, made by S. Plœssl, of Vienna, in 1845. With screw micrometer and other accessory apparatus. (Property of Prof. F. Cohn.)
Prof. Ferd. Cohn, of the Institute of Vegetable Physiology in the University of Breslau.

This microscope of Simon Plœssl was used by Prof. F. Cohn in all his researches during the years 1845–1862, and affords a historical proof of the degree of perfection attained by microscope manufacturers 30 years ago, as, at the time it was constructed, it was considered one of the best instruments on the Continent.

3531c. Goring's Operative Aplanic Engiscope, described and figured in Pritchard's "Microscopic Illustrations," 1830. An early form of compound microscope mounted with ball and socket universal joint, oval plane mirror and 2 in. condenser beneath the stage, black box for opaque objects, forceps, &c.

John Spiller.

3531d. Slit and Prisms used in the first form of Spectrum Microscope. *H. C. Sorby.*

As shown by the woodcut exhibited, the slit was placed some distance from the microscope, and the prism fixed below the achromatic condenser, so that an image of the spectrum was seen on the object examined under the microscope, and the relative powers of absorption for different rays observed. This form of apparatus has been entirely abandoned for many years, and replaced by the exhibitor's direct vision spectrum eyepiece, exhibited by Mr. John Browning, and by the exhibitor's binocular spectrum apparatus, exhibited by.Messrs. R. & J. Beck.

This apparatus is described in the exhibitor's paper in the "Quarterly Journal of Science," 1865, vol. ii., p. 198.

(*b.*) MICROSCOPES BY MODERN MAKERS.

3532. Microscope ; the first made on Joseph Jackson Lister's model, by James Smith, in 1839, but refused by the trade as being too great a departure from the old and approved model.

Joseph Beck.

3532a. Microscope with Large Compound Stand. Designed by Andrew Ross about 1832. With rotating stage and sub-stage added in 1851. $\frac{1}{12}$ inch objective of old form attached.

Ross & Co.

3532b. Microscope with Large Compound Stand on the Jackson slide principle. Designed by F. H. Wenham in 1873. With $\frac{1}{8}$ inch new patent objective attached. *Ross & Co.*

3532c. Microscope with Portable Folding Stand for travellers. Designed by F. H. Wenham in 1874. With 1 inch triple objective attached. *Ross & Co.*

3532d. Microscope with Portable Stand, with objective suitable for students. *Ross & Co.*

3533. Microscope, on the approved Jackson model, with a limb continued under the stage and planed out in one continuous groove to insure perfect concentricity of the optical and illuminating apparatus. Designed for and exhibited in the Great Exhibition of 1851. *R. & J. Beck.*

3533a. Microscope Stand, with modification of Jackson's limb, Wenham's binocular body, a concentric rotating stage, and Brown's iris diaphragm.   *R. & J. Beck.*

3534. " The Educational Microscope." The first cheap microscope supplied with English object glasses. Prepared for and exhibited in the Paris Exhibition of 1855 by Smith, Beck, and Beck.   *Joseph Beck.*

3535. " Universal Microscope." This instrument, made for the Exhibition of 1862, showed a new method of varying the object glasses and eyepieces without removing them, the mode of constructing a simple binocular body, and a loose lever as a slow motion.   *R. & J. Beck.*

3536. " The Popular Microscope." The first cheap binocular microscope brought out in this country, in 1864.
                                      R. & J. Beck.

3537. " The Economic Microscope;" a working instrument when the binocular arrangement is not required.
                                      R. & J. Beck.

3538. " Darwin's Dissecting Single Microscope."
                                      R. & J. Beck.

3538a. Sorby's Binocular Spectrum Microscope, with his new apparatus for measuring the wave length of every part of the spectrum.   *R. & J. Beck.*

3539. " Beck's Dissecting Single Microscope."
                                      R. & J. Beck.

3540. Stephenson's Erecting Binocular Microscope.
                                      J. W. Stephenson.

The primary objects of this microscope are, as the name implies, the erection of the image and the utilization of deep powers by the binocular. The binocular effect is produced by the use of two equilateral prisms, placed together at an angle of 4° 30′ ; the two edges in contact form a wedge by which the cone of light from the objective is divided, and, after internal total reflection, is laterally inverted. The divided pencil of light is then received on a plate of silvered glass at the polarising angle, the reflection from which completes the erection of the image. This plate, which is of black glass, rotates on its axis, the black side being instantly exchanged for the silvered side, when an analyser is required.

The advantages of this instrument are,—

1. The erection of the image.
2. The small angle (9°) at which the bodies converge, giving a convergence towards an imaginary point at a distance of 14 to 15 inches from the eye.
3. A horizontal safety stage, yielding to a slight pressure by the objective, with the bodies inclined at a convenient angle.
4. The immediate substitution of an analysing plate of black glass when polarised light is used.

5. Identity of illumination in each tube.

6. The use of the highest powers rendered possible by the projection of the prisms beyond the body of the instrument into the objective.

3540a. New Binocular Microscope. *Henry Crouch.*

This instrument has a concentric rotating stage provided with adjustments for centering to the highest powers that can be applied. The sub-stage is also of new construction, being detached from the stand by means of a horizontal slide, and leaving the space underneath the stage entirely clear for greater convenience in the use of ordinary oblique illumination.

3540b. Newly arranged Binocular Microscope.
Henry Crouch.

This is of somewhat similar construction to the above, but not so expensively finished ; with and without sub-stage.

3540c. New Premier Binocular Microscope for Botanical and Histological Work. *Henry Crouch.*

3540e. Students' Binocular Microscopes of Old and New Pattern. *Henry Crouch.*

New Educational Monocular for Botanical and Histological Work. *Henry Crouch.*

3541. Student's Microscope. *James How & Co.*

3542. Popular Binocular Microscope.
James How & Co.

3543. " Educational Microscope." With two object glasses, two eyepieces, condenser, &c., in mahogany box. Suitable for biological, histological, and physiological research.
M. Pillischer.

3544. " Student's " or " Educational Microscope." With three object glasses, three eyepieces, condenser, polarising apparatus, animalcule cage, stage-forceps, &c., in mahogany case. Suitable for clinical and biological research, &c. *M. Pillischer.*

3545. " New College Microscope." *James Swift.*

In this microscope the optical tube slides through a fitting lined with velvet for smoothness of action. The fine adjustment has direct central movement, and is so constructed as not to be deranged by constant use. The diaphragm is flush with the surface of the stage, thus leaving the tube-fitting free underneath for the use of apparatus, and, if required, the diaphragm can be used in conjunction with any of the stage accessories.

3546. " University Student's Microscope," in a cheap and efficient form, especially designed for medical and botanical students. *James Swift.*

3547. Crane Arm Binocular Microscope, with newly contrived concentric stage and adjustable object holder moving upon glass bearings. All the fittings of this microscope are made self-compensating for wear. *James Swift.*

3548. Student's Microscope in an alloy of German silver and aluminium. *James Swift.*

3550. Microscopes. *G. S. Wood.*

3551. Microscope with complex adjustments, searcher, and oblique condenser apparatus. *Dr. Royston-Pigott, F.R.S.*

This microscope is fitted with a peculiar hypocycloidal movement and traversing screws for very delicate observations. The condenser possesses wide rectangular movements combined with a unique oscillating oblique action for directing the minute image of a flame or the sun either directly or obliquely upon any desired point in the field of view, giving fine views of many difficult objects, and gorgeous diffraction phenomena with circular solar spectra. It is also fitted with Dr. Royston-Pigott's searcher for aplanatic images, by which much greater depth of focus is attained, and new powers of correcting chromatic and spherical aberration, by moving the searcher between the objective and the eye piece or ocular.

3552. Improved Microscope, with rotating body.
John Browning.

This instrument is contrived so as to combine the advantages of the English with the continental models ; it is especially adapted for dissecting purposes, as the body rotates with the stage ; objects may be examined with any power without losing their centricity.

3553. Stephenson's Binocular Microscope.
John Browning.

In this instrument, for the first time, the planes introduced by Mr. Stephenson for altering the direction of the ray, so that the microscope can be used with the stage in a horizontal position, have been introduced near the eyepiece in the separate bodies ; this arrangement will, it is believed, be found to possess considerable advantages. With Stephenson's binocular, objects may be examined with both eyes with the highest objectives.

3554. Microscope with Micro-Spectroscope.
John Browning.

The micro-spectroscope is intended for the observation of absorption bands in the spectra of solids or fluids, either by reflected or transmitted light. The instrument exhibited contains Mr. Sorby's most recent improvements.

3555. Pocket or Field Microscope, with two achromatic object glasses, contained in a leather case, measuring $7 \times 3 \times 1\frac{3}{4}$ in.
John Browning.

3556. New Portable Microscope. *John Browning.*

This powerful and complete instrument, fitted with a sub-stage, accessory apparatus, polariscope, &c., is contrived to fold on a hinged joint in such a manner that when set up it is the size of an ordinary microscope, but when closed, it packs in a case of which the outside measurement is 6 × 6 × 9.

3556a. Field Naturalist's Microscope.
Washington Teasdale.

The field naturalist's microscope was designed by the exhibitor specially to meet the difficulty of teaching, and making generally popular, the use of the microscope. Messrs. R. Field & Co., of Birmingham (makers of the Society of Arts prize microscopes), have undertaken to make and keep it for sale to the public, and to supply with each instrument a copy of the paper on " The Simple Microscope as an Educational Instrument," by W. Teasdale.

3557. Microscope Stand with Chain Movements, exhibited in 1851. *William Ladd & Co.*

Educational Microscope. Plain stage, with dividing English object glass 2-inch, 1-inch, ½-inch. In cabinet.
Edmund Wheeler.

Educational Microscope, with mechanical stage, with 1-inch 15°, and ¼-inch 70°, English objectives ; in cabinet.
Edmund Wheeler.

Dissecting Microscope, 3 powers.
Edmund Wheeler.

Dissecting Microscope, with glass stage and mirror, 3 powers. *Edmund Wheeler.*

Pocket Microscope, for collecting diatomaceæ, &c.
Edmund Wheeler.

Pocket Microscopes, for botany, entomology, &c.
Edmund Wheeler.

3557a. Polarising Microscope, large Des Cloiseaux pattern, capable of being placed either horizontally or vertically by means of a joint and clamp, with all requisite accessories for the complete study of the optical bi-refracting properties of natural or artificial crystals, such as a goniometer for measuring the distance between optical axes in air or in oil, at various temperatures ; ice-pan for oil ; copper stove with thermometers and spirit lamps ; the microscope frame and Nicol polarising prism, both being movable by means of a rack and pinion. *M. A. Picort, Paris.*

3557b. Polarising Microscope, large pattern, with all requisite accessories, the body of the microscope alone being movable on the stand. *M. A. Picort, Paris.*

3557c. Polarising Microscope, small pattern, mounted vertically with a small goniometer for measuring the distance between axes in air. *M. A. Picort, Paris.*

3660. Microscope Stand, medium size (No. 1b in the exhibitor's catalogue), with rack and pinion for the coarse adjustment, and with rotation round a vertical as well as horizontal axis, with five eyepieces, and Abbé's illuminating apparatus.
C. Zeiss, Jena.

3661. Set of Objectives, from 30 millimetres to 0·7 millimetre focal distance, in all 21 pieces. Some with the correction
for cover glass. The whole in case with lock and glass lid.

C. Zeiss, Jena.

This collection embraces a complete set of microscope objectives, as contained
in the catalogue of the exhibitor No. 21, 1874. Some of them are furnished
with an "adapter," so as to suit the length of the tube of English microscopes.
The threads are those of the "Microscopical Society's screw."

3559. Large Microscope, suitable for any research, with
rotating stage, and joint for inclining the body of the microscope.

R. Wasserlein, Berlin.

3560. Small Student's Microscope.

R. Wasserlein, Berlin.

3561. Microscope No. 1, combined with a photographic
apparatus and accessories.   *Seibert & Krafft, Wetzlar.*

3562. Microscope No. 3.   *Seibert & Krafft, Wetzlar.*

3563. Microscope No. 5.   *Seibert & Krafft, Wetzlar.*

3564. Microscope No. 7.   *Seibert & Krafft, Wetzlar.*

3565. Complete Microscope, with Abbé's illuminating
apparatus, magnifies from 20 to 1,100 diameters.

F. Schmidt and Haensch, Berlin.

3566. Microscope (specially adapted for mineralogical research), with polarisation apparatus.   *Ernst Leitz, Wetzlar.*

Height of the instrument, 1 foot 3 inches.

There is an arrangement for turning the tube round the optical axis, together
with the upper round stage, which is prepared (after the design and drawing
of Professor Möhl of Cassel) with a graduated rim.

The under-stage carries the vernier, and is provided with a circular
diaphragm, which, together with the polariser, is furnished with a convex lens.

The coarse adjustment is done by a rack and pinion, and the fine adjustment by a micrometer screw; with draw-tube, and plane and concave mirror
with vertical and horizontal movement. An eyepiece micrometer, the cross
lines of which can be easily used for estimating the angle of dichroism, and
angle measurement, by holding the eye-piece fast and turning the stage on
which the object lies. A revolving nosepiece for holding five objectives.
Analyser for fixing to the eyepiece, with a graduated circle.

A calcspar plate placed between the eyepiece and the analyser serves for
the stauroscopic examination of crystals; magnifying powers 60 to 2,800
diameters.

3567. Microscope, with nine objectives, four eyepieces,
and other apparatus.   *E. Leitz, Wetzlar.*

3568. Microscope, with three eyepieces and four objectives.

E. Leitz, Wetzlar.

3569. Microscope, with two eyepieces and four objectives.
E. Leitz, Wetzlar.

3570. Microscope, with two eyepieces and three objectives.
E. Leitz, Wetzlar.

3571. Microscope, with two eyepieces and three objectives.
E. Leitz, Wetzlar.

3572. Microscope, with one eyepiece and two objectives.
E. Leitz, Wetzlar.

These instruments are exhibited as specimens of accurate and elegant work, combining solidity and cheapness.

3572a. Microscope, large model, with arrangement for the study of rocks with polarised light. *A. Nachet, Paris.*

3572b. Microscope, with camera obscura, for photographing microscopic objects, arrangement of M. Aimé Girard.
A. Nachet, Paris.

3572c. Microscope for demonstration, which can be passed from hand to hand in histological lectures. *A. Nachet, Paris.*

3572d. Portable Microscope. *A. Nachet, Paris.*

3572e. Binocular Microscope, producing at will the effects of the stereoscope and the pseudoscope. *A. Nachet, Paris.*

3572f. Microscope for the study of the cornea and of the pupil. *A. Nachet, Paris*

3572ff. Nachet's first form of Binocular Microscope.
F. Crisp.

3572g. Nachet's later form of Binocular Microscope (both stereoscopic and pseudoscopic). *F. Crisp.*

3572h. Wenham's Binocular Microscope for high powers. *F. Crisp.*

3572i. Ahrens' Binocular Microscope, with double similar prisms. *F. Crisp.*

3572j. Ahren's Binocular Microscope, with prism of Iceland Spar (using the ordinary and extraordinary images).
F. Crisp.

3572k. Holmes' divided Object Glass Binocular Microscope (the object glass being cut in halves and each half forming one of the images). *F. Crisp.*

3572l. Tolles' Binocular Eyepiece. *F. Crisp.*

3572m. Beck's Binocular Simple Microscope.
F. Crisp.

3572n. Nachet's Triocular Microscope.   *F. Crisp.*

3572o. Nachet's " Grand Microscope renversé avec miroir argenté " (giving very large magnification from the long distance from object glass to ocular).   *F. Crisp.*

3572p. Chevalier's Universal Microscope.   *F. Crisp.*

3572q. Beale's Demonstrating Microscope.   *F. Crisp.*

3572r. Brown's Pocket Microscope.   *F. Crisp.*

3573. Microscope for Demonstration.
Prof. Recklinghausen, Strassburg.

The instrument is so made that it can be taken up in the hand like a telescope, placed against the window, and the light found by even an unpractised observer in a very short time. By using a condensing lens with the instrument, ample illumination is obtained; but faint daylight and even lamplight will suffice. The instrument can be conveniently used with a magnifying power of 400 diameters, especially as an arrangement (as in Hartnack's immersion systems) is fitted to the lower end of the tube and affords the most accurate adjustment.

The object is to demonstrate microscopic preparations to a large audience. This demonstration-microscope has this advantage over ordinary microscopes for demonstrations, that it can be passed quickly round a class. As the microscopic object under observation has to be vertical, it follows that the preparation must be mounted and the cover glass fixed, and moreover it must not float in a liquid.

The tube with adjustment stand is made by Hartnack and Pratzmowski, the apparatus for illumination and the table by Majer of Strassburg.

3574. Microscope, with a set of objectives and eyepieces, consisting of—

a.	Objective,	No. 5, magnifies	184 to	506	times.
b.	„	No. 6,	„	220 „	612	„
c.	„	No. 8,	„	366 „	1,012	„
d.	„	No. 9,	„	458 „	1,266	„

R. Winkel, Göttingen.

3575. Microscope constructed specially for the examination of microscopic sections.   *R. Fuess, Berlin.*

3576. Immersion Lens, with special contrivance for increasing the magnifying power from 300 to 1,000 times with the low eyepiece.   *R. Winkel, Göttingen.*

3577. Microscope (small model), commonly used.
Dr. Hartnack, Potsdam.

3578. Small Microscope, specially suited for mineralogical researches.   *Dr. Hartnack, Potsdam.*

3579. Large Microscope. *Dr. Hartnack, Potsdam.*

These instruments are sufficiently well known, and it is not necessary to bring forward their peculiarities, but the mineralogical microscope has been recently constructed, and the exhibitor requests mineralogists to direct their attention to it.

3579a. New Dissecting Microscope with three achromatic powers, support for arms, and rack focussing giving three movements. *Harvey, Reynolds, and Co.*

3580. Microscope for Dissecting, with forceps and plate for dissection, and simple Microscope with rod action.
Geneva Association for the Construction of Scientific Instruments.

The Geneva Association constructs three different patterns of microscopes. In the simplest form the focussing movement is worked direct by the hand acting upon a rod which supports the optical system, and which slides up and down through a groove. The light is given by a mirror placed below the stage. In the second pattern the focussing is effected by means of a rack. Lastly, in the third pattern, the most complete of all, a pair of forceps revolving around two rectangular supports is substituted for the plate. By this means an object can be successively examined on all sides, and without leaving the focus of the glass. The optical parts are the same as in the preceding.

The optical system of each of these microscopes is composed of three "Wollaston" doublets with inner diaphragms, of 1 inch, $\frac{1}{2}$ inch, and $\frac{1}{4}$ inch focal distance, giving magnifying powers of 9, 18, and 36 times, for 25 cm. length of vision.

The Geneva Association supplies besides to those who require higher powers achromatic lenses of short focus, giving great clearness and light. An achromatic lens of 2 millimetre focus produces a magnifying power of 125 times with perfect clearness.

3558. Small Microscope, with a Cylindrical Clip for Objectives.
Geneva Association for the Construction of Scientific Instruments.

This pattern of microscope is at once strong and simple. The system of the cylindrical clip is as follows :—

The objective is not screwed into the tube, but is held by a spring against a carefully turned bearer. To remove the objective, it is sufficient to draw it away transversely after having lowered it so as to exert a pressure in the direction of the axis. The objective is more easily placed in position than removed. The advantages resulting from this arrangement are :

1. A great saving of time to the observer, changes of magnifying power being rapidly accomplished.
2. The mechanical centering of the objective is better than that usually obtained with the screw.

3581. Large Microscope, with Reversing Action, and cylindrical clips for objectives.
Geneva Association for the Construction of Scientific Instruments.

40075. 3 M

This instrument, of which the general construction is that of Ross' English microscopes, has the following arrangements :—It may be used in different positions, vertical, horizontal, or oblique, and is specially applicable to the photography of microscopic objects. The illumination is effected by means of a mirror carried by a system of jointed rods, which limits its movement to an arc, the centre of which is the object to be observed. The best illumination is thus quickly secured, the observer not having to regulate at the same time the focal distance of the mirror and its lateral distance from the axis of the microscope. The condenser is moved by means of a rack, which serves to regulate exactly the position of the diaphragm. The whole condensing system, turning around an eccentric axis, can be very rapidly modified.

The stage turns independently of the tube. The focussing is effected, for coarse adjustment, by means of a rack, and for fine adjustment by means of a milled screw acting upon the tube supporting the objective, by the intervention of a lever that lessens the amplitude of the movement.

The tube supporting the objective is carried by a spring which serves to prevent the objects being crushed, when by mishap the objective is lowered too quickly. The milled screw for focussing is used as well for the micrometrical measurement of objects under observation as for focussing in the use of the objective in photography.

The objective is not screwed on to the tube, it is only pressed by spring clips against a steel bearing adjusted with the greatest care. To take off the objective, it is sufficient to draw it away transversely by pressing in the direction of the tube of the microscope. The objective is fixed as instantaneously as it is removed. The advantages resulting from these arrangements are the following :—1st. A great saving of time to the observer. 2nd. A mechanical centring of the objective much more perfect than that obtained by a screw. The defects of centring being immediately discovered may be partly corrected. 3rd. There is the easy choice of the side of the objective that gives the best effect when the oblique illumination is employed.

3582. Microscope, with durable body, with rotating movement round its axis, fine adjustment by means of steel prism, condenser, polarising apparatus, magnifying power from 10 to 1,200 diameters.   *W. Teschner (successor to Amuel), Berlin.*

3583. Microscope, with magnifying power from 20 to 1,000.   *W. Teschner, Berlin.*

3587. Polarising Microscope, with movable body and prism.   *M. A. Picort, Paris.*

3588. Polarising Microscope, with movable body.   *M. A. Picort, Paris.*

3589. Picort Vertical Polarising Microscope.   *M. A. Picort, Paris.*

3589a. Polarising Microscope, with lens for parallel light, and Nicol prisms for measuring the axes.   *Laurent, Paris.*

The illuminating part which is used in the case of crystals of one and of two axes, that is to say, for converging light, can be removed by hand and replaced by a lens set in a ring ; the whole objective part is also removed, and the apparatus is then used for parallel light. There is, moreover, the movable eyepiece for the sighting, and the flat tints are very uniform and bright. The field of view is very large, the two axes of the white topaz are seen. Compensation can be effected with a parallel or a perpendicular quartz.

3589b. Polarising Microscope on round foot.

L. Laurent, Paris.

Same model, but mounted knee-wise, upon a pillar with lengthening piece, and a tripod. This model can be placed horizontally and is then used for measuring the divergence of the axes of crystals, immersed in oil, heated, &c. For this purpose the small glasses at the objective and at the illuminator are removed, they are mounted so as to travel easily. The glass support is also removed and replaced by the tube that contains a Nicol. The distance between the two plates of the glasses of the objective and of the illuminator is 15 millimetres, which allows room for either an oil pan or a stove by means of a support bearing a divided dial (which is not exhibited). The principle is that of M. Des Cloizeaux, and the arrangement, especially for this last model, is due to the exhibitor.

3589c. Early Microscope. *John Waugh.*

3589d. Two Microscopes in cases.

W. Techner, 180, Friedrich Strasse, Berlin, W.

Polarising Microscope for projection.

L. Laurent, Paris.

Glass Sunk Cells for the microscope (4).

L. Laurent, Paris.

3590. Solar Microscope, with alum cell. *Laurent.*

Model, prepared for projections, at the request of M. Des Cloizeaux. It has already been used at the Sorbonne and at the Paris Museum.

The field of view has been made as large as practice allows, with especial regard to illumination. The gypsum and its variations through heat can be clearly seen. By the advice of M. Des Cloizeaux, the horizontal pincers have been replaced by vertical ones which are much more handy. An alum pan for stopping the greater part of the heat has also been added to the projection lantern, and it has then been possible to project platinocyanure of barium, and formiate of copper, without altering them.

3590a. Dr. Lionel Beale's Portable Microscope, for class purposes or for travelling. *Thomas Hawksley.*

(*c.*) ACCESSORY APPARATUS.

3591. Photographic Apparatus for the Microscope.

Seibert and Krafft, Wetzlar.

This apparatus can be connected to any microscope, which, for this purpose, is placed perpendicularly under the box in front of the apparatus, after which, in order to exclude unsuitable light, the cloth-bag, fitting tightly by means of a gutta-percha ring, is pulled over the tube. The adjustment of the picture is effected by means of the glass-plate contained in the box, which is provided with squares and the ocular-like microscope. The lines of the glass-plate must be placed downwards. The microscope is placed

3 M 2

on the glass-plate, and regulated by the upper lens in such a manner that the lines can be seen plainly; thereupon the picture is inserted, the microscope being used as ocular. In this manner the picture is sure to lie exactly on the lower surface of the glass-plate; consequently also on the prepared plate to be inserted afterwards. By applying the illuminating apparatus the plane mirror of the microscope should be used, the arrangement of which is easily understood. The achromatic condenser, accompanying the apparatus, consists of three achromatic lenses, and fits to microscope No. 3 (No. 3562) in the Exhibition.

3592. Microscopic Apparatus. Series of the more important pieces of apparatus and object glasses now supplied for use with the microscope.   *R. & J. Beck.*

3593. Object Glasses (4), showing the lenses before they are put into the cells.   *R. & J. Beck.*

3593a. New Stand on the Continental Model, the diaphragm being provided with centering screws so that it can be adjusted to each objective as applied.   *Henry Crouch.*

3593b. Microscope Objectives and accessory Apparatus.   *Henry Crouch.*

Suited to the stand described above, and for microscopic work generally, and mounting apparatus and materials.

3594. Object Glasses, Apparatus, and Accessories used in the different branches of microscopical research.

M. Pillischer.

3595. Microscope Lamp.   *M. Pillischer.*

3542a. Microscope Lamp.   *James How & Co.*

3542b. Tate's Air Pump.   *James How & Co.*

3542c. Selection of Transparent Photographs, for the **Lantern,** illustrating lectures upon geology, consisting of sections of strata, groups of fossils characteristic of the various sedimentary formations, restorations of extinct animals, &c.

James How & Co.

3595a. Fittings for Microscope.   *M. Lutz, Paris.*

3596. Apparatus to be used with the Microscope, for securing perfectly central illumination, manufactured by Wood of Liverpool.   *Rev. W. H. Dallinger.*

The object of the apparatus is to secure minute and delicate alterations in the position of the flame-image upon the mirror or prism, since it has been found by the exhibitor that perfectly central illumination can only be secured by having the image of the flame exactly under the optical axis of the sub-stage combination, after the latter has been made to coincide with the optical axis of the object glass. This will enable the microscopist to illuminate the whole field of a $\frac{1}{4}$-inch object glass through an aperture of $\frac{1}{100}$th of an inch in diameter. But it can only practically be done by a fine set of mechanical motions in rectangular positions giving perfect command of the position of the flame.

3597. Apparatus for Continuous Observation of Minute Organisms with the highest powers, by preventing evaporation of the fluid in which the organisms live.

Rev. W. H. Dallinger and J. J. Drysdale, M.D.

This apparatus was devised by Messrs. Dallinger and Drysdale for prosecuting their "Researches into the Life-History of the Monads." It is used upon the ordinary "mechanical stage" of the microscope, so as to admit of the continuous examination under the highest powers of the same drop of a putrefactive or other fluid without allowing it to evaporate. The glass cup to the right is to be filled with water. The capillarity of the linen carries over a constant supply of moisture ; part of this linen is included in the central cylinder into which the object glass projects, and the water thus constantly carried into this air-tight chamber causes the air therein to become so saturated that evaporation from the fluid under examination cannot take place. The highest powers may be used, because the india-rubber diaphragm yields instantly to the screw of the "fine adjustment."

3598. Microscopic mounting materials and dissecting instruments, consisting of :—

Wood-cutting instrument and chisel, instrument for cutting circles of thin glass, glazier's diamond, writing diamond, cell-making instrument, brass table and lamp, Page's forceps, case of dissecting instruments (containing four knives, two hooks, two points, three pairs of scissors, three pairs of forceps, and needle-holder), Valentin's knife, 1 oz. thin glass, 9 doz. slips (3 in. by 1 in.), 3 doz. wooden slips, 3 doz. cells, 200 labels, 5 capped bottles (containing Canada balsam, asphalte, gold-size, glycerine, and marine glue), bottle of Deane's medium, three stoppered bottles for containing chloroform, nitric acid, and liquor potassæ. *R. & J. Beck.*

3603. New Sub-stage Condenser, comprising achromatic condenser of high angle of aperture. *James Swift.*

The above is an effective spot lens, from 3 inch objective to ⅛ inch ; it has centering adjustments, contracting diaphragm, and complete polariscope, in which all the colours of the rainbow can be obtained. It also forms a complete substitute for every piece of apparatus used under the microscope.

3604. Low Angle Objectives. A series from 1 inch to ½ inch. *James Swift*

3605. Popular Achromatic and Tinted Condenser.
James Swift.

This apparatus combines the achromatic condenser, light modifier, spot lens, and polariscope.

3606. Microscopic Objects, illustrating every department of microscopy.

Machine for cutting sections, with knife.

Improved turn table.
Hot plate and lamp.
Spring compress, and spring clips.
Instrument for cutting circles in thin glass.
Glass slips and thin glass covers.
Labels for covering and naming objects.
Buff block for cleaning covers.
Centering card.
Case of dissecting instruments.
Knives, forceps, needles, &c.
Syringe and appliances for injecting.
Small air pump for mounting.
Lamp for microscope, with white cloud reflector.
Bottles and dipping tubes for collecting materials.
Collecting bottle, net, hook, and jointed rod.
Cements, varnish, liquids, and media for mounting.
Boxes, cases, cabinets, for finished objects.

Edmund Wheeler.

3606a. Micrographic Study of Paper Manufacture.

M. Aimé Girard, Paris.

3606b. Collection of Microscopic Slides.

John Waugh.

3607. Instrument for verifying Micrometers of Microscopes.

W. F. R. Suringar, Leyden.

Consists of a metallic frame, resting at each side upon four pillars fixed in a wooden stand through which pass three screws, two at the one side, one at the other side, for regulating the horizontal position of the instrument. Upon the frame slides another, the quick movement of which is regulated by a rack and pinion, and upon this a third, destined for bearing the scale, with slow motion, directed by an adjusting screw with milled head. For using the instrument, put the scale, divided into a certain number (fig. 35) of millimeters on the inner slide, or (in the case of microscopes with very small stage, such as the small Oberhäuser-microscope now shown), on the three extra supports on the outer side, regulate the screws of the small additional table destined for bearing the microscope, in such a manner that the stage of the microscope glides under the frame, and the divisions of the scale can be centered with the field of the microscope. After having adjusted the ends of the eyepiece micrometer in the microscope to the ends of the stage micrometer that is to be verified, seen through the microscope, place the microscope over the middle part of the instrument, move the scale till the first division of the scale can be seen in the field of the microscope, adjust and measure the first division, and subsequently all the others, taking care to measure always from the same side (*e.g.* the left) of the lines. Notice the number of entire divisions of the eyepiece micrometer, and note the tenth part of them, to which each division of the scale corresponds. After having measured them all, one by one, make the addition, and divide the amount by the number of divisions measured. The quotient indicates the mean value of each division of the scale expressed in parts of the eyepiece micrometer, viz., of the millimetre of the stage micrometer. This quotient, divided into one, expresses the exact value of the millimetre to be verified in parts of the scale used for comparison, and itself previously compared with a standard scale. The accuracy obtained is proportional to the square root of the number of divisions compared one by one, and can be relied on to the $\frac{1}{1000}$ part of a millimetre.

The instrument should be placed before a window, and at a distance not exceeding six feet from it, because otherwise the scale would not receive sufficient light for seeing the divisions clearly through the microscope.

A description of verifications made by the instrument, with cut, accompanies it.

3608. Apparatus for Microscopical Research in the open air. *Prof. Dr. Leonard Roesler, Klosterneuburg.*

3609. Re-agents for Microchemical Researches, in box. *Prof. Dr. Leonard Roesler, Klosterneuburg.*

3610. Instruments for Microscopical Researches. *Prof. Dr. Leonard Roesler, Klosterneuburg.*

3611. Instruments for Microscopical Researches on living organisms, adapted to their observation when under the influence of various gases and at different temperatures. *Prof. Dr. Leonard Roesler, Klosterneuburg.*

3612. Drawings illustrating preceding Apparatus. *Prof. Dr. Leonard Roesler, Klosterneuburg.*

3615. Kratometer for finding magnifying power and focal length of objectives. *Dr. Royston Pigott, F.R.S.*

This simple instrument is so contrived that whatever object glass is used, the actual power is ascertained at once, whatever be the length of body employed. A stage micrometer of lines ruled to 100ths and 1,000ths of an inch is viewed by the kratometer eyepiece, and the number of divisions of the stage micrometer embraced by ten of the kratometer, gives exactly the magnifying power, if multiplied by ten. This instrument at once determines the focal length and comparative powers of all object glasses submitted to this test.

3616. Microscopic Refractometer, for ascertaining the mean refractive index of plates of glass or lenses. *Dr. Royston-Pigott, F.R.S.*

A new refractometer for determining the refractive index of white light (or mean rays or line E in the solar spectrum) of small plates or lenses of refracting material. The instrument measures to the 100,000th of an inch, the thickness of a thin plate as covering glass ·004″ thick, and the distance which an image is refracted upwards.

A minute prism reflects the light through a small plano-convex lens fixed at the end of the measuring screw ; the prism is illuminated by solar or artificial light by a condenser, and reflected up the axis of the microscope. Differential toothed wheels measure the number of revolutions of the screw (nearly 100 threads to the inch), and indicators give the 100ths, 1,000ths, 10,000ths, and 100,000ths of an inch. The instrument detected 32 changes of colour in Newton's rings of contact between the central black spot in air, caused by a film half a millionth of an inch thick, and the last vanishing colour on separating the plano-convex lens from a plane surface with which it had been in contact. It has measured the refractive index to three places of decimals in thin flint glass 0″·0042 thick.

3616a. Sorby's Standard Interference Scale, for measuring the position of absorption-bands in spectra.

R. & J. Beck.

3616b. Sorby's Volute Diaphragm, for regulating the amount of light for the spectrum microscope.   *R. & J. Beck.*

3617. Early Form of Machine, with Knife, for making microscopical sections of wood, &c., devised by Andrew Pritchard, F.R.S.E., prior to 1835.   *Dr. Urban Pritchard.*

The block of wood for section, if large enough, is fixed in the movable triangular chamber by means of the little screw. The larger screw at the bottom gradually elevates the whole, and the knife, held in both hands, shaves off thin sections as the wood is raised. Should the piece of wood be too small to be placed in the triangular chamber, it must be glued on to a block of convenient size.

The whole machine is made to screw on to a bench.

3613. Freezing Apparatus. A simple form for preparing soft tissues for microscopical examination, consisting of a solid copper cylinder with wooden handle and felt cap to fit over cylinder.   *Dr. Urban Pritchard.*

Mode of use :—Immerse cylinder in mixture of ice and salt for a few minutes; then remove and wipe, place tissue to be cut, after being moistened with gum, on metallic end. Put felt cap over cylinder; tissue will be frozen and ready for cutting in two or three minutes.

3614. Glass Microtome for microscopical sections of hardened or frozen tissues.   *Dr. M. E. Mulder, Groningen.*

The hardened tissue, previously imbedded in a cylindrical mass of wax, consisting of a mixture of Stearine -   -   - 30

Hogs' lard -	-	-	24
White wax	-	-	16

is introduced into the glass tube of the microtome, which is polished at its upper margin.

Any rotation or movement of the wax cylinder, which is exactly of the same size as the tube, is prevented by the four pins. The wax cylinder can be moved up or down by means of a screw, and sections are made by passing a knife over the polished surface of the tube.

The thickness of the sections is indicated by the lines on the screw. (Every line is $\frac{1}{50}$ mm.)

When the microtome is used for freezing, the glass tube is unscrewed and replaced by the freezing box. The tissue, imbedded in a solution of gum arabic, is frozen by filling the freezing-box with snow or ice and salt.

3618. Apparatus for maintaining an even Temperature in Microscopic Observations.

Geneva Association for the Construction of Scientific Instruments.

This instrument is intended for performing microscopic operations in a perfectly even temperature by means of hot water circulation. A small spirit lamp is placed under the reservoir at the extremity of the apparatus. A thermometer placed inside the instrument serves to regulate the temperature.

3619. Camera for Microscope.
Geneva Association for the Construction of Scientific Instruments.

3620. Two Ross Compressors, for the Microscope.
(Ordinary models.)
Geneva Association for the Construction of Scientific Instruments.

3621. Compressor, by Schick, for the Microscope.
(Ordinary model.)
Geneva Association for the Construction of Scientific Instruments.

3622. Microtome, with inclined plane for producing microscopic sections of extreme tenuity and regularity.
Geneva Association for the Construction of Scientific Instruments.

3623. Microtome by Professor Wilhelm His.
Geneva Association for the Construction of Scientific Instruments.

Many instruments for making microscopic sections exist, but most of them have been distrusted by scientific men. This distrust was more or less justified by the imperfection of the apparatus known hitherto, and it was mostly preferred to operate simply by hand, without its being possible to calculate with positive accuracy the thickness of the section.

During the course of his studies upon vertebrated animals, Professor W. His, of Basle, found the absolute need of an instrument for rapidly effecting sections of positive uniformity in thickness.

It is from his suggestions that the Geneva Association for the Construction of Scientific Instruments has constructed a Microtome, which has now the complete approbation of the scientific men who have had occasion to use it.

The requirements were : 1, that the knife should be directed very securely while remaining independent of the apparatus ; 2, that the object to be sliced should be movable in a parallel line under the knife, and that the amount of the displacement should be read with great accuracy ; 3, that the object should be fixed firmly in any position.

The instrument consists of a table, 75^{mm} long and 60^{mm} wide, capable of being inclined at will upon its foot. The table has a small tongue which is movable in parallel line by means of a micrometric screw indicating $\frac{1}{100}$ths of a millimetre. To the tongue is affixed a stay, having a tightening screw, under which is placed the object to be sliced. Above the table is a kind of steel arch bearing a plane, perfectly adjusted, perpendicular to the table. It is upon this surface that the knife, of which one side is also adjusted with the greatest care, is made to slide by hand.

3624. Professor Vogelsang's Apparatus, serving to raise the temperature of Microscopic Objects.
From the Collections of the Royal Polytechnic School at Delft, Prof. J. Bosscha.

By this instrument it was shown that the air bubble in the liquid carbonic acid contained in quartz crystals ceased to be visible at a temperature of about 32° C. (Poggendorff's Annalen, vol. 137, table III., fig. 2.)

3625. Parallel Compressor, with micrometric screw, specially intended for the study of ova, and for observing the development of lower organisms.

Geneva Association for the Construction of Scientific Instruments.

The upper cup takes off to admit the dissecting trough. The pressure is effected by means of the side adjusting screw. By moving the dissecting trough with the hand a round body can be examined on all sides successively.

The dissecting trough can hold a good quantity of water, of which no portion comes in contact with the metal, a favourable condition for the preservation of live organisms.

3626. Dr. Burdon Sanderson's hot or cold Stage, for use with microscope, and with arrangement for heating by gas, hot water, or lamp.   *T. Hawksley.*

3626a. Boiler with incubating cells.   *T. Hawksley.*

3627. Freezing Machine for making microscopic sections, by Professor Rutherford.   *T. Hawksley.*

3628. Saccharometer, as accessory to the large microscope, from the same contributor.   *R. Wasserlein, Berlin.*

The saccharometer is to be regarded as an accessory to the stand of the microscope. By this combination of two instruments (as the saccharometer can be fitted to smaller stands) cheapness is attained. The chief aim of the maker is to furnish a useful instrument, which at the same time shall be specially suited to the wants of the medical chemist in estimating grape sugar (as in diabetes). For this purpose the scale of the instrument is graduated so as to indicate in whole numbers and tenths the per-centage of grape sugar. For other substances a calculation is necessary.

3629. Woolmeasurer, fitting into the Student's Microscope.   *R. Wasserlein, Berlin.*

This woolmeasurer, constructed by the exhibitor, is made at the express wish of Mr. Bohm, woolfactor at Leipzig. It affords the investigator the possibility of a complete command over stuffs composed of vegetable or animal fibres, as wool, silk, and all vegetable fabrics, so that they can be stretched and extended under the microscope; also in case the fibre is twisted, as in wool, it may be evenly stretched and rotated, so that its average diameter may be estimated. The apparatus allows the object which is thus stretched to be placed on a glass slide, so that it can be treated with acids, &c., and covered with a cover glass. All this can be done without preventing the manipulation described above. Threads up to a length of 1½ inch can be moved successively across the field of the microscope.

3630. Small Microtome, with an arrangement for the vertical elevation of the objects.   *F. Süss, Marburg.*

3631. Larger Microtome, for the section of the spinal cord, with vertical elevation and divided circle.   *F. Süss, Marburg.*

3632. Quadrant Microtome, for making microscopical sections at any given angle, with an arrangement for measuring the angle, and scale for showing the relative thickness of the section.
F. Süss, Marburg.

3633. Large Microtome, for cutting sections of the brain.
F. Süss, Marburg.

These microtomes are suited for embedding preparations. Samples of embedding preparations (soap mixtures), with instructions as to the use of the instruments, as well as the composition of the soap mixtures, are placed with the microtomes.

3634. Large Schiefferdecker's Microtome, with plate, two knives in case, glass bell jar, and brass mould for fixing preparations in an embedding mass. *F. Majer, Strassburg.*

3634a. Microtome. *Dr. Gasser, Marburg.*

This adaptation of the large Gudden microtome has been designed for embedding. The object and the enclosing mass being held fast by the screw at the top of the piston, and the often-condemned defect of repeatedly melting down in the case of large objects is avoided. The cylinder is slightly narrowed towards the top, in order that the object may more surely resist the knife which is guided by the hand on a polished steel surface. The instrument can be employed with any suitable liquid in the bowl, which can afterwards be drawn off through a gutta-percha tube, and the object being slightly screwed down, may be covered with a glass slide as a protection against drying up. In this way operations on the same object on several successive days may be carried on. The cylinder is of larger size than in the largest Gudden instrument, but can be further enlarged, if necessary, so as to enable sections of the entire human brain to be made.

3634b. Freezing Microtome, for enabling the microscopist to make thin slices of soft tissues. Invented by Professor Rutherford ; made by the exhibitor. *Gairdner.*

The freezing microtome was devised by Professor Rutherford for the purpose of enabling the microscopist to make thin sections of soft tissues and organs. The tissue is placed in the well ; a strong watery solution of gum arabic is poured around it, and a freezing mixture of ice or snow and salt is placed in the box. The sections are made with an ordinary razor. A full description is given in Rutherford's " Outlines of Practical Histology." (Churchill, London, 1875.)

3635. Smith's Microtome, modified by Schiefferdecker.
F. Majer, Strassburg.

3636. Brass Mould, for fixing preparations in an embedding mass. *F. Majer, Strassburg.*

3637. Microscopical Cutting Apparatus, for soft objects.
Prof. Jessen, Eldena, Pomerania.

3638. Fluid for keeping Parts of Plants, and preserving their colours in glass vessels.
Prof. Jessen, Eldena, Pomerania.

3639. Warm Moist Chamber, for observing microscopical objects at an elevated temperature during long periods of time. (Exhibitor's construction.)

Prof. Dr. Ferd. Cohn, of the Institute of Vegetable Physiology in the University of Breslau.

The apparatus in question combines the advantages of the moist chamber and the warm stage of the microscope. The object is laid in a drop of water on the underside of the glass plate in the centre of the cover, which has been previously removed ; the cover is then firmly fixed, and the whole apparatus placed on the stage of the microscope.

It is warmed to the desired temperature, which is indicated by a thermometer, by means of a small gas flame placed underneath the descending heating tube. The object is observed through the glass plate in the cover which takes the place of a cover glass. By regulating the flame, the temperature in the chamber, which is kept moist by the drop of water which cannot evaporate, can be maintained perfectly constant for a long time, and the development of microscopical organisms at a high temperature can be observed.

3640. Two Cases of Microscopic Objects.

Voigt & Hochgesang, Göttingen.

3640a. Cases of Microscopic Preparations.   *F. Enock.*

3641. Microtome with Preparations, made by W. Apel, mechanician to the University of Göttingen.

Prof. W. Krause, Göttingen.

This microtome is accurately described by Prof. Krause in Waldeyer's and De la Valette's Archiv für microscopische Anatomie (earlier in Max. Schultze's Archiv), 1875, vol. XI., p. 216, plate XIII. It cuts in a purely mechanical manner, more by drawing than by pressing the knife, whilst any shifting either of the preparation or of the cutter by the hand is excluded by the principle of the construction of the microtome. The thickness of the section to be made can be read off on a circular disc divided into degrees, which can easily be applied to the instrument. Any other (much longer) knife can be substituted for the one in the instrument ; the box in which the preparation is fixed can also be changed. The knife can be moistened by means of an " irrigator " with either alcohol or water, and the apparatus can be fastened to a wall so that the blade of the knife is horizontal. A microscopical preparation containing two sections of different thicknesses made with the above apparatus is enclosed for exhibition.

3642. Holle's Drawing Apparatus for the Microscope.

Cuno Rumann, Göttingen.

The construction of the apparatus has just been completed, and only one specimen has been made.

An account of its leading peculiarities will be found in the paper enclosed.

3643. Photograph of the Drawing Apparatus.

Cuno Rumann, Göttingen.

3644. Sectional View of Drawing Apparatus.

Cuno Rumann, Göttingen.

3645. Description of the apparatus of Prof. Grisebach, from the "Nachrichten von der Königl. Gesellschaft der Wissenschaften zu Göttingen."   *Cuno Rumann, Göttingen.*

3646. Dr. Thoma's Stages (*see also* X. 21).
                                        Rud. Jung, Heidelberg.

The three stages afford the means of making an exact microscopical examination of the tongue, the mesentery, and the interdigital membrane of the living frog, during the time that these parts are submitted to constant and determinate conditions by a continuous application of salt solutions of known concentration.

At the same time the apparatus allows continual infusions of different liquids, salt solutions, water, or colouring matters into the vena abdominalis mediana of the animal.

For directions for practical use see Virchow's Archives, vol. 65, page 36.

3647. Querschnitter, an Instrument for cutting under the Microscope.   *Prof. Dr. V. Hensen, Kiel.*

The nature and use of the apparatus are given in the accompanying treatise. The instrument only differs from that described in having the curved piece (Bügel) turned forwards.

3648. Microtome, with steel disc, water-trough, wedge-shaped knife, four moulds.   *Prof. K. Möbius, Kiel.*

3649. Two Microtomes, with steel disc and three moulds.
                                        Prof. K. Möbius, Kiel.

A written explanation accompanies the objects exhibited.

3650. Five Moist Chambers, for microscopical research (after the designs of Recklinghausen, Ludwig, Geissler, de Bary, and Klebs).   *Ch. F. Geissler & Son, Berlin.*

3651. Fritsch's Sliding Microtome.
                                        Prof. G. Fritsch, Berlin.

The microtome, of which (on account of some recent alterations) an incomplete account is sent, has on its left a movable box for the reception of the object, on the right a slide (Schieber) with eccentric disc and screw as a holder for the knife. Other knives will be found in the lid for use, according to the degree of resistance offered by the object.

The object to be cut must be so placed in the metal box that the part to be cut projects over the upper rim. The preparation is imbedded in a suitable mass, or is frozen in a mixture fit for cutting (*e.g.* Rutherford's). In the brown lacquered box are plates of spermaceti, and mixtures of the latter with cocoa-butter, suitably selected. Some elder pith is imbedded as a specimen.

After an advantageous position, as regards the object, has been given to the knife, and the box containing the embedded object has been fixed by means of the screws of the large movable box, so that it can be cut by a gentle elevation of the movable box, the screws of the disc and knife are sharply drawn across. Should the position of the object be too low, one of the metal plates furnished with the apparatus is laid under the box, and the screw of the movable box again turned up, but not too tight, in order to prevent the strain and pressure on the side of the box becoming too great. The work of preparing is automatic, so that when a clear section surface is obtained, the movable box, by means of the vernier attached, can be raised a definite fraction in height, and the object in it; the knife screwed up on the slide is guided by means of the horizontal slot, whereby the uppermost part is taken off as a thin section. After drawing back the knife to the other end of the slot (turned from the operator), and a fresh raising of the box, a second section is obtained in a similar manner, and so on until the object is converted into thin sections. With objects difficult to cut it is necessary to moisten the blade during the cutting of the section with weak spirit contained in a wash bottle, one tube from which overhangs the blade, whilst the other is kept in the mouth for blowing. If the objects resist very much, so that the blade tends to give, then the safety of the section will be secured by laying the curved piece of metal from the mahogany box under the larger screw of the eccentric disc, and a corresponding tightening of the adjusting screw placed over the free end of the knife. The movable box can (for the purpose of keeping the left hand for guiding) be easily fixed by means of a clamping screw through the slot.

Some sample sections are contained in the mahogany box.

3651a. Microscopic Section Cutting Machine.
G. Moritz Leyser, Leipzig University.

3652. Counting Micrometer, for enumerating microscopical forms. To this is added a figure under glass, representing the micrometer magnified eight times.
Prof. H. Welcker, Halle an der Saale.

3653. Sliding Tray, carrying microscope lamp and Hartnack microscope, for demonstrating objects to a class.
W. R. M'Nab, M.D.

This tray runs on two rollers partly covered with india-rubber, thus destroying vibration, and permitting the tray to be used on an ordinary table.

3654. Mica-plates for preserving botanical and anatomical preparations. (*See* Mineralogy.)   *Max. Raphael, Breslau.*

Mica-plates are conveniently employed for preserving hygroscopic anatomical preparations, such as of foliaceous mosses ("Laubmoosen"), &c., For this purpose, the mica is split halfway through, and the preparation placed, in a moist condition, between the plates, which close of their own account owing to the elastic nature of the mica.

Should the preparations be required for examination, the plates have only to be dipped for a second or two into water; they may thus serve for years for repeated microscopic examination. It is obvious that covering plates of mica may be used for all botanical microsopic purposes.

3654a. New adjusting and self-centering Shadbolt's Turn-table for making microscopic cells.
Harvey, Reynolds, and Co.

3655. Knife for cutting microscopic sections, by Dr. Gower, with divided adjusting screw.   *T. Hawksley.*

3656. Knife, with three blades, for cutting microscopic sections by Dr. Madox.   *T. Hawksley.*

3656a. Dr. Klein's Section Knife. *T. Hawksley.*

3657. Moist Chamber. *T. Hawksley.*

3658. Series of Glass Canulæ, and **Nozzles. Dissecting Instruments, Forceps, Scissors, Hooks, &c.**
T. Hawksley.

3662. Preparations of normal anatomy of the human body, injected and varnished, in 1809, by Prochaska.

The microscope and preparations were given by the author to the first Baron Larrey, with the description in Latin that is affixed to the instrument. *Baron Larrey, Paris.*

3662a. Microscopic Section Cutting Machine, with Freezing Cell. *Harvey, Reynolds, and Co.*

3662b. Williams' Freezing Microtome. *J. Swift.*

3663. Microscopical Double Knife, constructed by A. Stelzig, surgical instrument maker in Prague. With flexible curved blades. Compared with knives of older shapes, the parallelism of the edges is to a much greater extent preserved in this form.

The Imperial and Royal Pathological and Anatomical Institute of the University of Prague (Director, Prof. Edwin Klebs).

3664. Valvulotome. An instrument for making incisions in the valves of the heart, being a straight catheter, at whose front extremity, which is rounded, a small blade is attached, which by means of the screw inside the instrument can be opened, so that the free rounded end will project more or less. The edge of the knife is turned towards the catheter. After the introduction of the instrument into the jugular vein, or the carotid artery, the right or left ventricle of the heart can be easily reached. The blade having been protruded, the chordæ tendineæ of the heart are grasped and cut off by pulling the knife backwards into the catheter. Artificial disorders of the heart so produced have been observed for more than a year, and the circulation studied on cymographical curves.

(Prague Weekly Journal, '76. No. 2.)

The Imperial and Royal Pathological and Anatomical Institute of the University of Prague (Director, Prof. Edwin Klebs).

3665. Microscopical Warm Stage. The constant equality of the temperature is obtained by the increased bulk of the thick copperplate. The object is placed in the recess, and the centre aperture is closed by a glass disc. A broad, thick copper ring with central aperture for the object is placed on the upper surface of the disc, thus closing the aperture. The lateral movement of the microscope with the coppering is effected by a plate at the bottom of the apparatus on which the microscope is fixed. The same can be shifted in two directions, at right angles to one another by means of micrometer-screws. Four side-apertures admit the introduction of thermometers or liquids for keeping the chamber moist, or for the infusion of gas. A small gas-flame, which imparts heat to the adjoining piece, serves as a heat conductor.

> *The Imperial and Royal Pathological and Anatomical Institute of the University of Prague (Director, Prof. Edwin Klebs).*

3668. Microtome. A circular knife is set revolving by a turning-lathe. The preparations, which are embedded in a metal case, are moved forward by a micrometer screw, and passed across the edge of the rotating knife by a contrivance similar to that in a dividing engine.

> *The Imperial and Royal Pathological and Anatomical Institute of the University of Prague (Director, Prof. Edwin Klebs).*

3668a. Holman's Syphon Slide, invented to keep objects alive while under the microscope by a flow of cold water.

> *D. S. Holman, Philadelphia.*

3668b. Holman's Life Slide.

> *D. S. Holman, Philadelphia.*

By the cross groove causing live objects, such as vinegar eels, &c., to take up their position in it, they are enabled to be more readily examined.

3668c. Holman's Current Slide.

> *D. S. Holman, Philadelphia.*

3668a. Fifteen Frames, containing photographs of **Microscopical Sections.**   *Richard Daintree.*

II.—OPHTHALMOLOGICAL APPARATUS.

a. OPHTHALMOMETERS.

3692. Ophthalmometer, according to the directions of Professor Helmholtz.

> *Aug. Becker (Dr. Meyerstein's Astronomical and Physical Workshops), Göttingen.*

3693. Mirror Apparatus for the above, for measuring the radius of curvature of the different meridians, with balancing weight.

> *Aug. Becker (Dr. Meyerstein's Astronomical and Physical Workshops), Göttingen.*

The mirror apparatus is fixed into the opening of the index disc, which is placed at the further end of the ophthalmometer. The instrument must, when put into use, be so placed upon a stand that the long rods of the mirror apparatus clear the disc of the stand, when it is rotated. A weight, passing over the telescope, balances the instrument.

3686. Coccius' Double-refracting Ophthalmometer.
E. Stöhrer, Leipzig.

Design of Apparatus.—To measure the principal curved surfaces by means of two movable sources of lights and Iceland spar with doubly refracting prism of Iceland spar whose angle of dispersion is about 3 mm.

Practical Application.—The cornea, on which three points of light (the distance of the images of the respective lights being three millimeters) or four points of light (the distance of the images from the respective sources of light thrown on the cornea being $1\frac{1}{2}$ mm.) can be shown. In the last case the distance of the lights from each other is the half of the distance which three or four points of light show. When the position of the lamps is vertical, the Iceland spar must be turned round.

b. OPTOMETERS.

3710a. New Optometer, with double refracting lens of calc-spar, give double readings, and greater precision in determining the distance of sight. *Prof. Carl Wenzel Zenger, Prague.*

3714a. Perrin and Mascart's Optometer.
M. Roulot, Paris.

3713a. Perrin and Mascart's Optometer (small pattern).
M. Roulot, Paris.

3713. Dr. Badal's Optometer.
M. Roulot, Paris.

3697. Graefe's Binocular Optometer.
Dr. Weber, Darmstadt.

3710a. Optometer, for determining the condition, degree of refraction, and acuteness of vision.
William Laidlaw Purves, M.D.

The optometer consists of a disc of convex spherical, a disc of concave spherical, a disc of convex cylindrical, and a disc of concave cylindrical lenses. These move upon a central axis, so that any of the four discs can be superimposed upon another, and thus any combinations made which the particular lenses used can arrive at. The cylindrical lenses, having their axes all placed in the same direction, and at any one meridian of the instrument, the different powers are speedily brought before the eye, so as to act on the desired meridian of the eye, and on it only. (*Vide* "British Medical Journal" for January 1875.)

3710b. Boxwood Scale, on which the combinations are calculated: belongs to the optometer last described.
William Laidlaw Purves, M.D.

3710d. Instrument for the determination of the exact position of the **chief Meridians** of the **Eye.**
William Laidlaw Purves, M.D.

This instrument consists of three round apertures, one in the centre, and two at equal distances from the periphery in the same meridian, bored in a screen. When these, placed against a strong light, give major axes of diffusion circles, the disc is revolved till all the major axes are on the same line. This lens has an indicator attached, which points to the degree on a graduated circle in which the disc revolves. (*Vide* Graefe's Archiv, 1873.)

3710e. Test Card for determining the **Acuteness of Vision** in any **Meridian.**   *William Laidlaw Purves, M.D.*

This is a modification of the usual astigmatic test cards. By working the disc, and so bringing varied numbers of lines into any desired meridian, the acuteness of vision at that meridian is determined by the ability of the observer to count the number so placed.

3710c. Modified Stokes's Lens, for determining the degree and condition of refraction and acuteness of vision of any meridian of the eye, or of simple or combined lenses of less power than one-fifth.   *William Laidlaw Purves, M.D.*

This is a further modification of Snellen's modification of Stokes's lens. In this modification the power of the lenses are increased from one-tenth to one-fifth. The constant axis is obtained by means of a screw and wheels, instead of by springs, and, in addition to former modifications, the exact position of the axis of the whole lens is obtained by moving the whole lens in a graduated circle. As each cylindrical lens acts only in one meridian, and that meridian is determined by the position in which the lens is placed, any number of meridians may be determined successively. The power required in any meridian is calculated in the usual way. (*Vide* Donders' Accommodation and Refraction of the Eye.) By using a modification of Scheiner's experiment, and other methods, the power of single or combined lenses may be determined. (*Vide* Graefe's Archiv, 1873.)

808. Stokes's Lens, modified by Dr. Snellen. In one and the same mounting are set two plano-cylindrical lenses ($+12, -12$) capable of rotation in opposite directions. When their axes are parallel, their refraction becomes annulled ; when they are at right angles, they become added ; in intermediate positions they give a graduated series of cylindrical refractions. The axis is fixed for the same reasons as above.   *M. Crétès, Paris.*

3669a. Trial Box and collection of metrical glasses, for ophthalmology.   *A. Nachet, Paris.*

3717. Set of Lenses for Testing the Sight.
W. Campbell & Co.'s successor, J. Wohlers, Hamburg.

3715. Box of Lenses, containing 120 spherical, 36 cylindrical, 10 prismatic, and two trial eyepieces.   *M. Roulot, Paris.*

3715a. Box of Lenses, 120 spherical and trial eyepiece.
M. Roulot, Paris.

3715b. Box of Lenses, 36 cylindrical and eyepiece.
M. Roulot, Paris.

3715c. Box of Lenses, from 2° to 20°, square, cylindrical.
M. Roulot, Paris.

c. PERIMETERS.

3714b. Dr. Badal's Perimeter. *M. Roulot, Paris.*

3680a. Dr. Forster's Perimeter. *R. Sitte, Breslau*

This apparatus is used for determining the external limit and the colour limit of the defects of the field of vision. The position of the displaceable objects is read off by meridians and parallels. A more detailed description will be found in Zehender's Klinischen Monatschriften für Augenheilkunde, 1869, p. 411.

3711. R. Brudenell Carter's Perimeter for ascertaining the boundary of the field of vision and the area of the blind spot.
T. Hawksley.

d. PHOTOMETERS.

3695. Graefe's " Leuchtscheibe " (illuminating disc) ; from the property of the late Professor von Graefe.
Dr. Weber, Darmstadt.

3680. Dr. Förster's " Lichtsinnmesser."
R. Sitte, Breslau.

A source of light of constant intensity, but whose size can be varied, illuminates the object, which may be either black bands or large letters on a white ground. The source of light is behind a sheet of white paper, and shines only through it, the paper being illuminated on the other side by a standard candle burning at almost the same height. The size of the source of light depends on an adjustable diaphragm, whose centre always remains at the same point. The diagonal of the square aperture of the diaphragm is measured by a scale which moves with the plates of the diaphragm. The size of the illuminated surface $= \dfrac{d^2}{2}$, if half a square millimeter be taken as the unit. By means of this apparatus all quantities of light, from 1 to 2,500 units, can be estimated.

See Zehender's Klinische Monatsblätter fur Augenheilkunde, 1871, p. 337.

3720. Weber's Photometer and Chromometer.
Dr. Weber, Darmstadt.

3721. Weber's Chromoptometrical Apparatus.
Dr. Weber, Darmstadt.

The examination of the sense of colour with this apparatus rests upon the production of simultaneous contrasts, for which purpose a gray ring is placed upon the primary colour and both are covered with transparent paper. For the determination of the numerical size of the perception of colour, use is made of a gray disc which can be divided into sections, by means of which it can be determined in what relation the primary colour stands to the gray surface, in order to place upon it likewise the complementary colour. For determining

3 N 2.

the limit of the field of perception of colour, use is made of a number of concentric gray rings, for which the middle point of the prime-coloured surface
serves as a centre which is fixed upon by the eye under examination.

e. TONOMETERS.

3675. Graefe's Tonometer (Augendruckmesser).
Dr. Weber, Darmstadt.

Graefe's tonometer has only an historical value, since it was the first apparatus with which he endeavoured to measure the tension of the eye-ball
tegument ; it could not, however, be introduced into practice on account of
many inconveniences.

3676. Weber's Tonometer.   *Dr. Weber, Darmstadt.*

**3674. Tonometer, for Measuring the Tension and
Convexity of the Eye-ball.**
*Geneva Association for the Construction of Scientific In-
struments.*

This apparatus is specially used for glaucoma. It serves to produce a definite
deformation of the eye, which may be variable at the operator's will, according
to the most suitable circumstances for a special case.
The instrument is composed of a rod, of which the ivory-tipped end presses
upon and deforms the eye, while the other end acts upon a dynamometric
spring, the flexure of which, amplified by a catch and by the indicating needle,
shows in grammes upon the dial the pressure exercised upon the rod, and,
consequently, upon the eye.
An index marks the greatest oscillations of the principal needle.

f. PUPILLOMETERS.

3671. Pupillometer.   *Emil Stöhrer, Leipzig.*

g. EXOPHTHALMOMETERS.

3682. Cohn's Exophthalmometer, three specimens.
Prof. Dr. H. Cohn, Breslau.

These instruments possess a certain amount of interest, as they are almost
the first instruments which were made for measuring the prominence of the
eye from the orbit.   Dr. Cohn exhibited the first of the accompanying instruments, on Nov. 17th, 1865, before the Medical Society of Breslau, and at the
same time gave it the name of Ophthalmoprostatometer ; but later, in the
year 1867, he changed the name to Exophthalmometer. The second instrument Dr. Cohn has not publicly demonstrated, as he altered the principle
of measurement; nevertheless he determined to send the instrument, so
that one can judge of the reasons which induced him to regard the external
wall of the orbit which he had taken as his fixed point (and is still so taken
by some) as not suitable for the purpose.   The third exophthalmometer
with which the exhibitor finally carried out his measurements, was laid
before the Ophthalmic Congress, at Paris, on the 12th of August 1867.   Its
peculiarities and applications are described in the accompanying pamphlet.

3682a. Cohn's Exophthalmometer, with band.
Prof. H. Cohn, Breslau.

3682b. Cohn's Exophthalmometer, with small handles.
Prof. H. Cohn, Breslau.

h. OPHTHALMOMETERS.

3710f. Ophthalmoscope, for the determination of the refraction of the observed eye by the observer.
William Laidlaw Purves, M.D.

Behind the mirror are two discs containing numerous convex and concave lenses, which may be brought rapidly over the aperture of the mirror by means of a rack moved by the thumb of the hand which holds the ophthalmoscope. As the convex can be revolved over the concave lenses, or *vice versâ*, any combinations can be made which the number of the lenses employed will allow of.

3710g. Ophthalmoscope, for the determination of the refraction of the observed eye by the observer.
William Laidlaw Purves, M.D.

Behind the mirror two cylindrical lenses, $+\frac{1}{2}$ and $-\frac{1}{2}$, are revolved on the same method as in Stokes's lens, and thus the powers, which have been calculated and marked on the instrument, are brought into use instead of using a number of lenses as in the instrument before described.

3681. Cohn's Refracting Ophthalmoscope.
Prof. Dr. H. Cohn, Breslau.

In the year 1871 the exhibitor constructed this apparatus for examining as quickly as possible a number of scholar's eyes. In 1872 the instrument was described in the Klinischen Monatsbl. of Augenheilkunde, in the October number. Twenty-four glasses can be quickly pushed past behind one and the same mirror. As the disc which bears these glasses is centred above the mirror, the nose of the observer cannot come into contact with it. There is no necessity to shift three discs with glasses, as in Loring's Ophthalmoscope. If the mirror be placed in front of the eye to be examined, and the patient looks through the central hole at some letters in the distance placed on an inclined table desk (as Snellen's alphabet), the instrument can be used instead of a test lens, and thus the wearisome selection of the glasses is avoided; in the same way, by quickly turning the disc with the correction glasses, its refrangibility for an object can be removed.

3708. C. J. Oldham's First Ophthalmoscope, with one diaphragm at back of mirror to carry lenses. *T. Hawksley.*

3709. C. J. Oldham's Second Ophthalmoscope, with three diaphragms and 27 lenses. *T. Hawksley.*

3719. Metrical Ophthalmoscope, of Dr. de Wecker. In one zone are placed 20 convex glasses ascending gradually from 0·50 to 10 metrical dioptric. To obtain the concaves, a concave lens is interposed, more powerful than the maximum of the convex glasses ($10\frac{1}{2}$ dioptrics), and by turning the wheel a descending series of 0·50 to 10·50 concave dioptrics is obtained.
M. Crétès, Paris.

3714. Dr. Giraud Teulon's Binocular Ophthalmoscope.
M. Roulot, Paris.

3706. R. Brudenell Carter's Demonstrating Ophthalmoscope, showing the fundus oculi magnified to 12 inches in diameter.   *T. Hawksley.*

3707. Dr. Lionel Beale's Self - illuminating Demonstrating Ophthalmoscope. May be used in daylight.
T. Hawksley.

3696. Weber's Synamphophthalmoscope.
Dr. Weber, Darmstadt.

The apparatus serves for the simultaneous investigation of both eyes in the inverted image, and thus permits a comparison of the details of the background of the eye as to size, colour, &c. In using it a double source of light and a perforated reflector are required.

3707a. Prof. Laqueur's Instrument for measuring the Ophthalmoscopic Image. *Prof. Laqueur, Strasburg.*

3684. Apparatus for giving Instruction in the use of the Ophthalmoscope; also 18 water-colour drawings in a box, giving the typical aspects of the retina, choroides, and optic nerve in health and disease, from nature.   *Dr. Magnus, Breslau.*

This apparatus presents, in a very instructive way, the various laws of refraction of the eye, by shifting the lens which is contained in it; the apparatus can be used for demonstration with the greatest facility. Besides these physical problems, the apparatus has 18 water-colour drawings, painted by the exhibitor from nature, and showing the principal features of the healthy and diseased retina, choroides, and optic nerve, for the purpose of affording the student the opportunity of investigating ophthalmoscopically the pathological changes occurring inside the eye.

3673. Hermann's Blemmatotrope, for representing Listing's law of the positions of the eye.
Prof. Dr. L. Hermann, Zürich.

This apparatus can be arranged for vertical, horizontal, or oblique axes of rotation, by changing the position of the arcs which support the eye-balls. The same number that is marked by the arc on the edge of the fixed concave disc behind the eye-ball, must be crossed, on the equatorial ring of the eye, by the red (vertical) meridian. When the axis of rotation is vertical or horizontal, the blue (horizontal) meridian corresponds, during rotation, with the thin brass plate (visual plane); when the axis of rotation is oblique, the horizontal meridian and the visual plane will form an angle (the so-called Raddrehungs-Winkel).

3669c. Diplometer, by Dr. Landolt.   *A. Nachet, Paris.*

3689. Actinallactor, an instrument for demonstrating the persistence of light impression on the eye.
Prof. Buys-Ballot, Utrecht.

It consists of a black circular cardboard, that can be turned by a handle round an axle passing perpendicularly through its centre. Excentrically is adjusted a second axle, round which turns a rod containing four coloured wafers, or beads, or ignited pieces of German tinder, forming the same figures as are shown by the bead of Wheatstone's photometer. These figures can be changed by differently fitting the india-rubber band behind the black board. Made and invented by J. van Dreeven, Amanuensis of the Physical Cabinet of Utrecht University.

3685. Coccius' Polarising Ophthalmoscope.
E. Stöhrer, Leipzig.

Design of Apparatus.—To investigate the eye with perfectly polarised light, the images being either erect or inverted.

Practical Application.—To determine the defects of transparency on the iris, the cornea, and refracting media of the eye, with reference to the degree of nebulosity in the retina, to facilitate the estimation of the degree of atrophy of the optic nerve, and by removing all reflections to obtain a more distinct view of the fundus.

3710. Couper's Ophthalmoscope, with five discs and 45
lenses, two mirrors which are adjustable for right or left eye.
T. Hawksley.

3679. Micrometer, for measuring objects in the fundus oculi.
Prof. Laqueur, Strasburg.

"This apparatus is used for measuring in the living eye the real size of the "papilla," the diameter of the blood vessels of the retina, the size of the macula lutea, the distance of the latter from the external edge of the blind spot, also the dimensions of any extravasations of the retina.

For the method of using the apparatus, see Centralblatt der med. Wiss. 1873, No. 59.

It can also be used for determining the condition of the refrangibility of the eye in ophthalmoscopic examination, by placing the toothed part of the stems in front of the lamp which serves as a source of light.

An account of the apparatus is published in the Proceedings of the Ophthalmological Society of Heidelberg for 1875.

3687. Coccius' Micrometer for the Eye.
E. Stöhrer, Leipzig.

Design of Apparatus.—To measure the transparent and sensitive parts without contact.

Practical Application.—To observe any object in the cornea, scars, opacities, cloudiness of the pupil in diseases, the degree of movement of the iris, or the widening of the pupils by the inspiration.

3669. Instruments for Extraction of Cataract.
Dr. Adolph Weber, Darmstadt.

3678. Rose's Schistoscope, for the physiology of colour.
F. Schmidt and Haensch, Berlin.

3691. Apparatus for demonstrating the **Reverse Position of the Image** on the retina of the observer's own eye. (Old.)
Prof. Dr. Dove, Berlin.

3705. Dr. Archer Warwick's Endoscope. *T. Hawksley.*

3688. School Apparatus for the demonstration of the refraction of light in the eye.   *Physiological Institution, Prague.*

3690. Schematic Eye, on Dr. Kühne's principles.
Rud. Jung, Heidelberg.

About ten times the natural size. The cornea and lens are, as far as their optical values are concerned, as near as possible, imitations of the natural eye. The box is filled with water, containing a trace of quinine, and can be thus used for demonstrating the passage of the ray in the natural eye under differing conditions. By removing the lens the conditions of the eye after a cataract operation are represented. The plane cylindrical vessel filled with water can be used for illustrating astigmatism.

3669b. Schematic Eye, by Dr. Landolt. *A. Nachet, Paris.*

3694. Specimens of Transparent Drawing, showing cases of **Optical Illusion,** described in Prof.Helmholtz's Physiologische Optik.

> *T. A. Snyders, Lecturer at the Royal Polytechnic School, Delft.*

To be used in the projecting apparatus of M. Duboscq. The figures are drawn in Indian ink on sheets of unpolished glass, which are afterwards rendered translucid by a thin layer of varnish.

3712. Perrin's Artificial Eye Apparatus. *T. Hawksley.*

3716. Various kinds of Optical Glass. *Weslëin, Paris.*

3717a. Box with Specimen Spectacles for Oculists.
W. Campbell and Co.'s successor, J. Wohlers, Hamburg.

3683. Several Pairs of Cohn's Mica Spectacles, manufactured by M. Raphael in Breslau.

Prof. Dr. H. Cohn, Breslau.

In February 1868 the exhibitor examined the eyes of 1,283 metal workers and, in consequence of the enormous frequency of injuries to them, caused the first pair of mica spectacles to be made. They cannot splinter, and are extremely cheap. They are recommended for workmen who manipulate explosive materials. In the accompanying pamphlets is shown the gradual perfection to which these glasses have been brought by the manufacturers, Messrs. Max Raphael, Breslau, under Dr. Cohn's directions.

III.—WEIGHING AND MEASURING APPARATUS.

PROFESSOR A. CRUM BROWN'S PREPARATIONS AND APPARATUS.

3733. Skull with internal ear prepared for the measurement of the relative position of the planes in which the semicircular canals are situated.

Two saw-draughts are cut through the outer table of the skull, forming a large angle with one another. These saw-draughts pass through the mastoid part of the temporal bone, and are continued into the neighbouring parts of the skull ; two pieces of steel plate are cut so as to fit into the saw-draughts, and the position in which they fit indicated by marks on the plates and on the mastoid bone. The greater part of the mastoid and the whole of the petrous portion are then sawn out, and the steel plates fixed to the portion of bone thus removed. This portion can now be replaced exactly *in situ* by placing the ends of the steel plates into the portions of the saw-draughts remaining on the skull. The remaining portion of bone is now plunged into a bath of fusible metal, and placed under the receiver of an air pump. On exhausting, bubbles of air escape from the cavities in the bone, and, on re-admitting air, these cavities are filled with fusible metal. By repeating this operation ten or twelve times in different positions, all the air can be pumped out and the cavities completely filled with fusible metal. The bone is now removed from the bath of fusible metal, and the adherent metal removed. It is then placed in melted paraffin, so as to cover the steel plates, and the greater part of the mastoid portion. When the paraffin has solidified, the whole is placed in a vessel containing dilute hydrochloric acid. This dissolves the unprotected part of the bone, leaving a cast of the cavities in fusible metal. From this the casts of the mastoid cells are carefully removed. To the external part of the mastoid is now soldered a brass pin fitting into a socket in the large gonio-meter. The cast can then be brought successively into positions in which each of the canals lies in a horizontal plane. In each of these positions the skull is replaced by means of the steel plates, and a glass plate fixed to the skull in a horizontal plane. These plates are therefore parallel respectively to the planes of the canals ; and their relative position can be ascertained by means of the goniometer. *See* Proceedings of the Royal Society of Edinburgh, January 1874, and Journal of Anatomy and Physiology, viii. 327.

3734. Skull of Owl, prepared to show relative position of the semicircular canals of internal ear.

The spongy osseous tissue has been removed from the dense bones forming the canals, and these have been coloured so as to indicate the pairs of parallel canals.

3735. Skull of Crow, prepared to show relative position of the semicircular canals of internal ear.

3736. Skull of Heron, prepared to show relative position of the semicircular canals of internal ear.

3737. Cast, in plaster of Paris, of internal ears of the skate.

3738. Apparatus for measuring the relative **position** of the **planes** in which the **semicircular canals** of the **internal ear** are situated, with skull, illustrating its application.

The apparatus is simply a reflecting goniometer on a large scale. The telescope used with it is not sent, as it requires to be fixed at a considerable distance from the apparatus.

The mode of preparing the ear for observation is described above.

3739. Cast, in solder, **of human internal ear,** right side.

3722. Craniometer, by which measurements of crania may be taken rapidly and accurately.   *Prof. Struthers.*

It is essentially a glass box, the panes accurately ruled and fitted into a carefully made brass frame. Sides 9 inches square ; ends, bottom, and top,

9 inches by 7 inches. Bottom of strong plate glass. Top lifts out. Panes ruled both ways at distances of an inch. Middle division of each pane is halved by a median line, which is cross-marked at each ¼ inch. Skull placed so that middle line corresponds to middle line of panes. By looking through panes, and getting eye on corresponding lines of opposite pane, accurate measurements may be easily and rapidly read off of top, sides, front and back, or of interior of base. For outside of base, cranium is turned over and steadied. The cranium contained in the craniometer is a well-marked specimen of the scapho-cephalic form, with the usual early oblitera-tion of the inter-parietal suture. The instrument was made for Professor Struthers by Mr. P. Stevenson, philosophical instrument maker, Edinburgh. An account of this craniometer was given by Professor Struthers in the Edinburgh Medical Journal, 1863.

3722a. Craniometer, an instrument for taking measurements of the human cranium.   *Geo. Busk.*

The instrument is constructed on the principle of the common shoemakers' gauge, and consists of a straight stem about 12 inches long, having an arm jointed to it at one end, which can be erected so as to stand at a right angle, and a second arm which can be slid up and down the stem, and is also capable of being erected to a right angle, so as to stand exactly parallel with the former. The stem and arms are graduated on one side in inches and tenths, and on the other in centimeters and millimeters. The graduation of the stem, begins at the fixed arm, and that on the arms from the stem. In order to render the instrument capable of taking radial measurements, a conical peg can be slipped upon each of the arms, the points of which are inserted into each external auditory foramen. The radial distance from the centre of the foramen can thus be measured to any point on the periphery of the skull in the mesial plane, the distance being read off on the short arms.

3727. Dr. F. W. Spengel's Craniometer.
A. Wichmann, Hamburg.

An account of the fixing of the skull, as well as the use of the apparatus, is given in the accompanying " Correspondenz-Blatt der deutschen Gesellschaft für Antropologie, Ethnologie und Urgeschichte.

3728. Virchow's Portable Craniometer.
A. Wichmann, Hamburg.

3729. Virchow's Tactile Compasses (Æsthesiometer).
A. Wichmann, Hamburg.

3730. Virchow's Scale.   *A. Wichmann, Hamburg.*

3731. Kephalograph or Craniometer, to determine the shape and dimensions of the human skull.
Dr. P. Harting, Professor at the University of Utrecht.

This instrument, which is intended to determine the shape and dimensions of the human skull, consists of three principal parts or distinct instruments.

The first (Fig. 1., Pamphlet) is intended to fit round the head, and to trace its principal circumference.

The second (Fig. 2) may be fitted lengthwise, crosswise, or in any other direction required.

The principal use of the third (Fig. 3) is to determine the profile of the face.

For further particulars, see the monograph "Le Kephalograph," by Professor Harting.

3723. Balance for Physiological and Clinical Purposes, on polished oak.
The Brandenburg Balance Manufactory, Messrs. Kuhtz & Co., Brandenburg on the Havel.

These balances and weights are being extensively used in all scientific schools, and especially by practising physicians. They are available and almost indispensable for the investigation of the disturbances of nutrition in childhood ; for the estimation of the loss or gain of weight by the body during the course of lung disease, diabetes mellitus, acute diseases, &c. ; for the study of the normal growth of the body ; and for pathological and anatomical investigations in which weighings are required.

The balance carries 150 kilogrammes and turns with 0·25 grm. By means of the pointer attached to the beam and the ivory scale, a still smaller difference in weight can be detected.

3724. Set of Iron and Brass Weights, accurately adjusted.
The Brandenburg Balance Manufactory, Messrs. Kuhtz & Co., Brandenburg on the Havel.

2549. Drawings (4) of two apparatus for testing the **Products** of the **Perspiration** of **Animals ;** the larger of this apparatus is destined for horses, cattle, swine, and sheep, &c. ; the smaller serves for the same researches on poultry and rabbits, &c.
Prof. Dr. Ignaz Moser, Vienna.

3732. Manometer, with movable level of the mercury.
Physiological Institution, Prague.

The above is a quicksilver manometer, the legs of which communicate with a vessel containing mercury ; the bottom can be lowered or raised by means of a screw. The manometer allows, among other things, a very quick measurement of the pressure of the saliva, during secretion.

3740. Dynamometer for **Paracentesis Thoracis,** by Dr. Douglas Powell. *T. Hawksley.*

3741. Dr. Douglas Powell's Instrument for **measuring Thoracic Resilience.** *T. Hawksley.*

3741a. Instrument for the **Identification** of **Persons.**
Joseph Bonomi.

The normal proportion of the human frame is that the measure of the distance from the extremity of one hand to the extremity of the other when the arms are extended should be the same as that from the top of the head to the sole of the foot, and any departure from this normal proportion furnishes a means of individual identification.

The purpose of the instrument is to obtain these two measurements simultaneously.

It consists of two laths of wood fixed at a certain angle against a wall. Down the centre of each lath is a groove in which slides an index to show the required measurements. The third index below is for the purpose of ascertaining with great nicety the law of growth in the dimensions of the external divisions of the human figure.

IV.—CHEMICAL APPARATUS USED IN PHYSIO-LOGICAL RESEARCH.

3742. Closed Flask for Experiments in Abiogenesis, &c.
Prof. D. Huizinga, Director of the Physiological Laboratory of the University of Groningen.

The neck of the flask is surrounded by a glass cylinder closed below so that a circular groove is formed. When the liquid in the flask has been raised to boiling temperature, the flask is closed by the previously heated iron cylinder, the under surface of which, as well as the mouth of the flask, have been carefully ground so as to fit accurately. The circular grooves at the top and bottom of the cylinder are then filled with heated mercury. The steam now escapes through the axis of the cylinder. When ready for closing, after sufficient boiling, the lateral silver tube is heated by a pair of gas burners, and the bell-shaped iron cover is also heated and placed on the top of the cylinder so as to dip into the mercury of the upper groove. The flame under the flask is then removed. The air which now enters the flask must necessarily pass through the continually heated silver tube. If judged necessary, the inner width of this tube can be reduced by passing through it a platinum wire.

In experiments on abiogenesis a hatching apparatus is then applied to, but without touching, the flask, so that during the whole experiment the silver tube remains exposed to the flames of the burners.

The lateral iron bar with the movable weight serves for equilibrium, so as to secure a firm position for the whole on the mouth of the flask.

3743. Dialyser.
Prof. D. Huizinga, Director of the Physiological Laboratory of the University of Groningen.

A piece of parchment is glued on both sides of an ebony frame, and strips of the same material are glued about the edges, and thus is formed a perfectly tight bag. The whole is then exposed to diffused daylight, the superfluous bichromate removed, and the dialyser is ready for use. The glue used is a mixture of gelatine and bichromate of potash.

The dialyser can be tested thus :—

1. When the paper is dry and water poured in, no perceptible drops of moisture, even after long standing, ought to appear on the outer surface.
2. A fresh solution of hæmoglobine (diluted fresh blood) being poured in, and the dialyser placed in water, should the red hæmoglobine appear in the water the apparatus is defective.

For further particulars respecting this apparatus *see* "Archiv für Physiologie," Vol. xi. p. 392.

3744. Two Self-acting Filtering Glasses, for microscopical reagents. *Prof. Jessen, Eldena, in Pomerania.*

V.—THERMOMETRIC APPARATUS USED IN PHYSIOLOGICAL RESEARCH.

Thermometer, in the shape of a little frog; it contains balls of different densities and differently coloured, which sink in succession as the fluid within the instrument rises in temperature and consequently diminishes in density. It served, being tied to the arm, to determine the degrees of heat in fever.

The Accademia del Cimento.

The thermometer invented by Galileo underwent afterwards important improvements. Giov. Francesco Sagredo, a noble Venetian, and a very great friend of Galileo, wrote to him on the 9th of May 1613 as follows:—
" I have succeeded in reducing the instrument, which you have invented for
" measuring heat, into several very convenient and delicate shapes, inasmuch
" as the difference between the temperature of two rooms can be seen up to
" 100 degrees, &c."

It was Sagredo, who, in 1615, first hermetically closed the thermometer.

Other improvements were effected by the Grand Duke Ferdinand II. The following is what Padre Urbano Daviso says about them in his book entitled "Pratiche Astronomiche." Having first of all described a thermometer similar to Galileo's, he proceeds—" But His Serene Highness, not being satisfied with
" this invention, endeavoured to render it absolutely perfect. He constructed
" a little glass phial, of about the size of a musket-ball, with a neck of about
" the length of half a palmo, but so thin that it is barely possible to insert a
" millet-seed into it. Being filled with well refined spirit, either coloured
" or in a natural state, in sufficient quantity for it to reach up to the middle
" of the tube, and then the mouth being closed with a seal of hermete, it will
" show by means of the rising or falling of the liquid in the neck (which
" must have points marked on it for that purpose) the degree of greater or
" less heat or cold. And it has this advantage over the other instrument,
" namely, that the latter could only show the state of the air, whilst this one
" indicates, in addition to the temperature of the air, that of any liquid
" in which the bulb shall have been immersed; hence it will be possible
" to warm water, or a room, or a furnace to any degree, and to keep it
" in that state, or increase it, just as one pleases. Thus it will be easy to
" find out when a thing has been heated to the degree that was necessary for
" cooking it properly, operations from which, it may be said, that the
" chemical art has received its finishing touch. And in the same way it will
" be possible with instruments of this kind to find out the state of heat or
" cold of any place or province . . . . . And by this means it has been
" discovered that the water of wells and springs, as also cellars, grottoes,
" and other deep subterranean places, which in winter seem to our senses to
" be warmer than in summer, are, in fact, of the same temperature at both
" seasons. And we are, therefore, forced to say that the apparent difference
" comes from the surrounding atmosphere which affects our senses, and not
" from any change in the degree of heat or cold in the vaults."

As to the way of dividing them into degrees, the Diario dell' Accademia del Cimento says that the cinquantigrado " immersed in melting snow,

" descended to 13⅗ degrees,· although on some occasions the cold air at
" Florence reduced it to 7°.   And when exposed to the rays of the sun at
" midsummer, in the open air, and free from any kind of reflection, it rose
" to 43°, and at the same season, in the shade, to 34°." So that the difference between the cold of ice and the heat of the sun was divided into 30
degrees, and these were the points of comparison for the thermometers of
those days.

With regard to meteorological observations, they were instituted in the
year 1654, by the Grand Duke Ferdinand II. and continued by the Accademia
del Cimento.   They used to be taken at Florence at the Palazzo Pitti, at the
Giardino Boboli, and at the Convento degli Angeli ; and afterwards at
Cutigliano, Vallombrosa, Bologna, Patma, Milan, Warsaw, Innsbruck, &c.
Observations were usually made, at different hours of the day, of the state of
the thermometers exposed to the north and to the south ; the condition of
the sky, the direction of the wind, the barometric pressure, and the moisture
of the atmosphere.   Now, if these observations taken so long ago, be compared to recent ones, it will be seen that, after due corrections have been
made, or if observations be taken now with some of the best instruments of
the Accademia del Cimento, the meteorological conditions of Tuscany have
not changed.

It was with the thermometer No. 13 that the members of the Accademia del
Cimento carried out their first experiments on radiant heat, which showed
how both the heat of burning cinders and the cold of ice are reflected, according to the same law, by a concave mirror.   Illustrations of these experiments
are to be seen in the photograph of Galileo's Gallery, in the lunette on the
right hand side of the Tribuna.

Several thermometers, similar to No. 14, were placed by the Grand Duke
in his rooms, in order to secure an equable temperature in all of them.

As to thermometer No. 17, most important use was made of it for determining the temperatures of sick persons.   The various degrees of heat were
shown by the sinking of little balls of different colours in proportion as the
temperature rose.

At the present day we have returned to the cylindrical shape of ther-
memeter bulb, already used by the Accademia del Cimento.

3745. Clinical Thermometers, made upon Dr. Phillip's
principle.   *Francis Pastorelli.*

The ball and part of the stem are filled with mercury ; above the main
column, separated by an air speck, is a small mercurial index ; when heat is
applied to the ball the column and index are driven forward ; on cooling the
main column only recedes, the index remains, the upper end of which indicates
or measures the amount of heat applied.   Above the air speck the space is a
vacuum.

3745a. Dr. Clifford Allbutt's Clinical Thermometers ;
Fahr. and Cent. scales ; various patterns.

                                         Harvey, Reynolds, & Co.

The general introduction of the thermometer for ascertaining the temperature of the body in disease indicates one of the chief advances in the methods
of diagnosis.

Dr. Aitken used thermometers 10 inches long, and the instrument was
hardly met with beyond the wards of a few hospitals.   In 1867, Dr. Clifford
Allbutt requested Messrs. Harvey, Reynolds, and Co. to make instruments
with a chamber anterior to the bulb, reducing the length of the tube from

10 inches to 6 inches, then to 4 inches, and to 3 inches. From that time the use of the clinical thermometer has rapidly extended, until now it is found in the pocket of almost every medical practitioner.

3745b. Clinical Chart Forms for temperature, &c.

Harvey, Reynolds, & Co.

Quarto size for hospitals; octavo size for medical practitioners and their patients. Designed by Edward Casey, M.D., Windsor.

3745c. Registering Clinical Thermometers for the Pocket. *Harvey, Reynolds, & Co.*

The clinical thermometer, as introduced by Dr. Aitken, was made of the length of 10 inches. It was seldom used beyond the walls of a hospital, from its want of portability. In the year 1867, Dr. Clifford Allbutt, of Leeds, requested the exhibitors to make for him a thermometer of the length of 6 inches, the reduction in the length of the stem being compensated by making a small chamber in the bulb. This served for the retreat of the mercury at temperatures below 90° Fahr.

Instruments of the length of 4 inches and 3 inches were subsequently made, and the result of this portability has been the universal adoption of the clinical thermometer in medical practice in Great Britain. Beyond this, physicians often direct patients to be guided by its indications as to when to send for medical advice.

3746. Standard Thermometer, for physiological purposes, 0°–50°, divided into $\frac{1}{10}$° (Virchow).

Ch. F. Geissler & Son, Berlin.

3747. Standard Thermometer, for physiological purposes, 30°–45° (Heidenhain). *Ch. F. Geissler & Son, Berlin.*

3748. Two Pocket Maximum Thermometers, for the use of physicians, in a case and ebonite sheath.

Ch. F. Geissler & Son, Berlin.

3749. Geissler's Standard Thermometer, for determining the temperature of the skin, divided into $\frac{1}{10}$°.

Ch. F. Geissler & Son, Berlin.

3750. Geissler's Standard Thermometer, for determining the temperature of the ear. *Ch. F. Geissler & Son, Berlin.*

3751. Rosenthal's Electrical Thermometer, for determining animal heat. Constructed by Dr. A. Lessing, of Nuremberg. *Prof. Rosenthal, Erlangen.*

The electro-thermometer consists of a bundle of iron and German silver wires, which is contained in an elastic catheter, and can be easily introduced into any cavity of the body. It serves for the measurement of the temperature of the body in different places, and especially for lecture demonstrations. The resistance of the exhibited example = 0·25 Siemens' units. For introduction into the heart, e.g., by the jugular vein, a smaller sized instrument must be used.

3752. Circulation Thermometers, in pairs, 12″ long, divided from 35° to 45° centigrade. May be read to $\frac{1}{40}$ of a degree.   *T. Hawksley.*

3752a. Dupré's Thermometer, with spiral bulb and silver reflector, for ascertaining the temperature of the surface of the body.   *T. Hawksley.*

3752b. Pair of Clinical Thermometers.   *E. Cetti & Co.*

3752c. Three Clinical Thermometers, in cases.
                    E. Cetti & Co.

1037. Two Thermometers, on the plan of Virchow, for physiological investigation.
                Will. Haak, Neuhaus am Rennweg, Thüringen.

1042. Two Thermometers, for medical purposes, on the plan of Traube, divided in tenths from +25 to +45° C.
                Will. Haak, Neuhaus am Rennweg, Thüringen.

VI.—APPARATUS FOR INVESTIGATING THE FUNCTIONS OF CIRCULATION AND RESPIRATION.

3753. Hermann's Heart Pump.
                    Prof. Dr. L. Hermann, Zürich.

This apparatus, exhibiting the action of the auricle, may be applied in connexion with Weber's model of the circulation of the blood.   The auricle (the narrow chamber of the pump), without entrance valve, works in such a manner that the heart receives blood from the veins both during the systole and the diastole of the ventricle (the wide chamber).   When the play of the auricle is prevented, by shutting off its piston from the lever, the heart receives blood only during the diastole of the ventricle.

3957a. Prof. Rutherford's Model of the **Circulation** for explaining the **Blood Pressure** and the **Pulse.**
                    Prof. Rutherford.

Model of the circulation for explaining the blood pressure and the pulse. The tubes are filled with water.   The heart is represented by an elastic pump. The apparatus is fully described in the Journal of Anatomy and Physiology, Vol. VI., p. 249.

3754. Apparatus for Artificial Respiration.

Prof. Stricker, for the Institute for General and Experimental Pathology, Vienna.

The apparatus consists of clockwork set in motion by two steel springs, each five centimètres wide, and 598 centimètres long.   The clock moves bellows, through which an animal, narcotized for the experiment, can receive sufficient air to maintain circulation.

To give the clockwork (made by Siegfr. Marcus, of Vienna) a uniform movement, Professor Stricker has constructed a regulator the principle of which is that with increasing velocity in the rotation of a balance wheel, two rods are thrust from the plane of the wheel, which act against a brush and so diminish the velocity. As the position of the brush can be regulated, the checking of the velocity can be commenced at any time within certain limits.

The clockwork can be made to work the bellows from 12 to 160 times per minute.

3755. Lowne's Patent Portable Spirometer.

R. M. Lowne.

The measurement of the vital capacity is obtained by measuring the velocity of the expired current during the time of expiration, and the instrument is arranged so as to reduce the velocity of the current to cubic measure. The indications of the instrument are shown by means of hands revolving on a dial which denote the number of cubic inches expired.

3755a. Improved Portable Spirometer. *E. Cetti & Co.*

3755b. Hæmotachometer of Vierordt (1857).

Prof. Vierordt, Tübingen.

The stand to be used with the instrument is omitted.

3756. Pump for Transfusion of the Blood. Invented by Dr. Valentin.

Geneva Association for the Construction of Scientific Instruments.

This instrument is provided with a glass cylinder, thus rendering visible the presence of air in the pump. The piston also is made of a single piece of glass, which ensures the greatest possible cleanliness.

3756a. Blood Transfusor. *Collin & Co., Paris.*

3756b. Spiroscope. *Collin & Co., Paris.*

3756c. Aspirator of Liquids. *Collin & Co., Paris.*

3756d. Apparatus for delicate Injections.

Collin & Co., Paris.

3756f. Hematimeter, apparatus intended to determine the number of globules in the blood, by Dr. Hayem and A. Nachet.

A. Nachet, Paris.

3757. Sphygmograph of Vierordt (1853).

Prof. Vierordt, Tübingen.

The pad for fixing to the arm with the weights for equilibrating the instrument are omitted.

3758. Photo-sphygmograph, with magnesium lamp, condenser, and plaster cast.

Dr. S. Th. Stein, Frankfort on the Maine.

This apparatus is used for photographing the human pulse. See Vogel's "Photogr. Mittheilungen," September and October 1875. Specimens of the photographs, a normal and fever pulse tracing, accompany the instrument.

3759. Sphygmodynamometer, an apparatus for estimating the blood pressure in the radial artery of man.

Dr. C. Friedländer, Strassburg.

The apparatus is made for determining the variations of the blood pressure in the radial artery in man ; *e.g.*, the diurnal variations, the changes which the blood pressure suffers under the influence of poisons and various physiological and pathological conditions. Thus the apparatus is actually intended for purposes different from those of the sphygmographs which have been hitherto described, but it can also be used as a sphygmograph. A description of it will shortly be published in a physiological journal.

3760. New Sphygmograph (Sommerbrodt's).

Dr. Sommerbrodt, Breslau.

This sphygmograph transfers the movements of the wall of the blood vessel to a point moving up and down vertically by means of a one-armed weighted lever, and has a contrivance for fixing, which excludes, especially in researches on the radial artery, any vibration of the apparatus.

The performance of this apparatus in relation to the already known characteristics of the pulse traces surpasses in a striking way that of Marey's sphygmograph, which has hitherto been universally used, and which depends on the action of a spring. Besides, it has been possible, with the help of this apparatus, to make a series of new observations on the pulse curves given by the radial artery, and to discover some completely new features which it presents. The details of these observations are given in a paper published in March of the present year, which bears the following title :—" Ein neuer Sphygmograph und neue Beobachtungen an den Pulscurven der Radialarterie von Dr. Julius Sommerbrodt. Privatdocent, Breslau. A. Gosohorsky's Verlagsbuchhandlung (Adolf Kiepert).

3761. Riegel's Double Stethograph.   *Weber, Würzburg.*

3762. Sphygmograph.   *Weber, Würzburg.*

628. Portable Pneumatic Apparatus.

Prof. Dr. Weldunburg, Berlin.

This apparatus serves for condensing and rarefying air for medico-therapeutical and physiological purposes. It is possible by its means—

1. To inspire during its use compressed air whose degree of compression remains constant, but the compression of which can be varied to any extent by means of a known weight lying on it.
2. To exhale into rarefied air whose rarefaction is known and can be varied by means of the superimposed weight.
3. To inspire rarefied air.

These three methods of action not only produce a certain mechanical effect on the organs of respiration, but also affect the heart and the circulation. The apparatus is therefore adapted for studying the physiological action of compressed and rarefied air on the circulation and respiration.

Its principal applications, however, are to medico-therapeutical objects, for which its mechanical action renders it applicable ; as, 1, the inhalation of compressed air ; 2, exhalation into rarefied air ; and, 3, inhalation of rarefied air. As a remedy it is suitable for a series of lung and heart diseases ; as, for example, the exhalation into a rarefied atmosphere has been ascertained to be beneficial in many cases of emphysema of the lungs and asthma. The inhalation of compressed air in the treatment of pleuritis, bronchitis, phthisis, &c.

3763. Pneumatometer. *Prof. Dr. Weldunburg, Berlin.*

This instrument consists of a manometer, one limb of which is connected by means of a tube with a mask for the face. It is used for measuring the force of inspiration and the force of expiration in healthy and diseased individuals. In persons who suffer from diseases of the organs of respiration, the instrument not only gives the measure of the strength of inhalation and exhalation as such, but especially records the relative proportion in which each function is modified, so that in certain diseases the force of inhalation is principally or entirely affected ; in others, the force of exhalation ; and in others finally, both are affected. In this way the pneumatometer serves as a new physical expedient for the diagnosis of disease.

3764. Kymographion, for the registration of the pressure of the blood and of the respiratory movements on a blackened surface of 250 centimetres long and 23 centimetres high, which may be changed for another within half a minute.

Physiological Institution, Prague.

Accessory apparatus are used for the blackening and for fixing on the black. Two quicksilver manometers, one spring manometer, and a drum recorder. Several may be put into action, either singly or together. An electro-magnetic signal apparatus marks every single and every fifth second. Two other similar signal apparatus serve in connexion with two differently constructed electric double keys for marking the beginning or end of the electric shocks, or other effects. The rapidity of the movement of the blackened paper is variable within wide limits.

3765. Apparatus for Artificial Inspiration and Expiration. *Physiological Institution, Prague.*

Two pumps working synchronously, one of which is devoted to inspiration, the other to expiration. The number and depth of the respiratory acts are variable within wide limits. The apparatus allows also the employment of mixtures of gases and the preservation of the air expired without loss.

3766. Portable Injection Apparatus, with constant pressure, for artificially filling the blood and lymph vessels.

Physiological Institution, Prague.

By means of this apparatus every pressure between 2 and 300 mm. of mercury can be produced, and it remains unaltered during the whole continuance of the injection. The apparatus may be used for the simultaneous injection of two groups of vessels.

3767. Scheme of the Circulation.

Physiological Institution, Prague.

The apparatus consists of two glass cylinders, communicating by a horizontal tube, one representing the arterial, the other the venous, system. In the horizontal tube there is, in imitation of the capillary system, a sponge, in which the resistance to the current may be diminished or increased at will by means of a screw. A small pump brings the fluid with the desired force, or in the desired quantity, from the venous into the arterial system.

3768. Gas Sphygmoscope, by Sigismund Mayer.
Physiological Institution, Prague.

The apparatus consists of a small drum or capsule placed on the skin above a beating artery, through which common coal gas is allowed to flow towards a burner. Each pulsation causes a contraction of the flame, and shows, among other things, very plainly the dicrotism of the pulse.

3769. Apparatus for the Registration of the Respiratory Movements. *Physiological Institution, Prague.*

The above consists of an air-tight case containing a living animal, which breathes through a tube fastened in the trachea. This tube communicates with the atmosphere through the side of the chest. The air enclosed in the chest is moved in oscillations of pressure by the breathing of the animal, and these oscillations are registered on the kymographion.

The apparatus allows at the same time the registration of the pressure of the blood, the section of the nervi vagi, the irritation of any desired nerves, and the injection into the jugular vein, without opening the case.

3770. Apparatus for **Registering** the **Respiratory Movements.** *Physiological Institution, Prague.*

In this apparatus the animal inhales from a large air-tight closed receiver, which communicates with the trachea by a tube. The changes in the pressure of the air in the receiver, thus produced, are shown on the kymographion. The air in the receiver is frequently renewed.

3771. Apparatus for the **Demonstration** of the mechanical influence of the **Action of Breathing** on the circulation of the blood. *Physiological Institution, Prague.*

3772. Sphygmometer by Dr. Handfield Jones. *T. Hawksley.*

3773. Sphygmometer by Dr. Sibson. *T. Hawksley.*

3773a. Original Double Stethoscope, invented by Arthur Leared, M.D. *Dr. Burdon Sanderson, F.R.S.*

3773b. Double Stethometer, invented by Arthur Leared, M.D. *Dr. Burdon Sanderson, F.R.S.*

3774. Stethometer by Dr. A. Ransome. *T. Hawkesley.*

3775. Dr. A. Ransome's Stethograph for delineating the movements of the ribs during respiration. *T. Hawksley.*

3776. Dr. Burdon Sanderson's Kymograph, arranged for taking continuous ink tracing of the movements of the arteries. *T. Hawksley.*

3777. Dr. Burdon Sanderson's Kymograph with smoked cylinder for spiral traces. *T. Hawksley.*

3778. Dr. Oliver's Kymograph, with Sanderson's Cardio graph and Marey's Tambour, arranged for spiral traces, for use a the bed side. *T. Hawksley*

3779. Five hundred yards of Paper for continuous traces.
T. Hawksley.

3780. Dr. Burdon Sanderson's Cardiograph for recording the movements of the heart. *T. Hawksley.*

3781. Dr. Burdon Sanderson's Stetho-Cardiograph for recording the movements of the chest and the heart.
T. Hawksley.

3782. Marey's Sphygmograph, with improvements by Drs. Sanderson, Sibson, and Handfield Jones, for recording the motion of the pulse. *T. Hawksley.*

3783. Dr. F. Sibson's Cardiograph for right and left ventricles and apex of heart. *T. Hawksley.*

3784. Marey's adjusting Pneumatic Tambour.
T. Hawksley.

3785. Electro Marker. *T. Hawksley.*

3786. Electro Magnetic Marker. *T. Hawksley*

3787. Voit's Apparatus for investigating the gases given off in the respiration of small animals, exhibited in the Physiological Institute of Munich. *C. Stollnreuther, Munich.*

3788. Box containing **Apparatus** for the direct **Transfusion** of the **Blood** :—
1. Regulating canal formed by two tubes.
2. Three canals terminating in olive wood, shaped like a clarionet reed to insert in the arteries of animals.
3. Hollow and lancet.
4. Three canals of steel, reed-shaped, to fix in the regulating canal.
5. Two pincers for compression.
6. India-rubber tube for joining the regulating canals to the arterial canal.

More detailed explanations as to the mode of setting up the apparatus, and the different positions of the internal tube as to the regulating canal, will be found in the description of the apparatus, two copies of which, with sketches and plans, are appended.

Dr. Giuseppe Albini, Professor of Physiology in the Royal University of Naples.

4557. Drawings of Scientific Instruments in the University of Turin. *Director, Dr. Angelo Mosso.*

1. Physiological apparatus, for measuring the movements of the blood-vessels :—
    A. Pletismograph, and apparatus for producing artificial circulation in organs separated from the body.
    B. Pletismograph, to note the movements of the blood-vessels in man. (See two printed descriptions attached.)

3788a. Copy of "Ueber einen neuen Respirations Apparat."
Dr. Max Pettenkofer, Munich.

3788b. Mosso's Plethysmograph, modified by Kronecker, and photograph of the same.   *Prof. Gerald Yeo.*

3788c. Kronecker's Improved Manometer for frog's heart.   *Prof. Gerald Yeo.*

VII.—ELECTRICAL APPARATUS USED IN PHYSIO-LOGY.

3789. Hermann's Universal Commutator, with adjustable contact-springs.   *Prof. Dr. L. Hermann, Zürich.*

This apparatus may be turned by hand or by water power, and admits of many physical and physiological applications, *e.g.*: 1. Closing a simple electrical circuit ("interruptor"). 2. Change of direction in a simple electrical current ("inversor"). 3. The same in two circuits, for instance, in determining resistance in a polarisable conductor, by Wheatstone's method, under currents of changing direction. 4. Choice between opening or closing induction currents (Dove's "Disjunctor"). 5. Determination of the polarisation residuum in a conductor after shutting the current (like Siemens's "Commutator"). 6. Comparison of two residua of polarisation, by connecting the two conductors with the galvanometer, one opposite to the other. 7. Many physiological rheotomic inquiries.

3790. Hermann's Non-Polarisable Electrodes, of large form.   *Prof. Dr. L. Hermann, Zürich.*

By this apparatus a nerve may be connected with any number of non-polarisable electrodes in a small space, for instance, in the wet chamber of a myograph. The ends of the zinc wires, well amalgamated, receive a covering of clay, mixed with saturated solution of sulphate of zinc, and above this a small quantity of clay mixed with 0·6 per cent. solution of chloride of sodium.

3790a. Becquerel's Thermo-Electric Apparatus for Physiological Purposes.   *M. Ruhmkorff.*

3791. Electric Excitors.   *T. Hawksley.*

3792. Daniell's Constant Current Batteries for Clinical use.   *T. Hawksley.*

3793. Dr. Herbert Tibbitt's Constant Current Batteries and Induction Apparatus.   *T. Hawksley.*

3793h. Physiological Recording Apparatus, capable of moving at any speed from three inches to six feet per minute.
Dew. Smith.

3793a. Electro-Medical Pocket Case.
M. Trouvé, 6, Rue Thérèse, Paris.

This contains a battery, an induction coil, and all the accessories used in the practice of electro-therapeutics. (See pamphlet of Dr. Althean, translated by Dr. Davin, Paris, Delahaye.)

3793b. Large Electro-Physiological Apparatus.
Messrs. Trouvé & Onimus. M. Trouvé, 6, Rue Thérèse, Paris.

This apparatus, intended for electro-therapeutics, is at the same time very valuable for physiological experiments; it is the only one that produces induced currents strictly co-equal, whatever their number. In a given time, it shows instantaneously, and at will, the exact number of interruptions of the current corresponding with the electro-muscular shocks desired to be produced per second. It unites all the conditions requisite for making an exact study of the electro-physiological phenomena, and enables the progress of tetanus in a muscle to be observed.

3793c. Electric Exploring-Extractor.
M. Trouvé, 6, Rue Thérèse, Paris.

This serves to diagnose the presence in the organism of any foreign substance, metallic or otherwise, its nature, whether lead, copper, iron, wood, stone, &c.; the direction it has taken, its depth, and thus to facilitate its extraction.

3793d. Portable Electro-Therapeutic Apparatus, with equal and continuous Current.
M. Trouvé, 6, Rue Thérèse, Paris.

This apparatus, specially intended for electro-therapeutics, is composed of :—
1. A battery of 80 elements, simply moist.
2. An apparatus for graduating the tension of the battery, and altering the direction of the current.
3. A galvanometer intended to register the passage of the current; the whole enclosed in a very portable case. Each element is constituted similarly to battery No. 3793g.

3793e. Apparatus for Galvano-cautery, by G. Trouvé.
M. Trouvé, 6, Rue Thérèse, Paris.

This apparatus, specially intended for surgical operations, takes up a very small space, and is composed of movable elements enabling the practitioner to replace by himself, without the help of a specialist, the carbon and the zinc. The cauteries resulting therefrom are isolated by fuzed porcelain, which serves to resist all temperatures and all causes of deterioration.

3793f. Hermetical Battery, by G. Trouvé.
M. Trouvé, 6, Rue Thérèse, Paris.

This battery works only when reversed or placed horizontally; the zinc and carbon of which it is composed take up only the upper part of the sheath, made of hardened india-rubber; the other half, the lower, contains the stimulating liquid, water and bi-sulphate of mercury. This battery works the electro-medical case by the same inventor, as also his electro-physiological apparatus and his electric exploring extractor.

3793g. Moistened Battery, with constant Action.

M. Trouvé, 6, Rue Thérèse, Paris.

This battery is composed thus :—Between two discs, one made of copper the other of zinc, forming the two electrodes, are piled rounds of blotting paper. The lower half of these rounds is previously saturated with sulphate of copper, the other half with sulphate of zinc. It is continuously uniform, and lasts a long time. It is applied with great efficacy to electric clockwork, to telegraphy in general, and to all electrical apparatus having a resisting circumference. Moreover it forms part of several apparatus exhibited by the same inventor, particularly of the " military telegraph," the " apparatus with continuous current," &c., &c., &c.

VIII.—APPARATUS FOR INVESTIGATING THE FUNCTIONS OF MUSCLES AND NERVES.

3794. Muscle Telegraph. *T. Hawksley.*

3795. Apparatus for investigating the **action** of **Poisons** on **Muscles.** *T. Lauder Brunton, F.R.S.*

3796. Double Lever Apparatus for demonstrating the movements of the auricles and ventricles.

T. Lauder Brunton, F.R.S.

3797. Apparatus for demonstrating the influence of **Heat and Cold** and the **action of Poisons on the Frog's Heart.** *T. Lauder Brunton, F.R.S.*

The fact that heat accelerates and cold retards the pulsations of the heart is one of such fundamental importance, both in regard to a right understanding of the quick pulse, which is one of the most prominent symptoms of fever, and to a correct knowledge of the proper treatment to apply when the heart's action is failing, that for the last year or two the exhibitor has been accustomed to demonstrate it as a lecture experiment. The apparatus used is exceedingly simple, but it answers its purpose well, and by its means the pulsations of the frog's heart can be readily shown to several hundred persons at once. It was exhibited at the meeting of the British Medical Association in London more than two years ago, and a description of it appeared in the " British Medical Journal " for August 23, 1873 ; but as there is reason to believe that few physiologists have seen either the instrument or its description, it may not be amiss to add a few words regarding it here. It consists of a piece of tin plate or glass three or four inches long and two or three inches wide, at one end of which an ordinary cork, cut square, is fastened with sealing-wax in such a manner that it projects half an inch or more beyond the edge of the plate. This serves as a support to a little wooden lever about three inches long, a quarter of an inch broad, and one-eighth of an inch thick. A pin is passed through a hole in the centre of this lever, and runs into the cork, so that the lever swings freely upon it as on a pivot. The easiest way of making a hole of the proper size is simply to heat the pin red hot, and then to burn a hole in the lever with it. To prevent the lever from sliding along the pin, a minute piece of cardboard is put at each side of it, and oiled to prevent friction. A long fine bonnet straw or section of one is then fastened by sealing-wax to one end of the lever, and to the other end of the straw a round piece of white paper, cut to the size of a shilling or half-crown, according to convenience, is also fixed by a drop of sealing-wax. The pin, which acts as a pivot, should be just sufficiently beyond the edge of the plate to allow the lever to move freely, and the lever itself should lie flat upon the plate. Its weight, too, increased as it is by the straw and paper flag, would now be too great for the heart to lift, and so it must be counterpoised. This is readily done by clasping a pair of bull-dog forceps on the other end. By altering the position of the forceps the weight of the lever can be regulated with great nicety. If the forceps are drawn back as at *c*, Fig. 1, the flag is more than counterbalanced, and does not rest on the heart at all, while the position *a* brings the centre of gravity of the forceps in front of the pivot, and increases the pressure of the lever on the heart. The isolated frog's heart is laid under the lever near the pivot, and as it beats the lever oscillates up-

wards and downwards. If the tin plate be now laid on some pounded ice, the pulsations will become slower and slower, and if the room be not too warm the heart may stand completely still in diastole. On removing the plate from the ice the pulsations of the heart become quicker. If a spirit lamp be now held at some distance below it, the heart beats quicker and quicker as the heat increases, until at last it stands still in heat tetanus. On again cooling it by the ice its pulsations recommence. At first they are quick, but they gradually become slower and slower. On again applying the spirit lamp they become quicker, and by raising the temperature sufficiently the heat tetanus is converted into heat rigor. Then no application of cold has the slightest effect in restoring pulsation.

Not only the effects of heat and cold, but the effect of separating the venous sinus or the auricles from the ventricle can readily be shown with this apparatus, as well as the action of various poisons. The best for the purpose of class demonstration is muscaria. A drop of saline solution containing a little of the alkaloid being placed on the heart, it ceases to beat entirely. If a drop of atropia solution be now added, the beats recommence. The exhibitor has seen them do so on one occasion after they had entirely ceased for four hours. When used for demonstrating the action of poisons the wooden lever should be covered with sealing-wax, so as to allow every particle of the poison to be washed off it, and thus prevent any portion from being left behind and interfering with a future experiment. By attaching a small point to the end of the straw in place of the paper flag, tracings may be taken upon smoked paper fixed on a revolving cylinder.

3798. Rosenthal's Rotating Myographion, constructed by Th. Edelmann, of Munich. *Prof. Rosenthal, Erlangen.*

This myographion, lately constructed by the exhibitor, and not hitherto described, consists of a large glass plate, which is swiftly rotated round a horizontal axis, by a weight. As soon as the desired velocity is attained, the glass plate is slightly displaced parallel to itself in the direction of its axis. It makes exactly one revolution in its new position, and then returns of itself to its original position in order to keep on rotating. Whilst this one revolution is taking place the excitation and contraction curves are marked by a very simple mechanism. The time of duration can be exactly determined by a simultaneous tracing of a tuning fork curve. For investigations on reflex action, especially for measuring the time of the (so called by the exhibitor) "querlietung" (cross conduction), a contrivance is made by means of which two muscle curves can be simultaneously traced.

In order to determine very exactly the point at which the contraction curve commences, the movement of the muscle is very considerably magnified. As the form of the muscle curve given by this apparatus cannot be depended on, it will not serve for studying its actual details.

The time of revolution of the glass plate may vary from two to half a second. It was impossible to measure with any apparatus hitherto constructed a space of time of such length with sufficient rapidity.

The linear velocity of the instrument revolving at its quickest rate amounts to about 2,500 mm. in a second; values of 0·001 second can thus be easily measured.

3799. Methods of Physiological Experiments and Vivisections. Methodik der physiologischen Experimente und Vivisectionen, von E. Cyon ; with Atlas, published by Ricker, of Giessen and Petersburg, 1876.

Physiological Institute, Leipzig (Prof. Kronecker).

3800. Apparatus for demonstrating the Pulsations of the Frog's Heart, constructed by E. S. Stöhrer.

Physiological Institute, Leipzig.

3801. Canula for the frog's heart apparatus, constructed by Osw. Hornn.   *Physiological Institute, Leipzig.*

3802. Pendulum Commutator, constructed by Baltzer and Schmitz.   *Physiological Institute, Leipzig.*

3803. Double Myoscope for the examination and demonstration of the laws of muscular contraction.

Physiological Institute, Prague.

The apparatus allows two nerves to be placed in a damp chamber, at the same time in opposite directions, and to be traversed by the same electrical current. The contractions of the muscles connected with them may be read on two dials.

3803a. Dr. Sibson's Improved Gastric Canulæ.

3804. Professor Foster's Levers for recording the **Movements** of the **Muscles, Nerves, &c.**   *T. Hawksley.*

3805. Different Modifications of the **Spring Myograph** of Du Bois-Reymond.   *Prof. Theodore Schwann, Liége.*

A diapason or tuning fork is added, with a hammer to make it vibrate. By turning the handle placed on the left-hand side the hammer is freed in the first instance, and in the following moment the blackened glass plate. Excitation is caused by means of an induction coil. Previous to the freeing of the glass plate, and at the commencement of its motion, the current from the pile is interrupted, because it has to pass through the small insulated spring placed underneath the frame, and through the frame itself. But the latter has fixed to its under edge, on the right-hand side, a small ivory plate, which interrupts the metallic continuity. The excitement takes place just as the frame commences, in its motion, to touch with its brass surface this spring. An instrument fixed to the rod which carries the muscle enables the horizontal and the curved line of the muscle to be traced during the same experiment.

3806. Muscular Balance.
Prof. Theodore Schwann, Liége.

This apparatus was exhibited to the meeting of German naturalists at Jena in 1836. It is intended to demonstrate that muscular contraction takes place in accordance with the laws of elastic bodies. At that time the means of producing the continuous contraction of a muscle were not known. In order to determine the degree of contraction without a load or with increasing loads, the beam of the balance was lowered by means of a screw until the muscle could scarcely bear the load. "It was the first time," says M. Du Bois Reymond (Mem. de l'Acad. de Berlin, 1859, p. 79), "that an evidently " vital force was investigated in a similar manner to a physical force, and " that the laws of this force were mathematically expressed in figures." The description of the apparatus is also to be found in J. Müller's Physiology, edition of 1840, Vol. II., p. 59.

IX.—ANTHROPOLOGICAL APPARATUS.

3999. Anthropological Instruments.
Messrs. L. Mathieu and Son, Paris.

1. Anthropological truss.
2. Metrical measure, leather, divided into inches and centimetres.
3. Metrical measures, steel, divided into inches and centimetres (2).
4. Double metre, with articulated springs.
5. Two large squares, graduated, by M. Broca.
6. Facial goniometer   ,,   ,,
7. Median   ,,   ,,   ,,
8. Maxima frame.
9. Flexible auricular square, by M. Broca.
10. Plumb line.
11. Dermographic pencils, by Piorry.
12. Pocket dynamometers, by Mathieu.
13. Pneumometer, and pneumodynamometer.
14. Sphygmograph, by M. Marey.
15.   ,,   by Meurisse and Hy. Mathieu.
16. Anthropometer.
17. Graduated plane, by M. Broca.
18. Profilometer, by Sauvage.
19. Directing and exploring square, by M. Broca.
20. Meter, brass.
21. Caliper compasses, by M. Broca.
22. Micrometric caliper compasses, by M. Broca.
23. Cylometer   ,,   ,,

24. Compasses, with slides, of various shapes.
25. Endometer, by M. Mathieu.
26. Pachymeter, by M. Broca.
27. Craniostator, of the human cranium.
28.   „   for comparative anatomy.
29. Orbitostators, with orbital hands.
30. Bi-millimetrical rule.
31.   „   label, by M. Broca.
32. Craniophore, by M. Taupinard.
33.   „   by M. Broca.
34. Plane, for projections.
35. Craniograph, by M. Broca.
36. Stereograph   „
37. Suspensor, wood, with condylus plumb-line, by M. Broca.
38.   Do.   steel,   do.   do.   do.
39. Cephalometer, by " Antelme," modified by "Bertillon."
40. Millimetric ruler.
41. Tracing glass.
42. Diopter, by " Wirsig."
43. Camera, " Wollaston's."
44. Diagraph, by "Tavard," modified by Broca.
45. Designer, by Broca.
46. Pantograph.
47. Parietal toniometer, by M. Quatrefages.
48. Occipital level, by M. Broca.
49. Arteria basileris level, by M. Broca.
50. Occipital toniometer, with dial, by M. Broca.
51.   Do.   do.   rectangular, by M. Broca.
52. Caliper compasses, treble-armed, by M. Broca (2).
53. Auricular toniometer, treble-armed, by M. Broca.
54. Cephalic fan, by Second.
55. Rhinometer.
56. Intra-cranium holder.
57. Sphenoidal key.
58. Optical probe.
59. Turisk key.
60. Acoustic intra-cranium probes (2).
61. Optical-occipital probe.
62. Double disk for reconstructing compasses.
63. Cranioscope, by M. Broca.
64. Pivot, by Charles Bell.
65. Frame, by Pierre Camper.
66. Double frame, by Luca.
67. Frame, by Leach.
68. Craniometer, by Barclay.
69. Craniometer, by Bernard Davis.
70.   Do.   by Busk.
71. Toniometer, by Morton.
72. Apparatus, by Mantegazza.
73. Spindle.
74. Standard litre measure.
75. Half litre measure, graduated.
76. Funnel, with its operculum.
77. Double litre measure.
78. Shaft (*or handle*).

X.—SPECIAL COLLECTIONS.

PHYSIOLOGICAL AND OPHTHALMOLOGICAL INSTRUMENTS CONTRI-
BUTED BY THE PHYSIOLOGICAL LABORATORY AND OPHTHAL-
MOLOGICAL SCHOOL AT UTRECHT.

a. MICROSCOPY.

3951. Moist Chamber for micro-biological researches.

Prof. Engelmann, Utrecht.

Metallic box, length 80 mm., width 40 mm., depth 7 mm., thickness of the
plates 8 mm., bottom, a glass plate. Lid with central opening 15 mm.,
diameter closed hermetically by a glass cover.
Drop with object to be placed underneath the covering-glass. Some drops of
water on the glass bottom keep the chamber moist. The lid is pressed
against the upper surface of the chamber by two steel springs. If air-tight
closing is required the edges of the lid are greased. Each of the small sides
of the chamber is perforated by a metallic tube, 30 mm. long, 4 mm. wide,
for the entrance and escape of gases. To investigate the influence of
electric currents on the object, which is placed in the moist space filled with
air or any other gas, a lid of ebonite is used ; it is furnished with two metal
binding screws in which the wires are fastened. From each of the screws on
the inferior surface of the lid a movable platinum wire conducts to the drop.
In order to examine the influence of changes of temperature and of higher
temperatures a similar lid is used, but furnished with a single platinum wire
combining the two screws and passing through the drop. This wire is heated
by a galvanic current. By modifying the intensity and duration of the
current, fluctuations of temperature of any desired magnitude and velocity
can be produced in the preparation. To prevent evaporation of the drop
when heated it is also covered underneath with a glass.
(Onderzoekingen gedaan in het physiologisch laboratorium der
Utrechtsche Hoogeschool. Tweede Reeks, I. 1868, p. 140, &c., Pl. VIII.)

b. REGISTERING INSTRUMENTS.

3952. Set of Tympana for receiving the impressions of
movements and communicating them by air-transmission to the
cardiograph (modified stethoscope of König).

Prof. Donders, Utrecht.

On the rim of a hollow brass basin are fastened two elastic membranes, the
outer membrane convex, the inner membrane concave, owing to the air between them. Close to the rim of the basin a tube *a* conducts into the space
between the two membranes. In the centre of the basin (or elsewhere)
a tube *b* conducts into the space between the basin and the inner membrane. On sucking at the latter tube, *b*, the inner membrane falls against the
basin ; on closing now the first tube *a*, and ceasing to suck, the space between
the two membranes assumes the form of a bi-convex lens. With two india-
rubber lengths fastened on the tubes *c*, for the ears, these instruments are
excellent stethoscopes.
The special office of the tympana is to receive the impressions of movements (beating of the heart, pulsation of the carotid artery, respiration, &c.)
and to communicate them by air-transmission to the cardiograph. (Compare
the curves.)

3953. Chronoscopic Tuning Forks. Set of 4 tuning forks, with 10, 25, 50, and 100 vibrations per second, to be used as chronoscopes; they have a long, very pliable, flat spring, movable in its plane and perpendicularly on it, in order to be easily applied to the cylinder.   *Prof. Donders, Utrecht.*

3954. Direct registering Levers (large size) attached to lead wires.   *Prof. Donders, Utrecht.*

Very light registering levers with a pin (aluminium or cork) resting on the moving object (a contracting muscle, fontanelles, &c.). Two levers, one with a doubly bent pin, fixed on the same stand, can register two movements on the same cylinder, one immediately below the other (the contractions of the auricle and ventricle of the heart of a frog). Comp. Nuël. Onderzoekingen gedaan in het physiologisch laboratorium te Utrecht, Ser. 3, T. 11. p. 292, and Pflüger's Archiv. B. ix. S. 86.

3955. Apparatus for verifying the **Transmission** of **Movements** by **Air,** an india-rubber tube, and a **Marey's Cardiograph.**   *Prof. Donders, Utrecht.*

On a horizontal revolving axis is fixed an eccentrically ground brass disc, which moves a metallic spring placed underneath it. This spring registers its movements on a revolving cylinder, and at the same time presses on a tympanum, which communicates through an india-rubber tube with a Marey's cardiograph. If the cardiograph is made to register on the cylinder immediately under the spring, both curves may be compared. When the disc rotates slowly both curves are alike; with increasing velocity the difference augments. (Compare the curves; some are an imitation of physiological curves.)

3956. Brondgeest's Pansphygmograph. To register the movements of respiration, the beating of the heart, and the pulse of different arteries.   *Dr. Brondgeest, Utrecht.*

The little box contains : —
1. Cylinder, with internal spring, to be wound up by the hand.
2. Two Marey's sphygmographs to register on the blackened cylinder simultaneously respiration and pulse curve.
3. Little box with pens, to register, if desired, with ink on white paper.
4. A large tympanum, to register respiration, to be fastened with a tape round the chest or stomach.
5. Tympanum, with wooden plate and pin enclosed in a hoop, to be applied on the heart or on an artery.
6. Glazed papers for the cylinder.
7. Curves of respiration and pulse obtained simultaneously with the instrument.

c. CIRCULATION AND RESPIRATION.

3957. Scheme of the Circulation of the Blood (for instruction).   *Prof. Donders, Utrecht.*

An elastic bag with valves sucks water up from a vessel. By repeatedly pressing on the bag (with a lever) undulations are produced either in a glass or into an elastic tube, producing an interrupted or a continuous flowing out. On a long bent elastic tube the simultaneous existence of streaming and propagation of waves is further shown, and by means of two little springs on adjoining places of the tube the velocity of propagation of the wave is demonstrated, and if desired registered (also with different pressures indicated by a manometer). The tube at the same time offers an opportunity for auscultation before and behind a dilatation. (Compare Donders, Physiologie des Menschen. Leipzig, B. I. p. 78. Zweite Ausgabe, 1859.)

3958. Scheme of the relations of **Pressure** in the mechanism of **Respiration.** *Prof. Donders, Utrecht.*

A glass cylinder (thorax), closed underneath by an elastic membrane (diaphragm), at the top by a stopper with three openings: 1st, for a tube (trachea) with india-rubber bladder (lungs) ; 2nd, for a manometer, communicating with the space in the cylinder ; 3rd, for a tube to regulate the pressure of the air in the thorax. This pressure is made negative, as it is in reality ; which causes the air to enter the india-rubber bladder, and the diaphragm to become convex upwards. By pulling the diaphragm down (inspiration) the pressure in the thorax decreases and the lung is more distended ; the resistance of the lung is the negative pressure indicated by the manometer. Instead of an india-rubber bladder a fresh lung of a dog or of a rabbit may be used. The space in the cylinder outside the bladder is of no consequence, as not modifying the pressure on the inner surface of the wall.

d. Physiology of Muscles and Nerves.

3959. Electrodes, with moist chamber for physiological researches. *Prof. Donders, Utrecht.*

a. Non-polarisable, consisting of glass tubes, at the thin bent end having an aperture closed by moist salt-clay, continued in the lower part of the tube ; on this clay the nerve is spread. The tube is further filled with a solution of sulphate of zinc, wherein is a rod of amalgamated zinc.
1. Four of these electrodes on a stand between a vice reaching into a moist chamber of glass, two for a polarising current, two for a stimulating current on the extended nerve. (Compare Donders, Onderzoekingen gedaan in het physiologisch laboratorium, Ser. III., I. p. 1, 1873, and Pflüger's Archiv für Physiologie, Bv., p. 1.)
2. Four coupled with changeable distances.
b. Ordinary. A set of rods of zinc or platinum of various forms fitting into the same insulating glass tubes.

3960. Small round Muscle-Chamber, with non-polarisable electrodes and registering lever. *Prof. Engelmann, Utrecht.*

Bottom, an ebonite disc 50 mm. diameter. Lid, a glass shade perforated at the top, consisting of two halves, one furnished with a handle.
Electrodes, two bent glass tubes ; the vertical arm perforates the bottom of the chamber, and is at the top filled with clay impregnated with a solution of sulphate of zinc. Through the horizontal arm an amalgamated zinc rod soldered to a conducting wire is pressed into the clay. The openings of the tubes in the chamber are covered with salt-clay. Between them the preparation (heart, gastrocnemius, etc.) is put, whose contractions (increase in thickness) can be registered by a lever attached by its axis to a lead wire, and resting on the preparation by means of a vertical pin passing through the central opening in the lid of the chamber. The whole movable on a stand.

3961. Gas-Chamber for electro - physiological experiments particularly with living membranes. *Prof. Engelmann, Utrecht.*

The bottom. of the chamber is an ebonite disc, diameter 13 cm., thickness 13 mm., resting on three small feet. Lid, a glass shade, height 12cm., diameter 11 cm., with a broad flat ground rim resting in a groove of the ebonite disc; through the perforated neck a thermometer can be introduced.

On the bottom. of the chamber 6 binding screws, each separately in electrical connexion with corresponding binding screws on the outside of the chamber.

Inside the chamber, between the binding screws the object-support is placed, an ebonite plate furnished with :

1st. A central vertical vice to fix a horse-shoe cork frame upon which the animal membrane is spread.

2nd. On each side a glass pillar to fasten the electrodes.

To conduct the electric currents from the preparation small tubular non-polarisable electrodes are used; they are attached to lead wires and each connected by a very thin wire with one of the screws.

The other screws are to conduct electric currents, either to stimulate the object or to warm it by galvanism.

Two bent glass tubes, conducting through the bottom into the chamber, for the entrance and outlet of gases.

(Onderzoekingen gedaan in het physiologisch laboratorium der Utrechtsche Hoogeschool. Derde Reeks, II. 1873, p. 9, &c., Pl. I.)

3962. Isochronoscope.   *Prof. Donders, Utrecht.*

A brass lever has at one extremity a registering spring and at the other extremity a peg: On pressing the lever down with the hand, the spring registers the instant at which the peg, touching the mercury, closes the circuit (white electrodes). At this instant, the physiological or psycho-physical effects of this currrent, in all kinds of experiments of irritation, are registered on the same cylinder. Underneath the lever are two electro-magnets. When the current (green electrodes) passes, the lever is pulled down. This current is used either to note certain periods of time, as seconds, on the cylinder, or, by means of a more equal closing of the circuit, to obtain greater equability of the irritating current than can be obtained by the hand. The equable breaking of the circuit is then secured by means of a spring under the lever.

3963. Apparatus for the **Determination** of the **Co-efficient** of **Elasticity** of the **Living Muscles** in **Man.** (Compare Mansvelt, Elasticiteit der Spieren. 1860. Diss. inaug. Utrecht, 1863.)   *Prof. Donders, Utrecht.*

A vertical wooden board, sliding up and down in a stand with foot, has on its upper side a little sliding beam, and on its surface two recesses, which are the centres of two graduated arcs. With the face directed towards the profile of the board, the operator leans with the shoulder against an extremity of the cross beam and softly presses the internal condyles of the humerus into one of the recesses. The humerus has a vertical position ; the forearm is stretched either horizontally at $0°$, or above or below this position, and a weight suspended by a thread to a band round the wrist. On cutting this thread the forearm springs up, and the number of degrees is read off.

From different data the length of the muscle before and after the cutting of the thread and the action of the weight may be calculated, whence may be inferred the lengthening of certain muscles by definite weights at definite degrees of fatigue. One edge of the board is designed for the experiment with the left, the other for the experiment with the right, arm.

e. PHYSIOLOGICAL OPTICS.

3964. Double Spectacles, for the determination of refraction (Handbuch der Augenheilkunde von Saemisch und Graefe, III., p. 50). *Dr. Snellen, Utrecht.*

This is an opera-glass in the form of a pair of spectacles, consisting of a pair of negative glasses of one inch focal distance, and in front of these a pair of positive glasses of two inches focal distance. On changing the distance of the positive and negative glasses a hand is made to move on a disc furnished with a scale, which indicates the refraction of the system corresponding with the different degrees of sliding. To be had from R. Jung, optician, Heidelberg.

795. Phakometer (Snellen), for the determination of the power of lenses (by the method of placing object and image at equal distances from the lens). *Dr. Snellen, Utrecht.*

The object (points of light on ground glass) and the screen upon which the image is received are moved in a perfectly similar manner, but in opposite directions, each by a sliding steel spring.

For the determination of weak lenses, two auxiliary lenses, No. 2·75, are placed one on each side of the lens examined (at a distance of 24·33 mm. measured from centre to centre).

The screen which receives the image moves alongside a scale, marked at each double focal distance of the system of the three lenses obtained by calculation for the lenses commonly used in ophthalmological practice. Within $\frac{1}{20}$ " dioptrie " one can with sufficient accuracy estimate how much the lens examined differs from the powers marked on the scale.

The image having constantly the same magnitude, precise adjustment is easy. The screen carries a diagram of the image. If the image of the points of light cover this diagram, the centre of the glass will coincide with the diameter of the instrument. Not only the focus, but also the centre can be determined directly.

The scale may be controlled at any time by determining the strong glasses directly, the auxiliary lenses being removed. Then the double-focal distance is to be taken from the corresponding principal plane to the point where the image is formed.

The instrument as yet is only adapted for symmetrical (bi-convex) lenses. In order to determine plano-convex or periscopic glasses, it will be best to place two glasses of equal power one against the other, so as to obtain a symmetrical form. According to the calculated principal planes of this system, a scale has to be computed.

796. Lens of Stokes, with constant axis (Snellen); consisting of a negative and a positive cylindrical lens rotating equally, but in opposite directions. (Graefe's Archiv für Ophthalmologie, 1873, XIX. 1, p. 70.) *Dr. Snellen, Utrecht.*

Two cylindrical lenses (C $-\frac{1}{10}$ and C $+\frac{1}{10}$) centred one before the other can be made to rotate equally, but in opposite directions, about an axis perpendicular to the plane of the glasses, by means of two sliding steel springs. The principal meridians of the system here remain in the same direction.

The refraction in the principal meridians changes with the rotation proportionately to the sines of the angle between the axes of the cylinders. The scale which is to be read has been constructed on the rim of one of the glasses. For the determination of the refraction, parallel lines are viewed through this lens of changeable power, the lines are perpendicular to one of the principal meridians, and placed either at a distance or united with the lens in a tube. Manufactured by A. Crétès, optician, Paris.

797. Set of Spherical Lenses, metrical system (by Roulot).
Prof. Donders, Utrecht.

3965. Cylindrical Glasses (simple cylindrical, bi-cylindrical, and spherico-cylindrical) introduced by Donders (Astigmatismus en cylindrische glazen ; Utrecht, 1862). Metrical System by Roulot.   *Prof. Donders, Utrecht.*

To detect and correct astigmatism. By turning a very weak cylindrical glass before the eye in a plane perpendicular to the visual line, it will be found that no eye is absolutely free from astigmatism.

3966. Test-types, for the determination of the acuteness of vision. Williams and Norgate, London.   *Dr. Snellen, Utrecht.*

With letters and figures of definite magnitude are given, in metrical measure, the distances at which they exhibit themselves under an angle of five minutes. This angle is assumed to be the normal visual angle, and the sight is expressed by $V = \dfrac{d}{D}$, d being the distance at which the letters are recognised, and D the distance at which they exhibit themselves under an angle of five minutes.

3967. Set of Prismatic Glasses (by Roulot), introduced by Donders.   *Prof. Donders, Utrecht.*

Besides their application in investigating and compensating anomalies of the eye, they are used to demonstrate:

    a. The influence of the tendency to maintain binocular single vision on the movements of the eye.

    b. The maximum of divergence of the visual lines.

    c. The faculty of equal accommodation at different degrees of convergence.

    d. The apparent angle between equally directed meridians of the two eyes.

    e. The want of local sign (Local-zeichen) to distinguish the images of the right and left eye.

    f. The actual difference of those images in anisometropia.

    g. Pseudoscopy at double vision of horizontal lines.

3968. Microscope, to measure the depth of the chamber of the eye.   *Prof. Donders, Utrecht.*

By sliding the whole tube of the microscope we consecutively adjust for—*a*, surface of the cornea, *b* iris, and *c*, reflex image ; the reading from *a* to *b* is the apparent depth of the chamber, the reading from *a* to *c* furnishes the radius of curvature of the cornea, which enables us to calculate the real depth from the observed apparent depth. On the middle of the objective glass is pasted a very small piece of mirror : the eye observed, on looking in this mirror at the reflex image of a distant flame, is properly directed, and accommodated for distance, and shows to the observer the reflex image for the determination of the radius of curvature of the cornea.

3969. Phacoidoscope. (Anomalies, &c., p. 16.)
Prof. Donders, Utrecht.

A modification of the so-called ophthalmoscope, by means of which Cramer discovered the changes of form of the crystalline lens to be the true principle of accommodation. (Het accommodatie vermogen physiologisch toegelicht. Haarlem. 1853.) The three reflex images are seen in the eye under changeable angles, as the eye accommodates itself alternately to a distant and a near object.

3970. Meridian-ring of the Ophthalmometer. (Anomalies, &c., p. 361.) *Prof. Donders, Utrecht.*

A flat ring, 388 mm. diameter, upon which three small lamps can be moved; their reflexion-images have been used by Dr. Middelburg for the determination of the radius of curvature of the cornea in all meridians.

3971. Stenopæic Apparatus and narrow slit. (Compare van Wyngaarden, Archiv für Ophthalmologie, I. 1, p. 251.)
Prof. Donders, Utrecht.

3972. Metallic Plates with two openings, **for Entoptic Observation.** (Anomalies, &c., p. 201.)
Prof. Donders, Utrecht.

To determine the position as to depth of entoptic objects (muscae volitantes, black spots, pearly spots, &c.) in the humours of the eye, by the method "à double vue." Those of one eye are projected in the visual field of the other, and the distance of the double images measured with a pair of compasses (sliding compasses).

3973. Optometer. To determine the relative range of accommodation. (On the anomalies of accommodation and refraction of the eye. New Sydenham Society, London, 1864, p. 115.)
Prof. Donders, Utrecht.

An oblong quadrangular board on a stand. The board possesses three grooves with scales, in which a wire optometer or a point can be moved. On the width of the board are two half-rings supporting the glasses, whose relative distance may be regulated by means of screws. Each ring is movable in a circular groove, the centres of the circles coinciding with the centres of motion of the eyes. The position of the eyes can be controlled by small microscopes fixed on the sides of the board, and remains secured by the cheeks resting against two wooden rods. By this contrivance, the distance between the glass and the eye remaining unchanged, the visual line may at any degree of convergence coincide with the axis of the glass. This instrument is used to determine the play of accommodation at every degree of convergence, *i.e.*, curves of the nearest and farthest points, as functions of the convergence (relative range of accommodation).

3974. Simple Phænophthalmotrope (large size for demonstration), to demonstrate movements of the eye according to

3 P 2

the law of Listing, and the corresponding rotation. (Archiv für Ophthalmologie, XVI. 1, p. 154.)   *Prof. Donders, Utrecht.*

In an outer fixed ring can be turned another ring, representing the principal plane of axes, containing all the axes round which the eye moves out of the primary into any secondary position. Before turning on the intended axis, we put the arms of the cross vertical and horizontal; after the rotation these arms indicate the position of the vertical and horizontal meridians of the eye.

3975. Compound Phænophthalmotrope (large size, for demonstration).   *Prof. Donders, Utrecht.*

This instrument, as well as the simple phænophthalmotrope, No. 3974, demonstrates the movements of the eye according to the law of Listing; but can moreover put the line of fixation in any direction by rotation round a vertical and a horizontal axis (Seitenwendung und Erhebung of Helmholtz). The cross then indicates a position of the meridians different from the position obtained by rotation round a single axis according to Listing's law : the difference is the so-called " wheel-rotation" of Helmholtz.

The same instruments (small size) for private study of the movements of the eyes; they enable us to compare the direction of the after-pictures with the direction of the cross in corresponding movements. (Compare : Onderzoekingen physiologisch laboratorium, Utrecht. Cl. S. ii. D. iii. bl. 119, and Archiv für Ophthalmologie B. xvi. S. 160.)

3976. Simple Phænophthalmotrope, small size, for personal use.   *Prof. Donders, Utrecht.*

3977. Compound Phænophthalmotrope, small size, for personal use.   *Prof. Donders, Utrecht.*

3978. Snellen's Ophthalmotropometer, for determining the movements of the eye to right and left. (Handbuch der Augenheilkunde von Saemisch und Graefe, III. p. 236.)
            Dr. Snellen, Utrecht.

This instrument is intended to determine the movements towards right and left of each eye separately, and of both eyes with regard to each other.

It consists principally of two bars rotating about one axis; they form an angle which can be read directly on a graduated arc, attached to one of the bars.

The instrument is to be placed so that the point of rotation of the bars is perpendicularly below the point of rotation of one of the eyes.

For this purpose, the plate upon which the bars lie can slide forward and backward, and also to the right and to the left, independently of the points against which the head rests with the infra-orbital margins.

The eye under examination has to be fixed upon the middle of a telescope carried by one of the bars. By moving the plate to right or left the instrument is adjusted until the eye is in the middle of the field of view of the telescope. Then the bar with the telescope is turned some degrees, and the eye under examination has to be fixed again in the middle of the field of view; if the point of rotation of the eye does not coincide with the point of rotation of the instrument, the eye will now be seen out of the middle of the field of view. By sliding forward and backward until the eye appears again in the centre of the field of view, the instrument is carried into the proper position with regard to the eye.

The following can now be determined—

1. The mobility of each eye outward.
2. The limits of the lateral binocular vision.
3. The convergence of the two eyes in straightforward vision, and at any degree of lateral fixation.
4. The degree of deviation in strabismus.
5. The centre of motion of the eye.

3979. Ophthalmo-statometer, for the determination of the exorbital protrusion of the eye-ball. *Dr. Snellen, Utrecht.*

Two rods, united by a cross rod, are supported on the inner and outer margins of the orbit. From the cross rod a small crank can slide down until it presses softly against the eye-ball covered by the eye-lid. The thickness of the eye-lid is to be subtracted from the reading. The position of the top of the naked cornea may be read directly by means of a sight movable on one of the rods. (Handbuch der Ophthalmologie, Graefe und Saemisch III., p. 199.)

Manufactured by P. W. Hiele, Utrecht.

3980. Instrument for the determination of the **Movements** of the **Head** and **Eye,** in varying directions of the line of fixation. Donders and Ritzmann. (Compare Archiv für Ophthalmologie, XXI., 1, p. 131, 1875.) *Prof. Donders, Utrecht.*

A bent wooden rod has on one end a mouth-piece, on the other an arc (with a sliding sight), movable round an axis passing through the point of rotation of the eye, which point is at the same time the centre of curvature of the arc.

Over the axis of the arc the eye is fixed on a point *a*, then, without moving the head, on a point *b*, to which the sight is made to correspond ; now the eye is again fixed upon *a*, and with free movement upon *b* ; from the required displacement of the sight we learn how much the head has shared in the movement. If after the free motion the plane of the arc should not pass through *b*, we make it pass through this point, and read off how many degrees we have had to turn.

3981. Apparatus for verifying the **Laws** of **Donders** and **Listing** (hitherto undescribed). *Prof. Donders, Utrecht.*

A bent wooden rod has on one end a mouth-piece and on the other a coloured strip, movable about an axis passing through the centre of movement of the eye.

After fixation in the direction of the axis, directing the eye upon one extremity of the strip, the after-picture is seen in the prolongation of the strip (Law of Listing) in whatever way the line of fixation reaches the extremity (Law of Donders).

From experiments made with lateral inclination of the head, we may learn how far the associated rotary motion interferes with the law of Listing.

3982. Reflecting Lens, with a mouth-piece **for autoscopy of the eye.** Holländische Beiträge zu den anatomischen und physiologischen Wissenschaften, Düsseldorf und Utrecht, 1864, p. 384 ; and Mulder, Archiv für Ophthalmologie, Bd. XXI., Abth, 1, 1875. *Prof. Donders, Utrecht.*

Principally used to enable the eye, looking at a vessel of the conjunctiva, to observe its own rotations associated with lateral inclinations of the head. (The first application of the method consisting in uniting an instrument with the head, in order to secure the relative direction of the line of fixation.)

3983. Mouthpiece, with two vertical rods. Showing the rotations associated with lateral inclinations of the head and trunk, to be constantly equal for both eyes.

Prof. Donders, Utrecht.

The rods parallel to the apparent vertical meridians, at a distance of 70 mm., produce parallel half images near each other. In all movements of the head the images remain parallel, thus showing the rotations associated with lateral inclinations of the head and trunk, to be constantly equal for both eyes.

3984. Rod, with graduated arc to determine the degree of lateral inclination of the head.   *Prof. Donders, Utrecht.*

Seen with parallel visual lines, the rod shows double images of equal direction with the head (to be used in experiments of parallel rotations, &c.)

3985. Apparatus for determining the rotary motion connected with temporary and permanent lateral inclination. Donders and Mulder (Archiv. für Ophthalmologie, XXI. Abth. 1, p. 68, 1875.)

Prof. Donders, Utrecht.

This apparatus consists of (*a*) a head-holder, which, the head being fixed in the primary position, can revolve round a horizontal axis perpendicularly bisecting the basal line, and easily and quickly be secured in every position.

(*b*.) At a distance of some meters a round disc with horizontal light line of gas flames, situated in the principal plane of fixation, and possessing incisions on the rim for stretching diameters. After fixing the eye for some time on the middle of the light line, the gas is shut off, the head with the holder is turned and the latter secured, when the after-picture coincides with the stretched diameter. The difference between the inclination of the head and the inclination of the diameter is the rotary motion.

3986. Volkmann's Discs, united on one plate for the determination of the angle between meridians of apparent equal direction. The distance between the two axes must be equal to the distance of the parallel visual lines of the experimenter.

Prof. Donders, Utrecht.

3987. Volkmann's Discs of Glass, facilitating parallel position of the lines of fixation by looking at a distance.

Prof. Donders, Utrecht.

3988. Stereoscope with revolving mirrors for the determination of the angle between the meridians of apparent equal direction at different degrees of convergence.

Prof. Donders, Utrecht.

3989. Isoscope, to determine the angle of the meridians of apparent vertical and apparent horizontal direction. (Onderzoekingen physiologisch laboratorium, Derde Reeks, III. 2., 45 ; and Archiv für Ophthalmologie, B. XXI. 3. S. 100.)

Prof. Donders, Utrecht.

The instrument consists of a head-holder and a frame-apparatus. In the holder (on the principle of Hering) as the head turns round the basal line as axis, the eyes retain their places. The frame-apparatus can rotate round the same axis, and consists of a fixed frame and two pairs of movable frames, one pair for nearly vertical, one pair for nearly horizontal threads.

One of the vertical or one of the horizontal threads may have a fixed position; the direction of the other is obtained by turning the frame, until both the threads viewed with the eye appear to be parallel. The angle between them in this position is read on the graduated arc with vernier.

This instrument is used to determine the angle of the meridians of apparent vertical and apparent horizontal direction, either separately or simultaneously, at any degree of inclination of the plane of fixation, and at any degree of convergence of the lines of fixation, either symmetrical or asymetrical, and to investigate, at the same time, how lines or objects in the field of vision influence these angles.

3990. Perimeter Arc, with diagram of the plane of projection, for determining and registering the field of vision. (Handbuch der Augenheilkunde van Saemisch und Graefe, III. p. 57.)   *Dr. Snellen, Utrecht.*

A little below the centre of a movable metallic semicircle is a point for the infra-orbital margin to rest against; thus the eye under examination is kept as much as possible in the centre of the arc. Behind the arc a black board is placed, which is marked with meridians from 10° to 10°, showing, as it were, the projected arc with its divisions. The limits of the field of vision determined on the arc are mapped on the black board.

The scheme obtained represents clearly the limits of vision on each meridian.

To be had from D. B. Kagenaar, Utrecht. Its price is 22·50 f. (1l. 18s.)

3991. Cycloscope of Donders and Küster, to show the direction of the meridians, the great circles, the direction circles, and the parallel circles of the field of vision. (Archiv für Ophthalmologie, B. xxii. 1876.)   *Prof. Donders, Utrecht.*

This consists of a chair with head holder, and an arc with passing induction-sparks. The arc shows, by revolving round different axes, the directions of all meridians, all great circles, the direction circles of Helmholtz, and parallel circles by sliding up and down a stand. When in a dark room, the position of the head being well secured, one eye (the other being covered) is placed in the centre of curvature of the arc and fixed, in the primary position, upon a mark, which owing to its being rubbed with phosphorus throws out a faint light; the induction sparks we may fancy to be distant stars, and determine the direction in which they appear. On the bearing of such investigations, compare Helmholtz, Physiologische Optik, § 28, p. 550, and Krüster, *l. c.*

3992. Horopteroscope, for the determination of the line-horopter-plane, at different degrees of inclination of the plane of fixation and at different degrees of convergence.
    Prof. Donders, Utrecht.

The horopter-plane (a quadrangular plate) can revolve round the basal line of the head fixed by the mouth-piece; the relative distance of the two horizontal stripes, seen with or without crossed visual lines, determines the convergence; the horonter-plane is now made to incline until the horizontal stripes have one direction, and then turned round a horizontal axis passing through these stripes until the two perpendiculars on the first stripes are also parallel. (By this contrivance we find the position for close work with the hands, which, according to Donders, has in course of time determined the angle between meridians of apparent equal direction.)

3993. Spark-Stereoscope, with coloured lenses, showing stereoscopic vision independent of movement of the eye.   Archiv für Ophthalmologie, Band xiii. p. 33.   *Prof. Donders, Utrecht.*

A box with two apertures for the eyes.   A series of very small induction-sparks is used as object of fixation.   Between two electrodes attached to the sliding lid of the box, we may cause a large spark to pass at different distances before and behind the object of fixation.   In monocular vision we fail in recognising the place of the large spark with regard to the object of fixation ; but in binocular vision, at the first spark we judge correctly.   Between the apertures for the eyes are other electrodes.   The spark passing between these last two electrodes is out of sight, but its two reflex images are seen on a coloured lens ; one image is coloured the other not.   In binocular vision, their position with regard to the object of fixation is soon distinguished ; but without an object of fixation, many sparks are required for accurate estimation.   In general such coloured lenses are very convenient for demonstration and mutual comparison of the reflex images of concave and convex surfaces.

3994. Ptotostereoscope, to determine the relative value of the second eye for binocular vision, constructed upon the principle of Hering.   *Prof. Donders, Utrecht.*

The balls made to fall from the curved rod appear at the respective distances under equal angles, and seem to pass with equal velocity through the field of vision, so that every indication for the monocular vision is excluded, and the relative value of the second eye, for stereoscopic vision, can be deduced from the relation between the number of correct and incorrect cases (according to Fechner's method).

3995. Noematachometer.   (Onderz. physiol. laboratorium, Ser. 2, II., bl. 92, 1868).   *Prof. Donders, Utrecht.*

A vertical board with a central slit for the eyes, above a sliding electro-magnet with a long metallic prism suspended.   The prism has in front and behind a pair of sliding arms upon which a rod can rest, and at its back part two small sliding metal wire-frames, which can contain a very small piece of burning coal.   On closing the current the prism falls, and is received in a box with asbestos underneath the slit.   During the fall, the rods meet respectively a wooden and a metallic arm, and each produces its characteristic sound ; the frames with incandescent coal pass the slit one to the left, the other to the right, and one or other at pleasure coloured by a red glass.   Thus either two sounds, two lights, or a light and a sound, can be made perceptible with changeable difference of time (by sliding prism, rods, and frames), and in this manner we learn the time between two impressions, with increasing changes (correct and incorrect cases, Fechner) to perfect certainty, which enables us to judge about the priority of two impressions.

f. PSYCHICAL PHENOMENA.

3996. Noematachograph. Registering the time of psychical processes.   (Onderzoekingen physiol. laborat. Ser. 2, D., 1st. 21.) Archiv für Anat., u. s. w. 1868.   *Prof. Donders, Utrecht.*

The instrument consists of a horizontal cylinder, which, on being turned by the hand, moves in a spiral. The cylinder is covered with smoked paper, upon which a tuning fork with 250 vibrations per second, to be struck by a percussion hammer, registers its vibrations by means of a small steel spring. This tuning fork can slide and rotate on an axis by which it is suspended.

Without any further complication, the cylinder with tuning fork is sufficient for experiments with the simplified phonautograph, whose spring registers its vibrations next to those of the tuning fork. Two persons A. and B. sit before the phonautograph. A. shouts a short vocal sound, B. repeats it as quickly as possible; on the line registered by the spring the instants of the shouting of the sound by A. and B. are visible, and the time can be read on the adjoining chronoscopic curve. The experiments to be made are : a. B. knows which vocal A. will produce ; b. B. does not know which vocal (from two, three, or more) A. will produce ; c. B., not knowing which vocal he is going to hear, has to respond only upon a certain one, e.g., on a. The time for the intervening psychical processes may be deduced from the times ascertained in the experiments a, b, and c.

For other experiments the instrument is more complicated. At one end of the cylinder are two rings partly covered with copper, partly with ebonite. On each ring two pair of electrodes rest ; the lower break and close a constant current, the upper an induction current. The electrodes together are attached by a thick glass bar to a horizontal board following the cylinder, moving in a spiral, thus keeping the electrodes on the rings. The induction currents pass the tuning fork. In one series of experiments only the inner ring is used (the outer ring can be moved aside). On this ring quadrants of copper and ebonite alternate. The electrodes rest on copper, the tuning fork is struck, the cylinder moved half a turn ; at the instant the lower electrodes leave the copper, the induction current makes a hole in the paper on the chronoscopic curve, and a spark is made visible without being audible, or audible without being visible, or, the spark being neither audible nor visible, a shock is felt. The spark passes in a little box suspended on a cross-bar between electrodes with knobs, which are to be moved so far apart that the passing electricity constitutes only one spark. In order to see or to hear the spark, a Leyden jar is inserted in the circuit ; the shock is felt as unipolar conduction without an inserted Leyden jar.

On the stimulus being received, it is responded to by the hand moving a copper arm of a wooden cylinder with registering spring. By this contrivance, under similar objective and subjective conditions, the physiological time (the time between stimulation and response) can be determined alternately for impressions of vision, of hearing, and of touch at various places of the skin.

A dilemma may be put. Electrodes are applied right and left on the skin and the horizontal arm of the wooden cylinder struck in corresponding direction. In a similar manner, by a conventional response, a decision may be made between red and white light. In both cases response may be forbidden for one colour or for stimulation on one side. Instead of responding with the hand, the voice may be used acting on the phonautograph, and the physiological time determined in both cases. On the spark passing in the box, and making transparent letters visible, the physiological time for recognising these letters may be registered and compared with the time for simple light perception, etc.

The second ring has only a small conducting plate, upon which, at the rotation of the cylinder, the constant current is closed and broken, after contact with the electrodes of the induction current has followed. Thus, by sliding this second ring, sooner or later after the first spark, a second is made to pass, perforating the chronoscopic line, and at the same time, either visible in the second compartment of the box, audible, or sensible at another part of the skin. So we learn the time between two impressions on the same sense, or on two different senses, which enables us to judge about the priority.

In order to prevent, under all circumstances, the sparks from passing from one ring to the other, a somewhat greater distance between the rings is desirable.

PROFESSOR MAREY'S APPARATUS FOR OBSERVING AND REGISTERING PHYSIOLOGICAL AND MECHANICAL PHENOMENA.

3997. 1. Cardiac sounds for indicating the pressure in the cavities of the heart. A. Sound for the right heart. B. Sound for the left heart. C. Sound for showing negative pressures. D. Sphygmoscope. (The corresponding tracings are given in Plate I., 1–4.

2. Explorator for the heart of the rabbit. (Plates 2 and 2 quater.)

2A. Explorator for the human heart. (Plate II.)

Membrane explorator, for the pulsation of the heart, applicable to man or animals. By turning the regulating screw the button of the explorator may be made to project more or less, thus increasing or diminishing the sensitiveness of the instrument.

3. Explorator for the pulse (sphygmograph of transmission). (Plate III.)

4. Explorator for the respiratory movements. (Plate IV.)

5. Myographic pincers.

5A. Myographic pincers, another pattern.

6. Explorator for the investigation of to-and-fro movements, such as those of locomotion (of the horse), and of flight (of birds). (Plate VI.)

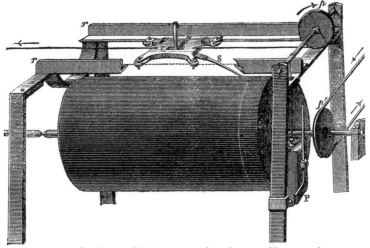

Apparatus for the graphical representation of any rectilinear motion.

7. Explorator for determining the rapidity of the flow of fluids.

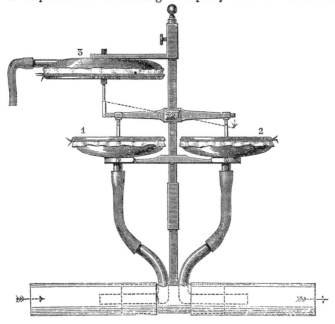

Model of an apparatus for the graphical representation of the motion of a liquid in a tube. or artery.

8. Explorator for determining the changes in the size of the hand.

8a. Explorator for determining the changes in the size of· the heart of the tortoise. (Plate 8 and 8 bis.)

9. Apparatus of " Donders " for verifying the correctness of tympana.

10. Myograph for transmission.

Sphygmograph for transmission placed in position on the wrist.

11. Myograph, direct, double.

12. Myograph, direct, single.

13. Myograph of the heart. (For these 3 apparatus, Plate 12 to 12³.)

14. Explorator of the movements of liquids in an elastic tube. (Plate 14 and 14 bis.) (*See opposite.*)

15. Explorator for investigating the pressure of the foot in human progression.

15A.
15B.   ,,   ,,   ,,   ,,   feet of the horse.
15C.

16. Apparatus for showing the action of muscular elasticity.

17. Tracing on paper transferred direct to wood.

18. Registering cylinder, with "Foucault" regulator, and rotatory hopper, by Marey.

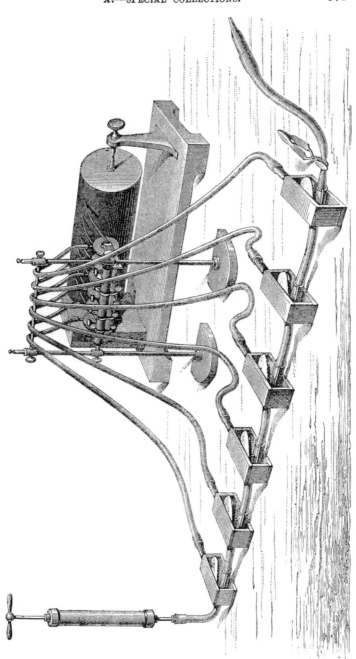

Series of explorators for investigation of a wave of fluid, attached to an equal number of recording lever membranes.   Arrangement for the experiment.

19. Tympanum and lever recording instrument.

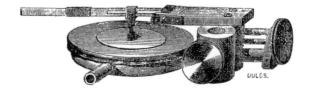

20. Hæmodromograph, by Chauveau.
21. Portable electric chronograph, giving 100 double vibrations per second.

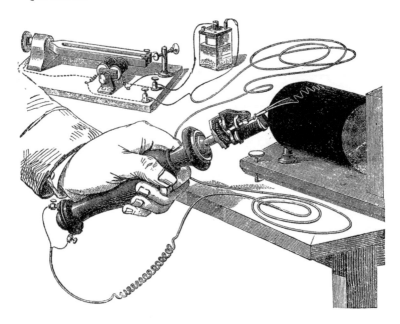

22. Tambours conjugués, for reproducing movements at a distance.

23. Pantograph of transmission, for reproducing movements at a distance. (*See opposite.*)

7. For demonstrating the laws of vision :

　　Large glass model of human eye ; stereoscope and geometrical slides, to show effect of binocular vision.

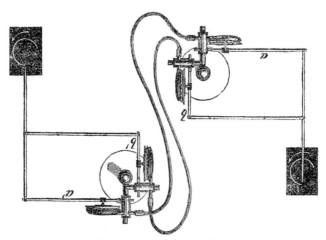

Pantographe of transmission.

8. Diagrams of—
   (*a.*) Shadow and penumbra.
   (*b.*) Lateral inversion of reflected image.
   (*c.*) Distance of image behind mirror.
   (*d.*) Refraction by prism.
   (*e.*) „   „   convex lens and formation of image.
   (*f.*) „   „   concave „   „   „
   (*g.*) The human eye.
      The foregoing set is contained in a strong case,
      24 × 18 × 9 inches, with fittings for facilitating
      the employment of the apparatus.
      Dr. Marey, Professor of the College of France, Paris.

XI.—DIAGRAMS, MODELS, PREPARATIONS, AND OTHER APPLIANCES FOR INSTRUCTION IN BIOLOGY.

3808. Recent Marine Mollusca. Typical collection, comprising 300 genera; to illustrate Woodward's "Manual of the Mollusca," and other works. *R. Damon.*

3809. Land and Fresh Water Mollusca. Typical collection, comprising 155 genera; to illustrate Woodward's "Manual of the Mollusca," and other works. *R. Damon.*

3810. British Marine Shells, illustrative of Dr. J. Gwyn Jeffreys's "British Conchology." 336 species. *R. Damon.*

3811. British Land and Fresh Water Shells, illustrative of Dr. J. Gwyn Jeffreys's " British Conchology." 122 species.
R. Damon.

3812. Skeletons. Two specimens of Manis or Pangolin, exhibiting a method of articulating skeletons, by which all the surfaces of the bones can be examined. Invented by Mr. James Flower, articulator to the Royal College of Surgeons.
Royal College of Surgeons of England.

In the larger specimen the parts of the skeleton are in their natural rela-tion, but any portion which may be required for examination can be removed without disturbing the rest, as is shown in the smaller specimen, which is divided into various segments, the individual bones of which are united by flexible wires.

3815. Case containing twelve **Injected Preparations** by the celebrated Lieberkühn (born 1711, died 1756) ; each prepara-tion is separately mounted in a brass magnifier with an ebony handle. *The Royal College of Surgeons of England.*

The object, which is opaque, is viewed with light concentrated upon it by means of a concave polished metal condenser (known as a "Lieberkühn"), and magnified by a simple lens adapted to a perforation in the centre of the condenser.

3813. Carbon Photographs of microscopic objects for pro-jection by the lantern, photographed by F. J. von Kolkoff.
Prof. D. Huizinga, Physiological Laboratory of the Uni versity of Groningen.

3816. Two small Cases with Lieberkühn's Preparations in microscopes.
Royal Museum at Cassel (Director, Dr. Pinder.)

The two collections of *Lieberkühn* have been made in the earlier part of the 18th century. According to *Harting,* the Royal College of Surgeons of England possesses a similar one.

3816a. Microscopical Preparations. No. 1–289.
I. History of the development of the Cephalopoda.
II. Microscopical anatomy of the Tunicata.
III. Histology of the nervous system of the Vertebrata.
Dr. Michel Oussoff, Russia.

3816b. Microscopic Preparations.
Zoology.—Preparations of tissues in their normal, and in their morbid state.
Vegetative physiology.—Elementory organs, cells, vessels, com-posite organs, epidermis, stems, roots, organs of reproduction.
Medicinal substances.— Stems, roots, barks, seeds, salts used in pharmacy.

Alimentary products.—Flours, feculæ, prepared with the object
of defining commercial adulterations.

Textile products.—Prepared with the object of defining commercial frauds.   *Eugène Bourgogne, Paris.*

3816c. Large Sections, for the **Microscope,** of various
Anatomical Tissues, showing variety of methods of histological staining.   *Drs. Hoggan.*

3818. Models exhibiting the **Anatomical Symptoms** of
Cattle Disease.   *A. Th. Ferhaar, Utrecht.*

No. 1.—Three cows' eyes :
  A. A healthy eye, of which the sclerotic and the accessory eyelid
     are pale.
  B. A healthy eye, in which the same parts are of a brown colour :
     this is considered as a distinctive mark that the animal is not liable
     to catch the disease.
  C. The eye of a cow attacked by the cattle disease. The red colour
     of the above-named parts may be noticed.
No. 2.—The extremity of the lower jaw and the inner surface of the lower
  lip of a cow attacked by the cattle disease. In these an irregular
  swelling and the difference of colour, as well as the excoriated and
  lacerated places on the gum, are to be noticed. A part of the same
  jaw seen inside, having the same symptoms.
No. 3.—The cartilaginous septum nasi of a cow that has died of the cattle
  disease, seen on the right side. The red colour and violent swelling of
  the mucous membrane and the great quantity of thick mucus are particularly remarkable ; on the edge of the nostril an affection of the skin
  and dry crusts of mucus are to be noticed.
No. 4.—The right side of the larynx and a portion of the windpipe of a
  cow that has died of the cattle disease. The red colour and the swelling of the pyogenic membranes are to be seen.
No. 5.—A portion o the œsophagus (gullet) of a cow that died of the
  cattle disease. One part of the inner membrane is nearly healthy, the
  other very red and swollen.
No. 6.—The palate of a cow that has died of the cattle disease.
No. 7.—The end of the abomasum and the beginning of the small intestines
  of a cow that has died of the cattle disease. The violent swelling, the
  red colour, the small pyogenic cavities, and the black crusts of blood
  on the mucous membrane of the stomach are remarkable.
No. 8.—A piece of the small intestines of a cow that has died of the cattle
  disease, opened and seen inside. The red colour, and the swelling of
  the mucous membrane, in which Peyerian glands swollen and empty are
  seen.
No. 9.—A piece like No. 8. Several black spots may be noticed.
No. 10.—A piece like No. 8 and No. 9. Besides the swollen glands the
  difference of colour may be seen.
No. 11.—A piece like Nos. 8, 9, and No. 10. Here and there a difference
  of colour is apparent.
No. 12.—A piece like Nos. 8, 9, 10, and 11. The mucous membrane appears
  thinner and duller.
No. 13.—The end of the small intestines and the cæcum of a cow that
  has died of the cattle disease, opened and seen inside. The much-
  swollen folds, the redness of the mucous membrane, as well as the black

spots, especially at the point of transition from the small intestines to the large, are remarkable.

No. 14.—The end of the rectum of a cow that has died of the cattle disease, opened and seen inside. The folds of a black colour, and the considerable swelling and redness of the mucous membrane are remarkable.

No. 15.—A piece of the skin of a cow that has died of the cattle disease, taken from the inner surface of the leg. Grey crusts covering the diseased parts of the skin, and between these the fissures in the true skin, are to be noticed.

3819. Skull of Sheep (*Ovis aries*). The bones have been disarticulated and mounted in segments, according to their morphological relations, so that each segment may be removed for lecture purposes. *E. T. Newton, Museum of Practical Geology.*
The following is the order in which the bones are grouped:—

BRAIN CASE.—1st Segment. Basioccipital ankylosed, with two exoccipitals supraoccipital.

2nd Segment. Basisphenoid ankylosed with two alisphenoids; two ankylosed parietals.

3rd Segment. Presphenoid ankylosed, with two orbitosphenoids; two frontals.

NASAL REGION.—Mesethmoid ankylosed, with two superior and two middle turbinals and vomer; two free nasals.

AUDITORY REGION.—Right side. Squamosal, tympanic, and periotic mass. (The periotic bones being ankylosed.)
Left side. Ditto ditto.

FACIAL REGION.—1. The two premaxillæ.
2. Right side. Maxilla, jugal, lachrymal, and inferior turbinal.
3. Left side. Ditto ditto.
4. Right side. Palatine and pterygoid.
5. Left side. Ditto ditto.

FIRST VISCERAL ARCH.—Right side. Ramus of lower jaw and malleus. (The latter, for convenience, is mounted separately).
Left side. Ditto ditto.

SECOND VISCERAL ARCH.— The hyoid, consisting of basihyal, two ceratohyals, two epihyals, two stylohyals, and two tympanohyals. (One of the tympanohyals is left attached to the tympanic bone). The two thyrohyals belong to the third visceral arch, but are mounted here for convenience.

The incus and stapes of each side are mounted separately on account of their small size; the former belongs to the second visceral arch, and the latter is most probably a part of the auditory capsule.

3820a. Collection of **Entomological Preparations** for the **Microscope,** comprising dissections of insects and whole insects, mounted in Canada balsam. *Frederic Enock.*

Some of these preparations are mounted in a deep cell, and by being treated in a peculiar manner their natural form is preserved, and the *internal* structure of muscles and tracheæ seen in their natural positions.

3821. Objects exhibited by Professor Moser.

The geometrical and perspective drawings, I., II., and III., represent an apparatus for the determination of the products of the respiration and perspiration of animals. The apparatus is arranged at the I. R. Chemico-Agricultural Experimental Institution at Vienna, for experiments on horses, cattle, sheep, and pigs. The plan is that of Professor M. Pettenkofer, of Munich, excepting that for the aspiration of the air, which is the same as that adopted in a smaller apparatus by Professor C. Voit, of Munich. The parts are as follows:—

A, chamber for the reception of the animal during the experiment.

B, tube of tin, for the passage of the gaseous products from the chamber.

C, a vessel in which the gaseous products are saturated with watery vapour.

D, a gasometer in which the volume of the gases is determined.

E, tube through which the measured gases escape.

F, engine (hot air) which, by turning the revolving drum of the gasometer aspirates the gases through the parts A, D, and, as a consequence, draws fresh air in through the openings in the back of A'.

To ascertain the quantities of carbonic acid, aqueous vapour, and methan (marsh gas) evolved by the animal, the composition both of the gases passing from the chamber A through the tube B and of the fresh air is determined, with regard to their contents of these substances, and the difference shows the quantities of the gases evolved.

For these examinations the apparatus shown on the table G is employed. Aspirating and forcing pumps ($c, c, c,$) which close with mercury, draw small quantities either of air from B through the tubes $a, a,$...or of fresh air through the tubes $b, b,$...and force them through the absorption apparatus $d, d,$... and $d1, d1,$ or $d2, d2,$ to the gasometers $e, e, e,$...where they are measured.

The motion of the pumps is effected by the engine F by means of the mechanism $o, p, q, r, s.$ To every absorption apparatus there appertain two pumps which alternately aspirate and force, and four valves with mercury joints (f, f, f...).

The carbonic acid is absorbed by a test solution of baryta in the tubes $d, d,$ ...and $d1, d1$....

The aqueous vapour is determined in the flasks $d2, d2,$...which contain pumice stone saturated with sulphuric acid.

The carburetted hydrogen is oxidated by passing over ignited oxide of copper. For this purpose a portion of the air to be analysed is drawn by the pumps through the tubes $g, g,$ filled with oxide of copper, and placed upon a furnace H. The products of the oxidation, carbonic acid and water, are then conveyed to the absorption apparatus and gasometers.

The drawing IV. represents a similar apparatus, but of smaller dimensions, which also is employed at the I. R. Chemico-Agricultural Expermental Institution at Vienna. Its construction is in the main the same, excepting that the ventilation is effected by the exhausting fanner K. The motion of K and of the pumps ($c, c,$) is effected by the mechanism L, which is worked either by a man or by a small water wheel. (The other signs on this drawing IV. correspond with those of the drawings I., II., III.)

The smaller apparatus is used for experiments with poultry, rabbits, and other small animals.

3822. Photographs of Microscopic Objects. Enlarged from negatives by Dr. Maddox, for illustration by means of the lantern. *James How & Co.*

3823. Mounted Preparation to show a method of putting up dried membranous specimens for museums.

Prof. Struthers.

The glass jar is not essential. The wire loop displays and also protects the preparation, and the tube enables it to be turned round on the stand,

3 Q 2

or to be taken out and held up by the stem for demonstration. Curators of museums may remove the cover and examine the mounting. The specimen is a preparation of the human pyloric "valve."

3823a. Specimens of Linnæus's MS.
The Linnæan Society.

" Iter Dalekarlicum," a MS. journal of a tour made at the suggestion of Governor Reuterholm, in the year 1734, through the provinces of East and West Dalecarlia, by Linnæus himself, accompanied by several students from the University of Upsala, who were devoted to the study of natural history, each student undertaking to record daily his observations on the particular branch assigned to him.

The journal, which is chiefly in Swedish, and does not seem ever to have been published, is illustrated by pen and ink sketches, and accompanied by one engraved and two MS. maps of the district traversed.

3823b. Specimens from Linnæus's Collections.
The Linnæan Society.

a. Zoological.

1. *Mya margaritifera*, and portion of shell with artificial pearls.

In his letter to Haller, dated Upsal, Sept. 13, 1748, Linnæus writes, " At " length I have ascertained the manner in which pearls originate and grow " in shells,* and I am able to produce, in any mother-o'-pearl shell that " can be held in the hand, in the course of five or six years, a pearl *as large* " *as the seed* of common vetch."

2. *Testudo pusilla*, L., native of the Cape of Good Hope. Both specimens named in Linnæus's own hand.

3. *Scomber Chrysurus*, native of Carolina.

* " For this discovery the illustrious author was splendidly rewarded by " the States of the Kingdom."—*Haller.* Specimens of pearls so produced by art in the *Mya margaritifera* are in the Linnæan Cabinet.

" The shell appears to have been pierced by flexible wires, the ends of " which perhaps remain therein."—(Smith's Selection of the Correspondence of Linnæus, II., p. 428.)

In his Memoir of Linnæus, in Rees's Cyclopædia, Sir James Smith, after mentioning that his patent of nobility was confirmed by the Diet in 1762, goes on to state that that august body honoured him with a still more solid reward, upwards of 520*l.* sterling, for his discovery of the art of producing pearls in the river mussel by wounding the shells; but the practice does not seem to have been prosecuted to any great extent.

b. Botanical.

(All glued on paper, poisoned to protect them from insects, and named in Linnæus's own hand.)

1. *Linnæa borealis*, Gron. Named in honour of Linnæus, by his pupil Gronovius.

2. *Sibthorpia Europæa*, Linn.

3. *Browallia demissa*, Linn.

4. *Hebenstretia capitata*, Thunb.

5. *Erinus capensis*, Linn.
6. *Buchnera Asiatica*, Linn.
7. *Pedicularis tuberosa*, Linn.

3823c. Dried Specimens of Plants from the Society's British Herbarium. *The Linnæan Society.*

1. Ranunculus, 17 sheets.
2. Dentaria, 3 sheets.
3. Matthiola, 2 sheets.
4. Linum, 7 sheets.
5. Fragaria, 4 sheets.
6. Oxalis, 2 sheets.

3823d. Specimens of the Linnæan Society's Publications. *The Linnæan Society.*

1. TRANSACTIONS. 4to. (*Illustrated.*)
   Vol. 19, Part 2, containing papers by—
   Hope on rare and beautiful insects from Silhet.
   Blackwall on British Spiders.
   Quekett on Ergol.
   Forbes on the *Ophiurida* of the E. Mediterranean, &c.
   Vol. 22, Part 4, containing—
   Hooker on the Pitchers of *Nepenthes*, with account of new species from Borneo.
   Hooker on new *Balanophoreæ*.
   Bentham on *Henriquezia*.
   Seemann on *Camellia* and *Thea*.
   Cobbold on new *Entozoa*, &c.
   Vol. 24, Part 1.
   Hooker on *Welwitschia*.
   Vol. 26, Part 4.
   Williamson and *Carruthers* on Fossil *Cycadeæ*.
   Vol. 27, Part 1.
   Welwitsch. Sertum Angolense: s. Stirpium . . . in itinere per Angolam et Benguelam observatarum descriptio, *Iconibus illustrata*.
   Vol. 29, Part 1.
   Col. Grant. Botany of the Speke and Grant expedition from Zanzibar to Egypt.
   TRANSACTIONS. SECOND SERIES; in which the two sections, *Zoology* or *Botany*, are published separately.
   Zoology.
   Vol. 1, Part 1, containing—
   Willemoes Suhm, on Atlantic Crustacea from the "Challenger" Expedition.
   Parker on the Morphology of the skull in the Woodpecker, &c.
   Allman on *Stephanoscyphus mirabilis;* the type of a new order of Hydrozoa.
   Botany, Vol. 1, Part 1.
   Miers on *Napoleona*, *Omphalocarpum*, and *Asteranthos*.
   Miers on *Auxemmeæ*.
   TRANSACTIONS. Specimens of papers, from the later volumes, printed off for separate sale.
   Botanical.
   1. *Welwitsch* and *Currey.* Fungi Angolenses (from Vol. 26, Part 1).
   2. *Scott* Tree Ferns of British Sikkim (from Vol. 30, Part 1).

Zoological.
    3. *Trimen.* Mimetic Analogies among African Butterflies (from Vol. 26, Part 3).
    4. *Murie* on the three-banded Armadillo (Vol. 30, Part 1).
  2. JOURNAL, 8vo. The two sections, *Zoology* and *Botany,* published separately. The papers illustrated when necessary.
    Zoology, Nos. 60–62, containing—
      Allman, on new genera and species of *Hydroida.*
      Cobbold on the large Human Fluke (*Distoma crassum*), &c.
    Botany.
      No. 74. *Berkeley* and *Broome.* Fungi of Ceylon.
      No. 82. Contribution to the Botany of H.M.S. "Challenger."
    Supplement to Vol. 5.
      Anderson. Florula Adenensis; account of the flowering plants hitherto found at Aden.

3824. Entire Plant of *Welwitschia mirabilis,* Hook, fil. Discovered by Dr. Welwitsch, in South-west Africa, in 1859.
                                 Museum, Royal Gardens, Kew.
  This plant is the subject of a memoir by J. D. Hooker, M.D., P.R.S., in the Transactions of the Linnæan Society, Vol. XXIV.

3825. Series of Fruits illustrating the variety of form in the **Dalbergieæ,** a tribe of **Leguminosæ.**
                                 Museum, Royal Gardens, Kew.
  Described by George Bentham, Esq., F.R.S., &c., in Supplement to Vol. IV. of the Journal of the Linnæan Society.

3826. Fruits illustrating the **Rubiaceæ,** a very large order of tropical trees and shrubs.   *Museum, Royal Gardens, Kew.*
  The genera of these have been revised and published in Part I., Vol. II. of the " Genera Plantarum," by G. Bentham, F.R.S., and J. D. Hooker, M.D., P.R.S.

3827. Series of Fruits, showing the variation in form and size of the different species of **Eucalyptus,** a genus of shrubs or trees, commonly known as gum trees, almost exclusively confined to Australia.   *Museum, Royal Gardens, Kew.*
  The species have been described by George Bentham, Esq., F.R.S., in Vol. III. of his " Flora Australiensis."

3828. Fruits and Seeds of plants belonging to the natural order **Bignoniaceæ,** a family of twining or climbing plants or trees. Found chiefly in the tropics.   *Museum, Royal Gardens, Kew.*
  A revision of the genera of this order will appear in the forthcoming volume of the " Genera Plantarum," by G. Bentham, F.R.S., and J. D. Hooker, M.D., P.R.S.

3829. Collection of Fruits illustrating the **Australian Proteaceæ.**   *Museum, Royal Gardens, Kew.*
  The collection consists chiefly of species of *Banksia* and *Hakea,* both of which are limited to Australia, and have been described by George Bentham, Esq., in the fifth volume of the " Flora Australiensis."

3829a. Model of the **Physiology** of a **Grain** of **Wheat** (30 times magnified).   *Doctor Auzoux, Paris.*

3829b. Model of the **Physiology** of a **Small Ear** of **Wheat** (belonging to the Conservatoire des Arts et Métiers).
Doctor Auzoux, Paris.

3829c. Model of **Anatomy** of the **Silk Worm** (belonging to the Conservatoire des Arts et Métiers).
Doctor Auzoux, Paris.

3829d. Model of **Anatomy** of the **Egg** of the **Epyornis.**
Doctor Auzoux, Paris.

3830. Pathological Preparations. A complete set of 24, from the human subject, mounted for the microscope.
Arthur C. Cole & Son.

3831. Physiological Preparations. A complete set of 24, from man and the lower animals. *Arthur C. Cole & Son.*

3832. Educational Preparations. A complete set of 24, for medical students and schools. *Arthur C. Cole & Son.*

3833. Vermiform Appendix, Cæcum, and parts of Small Intestine and Colon of Rabbit (*Lepus cuniculus*), with the Peyerian and also some of the mesenteric glands in connexion with them injected.
The Anatomical Department, University Museum, Oxford.

The ileum and colon lie side by side, the former describing a siphon-shaped curve before ending in a dilatation, known as the " *sacculus rotundus,*" and homologous with the aggregation of Peyerian follicles situated in man just anterior to the ileo-cæcal valve. The *sacculus rotundus* occupies the centre of the preparation; above and a little to the left of it another somewhat similar aggregation of Peyerian follicles is seen just beyond the ileocolic valve. The colon curves backwards externally to and concentrically with the ileum. The cæcum consists of two portions : one larger in calibre, thinner in walls, and sacculated spirally; the other, the homologue of the human " appendix vermiformis," smaller in calibre, but with much thicker walls. Prepared by Mr. Robertson.

3834. Appendix Vermiformis of Rabbit (*Lepus cuniculus*), with a small portion of the cæcum and a larger portion of the small intestine left in apposition with it.
The Anatomical Department, University Museum, Oxford.

The lacteal system has been injected with Berlin blue by the insertion of the point of a syringe into the lacunar system. The injection is seen to have passed into some of the adjacent mesenteric glands. Prepared by Professor Rolleston.

3835. One Longitudinal and two Transverse Sections of the Appendix Vermiformis of Rabbit (*Lepus cuniculus*).
The Anatomical Department, University Museum, Oxford.

The blood vessels have been minutely injected with carmine and the lacteals with Berlin blue. Prepared by Mr. Robertson.

3836. Right Half of the Skull of a Sturgeon (*Acipenser sturio*), macerated, and mounted in the ordinary way.
Prepared by Mr. Robertson.
The Anatomical Department, University Museum, Oxford.

3837. Left Half of the Skull of a Sturgeon (*Acipenser sturio*), stained to show the cartilaginous cranium.

The Anatomical Department, University Museum, Oxford.

This preparation was placed for several hours in staining fluid, which has tinted the cartilaginous cranium whilst leaving the bony elements with their natural white colour. Prepared by Mr. Robertson.

3838. Skull of a Sturgeon (*Acipenser sturio*), stained to show the cartilages.

The Anatomical Department, University Museum, Oxford.

The opercular bones of both sides have been removed, also the shoulder girdle, parietal and frontal bones of the left side, to show the cartilaginous cranium. Prepared by Mr. Robertson.

3839. Set of Wax Models, to illustrate the **Development** of the **Trout** (*Trutta fario*), containing 21 preparations, including 7 sections. The preparations can be viewed by transmitted light. Magnified 30 diameters.

Dr. A. Ziegler, Freiburg, Baden.

This series is the latest work of the exhibitor, and the 21st series of his scientific wax preparations for illustrating biological facts, and particularly the changes during development. These wax models are copied from nature, and enlarged so as to be useful for teaching purposes. Compare the accompanying prospectus and account of physiological preparations in wax for the use of English readers.

A full account of the development of the trout's egg will be found in Zeitschrift f. wiss. Zool., Vol. 22, Part IV., Plates 32 and 33, and Vol. 23, Part I., plates, 1, 2, 3, and 4, in papers by Professor Oellacher of Innsbruck.

3839a. Collection of **Preparations** for the **Study** of the **Embryology** of the **Sterlet** (*Acipenser ruthenus*), from the segmentation of the egg to the one year old fish.

Ph. Owsiannikoff, University of St. Petersburg.

The first artificial fecundation of the eggs of this fish was made by M. Ph. Owsiannikoff, in the year 1869, at Simbirsk on the Wolga, and in the same year a small quantity was brought by him to St. Petersburg. The exhibited embryos presented have grown up in an aquarium in the study room of M. Owsiannikoff, who wishes to present them to the South Kensington Museum.

3839b. Preparations on the **Embryology** of the **Sterlet** (*Acipenser ruthenus*).

Ph. Owsiannikoff, University of St. Petersburg.

The time is fixed approximately, as at an early stage temperature and quantity of food have a great influence on the development. No. 1. First day, the hole of Rusconii forms the half of the egg. No. 2. Second day, morning, the hole of Rusconii is smaller. No. 3. Second day, evening, the hole of Rusconii is very small. No. 4. Fourth day, the hole of Rusconii has disappeared. No. 5. Fifth day. No. 6. Sixth day. No. 7. Seventh day. No. 8. The third and fifth days after their coming out from the egg. No. 9. About a week after their coming out. Nos. 10, 11, 12, 13. From two to four weeks. No. 14. About six weeks ; bastard of an Acipenser Güldenstaedtii (soft roe), and of an Acipenser ruthenus (hard roe). No. 15. About two months. No. 16. About a year ; reared in an aquarium. If there is plenty of food the fish can attain about double this size in the same period.

3933b. Collection of Preparations on the Embryology of the Sterlet ((*Acipenser ruthenus*).
Ph. Owsiannikoff, University of St. Petersburg.

Most of the preparations represent tranverse sections of the eggs or the embryos, hardened in chromic acid, and partly coloured with carmine.

1. Ripe but not fecundated egg.
2. Embryo, taken from the egg thirty-seven hours after fecundation.
3. Embryo, about forty hours after fecundation.
4. Transverse section of an embryo of thirty-seven hours.
5. Transverse section of an embryo of forty-eight hours.
6. Section of the brain and the eye-vesicle of an older embryo.
7. The three embryonal plates.
8. Section through the nerve-ganglion ; the rudimentary gills are to be seen.
9. This section represents the cells that form the medulla spinalis, chorda dorsalis, and other parts.
10. Section through the head of a sterlet just coming out from the egg-shell.
11. Sterlet about three days after coming out.
12. Sterlet about a week.
13. Some longitudinal sections of a one week's sterlet. We see the eyes, the heart, the intestinal canal, particular nervous organs of the surface of the head.
14. The segment.
15, 16. The spinal furrow and the first plates.
17. The closing of the spinal furrow ; the plates are not completely formed.
18. Longitudinal section through an embryo in the egg-shell. The interesting feature of this preparation is the closing of the intestinal canal.
19. Transverse section of an embryo, formation of muscles, construction of the spinal chord, wolffian body, &c.
20. Preparation of a coregonus, showing the central nervous system.
21. Section of an egg of petronzyon fluviatiles, showing the formation of the intestinal canal.

3933a. Collection of Preparations of the Nervous System of Molluscs.
Ph. Owsiannikoff, University of St. Petersburg.

1-3. Whole nerve-ganglion and two sections, from a thetis.
4. Aeolis.
5, 6. Gastropteron.
7. Bulla, with eyes.
8. Umbrella.
9. Pleurobranchus.
10. Pleurobranchus, section of a nerve-ganglion.
11. Aplysia.
12. Lobulus electricus (Rajae). The section is made of a brain preserved for three years in spirit, the cells and fibres are still seen very well.
13. Dentalium.

3933c. Eight Drawings illustrative of the minute structure and reproductive process of the Mushroom Tribe.
W. G. Smith.

These drawings show the reproductive process, the minute structure, and every individual cell of *Coprinus radiatus*, one of the smallest members of the mushroom tribe. The fungus, natural size, is shown growing upon fragments of straw or dung on drawings Nos. 1, 2, 7, and 8. The enlarged details are magnified from 25 to 3,000 diameters by the aid of the camera-lucida attached to the eye-piece of the microscope. Height of the full grown fungus $\frac{2}{3}$ of an inch. Number of cells entering into the composition of one fungus, 25,760,000. Weight of one fungus, $\frac{1}{150}$ of a grain.

3933d. Four Glass Cases, containing a series of **dissections of Insects** of various orders. *Professor Westwood.*

3933e. Glass Case, containing 39 microscopic slides and dissections of insects. *Professor Westwood.*

3840. Original Cast of the trunk of a boy, the thorax opened to show the natural position of the heart after removing the lungs.
Steger, Anatomical Institute, University of Leipzig.

3841. Cast, with the **Abdominal Cavity** opened to show the natural form and position of the stomach, liver, and kidneys, &c.
Steger, Anatomical Institute, University of Leipzig.

3842. Several Small Original Casts.
Steger, Anatomical Institute, University of Leipzig.

3843. Histological Photographs.
Steger, Anatomical Institute, University of Leipzig.

3844. Photographs to illustrate the **Changes** during **Development** of the chick.
Steger, Anatomical Institute, University of Leipzig.

3845. 25 Photographs on Glass, for the magic lantern, for instruction in physiology (anatomy, physiology, biology, and development). *Dr. S. Th. Stein, Frankfort on the Maine.*
These slides have been photographed from nature.

3846. Three Large Diagrams.
Dr. S. Th. Stein, Frankfort on the Maine.
These contain—
1. 35 physiological photographs. (The whole taken from nature.)
2. The application of photography to physiology and medicine.
3. Thermophotography.

3847. Diagrams, for botanical teaching, by L. Kny. 1–20. *Prof. Kny, Berlin.*
These diagrams are published by Wigand, Hempel, & Co., Berlin.

3849. Model of the Mechanism of the Accommodation of the Eye. *Prof. Beetz, Munich.*
This model is made by Tauber of Leipzig from a design of Dr. Beetz, and is explained by a drawing and description which accompany the apparatus.

3850. Three Wall Charts, with representations of **the Technical Geography of Plants.**

Prof. Jessen, Eldena, in Pomerania.

3851. Apparatus for representing the mechanical principle of the **Circulation** of **Sap** in **Cells.** Rough model.

Prof. Jessen, Eldena, in Pomerania.

3852. Plan of the **Institute of Vegetable Physiology of the University of Breslau,** with an account of its arrangements for instruction in the biology of plants.

Prof. Dr. F. Cohn, Institute of Vegetable Physiology, University of Breslau.

The Institute of Vegetable Physiology in the University of Breslau was founded in the year 1866, and was intended (*a*) for experiments for the elucidation of the biology of plants, also for preparing all kinds of demonstrations for the lectures of its director, Professor Ferdinand Cohn, and to found for this purpose collections of apparatus, models, normal and pathological plant forms, herbaria, &c.; (*b*) to make the students, by means of courses of instruction on histology, familiar with the use of the microscope for phytobiological researches, and to afford them, by means of a collection of microscopical preparations, rich material for observation ; (*c*) to secure for original researches on the biology of plants the necessary rooms, instruments, and other assistance, as well as direction, and to publish the same as far as possible in the scientific publication of the institute, "Beiträge zur Biologie der Pflanzen ; Breslau, Verlag von Max Müller."

3853. Six Models, prepared by A. Lohmeyer, and presented to the collection of botanical models in the Phytophysiological Institute.

Vaucheria sessilis, development of the Sexual Organs.

Vaucheria sessilis, development of the Zoospores.

Eurotium aspergillus, development.

Models of Antherozoids.

Models of Zoospores.

Cypripedium Calceolus, model of the flower.

Prof. Dr. F. Cohn, Institute of Vegetable Physiology, University of Breslau.

M. A. Lohmeyer prepared, during the years 1862-67, by the suggestion and under the direction of Professor F. Cohn, an extensive collection of botanical models comprising more than 350 specimens, the first and most complete of its kind which illustrates the structure of flowers and fruits of the phanerogamic families of plants, with especial attention to officinal plants and the development of the cryptogams. Out of this sprang the collection of botanical models made for sale by Brendel, and exhibited in his name.

3860. Glazed Frame, with movable back for demonstrating biological preparations and drawings. The development of rusts. (Æcidium Berberidis, Puccinia graminis, Uredo linearis.)

Dr. F. Cohn, Institute of Vegetable Physiology, University of Breslau.

For lecture-demonstrations of herbarium specimens of morphological preparations, and collections of microscopical drawings and natural objects which
elucidate the diseases, &c. of plants, the glass frames with movable back, which
are in common use at the Institute of Vegetable Physiology, are exceedingly
well adapted.

3861. 12 Botanical Models of Cultivated Plants.
Robert Brendel, Berlin.

3862. 5 Botanical Models of Fruit Trees.
Robert Brendel, Berlin.

3863. 5 Botanical Models of Forest Trees.
Robert Brendel, Berlin.

3864. 45 Models of Flowers.   *Robert Brendel, Berlin.*

These 67 botanical models were constructed to illustrate the structure of
flowers, of cultivated and wild plants, and the natural orders. They include—

1. Cultivated mono- and dicotyledonous plants, *e.g.*, families of the Cruci-
   feræ, Lineæ, Leguminosæ, and Gramineæ.
2. Fruit trees of the families Ampelidæ, Pomaceæ, Rosaceæ, Ribesiaceæ,
   and Amygdaleæ.
3. Forest trees, from the families of the Coniferæ (Abietineæ Taxineæ),
   Salicineæ, and Cupuliferæ.
4. Models of the inflorescence of the principal German orders, including
   Ferns and Equisetaceæ.

The botanical models (which are durably made of gutta-percha, papier
mâché, &c., held together with cane, wires, &c., and painted true to nature in
oil colours) are intended to illustrate lectures on the artificial and natural
orders, regardless of the seasons, and in systematic order. As the models are
made on a more or less enlarged scale, they facilitate the recognition of all
the fine and even smallest organs, and the comprehension of the distinguishing
characteristics of the structure of flowers by comparison with living plants ;
besides, many can be taken to pieces, and are represented in sections. The
models, as a means of botanical teaching, were first prepared at the suggestion
and under the guidance and direction of Professor F. Cohn, by M. Lohmeyer,
who died in 1872, and presented his collection of over 300 models to the
Institute of Vegetable Physiology in the University of Breslau. The models
supplied by the exhibitor are now prepared directly from nature.

3864a. Six Photographs (in frame with glass)**, for demon-
strating the consequences of exterior injuries to trees,**
illustrative of the publication and atlas, " Über die Folgen äuperer
Verletzungen der Bäume."
*Royal Botanical Garden and Museum of the University of
Breslau, Prof. Dr. H. R. Göppert, Director.*

3864b. Seven Photographs (in frame with glass)**, for de-
monstrating the effect of frost on trees.**
*Royal Botanical Garden and Museum of the University of
Breslau, Prof. Dr. H. R. Göppert, Director.*

3864c. Four Photographs (in frame with glass)**, for de-
monstrating the deformed growth of trees.**
*Royal Botanical Garden and Museum of the University of
Breslau, Prof. Dr. H. R. Göppert, Director.*

3864d. Three Photographs for demonstrating the formation of the veins and branches of trees, according to the publication " Über Maserbildung."
Royal Botanical Garden and Museum of the University of Breslau, Prof. Dr. H. R. Göppert, Director.

3864e. Three Photographs and Sections of Wood (in frame with glass). Originals, illustrative of the publication, " Über Zeichen und Schriften der Bäume."
Royal Botanical Garden and Museum of the University of Breslau, Prof. Dr. H. R. Göppert, Director.

3864f. Scene from the Primeval Forests of the Bohemian Mountains.
Royal Botanical Garden and Museum of the University of Breslau, Prof. Dr. H. R. Göppert, Director.

3864g. Photograph of the coal profile erected in the Botanical Garden.
Royal Botanical Garden and Museum of the University of Breslau, Prof. Dr. H. R. Göppert, Director.

3864h. Section of Wood for demonstrating the process of grafting and improved growth, according to the publication, " Über das Veredeln der Obstbäume."
Royal Botanical Garden and Museum of the University of Breslau, Prof. Dr. H. R. Göppert, Director.

3864i. Zinc Model, representing the **Rafflesia Arnoldi.**
Royal Botanical Garden and Museum of the University of Breslau, Prof. Dr. H. R. Göppert, Director.

3864j. Atlas, with Photographs and Inscription, Austellung des Botanischen Gartens."
Royal Botanical Garden and Museum of the University of Breslau, Prof. Dr. H. R. Göppert, Director.

3864k. Göppert on Medicinal Plants, Botanical Museums, Formation of Veins, Inscriptions in Trees, Effects of Internal Injury, Process of Grafting.
Royal Botanical Garden and Museum of the University of Breslau, Prof. Dr. H. R. Göppert, Director.

3865. Six Models of Radiolaria, in papier mâché.
V. Fric, Prague.

3866. Twelve Types of Foraminifera, recent and fossil, representing the system of Carpenter, with a drawing of each on a large scale. *V. Fric, Prague.*

3867. Model of a Human Leg, with sections at intervals ; for use in giving anatomical and surgical instruction.
Dr. Pansch, Kiel.

This model has been moulded by the exhibitor accurately from nature, by 13 sections in successive planes through a firmly frozen leg. The advantages of such models for teaching, in lieu of the imperfect representations which have hitherto been used, are self-evident.

Models of other parts of the body similar to this have already been taken in hand by the exhibitor.

3868. Section of Head.

Rammé and Sodtmann, Hamburg.

By these anatomical models in papier mâché, the exhibitors intend to give a general insight into the structure of the human body to scholars and intelligent people. The aim of the exhibitors has been to combine accuracy with the greatest possible cheapness and durability.

Some of the models are taken from drawings by Dr. Simon, and some from those by Dr. Dehn.

3869. Head, with Muscles.

Rammé and Sodtmann, Hamburg.

3870. Model of the Ear enlarged, without muscles.

Rammé and Sodtmann, Hamburg.

3871. Eye enlarged. Perpendicular section.

Rammé and Sodtmann, Hamburg.

3872. Larynx, &c., natural size.

Rammé and Sodtmann, Hamburg.

3873. Lung. Perpendicular section.

Rammé and Sodtmann, Hamburg.

3874. Larynx and Lungs, in separable parts.

Rammé and Sodtmann, Hamburg.

3875. Model of Thorax, in separable parts.

Rammé and Sodtmann, Hamburg.

3876. Digestive Apparatus.

Rammé and Sodtmann, Hamburg.

3877. Perpendicular Section of Skin.

Rammé and Sodtmann, Hamburg.

3878. Biological Preparations for teaching zoology :—

1. Sphegidæ. Family of the carnivorous wasps, showing the rapacious manner of life and dwelling, with their victims.

2. Megachile betulina, a bee which cuts off a portion of birch leaf, showing its dwelling, with the piece of the leaf bitten off.

3. Bombus lapidarius, the mason bee, with construction of comb.

4. Myrmecoleon formicarius, the antlion, its funnel shaped-pit, larvæ, cocoons, &c.

5. Locusta viridissima, deposition of eggs and development.

6. Insects in amber, and their origin.

Prof. Dr. H. Landois, Münster, Westphalia.

The preparations which have been sent are intended as specimens for zoological instruction in the higher class educational establishments. They can be continually renewed by any teacher, and are not liable to destruction, and will raise the standard of the objects concerned in teaching. They attracted the notice of the judges in the Vienna Exhibition, and gained a medal as a mark of merit.

These and the following preparations were constructed for the exhibitor by Rudolph Roch, of Münster, manufacturer of anatomical preparations.

3878a. Two Preparations. Heads of children.

Ph. Owsiannikoff, University of St. Petersburg.

These preparations are taken from the Anatomical Museum of the Academy of Science at St. Petersburg; they were bought by the Emperor Peter the Great, in the year 1717, from the eminent anatomist of Holland, Ruysch (1638–1731), with his whole excellent collection.

The lessons of anatomy, the nervous system especially, must be demonstrated by preparations. Most of these preparations in spirit cannot be long preserved. The preparations of Ruysch, though very old, are extremely well preserved. The collection of preparations by Ruysch at the Anatomical Museum of the Academy of Science at Petersburg is unique in Europe.

3879. Horizontal Section of the Eye-ball through its meridian in a child six months old.

Anatomical Institute, Erlangen (Prof. Dr. B. Gerlach).

3880. Horizontal Section of the Tympanic Cavity and the cochlea of a child 10 months old.

Anatomical Institute, Erlangen (Prof. Dr. B. Gerlach).

3881. Frontal Section of the Head of a Child six months old through the cavities of the pharynx and mouth.

Anatomical Institute, Erlangen (Prof. Dr. B. Gerlach).

3882. Median Sagittal Section through the Head of a Child four months old.

Anatomical Institute, Erlangen (Prof. Dr. B. Gerlach).

3883. Lateral Sagittal Section through the Head of a Child four months old.

Anatomical Institute, Erlangen (Prof. Dr. B. Gerlach).

3884. Horizontal Section through the Neck of a Child six months old.

Anatomical Institute, Erlangen (Prof. Dr. B. Gerlach).

3885. Horizontal Section through the Neck of the same Child somewhat deeper.

Anatomical Institute, Erlangen (Prof. Dr. B. Gerlach).

3886. Three Longitudinal Sections through the Heart of a Child 32 days old.

Anatomical Institute, Erlangen (Prof. Dr. B. Gerlach).

3887. Two Sections through the Larynx of a Child six months old.

Anatomical Institute, Erlangen (Prof. Dr. B. Gerlach).

3888. Two Sections, one longitudinal, one transverse, **through the Testicle of an Adult.**

Anatomical Institute, Erlangen (Prof. Dr. B. Gerlach).

3889. Median Longitudinal Section through the Uterus of a virgin 20 years old.

Anatomical Institute, Erlangen (Prof. Dr. B. Gerlach).

3890. Cross Section through the Arm of a Child six months old, in the upper third.

Anatomical Institute, Erlangen (Prof. Dr. B. Gerlach).

3891. Cross Section of the Fore-arm of a Child six months old, in the upper third.

Anatomical Institute, Erlangen (Prof. Dr. B. Gerlach).

3892. Cross Section of the Fore-arm of a Child six months old, in the lower third (a glycerine preparation).

Anatomical Institute, Erlangen (Prof. Dr. B. Gerlach).

All these sections were prepared by the exhibitor for the purpose of demonstrating to a large audience, by means of the magic lantern (sciopticon), certain anatomical features, especially with regard to their topographical relations.

3893. Models of the Blood Corpuscles, for illustrating their form and size.   (Magnified 5,000 times.)

Prof. Dr. H. Welcker, Halle an der Saale.

3894. Model of the Bones of the Arm of Man, for demonstrating supination and pronation.   Also an osteological preparation to which the bone of the fore-arm is attached by a steel wire.   *Prof. Dr. H. Welcker, Halle an der Saale.*

3895. Orchitis interstitialis.

Pietro Toninetti Pathological Institute, Berlin (Director, Prof. Dr. Virchow).

By means of this fluid, discovered by the exhibitor, the objects were kept perfectly fresh, and when prepared in the dry way by the exhibitor retained their fresh appearance.

3896. Normal Heart.

Pietro Toninetti Pathological Institute, Berlin (Director, Prof. Dr. Virchow).

3897. Lipoma subcutaneum.

Pietro Toninetti Pathological Institute, Berlin (Director, Prof. Dr. Virchow).

3898. Glioma cerebri.

Pietro Toninetti Pathological Institute, Berlin (Director, Prof. Dr. Virchow).

3899. Slaty discolouration of the mucous membrane of the large intestine.

Pietro Toninetti Pathological Institute, Berlin (Director, Prof. Dr. Virchow).

3900. Two Kidneys.

Pietro Toninetti Pathological Institute, Berlin (Director, Prof. Dr. Virchow).

3901. Two Normal Brains.

Pietro Toninetti Pathological Institute, Berlin (Director, Prof. Dr. Virchow).

3902. Bullock's Heart and Lungs.

Pietro Toninetti Pathological Institute, Berlin (Director, Prof. Dr. Virchow).

3903. Histological Preparations for Teaching.

C. Rodig, Hamburg.

I. Collection of 150 drug preparations, in accordance with the pharmaceutical atlas of Professor Berg, adapted for pharmaceutical instruction.
II. Sixty anatomical plant preparations for teaching botany.
III. Algæ, fungi, and mosses.
IV. Cereals in sections for agricultural teaching.

3904. Sketch of an apparatus for investigating the **Influence of Temperature** on the life of **Plants** and **Animals.**

Dr. W. Velten, Physiologist, I. R. Station for Experiments relating to Forests, Vienna.

The apparatus consists of a box of zinc, with double partitions, the upper and sides perpendicular thereto being replaced by parallel glass plates. The space between the partitions must be filled with fluids, while that in the centre is destined for the objects to be experimented on. The whole is surrounded by a wooden cover suited to receive a refrigerator. The box is heated from below, and the temperature retained constant by means of a thermo-regulator brought into connexion with the apparatus. At the sides are openings into which the hands should be placed, when encased in india-rubber gloves only, in order to work without great change of temperature in the apparatus itself. By means of the same apparatus the influence of various coloured light can be determined as well as that of gas, &c. at different degrees of temperature.

3904a. Skeleton of a Dog, disarticulated and mounted in such a manner that every bone can be separately removed for examination. *Prof. Huxley, F.R.S.*

3904b. Typical Parts of the Skeletons of a Cat, a Duck, and a Codfish, disarticulated and mounted, to show the chief modifications of the vertebrate endoskeleton.

Prof. Huxley, F.R.S.

3904c. Series of Fifteen Dissections, illustrating the anatomy of the edible **Frog** (Rana esculenta).

Prof. Huxley, F.R.S.

3904d. Series of Sixteen Diagrams, illustrating the anatomy of the **Frog.**   *Prof. Huxley, F.R.S.*

3904e. Exoskeleton of the Common Lobster, disarticulated and mounted.   *Prof. Huxley, F.R.S.*

3905. Botanical Class Diagrams, used by the Professor of Botany, Royal College of Science. Diagrams of Closterium and Euastrum; also of Cycas circinalis, illustrating drawings used in the botanical class in the Royal College of Science.

W. R. M'Nab, M.D.

3906. Models of Monocotyledonous Embryos (8), prepared by Dr. Ziegler, of Freiburg, in Breisgau, part of a set of wax models used in the botanical class in the Royal College of Science.   *W. R. M'Nab, M.D.*

3907. Models of the flowers of Monocotyledonous and Dicotyledonous plants; part of a set of models prepared by Robert Brendel, in Breslau, used in the botanical class in the Royal College of Science.   *W. R. M'Nab, M.D.*

3908. Diagram of the Myxastrum Radians, a non-nucleated Protozoon.   *Prof. A. Leith Adams, F.R.S.*

3909. Diagrams of the Eucecryphalus Schulzei and Heliosphera Inermis.   *Prof. A. Leith Adams, F.R.S.*

The above are two forms of Radiolaria, showing their siliceous basket-like skeletons, and the cellform bodies of a bright yellow colour, containing starch. The nuclei and pseudopodia with granules are also represented.

3910. Case of **Specimens,** illustrating the **Domination of one Plant over another** in the mixed herbage of grass land, under the influence of different manures, each applied year after year on the same plot.   *John Bennet Lawes.*

The experiments were made in Mr. Lawes' park, at Rothamsted, near St. Albans, commencing in 1856, at which time the character of the herbage was apparently pretty uniform over all the plots, and there were 50 species or more growing together. There are about 20 experimental plots, from a quarter to half an acre each; two being left continuously without manure, and each of the others receiving its own special manure year after year. Under this varied treatment, changes in the *flora*, so to speak, became apparent even in the first years of the experiments; and three times since heir commencement, at intervals of five years, namely, in 1862, 1867, and 1872, a carefully averaged sample of the produce of each plot has been taken, and submitted to careful botanical separation, and the per-centage by weight of each species in the mixed herbage determined. Partial separations have also been made in other years. The specimens exhibited in the case show the botanical composition of the herbage on 12 selected plots, in the 17th season of the experiments, 1872 ; and the quantities represent the relative proportion by weight in which each species was found in the mixed produce of the plots.

The mean produce of hay per acre per annum has ranged on the different plots from about 23 cwts. without manure, to about 64 cwts. on the plot the most heavily manured.

The number of species found has generally been about 50 on the unmanured plots, and has been reduced to an average of only 20, and has sometimes been less, on the most heavily manured plots.

Species belonging to the order *Graminaceæ* have, on the average, contributed about 67 per cent. of the weight of the mixed herbage grown without manure, about 65 per cent. of that grown by purely mineral manures (that is, without nitrogen), and about 94 per cent. of that grown by the same mineral manures with a large quantity of ammonia-salts in addition.

Species of the order *Leguminosæ* have, on the average, contributed about 8 per cent. of the produce without manure, about 20 per cent. of that with purely mineral manures, and less than 0·01 per cent of that with the mixture of the mineral manures and a large quantity of ammonia-salts.

Species belonging to various other orders have, on the average, contributed about 25 per cent. of the produce without manure, about 15 per cent. of that with purely mineral manures, and only about 6 per cent. of that with the mixture of the mineral manures and a large amount of ammonia-salts.

Not only the amount of produce, but the number and description of species developed, have varied very greatly between the extremes here quoted, according to the particular character or combination of manure employed, as is strikingly illustrated by the arrangement of the specimens in the case.

3910a. Photographs under glass, for demonstrating the consequences of exterior damages to trees, the influence of cold on trees, and modes of protection of trees.

Royal Botanical Garden and Museum of the University at Breslau (Prof. Dr. H. R. Göppert).

3911. E. J. Spitta's working Model of the larynx.

T. Hawksley.

3912. 106 Original Water-Colour Drawings, by Wolf, illustrating the new and rare animals exhibited in the Society's Gardens. *The Zoological Society of London.*

(Uprights.)

1. Orang-Outang	-	-	- *Simia satyrus.*
2. Chimpanzee	-	-	- *Troglodytes niger.*
3. Hoolock Gibbon	-	-	- *Hylobates hoolock.*

(Longs.)

4. Ashy-black Ape	-	-	- *Macacus ocreatus.*
5. Black-fronted Lemur	-	-	- *Lemur nigrifrons.*
6. Lioness and Young	-	-	- *Felis leo.*
7. Painted Ocelot	-	-	- *Felis picta.*
8. Ocelot	-	-	- *Felis pardalis.*
9. Ocelot	-	-	- *Felis pardalis.*
10. Egyptian Cat	-	-	- *Felis chaus.*

11. Leopards	-	-	-	-	*Felis leopardis.*
12. Viverrine Cat	-	-	-	-	*Felis viverrina.*
13. Canadian Lynx	-	-	-	-	*Felis canadensis.*
14. Serval	-	-	-	-	*Felis serval.*
15. Caracal	-	-	-	-	*Felis caracal.*
16. Caracal	-	-	-	-	*Felis caracal.*
17. Eyra	-	-	-	-	*Felis eyra.*
18. Cheetah	-	-	-	-	*Felis jubata.*
19. Indian Civet	-	-	-	-	*Viverricula indica.*
20. Ratels	-	-	-	-	*Mellivora capensis & M. indica.*
21. American Skunk	-	-	-	-	*Mephitis mephitica.*
22. Fennec Fox -	-	-	-	-	*Canis cerdo.*
23. Azara's Fox -	-	-	-	-	*Canis axaræ.*
24. Bassaris	-	-	-	-	*Bassaris astuta.*
25. African Elephants	-	-	-	-	*Elephas africanus.*
26. Walrus	-	-	-	-	*Trichechus rosmarus.*
27. Southern Sea-lion	-	-	-	-	*Otaria jubata.*
28. African Rhinoceros -	-	-	-	-	*Rhinoceros bicornis.*
29. Andaman Pig	-	-	-	-	*Sus andamanensis.*
30. Red Potamochere	-	-	-	-	*Potamochærus africanus.*
31. Bosch-Vark -	-	-	-	-	*Potamochærus penicillatus.*
32. Collared Peccary	-	-	-	-	*Dicotyles torquatus.*
33. Hippopotamus	-	-	-	-	*Hippopotamus amphibius.*
34. Hippopotamus	-	-	-	-	*Hippopotamus amphibius.*
35. Panjaub Sheep	-	-	-	-	*Ovis cycloceros.*
36. Thar Goat -	-	-	-	-	*Capra iemlaica.*
37. Persian Gazelle	-	-	-	-	*Gazella subgutturosa.*
38. Eland (young)	-	-	-	-	*Oreas canna.*
39. Chinese Tailed Deer	-	-	-	-	*Cervus davidianus.*
40. Mantchurian Deer -	-	-	-	-	*Cervus mantchuricus.*
41. Formosan Deer	-	-	-	-	*Cervus pseudaxis.*
42. White-tailed Deer -	-	-	-	-	*Cervus leucurus.*
43. Japanese Deer	-	-	-	-	*Cervus sika.*
44. Rusa Deer -	-	-	-	-	*Cervus rusa.*
45. Swinhoe's Deer	-	-	-	-	*Cervus swinhoii.*
46. Persian Deer	-	-	-	-	*Cervus maral.*
47. Pudu Deer -	-	-	-	-	*Cervus pudu.*
48. Large-eared Brocket	-	-	-	-	*Cervus auritus.*
49. Alpacas	-	-	-	-	*Auchenia pacos.*
50. Great Ant-eater	-	-	-	-	*Myrmecophaga jubata.*
51. Great Ant-eater	-	-	-	-	*Myrmecophaga jubata.*
52. Two-toed Sloth	-	-	-	-	*Cholopus didactylus.*
53. Thylasine	-	-	-	-	*Thylacinus cynocephalus.*
54. Broad-fronted Wombat	-	-	-	-	*Phascolomys latifrons.*
55. Common Wombat	-	-	-	-	*Phascolomys wombat.*
56. Mantell's Apteryx	-	-	-	-	*Apteryx mantelli.*
57. Mooruk	-	-	-	-	*Casuarius bennetti.*
58. Rhea (young)	-	-	-	-	*Rhea americana.*
59. Painted Spur-fowl	-	-	-	-	*Galloperdix lunualso.*
60. Caspian Snow Partridge	-	-	-	-	*Tetraogallus caspius.*
61. Sœmmerring's Pheasant	-	-	-	-	*Phasianus sœmmerringi.*
62. Ring-necked Pheasant	-	-	-	-	*Phasianus torquatus.*
63. Japanese Pheasant -	-	-	-	-	*Phasianus versicolor.*
64. Horned Tragopan	-	-	-	-	*Ceriornis satyra.*
65. Rufous-tailed Pheasant	-	-	-	-	*Euplocamus erythrophthalmus.*
66. Vieillot's Fire-back -	-	-	-	-	*Euplocamus vielloti.*

67. Swinhoe's Fire-back - - *Euplocamus swinhoii.*
68. Horsfield's Kaleege - - - *Euplocamus horsfieldi.*
69. Siamese Pheasant - - - *Euplocamus prœlatus.*
70. Lineated Pheasant - - - *Euplocamus lineatus.*
71. Weka Rails - - - - *Ocydromus australis.*
72. Brush Turkey - - - *Talegalla lathami.*
73. Black-necked Swan - - - *Cygnus nigricollis.*
74. White-winged Casarca - - *Casarca leucoptera.*
75. Upland Goose - - - *Chloëphaga magellanica.*
76. Ashy-headed Goose - - - *Chloëphaga poliocephala.*
77. Indian Tantalus - - - *Tantalus leucocephalus.*
78. African Tantalus - - - *Tantalus ibis.*
79. Shoebill - - - - *Balœniceps rex.*
80. Egrets - - - - *Ardea candidissima & A. garzetta.*
81. Mantchurian Crane - - - *Grus montignesia.*
82. Kagu - - - - - *Rhinochetus jubatus.*
83. Angola Vulture - - - *Gypohierax angolensis.*
84. Bataleur Eagle - - - *Helotarsus ecaudatus.*
85. Schlegel's Clotho - - - *Clotho rhinoceros.*
86. Clotho - - - - *Clotho nasicornis.*

(Uprights.)

87. Stanger's Monkey - - - *Cercopithecus stangeri.*
88. Ocelot - - - - *Felis pardalis.*
89. Ocelot - - - - *Felis pardalis.*
90. Clouded Tiger - - - *Felis macrocelis.*
91. Norwegian Lynx - - - *Felis lynx.*
92. Binturong - - - - *Arctictis binturong.*
93. Syrian Bear - - - - *Ursus syriacus.*
94. Aoudad - - - - *Ovis tragelaphus.*
95. Markhoor - - - - *Capra megaceros.*
96. Bar-tailed Pheasant - - - *Phasianus reevesi.*
97. Australian Mycteria - - *Mycteria australis.*
98. Saddle-billed Stork - - *Mycteria senegalensis.*
99. Rhinoceros Hornbill - - *Buceros rhinoceros.*
100. Concave-casqued Hornbill - - *Buceros bicornis.*
101. Iceland Falcon - - - *Falco islandicus.*
102. Saker Falcon - - - *Falco sacer.*
103. Lanner Falcon - - - *Falco lanarius.*
104. Greenland Falcon - - - *Falco grœnlandicus.*
105. Spotted Eagle - - - *Aquila nœvia.*
106. Green Boa - - - - *Xiphosoma caninum.*

3913. Skeleton of a Man, mounted (after Beauchène).
Tramond, Paris.

3914. Skeleton of a Man, for practising dislocations.
Tramond, Paris.

3915. Human Head, with worn jaw (after Beauchène).
Tramond, Paris.

3916. Child's Head, showing first and second dentitions.
Tramond, Paris.

3917. Head of Python (after Beauchène). *Tramond, Paris.*

3918. Head of Tortoise (after Beauchêne).

Tramond, Paris.

3919. Skeleton of Young Chimpanzee. *Tramond, Paris.*

3920. Skeleton of Gibbon. *Tramond, Paris.*

3921. Skeleton of Crane. *Tramond, Paris.*

3922. Skeleton of Adult Beaver. *Tramond, Paris.*

3923. Model, showing **Muscles** and **Nerves.**

Tramond, Paris.

3924. Entire Head, showing **Nerves, Vessels, Sinus, &c.** *Tramond, Paris.*

3925. Anatomy of the Eye, in 13 pieces. *Tramond, Paris.*

3926. Two Fœtuses. *Tramond, Paris.*

3927. Complete Circulation in the Fœtus.

Tramond, Paris.

3928. Skeleton of Rattlesnake. *Tramond, Paris.*

3929. Lymphatic Vessels, natural size (vulva).

Tramond, Paris.

3930. Lymphatic Vessels, natural size (tongue).

Tramond, Paris.

3931. Lymphatic Vessels, natural size (leg and foot of a child). *Tramond, Paris.*

3932. Lymphatic Vessels, natural size (portion of the large intestine). *Tramond, Paris.*

3933. Lymphatic Vessels, natural size (half the head of a fœtus). *Tramond, Paris*

XII.—MISCELLANEOUS APPARATUS.

3934. Heliopictor, an automatic photographic apparatus for producing zoological and physiological photographs, with all necessary apparatus, in a box.

Dr. S. Th. Stein, Frankfort on the Maine.

See Vogel's "Photögr Mittheilungen," 1873.

3934a. Polarising Apparatus of **Hoppe-Seyler.**

F. Schmidt and Haensch, Berlin.

This apparatus is especially suited to the wants of medical men, for estimating sugar or albumen in urine. It is distinguished from the apparatus of Soleil, in giving the quantity of sugar in per-centages and tenths, by direct inspection of scale.

3935. Germinating Apparatus, for the simultaneous germination of a large number of seeds, or for the cultivation of microscopical organisms at a constant degree of temperature and moisture. (Exhibitor's construction.)

Prof. Dr. F. Cohn, Institute of Vegetable Physiology, University of Breslau.

This apparatus satisfactorily replaces the germinating apparatus of Nobbe. The seeds are soaked for 24 hours in water, then 200 are placed in each earthenware dish and covered over. The enclosed tin dishes are then filled with water, which keeps in the porous clay dishes the moisture requisite for the development of the seeds. The space between the double walls of the chamber is now three-quarters filled with water (the height of the water can be seen by the glass gauge), and a small gas flame, governed by a Bunsen's regulator and placed underneath the germinating chamber, keeps the temperature very constant. This apparatus is used for the examination of the germinating power of agricultural seeds, which takes place at the seed control station (Samen-Controll Station) connected with the Institute of Vegetable Physiology. Similar apparatus of different sizes are used in this station continually for the culture of plants, especially microscopical growths, at a constant temperature. They were employed especially by Prof. Cohn in his researches on Bacteria.

3935a. Cultivation Apparatus, with admission of air.

The Imperial and Royal Pathological and Anatomical Institute of the University of Prague (Director, Prof. Edwin Klebs).

The circular vessel is filled to half of its height with a mixture of water, glycerine, and sulphuric acid (or sulphate of copper), and the glass globe dipped into the cohesive fluid. The little round vessel contains the substance for cultivation, mostly gelatine. The open neck of the globe is wrapped round with wadding, and on this is placed the small, broad-brimmed, thick-shelled globe. The air in the globe, as well as the substrata of cultivation, are purified by the infusion of a powerful jet of steam. The introduction of germs, as well as the removal of samples, is effected under a "spray" of permanganate of potash.

Among 70 preparations by this method, only a very few showed a spontaneous development of fungi, whereas "Hyphomycetes" and "Schistomycetes" thrive excellently in them after implantation of their germs.

3935b. Apparatus for the cultivation of fungi.

The Imperial and Royal Pathological and Anatomical Institute of the University of Prague (Director, Prof. Edwin Klebs).

Two glass chambers for microscopical observation :—
a. Filled nearly full with isinglass gelatine. Implantation of "Microsporon septicum" of the year 1872. Ring-shaped progressive development of the "Schistomycetes" round a capillary tube, which contained the germs.

b. Filled half to its height with isinglass gelatine. The vegetation of the "Microsporon sept." ceased early, and the gelatine has to some extent become liquid.

"Hyphomycetes" did not make any appearance, although they thrive in such chambers. Without importation of germs, these chambers, with their contents, will remain free from any formation of fungi.

3936. Apparatus for demonstrating Knight's Experiment on the influence of gravity on the direction of the growth of roots and stems of budding plants. (Exhibitor's construction.)

Prof. Dr. F. Cohn, Institute of Vegetable Physiology, University of Breslau.

Some seeds (by preference Pisum sativum and Zea maize) are soaked for 24 hours in water, and then attached by long needles, which must not pass through the radicle or the plumula, but only through the cotyledons or the endosperm, radially to the circumference of a disc of cork. The apparatus is set in action by connecting the caoutchouc tube of the cover with the water supply or a water reservoir. By regulating the strength of the stream of water the rate of the waterwheel can be increased or diminished. The splashing of the water furnishes sufficient moisture for the germination of the seeds, from which all the roots are developed centrifugally, whilst the stems grow in a centripetal direction. An apparatus similar to this was employed by Prof. Ciesieboki in the Institute of Vegetable Physiology for his researches on the bending downwards of the root (see Cohn's Beiträge zur Biologie der Pflanzen, Vol. I., part 2). The observations can be made through the glass window, which is, however, generally closed by a shutter in order to keep out the light.

3937. Apparatus for Observing the Velocity of Growth in Plants (constructed by Prof. Reinke).

Institute of Vegetable Physiology of Göttingen (Director, Prof. Grisebach).

3938. Apparatus for Registering the Growth of Plants.
E. Stöhrer, Leipzig.

The principle of the apparatus is, that the making and breaking of a galvanic current sets in motion a lever which works between the poles of an electro-magnet, and carries a writing point at its extremity. This pen is pressed against and marks, on a drum of paper (which is turned round once in 24 hours by clockwork) a circular line. The portion of the plant under observation is brought into connexion with the pen by means of a delicate system of levers. A growth of even $\frac{1}{2}$-1 millimetre of the plant will cause the making and breaking of the current, the vibrator will thus move, and on the circular line will be recorded curved tracings which indicate the growth of the plant.

The second registering apparatus of which only a photograph is sent is intended for researches on the turning down and pressure of the root. A sheet of paper divided into minutes and hours is applied to a large brass cylinder, which is turned round on its axis daily or hourly by clockwork. The oscillations of the column of water are written on the paper by means of a cork float bearing a glass pen filled with anilin solution.

3939. Apparatus for Researches on the Physiological exchange of Material in the Sheep.

Prof. Henneberg, Agricultural Research Office, Station of the Hanoverian Agricultural Society, Göttingen.

It is necessary for researches on the total exchange of material in an animal to collect completely and separately the fæces and the urine. The stall exhibited serves to illustrate the method in which the experiment was carried out at the Research Station at Göttingen-Weende, when a sheep formed the subject of investigation. The single pen stall was erected in the chamber of Pettenkofer's respiration apparatus on those days when, in addition, the products of respiration were determined.

See Neue Beiträge zur Begründung einer rationellen Fütterung der Wiederkäuer, edited by Henneberg, Göttingen, 1870–71.

3940. Mercurial Air Pump, for the analysis of blood-gases.
Ch. F. Geissler & Son, Berlin.

The taps of the air pumps and other apparatus must be turned gently after they have been smeared slightly with lard until they appear transparent and clear.

3941. Sample Collection of Sections of Wood.
Dr. H. Nördlinger, Hohenheim, Württemberg.

The volume, which is the seventh published, contains, as well as the six preceding, 100 species of woods botanically determined. Although following as far as possible the Linnæan nomenclature, it introduces, however, about 50 new genera. It gives also an extraordinary variety of the inner structures of trees. There are specimens amongst them of the Suæda of the Mediterranean, and Atherstonia of the Cape, which can, on account of their abnormal structure, be only classified with difficulty.

Especially interesting is the comparison of the species of different genera of woods coming from different climates.

3942. Lucæ's Drawing Apparatus, modified by Spengel.
A. Wichmann, Hamburg.

3943. Orthoscope. *A. Wichmann, Hamburg.*

3944. Two Large Sachs' Vegetation Flasks, for coloured light, 500 mm. high, and 250 mm. across.
Warmbrunn, Quilitz, & Co., Berlin.

3945. Two Smaller Flasks, entirely of glass.
Warmbrunn, Quilitz, & Co., Berlin.

3946. Kronecker's Warm Chamber for Digestion.
Warmbrunn, Quilitz, & Co., Berlin.

3947. Ludwig's Strom ühr or Hæmodromometer.
T. Hawksley.

3949. Set of Physiological Operating Instruments.
T. Hawksley.

3949a. Table, with—

(a.) Fifteen pieces of the original burners which Middeldorpf, the discoverer of galvano-caustic, employed.

(b.) Collection of nine different uterus cauterizers. Wen constrictor. Bow cutter, with five adaptable tubes. Appliance for cauterizing the almond glands. Sliding corrector.

(*c.*) Universal handle for four porcelain cauterizers. Intestine cauterizer. Tooth cauterizer.

(*d.*) Tonsil cutter, for cauterizing the almond glands.
> *Kgl. chirurgische Klinik der Univ. Breslau (Prof. Dr. Fischer).*

3949b. Case, with two cautery handles, and six small cauteries for dentists.
> *Kgl. chirurgische Klinik (Prof. Dr. Fischer), Breslau.*

3949c. Case, with larynx instruments.
> *Kgl. chirurgische Klinik (Prof. Dr. Fischer), Breslau.*

3949d. Complete Case.
> *Kgl. chirurgische Klinik (Prof. Dr. Fischer), Breslau.*

3949e. Case, containing the most necessary cauteries for practical surgeons.
> *Kgl. chirurgische Klinik (Prof. Dr. Fischer), Breslau.*

3949f. Winterich's Lamp. *Friedrich Heller, Nuremberg.*

The instrument is very convenient for illuminating the throat.

3950. Hatching Apparatus at a uniform temperature; length 57 c., width 23, height 70 centims.

The apparatus was made in 1850, but not described. In the interior of the hatching apparatus is placed a metallic thermometer, the vertical axis of which passes through the bottom, and carries, underneath the hatching apparatus, a horizontal branch. At the end of the latter there is a chimney which, at a given temperature, is brought over the flame and diverts the hot air from the chimney which is placed in the water of the apparatus.
> *Prof. Théodore Schwann, Liége.*

3950a. Apparatus for Aërating the water in Aquaria.
> *Prof. U. J. Van Ankum, Director of the Zoological Laboratory of the University of Groningen.*

3950b. Elevations and Plan of the Laboratory for Biological Research, presented to the Royal Gardens, Kew, by T. J. Phillips Jodrell, Esq.   *H.M. Board of Works.*

3950c. Model of a Cupboard 3′.11″ high; 2′ broad; 1 deep, enclosing 2 cylinders.

1. Complete, with 4 circular shelves.
2. In section, to show the internal structure of the whole.
> *Laur Esmark, Christiania.*

The shelves are made to save room in museums in arranging specimens preserved in spirits of wine. The cylinders are to be turned round, so that the glass jars, which are placed along the periphery of the circular shelves, can be easily seen or taken out, instead of their being arranged as is generally the case in museums in a straight line, taking double or treble space. In some glass jars is to be seen the manner in which the specimens are fixed to glass threads, which prevent their coming in contact.

3950d. Page's Gas Regulators (2). *E. Cetti and Co.*

3950e. Geissler's Gas Regulator. *E. Cetti and Co.*

SECTION XIX.—EDUCATIONAL APPLIANCES.

South Gallery, Room A.

4000. Apparatus for **Instruction** in **Physical Science.**

Aug. Bel and Co.

1. Thomas's apparatus for showing pressure produced by dilation of liquids.
2. Daniel's hygrometer, for measuring the hygrometic state of the air.
3. Gravesande's hygrometer, for showing the expansion of metal by heat.
4. Bar of iron and copper for showing unequal expansion. 5. Apparatus to show the formation of vapour in a vacuum. 6. Air pump for producing a vacuum. 6a. Receiver for ditto. 7. Air pump. 7a. Receiver for ditto. 8. Air pump with glass barrel. 8a. Receiver for ditto. 9. Long glass tube mounted to show that heavy and light articles fall with equal rapidity in a vacuum. 10. Wire cage. 11. Heron's fountain for showing the elastic force of compressed air. 12. Water hammer (singing). 13. Haldat's apparatus for showing that the pressure of liquids depends upon their height and the surface of the bottom of the columns, and not upon the capacities of the vases. 14. Bunsen's spectroscope, for spectral analyses. 14a. Burners for ditto. 14b. Platina wire holder for ditto. 15. Hofmann's spectroscope (2). 16, 17. Polariscopes (5). 18. Magnesium lamp. 18a. Phosphorescent tube. 19. Newton's coloured disc for producing white light by rapid rotation. 20–23. Bichromate of potash batteries. 23a, 23b. Ditto, large. 24. Grenet's battery. 25. Induction coil. 26, 27. Apparatus for decomposition of water. 28. Galvanometer. 29. Apparatus for rotation of liquids by the electric current. 30. Ditto with hollow magnets. 31. Apparatus to show the attraction current. 32. Roget's spiral, for showing ditto. 33. Apparatus to show that the electric spark will not pass in a complete vacuum. 34. Astatic needle. 35. Secondary current apparatus. 36. Telegraph (two parts). 37. Smee's battery, six cells. 37a. Grove's ditto, five cells. 37b. Bunsen's battery. 37c. Set for electrotyping. 38. Set of mechanical powers. 39. Endless screw. 40. Centrifugal machine, four parts. 40a. Double cone. 40b. Electroscope. 41. One sportsman. 42. Sparkling jar. 43. Electroscope. 44. Whirl. 45. Egg. 46. Set of seven bells. 47. Apparatus for making a hole in a sheet of glass by the electrical discharge. 48. Diamond jar. 49. Henley's table. 50. Discharging rod. 51. Hand spiral. 52. Two brass plates for pith figures. 53. Electrical swing. 53a. Holtz electrical machine. 53b. Leyden jar, with movable coating. 53c. Electroscope. 53e. Jar with pith balls. 53f. Discharging table. 53g. Ditto rod. 54. Geissler's tube for showing direction of the current. 54a. Stand for ditto. 55–66. Vacuum tubes and stands. 67. Box of five tubes. 68. Tube and stand. 69. Ditto. 69a. Magneto engine for revolving vacuum tubes, &c. 70. Copper hot water funnel. 71. Still and condenser. 72. Small japanned pneumatic trough. 73. Large ditto. 74. Copper water bath, nickel plated. 75. Galvanised iron press. 76. Enamelled cast-iron air bath. 77. Bunsen's filter pump. 78. Ditto, modified by Dr. Frankland. 79. Copper oxygen retort and stand. 80–81a. Berzelius spirit lamp. 82. Revolving test tube stand. 83. Twelve holes and pegs, black. 84. Ditto, with glass pegs. 85. Ditto, in two stages. 86, 87. Test

tubes in two sizes for the above stands. 88. Bunsen's mercury trough. 89. Mahogany funnel stand for two. 90. Ditto, for one. 91. Stand for eight burettes. 22. Eight burettes for ditto. 93. Universal holder. 94. Two Geissler's burettes for ditto. 95. Universal holder. 96. Retort for ditto. 97. Bunsen's universal holder. 98. Table stand. 99–102. Bunsen's burner. 103. Steatite ditto. 104–106. Bunsen's ditto. 107, 108. Bunsen's blast burner. 109, 110. Bunsen's gas blowpipe. 111. Oxygen hydrogen burner. 112. Hoffmann's apparatus to show that 2 vols. of H. & 1 vol. of O. produce H. 2 O. 2 vol. 113. Burette stand for ditto. 114. Ozone tube. 115. Two stands for ditto. 116–118. Chemical balance. 119. Plattner's ditto. 120. Set of 1 gramme weight. 121. Ditto 50 ditto. 122. Ditto 100 ditto. 123. Ditto 6,000 grain. 124. Ditto 100 gramme weight, common. 125. Ditto 1,000 ditto. 126. Retort stand, galvanised foot. 126. Ditto, brass rings, clamp, and block. 127. Ditto, triangular foot. 128. Large retort stand, two clamp burner. 129. Bunsen's apparatus for determining the specific gravity of gas by effusion. 130. Glasshilig's condenser stand. 131. Aug. Belsder's crucible furnace (according to Perrot, Geneva). 132. Ditto, smaller (according to Forgnigum and Leslere). 133. Glasser's combustion furnace. 134. Hofmann's ditto. 135. Kekuli ditto. 136. Dr. Würtz ditto. 137. Large Bunsen's burner. 138. Tripod for ditto. 139. Funnel with stop-cock. 140. Set of bottles for mounting microscopic objects. 141. 5 decigallon jars. 142. 1,000 cc. ditto graduated jar. 143. 700 ditto. 144. 500 ditto. 145. 300 ditto. 146. 200 ditto. 147. 100 ditto. 148. 25 ditto. 149. 1,000 ditto, stop. 150. 500 ditto. 151. 300 ditto. 152. 75 ditto. 153. 25 ditto. 154, 155. Two cylindrical jar stops. 156, 157. Two cylindrical jars, plain. 153. Hydrogen gas bottles. 159. Eprouvette. 160–162. Three Woulff's bottles, assorted. 163, 164. Two Woulff's bottles. 165. One ditto, tubulated. 166–170. Five test glasses, assorted. 171. Lixiviating jar. 172–175. Four capsules. 176. Parting flasks. 177, 178. Two conical beakers. 179–182. Four flasks, assorted. 183–185. Three Berlin basins. 186, 187. Two conical flasks. 188. 750 cc. flask. 189. Set of three beakers. 190. Ditto four ditto, spouted. 191. Set of nine glass basins. 192. One decigallon flask. 192a, 193. Two basins. 194. 250 decim flask. 195. One retort. 196, 197. Two retorts. 198–200. Three receivers. 201. Bottle flasks. 202. Jar. 203, 204. Two specimen bottles. 205, 206. Florentine receiver. 207. French balloon. 208–214. Seven bottles, assorted. 215. Percolator. 216. Gas bottle. 217. Water bottle with tap. 218–221. Four funnels. 222, 223. Two spirit lamps. 224. Apparatus to prove that the mixing of chlorine with hydrogen into hydrochloric acid does not alter the volume, therefore one volume of hydrogen and one volume of chlorine produce two volumes of hydrochloric acid. 225. Geissler's potash bath. 226. Carbonic acid apparatus. 227–231. Five blow-pipes. 232. Spee gear, bottle in case. 233. Mercurial trough. 234. Ditto, ditto. 235. Pestle and mortar. 236, 237. Two Berlin capsules. 238. Desiccating pan. 239. Porcelain plate twelve cavities. 240. Copper condenser and stand. 241. Complete set of apparatus according to Gay Lussac for assaying silver by wet process. 242. Copper gas holder. 243. Vertical galvanometer for lectures.

SOUND.

301. Eight pieces of wood sounding the scale. 302. Opening pipe allowing a view of the interior of the pedal. 303. Wertheim's apparatus, new model. 304. Organ pipe with glass facing, and showing vibration of sound. 305. Organ pipe with changeable openings. 305 *bis*. Four organ pipes, all containing the same quantity of air ; one cylindrical, one cubical, one spherical, and one tetrahedral. 306. Three organ pipes, all containing the same quantity

of air; one prismatical, one straight conical, one reversed conical. 307. Organ pipe with glass furnace, according to Mr. Bourbouze, showing three different vibrations with reflector. 308. Five different sounding boards in wood, and one of wire. 309. Two flat wires for horizontal vibration. 310. Small support with movable bridges to fix the wire or sounding boards. 311. Sinometer, according Mr. Barberau. 312. Circular membrane in caoutchouc with changeable tension. 313. Ditto of paper, 30 cent. diameter. 314. Ditto, ditto, 30 cent. thick. 315. Two tuning forks in C, mounted upon wooden box adjusted to give four vibrations in the second. 316. Two tuning forks in C, mounted between magnetic poles, and intermitting tuning fork in C. 317. Vibration meter according M. Duhamel. 318. Apparatus showing the vibration through liquids. 319. Small apparatus, showing the vibration of vocals without movable reflector.

EXAMPLES OF ELEMENTARY PHYSICAL APPARATUS.

The pieces of apparatus in this collection were made by the pupils Messrs. Boughton and Owen in the Physical Laboratory at the Science Schools, South Kensington. They are intended to show what can be done by Science students in the matter of making their own apparatus, and to indicate that neither great expense nor great manipulative skill is required to put such students in possession of instruments capable of illustrating the ordinary laws and phenomena of physics.

Frederick Guthrie.

PROPERTIES OF AIR AND WATER.

4000a. Cardboard Mercury Tray.

4000b. Glass Millimetre Scale.

4000c. Boyle's Tube.

4000d. Siphon Barometer.

4000e. Water Hammer.

4000f. Specific Gravity Bottle.

4000g. Apparatus for estimating density of Mercury.

SOUND.

4000h. Monochord.

4000i. Compound Motion Spring.

4000j. Sensitive Flame.

4000k. Singing Flame Apparatus.

4000l. Singing Flame Apparatus [octave higher].

HEAT.

4000m. Differential Thermometer.

4000n. Alcohol Thermometer.

4000o. Mercury Thermometer.

4000p. Bulb for determining Expansions.

4000q. Apparatus for measuring Absolute Expansion of a Liquid.

4000r. Flask with Delivery Tube.

4000s. Conductivity Cones.

4000t. Hygrometer.

4000u. Ventilation Apparatus.

LIGHT.

4000v. Pin Hole Camera.

4000w. Instrument for measuring Vertical Heights.

4000x. Circle Dividing Board.

4000y. Multiple Image Apparatus.

4000z. Concave Mirror.

4000aa. Convex Mirror.

4000ab. Three Glass Bulbs and Beakers.

4000ac. Glass Cell.

4000ad. Bisulphide of Carbon Prism.

4000ae. Spectroscope.

4000af. Polariscope.

FRICTIONAL ELECTRICITY.

4000ag. Glass Tube.

4000ah. Amalgamed Silk Rubber.

4000ai. Balanced Glass Tube.

4000aj. Pointed Support for Glass Tube.

4000ak. Pith Balls with Insulating Threads.

4000al. Insulating stand for Insulating Threads.

4000am. Proof Plane [glass].

4000an. Proof Plane [sealing wax].

4000ao. Proof Plane [ebonite].

4000ap. Insulated Flannel Rubber.

4000aq. Gold Leaf Electroscope.

4000ar. Shellac Rod.

4000as. Flannel Rubber for Shellac Rod.

4000at. Sealing Wax.

4000au. Flannel Rubber for Sealing Wax.

4000av. Cylindrical Conductor on Insulating Stand.

4000aw. Conical Conductor on Insulating Stand.

4000ax. Spherical Conductor with Insulating Thread.

4000ay. Spherical Conductor with Insulating Thread.

4000az. Insulating Stand for Spherical Conductor.

4000ba. Insulating Stand for Spherical Conductor.

4000bb. Insulating Plate.

4000bc. Condenser.

4000bd. Fulminating Plate.

4000be. Bennett's Multiplying Condenser.

4000bf. Leyden Jar Discharger.

4000bg. Oval Conductor on Stand.

4000bh. Electrophorus Cover.

4000bi. Reversible Net.

VOLTAIC ELECTRICITY.

4000bj. Coil of Wire on Glass Tube.

4000bk. Zinc and Copper Plates.

4000bl. Astatic Galvanometer.

4000bm. Daniell's Cell.

4000bn. Wheatstone's Bridge.

4000bo. Rheocord and set of Resistance Coils.

4000bp. Electric Bell.

4000bq. Quadrant Electrometer.

4000br. Thermopile.

4000bs. Water Decomposition Apparatus.

MAGNETISM.

4000bt. Two-bar Magnets and Keepers.

4000bu. Magnetized Needle.

4001. Collection of **Models** and **Apparatus** for teaching **Physics, Sound, Light, Heat, Electricity,** and **Chemistry.**
Matthew Jackson.

Tate's air pump.
Magdeburg hemispheres.
Glass globe for weighing air.
Gay-Lussac's apparatus for vapour densities.
Dumas' apparatus for vapour densities.
Electrical alarum.
Savart's toothed wheel apparatus.
Syren.
Tuning forks.
Hopkins' fork tube.
Glass reflectors on stands.
Set of lenses and half lenses, with holder on card.
Fergusson's pyrometer.
Gravesand's ring and ball.
Sir H. Davy's apparatus.
Thermopile.
Cylinder machine.
Plate machine.
Bertsch machine.
Holtz machine.
Winter's machine.
Volta's condensing electroscope.
Aurora apparatus.
Bichromate battery.

Zinc plates for Bunsen's battery.
Tangent galvanometer.
Set (11) of Hofmann's tubes with stands (1, with 2 stands) and 1 span stand.
Winkler's gas apparatus.
Imp. sulph. hydrogen apparatus. 3 articles.
Ammonia apparatus.
Collection of small instruments (18) on cardboard.
Organic analysis apparatus. 30 pieces.
Desiccators. Different patterns.
Distilling apparatus for carbon bisulph. (4 articles.)
Assortment of 12 test glasses.
Distilling apparatus. (4 pieces.)
Set of 71 graduated and other glass instruments.
Carbonic acid apparatus. 7 portions in card boxes.
Arrangement of gas burners.
Bunsen's water bath and stand.
Sulph. hydrogen apparatus.
Cleaning brushes.

Wide tube, graduated tube, and thermometer for Gay-Lussac's apparatus.

Bulb and thermometer for Dumas' apparatus.

Hofmann's tube for steam, with glass worm and stand.

Cabinet with 26 bottles.

Glass tube, two nitrogen bulbs, and piece of india-rubber tubing for organic analysis apparatus.

Four weighing bottles for desiccators.

Twelve small tubes for Eggertz's carbon apparatus.

One nest of six beakers with spouts.

One nest of 12 beakers without spouts.

Two tin sand baths.

One Bink's burette 500 cc., on wood stand.

Ritchie's photometer.

Glass carbon and porous cells for Bunsen's battery, 10 each.

Wooden burette stands (5).

Stand for eight burettes.

Bath for Eggartz's carbon test.

Bunsen's clamp stand.

Regulator for Bunsen's water bath.

4026. Th. Müller's Di-Electric Machine, with accessory apparatus, viz. :—

Electric tuft.
Electric turbine.
Electric puppet-dance.
Electric peal of three bells.
Lightning tube.
Lightning conductor.
2 electric excitors.
Electric pistol.
2 condenser jars.

Kgl. Preuss. Oberbergamt, Breslau.

This machine has of late been much used in elementary schools in mining districts of Germany for instruction of the children in some parts of their future calling. With 12⅜ inches diameter of disc, it gives sparks over 4 inches long. The machine can be depended on even in unfavourable weather. It is simple in construction, and the peculiarity of it consists in the collection of the positive and negative electric currents in two separate Leyden jars of special form, which again are so connected together that the opposite electric currents compensate each other. Thus the loss of electricity is reduced to a minimum, and strong actions are obtained with the excited glass surface.

4002. Calculating Disc, constructed on the system of Professor Sonne. Specimen for demonstration, size 50 centimeters, with rough division ; for instruction in schools.

Landsberg and Wolpers, Hanover.

4002a. Universal Lever, for lecturing purposes.

C. Chzechovicz, Russia.

Apparatus consisting of a set of levers and parallelograms, intended to illustrate the principle of elementary statics, viz. :—

The combination of forces applied to a body.

Combination of parallel forces.

Testing of the laws of the lever of first and second class.

Examination of the properties of the balances.

Experimental proof of the proposition that every part of a decomposed force produces a corresponding effect.

4002b. Apparatus for showing the **Propagation** of **Wave-motions.** *C. Chzechovicz, Russia.*

For lecturing purposes, and intended to show the propagation of wave-motion by a system of bi-filar pendulums suspended on movable levers. By changing the plane of suspension, one is enabled to reproduce waves with plane vibrations as well as with elliptical ones.

4002c. Model of a Plummer-block, for use in classes.

4002d. Model of a Chain-holder, for use in classes.

4002e. Group, with photographs of models, for use in classes.
M. J. Prugger, Munich.

4002f. Glenny's Diagrams of Building Construction (10). *Chapman & Hall.*

4002f. Bristow's Table of British Strata.
Chapman & Hall.

4002f. Unwin's Machine Details (16).
Chapman & Hall.

4002f. Patterson's Zoological Diagrams (10).
Chapman & Hall.

4002f. Goodeve's Steam Diagrams (15).
Chapman & Hall.

4002f. Etheridge's Diagrams of Fossils (6).
Chapman & Hall.

4002f. Anderson's Diagrams (2). *Chapman & Hall.*

4003. Arithmometer, by Martinot, for imparting the knowledge of metrical arithmetic and practical geometry.
Alphonse Martinot, Belgium.

It is composed of one thousand small cubes of the size of one centimetre. Thirty-two of these are isolated, nine hundred and thirty-two others are united in several groups, forming forty-two pieces. There are ten pieces of two cubes, ten pieces of five cubes, thirteen pieces of ten cubes, bound together in the shape of reglets, or scale boards, one piece of twenty cubes, one piece of fifty cubes, and, finally, seven pieces of one hundred cubes, bound together like plane tables.

The apparatus is completed by four racks, into which, not only the isolated cubes, but the greater number of the groups of cubes may be bracketed so as to form, when united, the cubic decimeter, *i.e.*, the 1,000. It is easy to understand what intuitive power this apparatus places at the disposal of the professor, for teaching children, and for making them understand, in what may be called a material way, the principles of numeration, construction, and deconstruction of numbers, the four fundamental rules of arithmetic, fractional decimals, the formation of the square and cube of numbers, the extraction of their roots, and the generation of metrical measures, &c., &c. (Extract from the Report of Mr. André Van Hasselt, Inspector of the Normal Schools of Belgium.)

4003a. Machine for Arithmetical Instruction.
Antoine Arens, Frere Marianus, Rue de Bruxelles, Namur.

APPARATUS FOR INSTRUCTION IN PHYSICAL SCIENCE, CONTRIBUTED BY THE COMMITTEE OF THE PEDAGOGICAL MUSEUM, RUSSIA.

APPARATUS FOR INSTRUCTION IN PHYSICAL SCIENCE IN MILITARY SCHOOLS OF THE HIGHER AND LOWER GRADES IN RUSSIA.

4004. Bauler's Inclined Plane illustrating the acceleration of motion.

4005. Apparatus illustrating the **Expansion** and **Fusion** of **Bodies** from **Heat,** constructed under the direction of the Physical Section of the Pedagogical Museum.

4006. Slides, with weights, for illustrating the force of **Cohesion,** constructed under the direction of the Physical Section of the Pedagogical Museum.

4008. Wheel with Graham's Anchor. By Lermontoff.

4009. Bauler's Lamp, with Gasometer. By Lermontoff.

4011. Caoutchouc Electrical Machine. By Kresten.

4012. Apparatus for preparing Oxygen. By Kresten.

MATHEMATICS.

A.—ARITHMETIC.

4019a. Sets of pegs for integers and fractions.

4019b. Kanaeff's arithmetical box.

4019c. Kanaeff's arithmetical box, coloured.

4019d. Wooden blocks divided into cubes for the **comparison of numbers.**

4019e. Arithmetical boxes, made by the St. Petersburg School Workshop.

4019f. Zitowitch's arithmetical box.

4020. Paulson's abacus for integers.

4020b. Kanaeff's abacus for integers and fractions.

4020c. Model of Ilyin's abacus.

B.—GEOMETRY.

4028a. Collection of 6 geometrical forms. Made by the St. Petersburg School Workshop.

4028b. Strookoff's collection of 7 geometrical forms.

4029a. Collection of 16 geometrical forms prepared by the inmates of a Russian prison.

4033a. Cube of three.

4033b. Cube of a binomial quantity.

4033c. Square of a trinomial quantity.

Cube of a trinomial quantity.

Two-plane angle made of cork.

The Pythagorean theorem.

Four-sided geometrical figures.

Six Polygons.

Boldt's geometrical apparatus.

Zitowich's apparatus for determining the cubic content of geometrical bodies.

Ogerovsky's geometrical puzzle.

Ogerovsky's trigonometrical diagram.

C.—GEOMETRICAL DRAWING.

Shildknecht's apparatus for demonstrating the theory of projections.

Strookoff's collection of school apparatus for geometrical drawing.

Pantograph.

Ellipsograph.

Specimens of appliances for geometrical drawing, prepared by pupils of one of the military schools.

PHYSICS.

Lermontoff's wheel with Graham's anchor.

Van der Fleet's apparatus for illustrating Marriotte's law.

4012. Kresten's apparatus for preparing oxygen - - - - - Not included in the list as not belonging to educational appliances.

Schwedoff's electrical machine - -

Tepploff's two electrophore machines -

4009. Baulev's lamp, with gazometer -

GEOGRAPHY.

N.B.—These appliances were exhibited at the International Geographical Exhibition in Paris (1875), and the Museum was awarded a " Lettre de distinction " for them.

A. PROPEDEPHTICAL COURSE OF GEOGRAPHY.

Semenoff's pictures.

Simashko's ideal relief and section of the earth.

Relief of the sea bottom.

Michailloff's globe.

Class Plan of **St. Petersburg.**

Plan of the **Environs** of **St. Petersburg.**

Plan of **Moscow.**

Ilyin's Nets for **Drawing Maps.**

Ilyin's Nets with the **Outlines** of **Continents.**

Michailoff's System of **Drawing Maps** on oil cloth.

System of preparing **Relief Maps** in wax.

Class Orographical Map.

B. Physical Geography.

4049. Phenomena of nature in pictures.

Girotovsky's geyser.

       ,,   icebergs.

       ,,   attol.

Michailoff's gletchers.

       ,,   sand spouts.

Teich's Russian obos (train of carts).

       ,,   bargemen-work on Wolga.

Karagin's steppes and deserts (two pictures).

       ,,   sea of Aral.

       ,,   A caravan in the steppes.

       ,,   Barkhanes (sand-hills in Turkistan).

       ,,   the steppes in middle Asia.

Albums with views.

C. Political Geography.

Ilyin's school geographical atlases.

Ilyin's detailed school atlas of the Russian empire.

Ilyin's atlases of the Russian empire for elementary schools.

Ilyin's wall maps of the five parts of the world.

Ilyin's wall maps of the Russian empire.

Ilyin's detailed atlas of the Russian empire.

Ilyin's detailed atlas of Asiatic Russia.

Ilyin's post map of the Russian empire.

Ilyin's railway map of the Russian empire.

Ilyin's map of Finland.

Ilyin's maps of Poland.

Ilyin's map of Turkistan government.

Ilyin's map of Caucasus.

D. Ethnography.

Specimens of **Models** showing the national costumes of the different inhabitants of Russia.

4058. Types of races of man (models made of papier-mâché).

Busts of large size.

Busts of middle size.

Busts of small size.

Frames with busts in bas-relief.

Svetchnikoff's ethnographical map of Russia.

Mirkovitch's ethnographical map.

E. Mathematical Geography.

Model of Foucault's apparatus.

Tellurium.

4079. Kachovsky's apparatus for explaining the seasons.

Kachovsky's apparatus for explaining the formation of the ecliptic.

4080a. Apparatus for explaining the phases of the moon (without a lamp).

Kachovsky's apparatus illustrating the motion of the upper and lower planets.

Brouns' astronomical atlas.

Petchorin's atlas of mathematical geography.

Mathematical Instruments.

4019. Board, with set of pegs, for exercises in studying numbers from 1 to 1,000. Made by the St. Petersburg Workshop for School Apparatus.

4020. Arithmetical Boxes, for exercises in numbers. Made by the St. Petersburg Workshop for School Apparatus.

4021. Kachovsky's Class Abacus, for integers and fractions. Made by the St. Petersburg Workshop for School Apparatus.

4022. Kachovsky's Abacus, constructed on the system of the St. Petersburg Workshop of School Apparatus.

4023. Nomansky's Abacus, for integers and fractions.

4024. Russian Trade Abacus.

4025. Collection of Measures, for use in schools. (Long, superficial, and solid measures, weights, &c.)

GEOMETRICAL APPARATUS.

4027. Board, exhibiting square foot divided into square inches. Made by the St. Petersburg Workshop for School Apparatus.

4028. Cubic Quarter Arshin, divided into cubic vershoks. Made by the St. Petersburg Workshop for School Apparatus.

4029. Collection of Geometrical Forms, with sections. 17 models. Made by Stroukoff.

4030. Collection of 37 geometrical forms intersecting each other. Made by Stroukoff.

4031. The Cube of two, the cube of three, the cube of a binomial and a trinomial quantity. Made at the St. Petersburg Workshop of School Apparatus.

4032. Large Cone, with sections and ·hemispheres for determining the focus of curves. Made by Stroukoff.

4033. Sphere with sections. Made by Stroukoff.

4035. Apparatus for demonstrating the theory of projections. Made by Schildknecht.

4038. Collection of class school apparatus for geometrical drawing. Made by Stroukoff.

4039. Manuals and Exercises on all branches of mathematics and geometrical drawing ; treatises on the method of instruction in these subjects, adopted in the military schools of Russia. Published by Fenoult.

MAPS, DIAGRAMS, MODELS, AND OTHER APPLIANCES FOR TEACHING GEOGRAPHY AND ASTRONOMY.

4040. Apparatus illustrating the **Motion** of the **Planets** in general, and demonstrating the periods of the transit of Venus in particular, by Kachovsky.

4042. Geographical Pictures on glass for the magic lantern, by Ermolin.

4043. Atlas of Relief Maps. Published by the Juvenile and Pedagogical Library in Moscow.

4044. Relief Map of Asia. By Shulgin.

4045. Model in Relief of an Alpine country. Prepared by pupils.

4046. Orographical Maps of the World. By Simashko.

4047. Maps of the five divisions of **the World,** By Ilyin.

4048. Charts (two), showing the distribution of **Forests** and **Mineral Wealth of Russia.** Prepared by the Statistical Committee.

4049. Phenomena of Nature, in pictures :—
Givotovsky's Geyser, Toundra, Icebergs, and Attols.
Michailoff's Glaciers and Sand Spouts.
Teich's Russian Obos (train of carts), and Bargemen at work on the Volga.
Karasin's Steppes and Desert, two pictures; Sea of Aral, a Caravan in the Steppes, Barkhanes (sand hills in Turkistan), and The Steppes in Middle Asia, two views.

4050. Specimens of **Ermolin's Geographical Pictures** on glass for the magic lantern.

4052. Ilyin's Aids for Pupils in **Map Drawing,** with the outlines of the continents filled in.

4053. Terrestrial Globes of different dimensions, prepared by the Juvenile and Pedagogical Library in Moscow.

4054. Album with views of Lapland (travels of Nemirowitch Danehenks).

4055. Album with views of Polessie (the marshes of Pinsk). Views drawn according to a project for draining the marshes, by Colonel Gilinsky, with charts and plans.

4056. Photographic Views of Amoor, Western and Eastern Siberia, and Ural, five large vols. in folio.

4057. Atlas of **Physical Geography,** by Zooeff.

4058. Types of Races of Man. Models made of papiermâché (busts of large, middle, and small dimensions, busts in bas-relief, statues, etc.). Prepared by Heiser and Female Workshop.

4059. Races and Tribes of Man. Drawings, published by the Juvenile and Pedagogical Library in Moscow, and Ilyin's Chartographical Establishment.

Only specimens will be shown, the Museum being obliged to send the collections to the American Exhibition in Philadelphia.

4060. Ethnographical Charts of Keppen, Mirovitch, and Sweshnikoff.

4061. Ethnographical Album, by Pauli.

4064. Atlas of Russia, by Ilyin.

4065. Atlas of Russia in detail, by Ilyin.

4066. Rothstein's Atlas of the Russian Empire.

4067. Geographical Atlas (of all countries), by Ilyin.

4068. Atlas of the five parts of **the World,** by Ilyin.

4069. Jordan's Geographical Atlas.

4070. Illustrated Geographical Atlas, by Limberg.

4071. School Atlas, as projected by Colonel Ponlikoffsky.

4072. Geographical Wall-maps, of Ilyin, Rothstein, Sheveleff, Lebedeff, Zooeff, and Michailoff.

4073. Elementary Course of Geography. Systematically arranged in drawings, plans, and maps, prepared by Lapehenko and Michailoff.

4074. Apparatus for teaching the Blind. Collections of plans, maps, manuals, and models for teaching geography to the blind. Prepared and edited by General Grigorieff, director of the Institution for Blind Children in St. Petersburg.

4075. Manuals on all branches of **Geography,** adopted in the Military Schools of Russia; published by Fenoult.

4077. Atlas, by Petchorin.

4079. Apparatus for explaining the seasons (formation of the ecliptic), by Kachovsky.

4080. Apparatus for explaining the phases of the moon, by Kachovsky.

4081. Armillary Sphere, by Penkin.

4082. Armillary Sphere, by Koualsky.

NATURAL HISTORY.

4089a. Charts illustrating natural history, issued under the direction of the Imperial Economical Society. Published by Fenoult.

4089b. Arend's School Atlas of Natural History. Published by Fenoult.

1. SCHUBERT'S ZOOLOGICAL WALL CHARTS. Published by Fenoult.
2. MICHAILOFF'S SCHOOL ATLAS OF ZOOLOGY. Published by Fenoult.
3. MEINHOLD'S ANATOMICAL TABLES. Published by Fenoult.
4. CARUS' TABLES OF COMPARATIVE ANATOMY. Published by Fenoult.
5. BRANDT'S ATLAS OF COMPARATIVE ANATOMY. Published by Fenoult.
6. BOCK'S ATLAS OF COMPARATIVE ANATOMY. Published by Fenoult.

4089c. Specimens of prepared Skeletons and their parts : the skeleton of the bat, of the cat ; cervical and lumbar vertebra in section; horn of the bull with the core ; leg of the horse ; hoof and skull of the same. Prepared by Strembitsky.

(NOTE.) The committee consider it unnecessary to send complete school collections of skeletons and stuffed animals, and prefer sending a few specimens only, in order to show the mode of preparation, the price of articles and their use in teaching.

4089d. Collection of Furs. Prepared by Fenoult, and at the St. Petersburg Workshop of School Apparatus.

4089e. Specimens of **stuffed Mammalia** (cat and bat). Prepared by Fenoult, and at the St. Petersburg Workshop of School Apparatus.

ANATOMICAL MODELS, A COLLECTION PREPARED BY STREMBITSKY IN ACCORDANCE WITH THE INSTRUCTIONS OF THE NATURALISTS OF THE PEDAGOGICAL MUSEUM COMMITTEE.

4089f. Thorax (with the same details as Bock Stagger's).

4089g. Vertical section of the Head, with removable larynx.

4089h. Vertical section of the Cerebrum.

4089i. Human Mask, the mouth open.

4089j. Lungs, Heart, and Larynx.

4089k. Heart.

4089l. Larynx, from behind.

4089m. Larynx, with soft movable epiglotis and elastic communication of the hyoides with the thyroid cartilage.

4089n. Organ of Hearing.

4089o. Eye.

4089p. Spinal Cord and Cerebellum.

4089q. Lower Jaw (organization and progressive development of teeth).

4089r. Vertical section of the Skin.

4089s. Models of Joints (three).

4089t. Complete Model of a man. (2½′ high.)

4089u. Model of a human **Tooth.**

4089v. Models showing the anatomy and physiology of animals ; stomachs of ape, kangaroo, tiger, armadillo, seal, manatee. Prepared by Strembitsky.

4089w. Heiser's Models of mammalia (papier-mâché). Prepared by Schindhelm.

4089x. Birds. Specimens of skeletons and their parts ; the skeleton of the turkey, with wing and tail feathers ; skeleton of the fowl ; sternum, wing, foot, skull of the same. Prepared by Strembitsky.

4089y. Stuffed Birds, two specimens (fowl and linnet). Prepared by Strembitsky.

4089z. Model of digestive organs of a Goose. Prepared by Strembitsky.

4089aa. Skeleton of a Frog, Skull of a Poisonous Snake. Exhibited as specimens of skeletons of amphibia and reptilia. Prepared by Strembitsky.

4089ab. Frog in different stages of development. Exhibited as a specimen of preparation of amphibia in alcohol. Prepared by Strembitsky.

4089ac. Frog. Exhibited as specimen of stuffed amphibia. Prepared by Strembitsky.

4089ad. Collection of preparations in wax, illustrating the development of the frog. Prepared by Strembitsky.

4089ae. Skeleton of the Perch. Exhibited as specimen of skeletons. Prepared by Strembitsky.

4089ah. Specimen of stuffed fish. (Sturgeon.) Prepared by Strembitsky.

4089al. Dissected Insects, two specimens (a beetle and a locust). Prepared by Strembitsky.

4089am. Class collection of Insects, containing 25 typical forms. Prepared by Strembitsky.

4089an. Class collection of Insects on coloured plants, prepared by Vaviloff and the pupils of one of the military gymnasia.

4089ao. Class collections of Insects, the insects stuffed and arranged in characteristic postures on natural branches. Prepared by Strembitsky.

4089ap. Specimens of a class collection of insects in separate glass boxes (idea of St. Hilaire).

4089aq. Larvæ of different butterflies, dried (the colour retained). Prepared by Strembitsky.

4089ar. Model of the head of *Œstrus* with movable parts of the mouth (greatly enlarged). Prepared by Strembitsky.

4089as. Model of the head of *Libellula* with movable parts of the mouth (greatly enlarged). Prepared by Strembitsky.

4089at. Model of the head of a spider with movable parts of the mouth (greatly enlarged). Prepared by Strembitsky.

4089au. Model of the head of a butterfly with movable parts of the mouth (greatly enlarged). Prepared by Strembitsky.

4089av. Model of the leg of an insect, joints movable (greatly enlarged). Prepared by Strembitsky.

4089aw. Wax Model of the larva of *Meloloutha* (greatly enlarged). Prepared by Strembitsky.

4089ax. Model of *Bombyx mori,* in three stages of its development (vulcanized india-rubber). Prepared by Strembitsky.

4089ay. Collection containing six typical forms of Crustacea. Preserved in alcohol by Strembitsky.

4089az. Specimen of *Astacus fluviatilis.* Prepared in a special manner by Strembitsky.

4089ba. Collection of Crustacea, 10 typical forms, living in Russia (in form of a table). Prepared by Strembitsky.

4089bb. Collection of 5 typical specimens of Vermes in alcohol. Prepared by Strembitsky.

4089bc. Collection of three specimens of Mollusca in alcohol. Prepared by Strembitsky.

4089bd. Collection of 10 artificial specimens of Mollusca, representing them in motion to show the form of their bodies. Prepared by Strembitsky.

4089be. Collection of 17 artificial specimens of Mollusca. Prepared according to the instruction of Professor Owsiannikoff by Strembitsky.

4089bf. Sea Urchin, in alcohol. A specimen of the mode of preserving Echinodermata, Acalepha, Polypi, and Polycistina. Prepared by Strembitsky.

4089bg. Botanical Wall Charts. Prepared by Givotovsky.

4089bh. Botanical Hand Atlas. Published by Givotovsky.

4089bj. Botanical Atlas, by Schubert. Published by Fenoult.

4089bk. Wall Botanical Charts, one specimen of a collection. Coloured on oil-cloth. Prepared in the Military Gymnasium of Orenburg.

4089bl. Dendrological Collection. Illustrations of the trees of Russia. Prepared by Stolpiansky.

4089bm. Dendrological Collection. Illustrations of the trees of Russia. Prepared by Stroukoff.

MODELS AND OTHER APPLIANCES FOR TEACHING MINERALOGY AND CRYSTALLOGRAPHY.

4084. Class Collection of Minerals, with apparatus for their investigation (Heard's system), prepared by Latkin.

4085. Glass Crystals, with axes, illustrative of crystallography, prepared by Skibinevsky.

4086. Glass Crystals, intermediate forms, illustrative of crystallography, prepared by Skibinevsky.

4087. Crystals of Tin (specimens), prepared by Colonel Vonder-Weld.

4088. Models of Crystals in Wood, illustrative of crystallography, prepared by Stroukoff.

4089. Aquarium, Terrarium, and various apparatus for aiding pupils in forming zoological, botanical, and mineralogical collections. From the St. Petersburg Workshop of School Apparatus.

4090. Teacher's Portable Lecture Set, for demonstrating the principles of the science of **Sound,** comprising the following apparatus :—

1. For demonstrating the mode of Propagation of Sound.

  Model to show propagation of wave by the oscillation of a row of particles. The difference between the waves of sound and light is shown by the production of both simultaneously in close proximity. Row of marbles in grooved board.

2. For demonstrating the Necessity of a Conducting Medium.

  Air pump, 6-inch plate, with open receiver fitted with collar and sliding rod. Small bell on cushion.

3. For demonstrating the Reflection and Refraction of Sound.

  Tube, 2 inches diameter, and 6 feet long, with mouthpiece and trumpet-shaped end ; pair of 18-inch tin reflectors ; goldbeater's skin balloon.

4. For demonstrating the Lengths of Waves, Regularity, Music, and Noise.

  Tuning-fork, with steel points set in frame with groove and smoked glass slides ; whirling table with toothed wheel and card and perforated disc ; Trevelyan rocker and lead block.

5. For demonstrating the laws of the Vibrations of Strings.

  Iron frame, with stretched piano wire to show need of sounding board ; sonometer with four wires (two stretched by keys and two by weights), graduated decimal scale, movable bridges, paper riders, and extra strong bow ; elastic cord, 15 feet long, with attachment for whirling table.

6. For demonstrating the laws of the Vibrations of Air-columns.

  C, tuning-fork and glass jar ; glass tube, 1 inch diameter, 2 feet long, open at both ends ; set of 13 organ pipes with bellows and wind chest ; pipe, with sliding stopper for beats ; pipe, with glass slide and paper tray for showing position of node ; metal reed pipe ; harmonium reed large model ; burner for sounding flame, and set of glass tubes.

7. For demonstrating the laws of the Vibrations of Bells.

  Bell glass, 8 inches diameter, and squares of blotting paper ; 3-inch bell on stand ; 3-inch bell mounted for heating.

8. For demonstrating the laws of the Vibrations of Plates.

  Set of two glass and two brass plates, 6 inches diameter, with stand and thumbscrew ; box of silver sand.

9. For demonstrating the laws of Longitudinal Vibrations of Rods.

  Glass tube, $\frac{1}{2}$-inch diameter, 3 feet long, sealed at one end ; lycopodium ; brass rod, $\frac{1}{2}$-inch diameter, 3 feet long ; flannel, leather, and rosin.

10. For illustrating the production of Vocal Sounds.
Wooden funnel, with stretched membrane.
11. Diagrams.
    (*a.*) Interference of sound.
    (*b.*) Section of organ pipe.
    (*c.*) Formation of nodes on a string.
    (*d.*) Human ear.
    (*e.*) Human larynx.
                Matthew William Dunscombe, Bristol.

4091. Teacher's Portable Lecture Set for demonstrating the principles of the science of **Light,** comprising the following apparatus :—
1. For demonstrating the Velocity of Propagation.
Movable diagram showing Roemer's observation.
2. For demonstrating the Undulatory Theory.
Model showing propagation of wave by series of rotating particles. The difference between the vibrations of waves of light and sound is shown by the production of both simultaneously in close proximity.
3. For demonstrating the Law of Inverse Squares and Relative Intensities of Lights.
Two square frames, one four times the area of the other, fitted at the corners to four wires, representing rays of light diverging from a point; Wheatstone's photometer, a new and highly improved form ; discs of paper in frame with oiled spot for Bunsen's method.
4. For demonstrating the Laws of Reflection :
(*a.*) By plane mirrors.
Plane mirror and candle; pair of mirrors fringed for showing principle of kaleidoscope ; movable model (quite new), showing law of angular velocity in ray reflected by rotating mirror.
(*b.*) By curved mirrors :
Convex and concave silvered glass mirrors, carefully worked ; semi-cylindrical mirror to show caustic curves; three movable models illustrating the properties of the foregoing.
5. For demonstrating the Laws of Refraction :
(*a.*) At plane surfaces.
Equi-angular glass prism on jointed pillar; disc painted with lines showing paths of incident and refracted rays, for immersion up to the centre in water; movable models to show the law of lines and passage of a ray through a sheet of glass with parallel faces ; right-angled glass prism on pillar, to show total internal reflection.
(*b.*) Three-inch diameter double convex lens on pillar and

40075.   3 T

ground-glass screen ; set of six semi-lenses ; movable models to show refraction by convex and concave lenses.

6. For demonstrating the Laws of Chromatics :

Set of coloured discs to fit whirling table ; achromatic pair of prisms on jointed pillar.

Matthew William Dunscombe, Bristol.

4092. Geography Teaching Apparatus :—

1. Physical and hypsometrical wall map of Belgium, 1·75 metre by 1·60.

2. Hypsometrical wall map of Europe in three colours, 2 metre by 1·75 metre.

3. Physical, political, and commercial wall map of the World, in three colours, 2 metre by 1·75 metre.

4. Oil cloth board map of Belgium and Europe for chalk exercises, 1·30 metre by 1·75 metre.

5. Hypsometrical submergible relief, for facilitating the study of maps, painted plaster, 0·35 metre by 0·30.

6. Typical landscape in relief, painted plaster, for studying geographical nomenclature, 0·70 metre by 0·60 metre.

7. Several classical atlases (chromolithographic).

8. Series of hartographical exercises, for the use of students.

9. Manuals for masters and pupils.

Alexis M. Gochet, Professor at the Normal School of Carlsbourg (Belgian Luxemburg).

4092a. From Mottershead & Co., Manchester.

Figure plates.	Specimen of crystallized bismuth, with glass shade.
Thick horse shoe magnets.	
Pair of magnets.	Combined Holtz and Bertsch electrical machine.
Tate's air pump and receiver.	
Small air pump and receiver.	Pair bismuth and antimony.
Concave and convex mirrors.	Bar to show different expansion of metals.
Bottle prism.	
Sets of phosphorescent powders.	Voltameter.
Copper drying oven, with regulator.	Oersted's experiment.
Copper still and condenser.	Jewsbury's reflecting polariscope without Nicol's prism.
Liebig's condenser.	
French blast spirit lamp.	Bobbin to show induction.
Books of chemical labels.	Madgeburg hemisphere.
Electro-magnet.	Electroscope, with condenser.
Galvanometer.	Vulcanite electrophorus.
Bladder frame and weights.	Rammelsberg's air bath.
Prism mounted on stand.	Set of apparatus and chemicals according to Roscoe's primer.

4092b. Apparatus from J. J. Griffin & Sons.

Oil lamp furnace, with Griffin's blowing machine for the fusion of metals at high temperatures.	Gas burner, 16 jets, and a selection of internal fittings.
	Iron tripod stand.
Griffin's patent blast gas furnace, with	P. P. cylinder and crucible.
	Fireclay ventilator.

Fireclay cone.
„ plates.
P. P. crucible and cover.
Blowing machine for do., with weights.
Blast gas burner, 26 jets, with stand and crook.
Gas crucible furnace complete, large size.
Gas crucible furnace, complete, small size, and P. P. pots.
Gas muffle furnace, complete, large size.
Gas muffle furnace, complete, small size.
Large muffle and bottle necked.
Small „ „
Large clay atmopyre.
Small „
Bone ash cupels.
Plattner's gold assay apparatus.
Pair cupel tongs, with shields.
„ tongs straight, No. 126.
„ „ bent, No. 127.
„ „ charcoal tongs bent, No. 120.
„ „ bent, No. 125.
„ „ bow, No. 128a.
„ „ basket, No. 128.
„ „ bow, No. 124.
„ crucible tongs, No. 4,652.
„ flask tongs, No. 4,576.
Set (4), roasting dishes.
„ (3), cornet pots.
„ (5), scorifiers.
Wrought iron crucible.
„ „
Iron cupel mould.
Boxwood cupel mould.
Ingot mould, conical form on foot.
„ „ „
„ with two holes.
„ „ three divisions.
„ for 12 rods.
Glass parting flask.
Mallet.
Pair boxwood crucible moulds, with three stamps.
Cupeling furnace, with chimney.
Muffle for do.
B. P. tube for do.
Blowpipe set (pocket).
„ in one tin case.
„ in two tin cases in mahogany box.
„ in three tin cases in mahogany box.

Major Ross's blowpipe set (with extras).
Wire stands.
Aluminum plate.
Glass beakers.
G. S. blowpipe with platinum nozzle and stand.
Brass blowpipe lamp.
„ stand.
„ blowpipe.
Stand.
Brass gas blowpipe.
Stand.
Brass Bergmann's blowpipe.
„ German blowpipe.
Stand.
Portable G. S. blowpipe with stand.
Brass „ „
Tin laboratory lamp.
Brass blowpipe on stand.
„ gas blowpipe on stand.
Blowing apparatus.
Tin blowpipe lamp.
Brass stand.
„ pocket blowpipe lamp.
Pair iron tongs, with spoon.
„ brass „
„ „ crucible tongs.
„ „ tongs, long points.
Platina spoon and cover with handle.
Pair brass tongs with ivory points.
„ platina tongs.
Piece platina foil.
Pair G. S. crucible tongs, with plat. tips.
Patina crucible and cover.
Set of three charcoal borers, with charcoal.
Set of Plattner's cupel supports.
„ lead measure.
Agate mortar and pestle. 2 inch.
„ „ 2½ inch.
Plattner's roasting furnace.
„ brass sieve.
Platina spoon and handle.
Platnum wire holder and wire.
Plattner's brass scoop.
„ capsule support.
„ crucible support.
„ „ mould, with boxwood stamps.
„ mould for square blocks.
Albata spoon.
Ivory spoon.
Steel spatula.
Flat file.

Anvil.
Crushing mortar.
„   „   with base.
Blowpipe hammer.
Triple lens.
Single lens.
Coddington lens.
Magnetic and electric needle in case.
„   chisel.
Berlin porcelain cups.
Brass spirit bottle.
Griffin pastille mould, boxwood, and two stampers.
Griffin pastille mould, iron.
Support for sublimator.

Charcoal block and support.
Colour plate and support.
B. P. cup and wire holder.
Cupel support and two cupels.
Set of three watch glasses.
Bunsen's burner and blowpipe jet.
Box powdered glass.
Brass stand.
Wooden blocks.
Set of 100 minerals.
Glass case for do.
Set of 100 blowpipe minerals.
„   6 Von Kobell's minerals.
„ 120 fragments minerals.
„   9 Mohr's degree of hardness.

4092c. " **Educational Series.**" Physical and Chemical Apparatus.   *J. J. Griffin & Sons.*

Voltaic Electricity and Magnetism.

Thermo-electric Pile, simple form, for school use.
Thermo-electric Pile, very delicate, for investigations.
Galvanometer, vertical needle, simple, for school use.
Galvanometer, astatic, delicate, for use with thermo-piles.
Galvanometer, astatic, for lecturers, with long needle.
Astatic needle, simplest form, on support for demonstration.
Tangent Galvanometer, working model of.
Solenoid, single, with soft iron core, to explain electro-magnets.
Solenoid, double, with soft iron core, to explain induced currents and induction coils.
Pair of *Induction Planes,* copper wires coiled on a plane to exhibit induced currents.
Induction Coil, school form, to give shocks.
Ruhmkorff's Induction Coil, with Grove's battery.
Ruhmkorff's Induction Coil, with Bunsen's battery.
Rheostat, for interposing known resistances.
Electrolysis of water, Buff's apparatus, delivering the gases separately.
Smee's Galvanic Battery.
Model of *Cannon,* fired by electricity.
Faraday's Needle, rotating round a current parallel to itself.
Barlow's Stellar Wheel, rotating between the poles of a magnet.
Sturgeon's Copper Disc, rotating between the poles of a magnet.
De la Rive's Floating Battery, carrying a magnetic solenoid.
Electro-magnetic set of Apparatus, to exhibit the experiments of *Faraday, Ampère, Ritchie, &c.*
Electro-magnetic Machine, with *Siemens' Armature,* for heating wires, firing fuses, decomposing water, &c.
Horse-shoe Magnet, large and powerful, for use on lecture table.
Bar Magnet, suspended in cradle for lecture demonstrations.
Dipping Needle with universal motion.
Ritchie's Rotating Magnet, simplest form.
Soft-iron Discs, Stars, Rods, &c., for illustrating *magnetic induction.*

Hydrostatics and Hydraulics.

Pascal's Apparatus, proving that the pressure depends on the height of the column and the area of the base, and not upon the capacity of the vessels.

Holdat's Apparatus, for proving that the pressure depends on the height of the column and the area of the base, and not upon the capacities of the vessels.

Glass Cone with movable base for determining the pressure on the base of a cone, having a movable bottom.

Apparatus to prove that water presses equally in all directions if pressure be applied to any part of it.

Hydrostatic Bellows, for lifting a heavy weight by the pressure of a column of water.

Bramah's Hydrostatic Press, glass model.

Bramah's Hydrostatic Press, metal model of some power for school use.

Force Pump of Glass, with glass valves and pistons, to facilitate inspection.

Lift Pump of Glass, with glass valves and pistons, to facilitate inspection.

Fire Engine, tin-plate working model of, with double barrels and air chamber.

Hero's Fountain, in which the water jet is caused by the water column compressing the air.

Fountain, intermitting, glass model.

Spring, intermitting, glass model.

Syphon containing a fountain, glass model.

Appold's Centrifugal Pump, which raises water by the rapid rotation of a small water-wheel.

Archimedean Screw, glass model, to raise water.

Set of *Glass Tubes,* to exhibit the rise of water by capillary attraction.

Simple Glass Vessel to prove that water rises to a level in all communicating vessels.

Vessel to show that water rises to a level in all communicating vessels, superior construction for the lecture table.

Water Wheels, Set of *Models,* showing *Overshot, Undershot,* and *Breast Wheels.*

Physical Apparatus. Cheapest Set of Physical Apparatus for use in Elementary Schools. Prof. Balfour Stewart's set of apparatus, described in his Science Primer, No. III.

Description of Apparatus.

No. of
Experiment.

1, 2.—Tin pan, with peas.

3.—Iron plate with four strings.

4.—Balance to carry 2 lbs. in each scale; beam two feet long.
Piece of metal weighing 200 grains.
Set of weights, 600 grains to $\frac{1}{2}$ grain.

5. —2 lbs. mercury in a bottle.
Two pieces of glass two inches square.

6.—Apparatus unnecessary.

9, 10.—Beam of wood.
Two 4-lb. weights.

15.—Plumb-line.
Stoneware dish for mercury.

16.—Tube for showing level of water.

17.—Metal cylinder with two tubes and stoppers.
Tube with movable bottom and cord.
Water-jar for tube.
Indigo solution.

18, 19.—Substance weighing 1,000 grains, same specific gravity as water.

No. of
Experiment.
20.—Hollow brass cylinder.
  Bucket to contain it.
  Apparatus for attaching the bucket to balance.
21.—See Experiment 18.
22.—Block to illustrate Flotation.
24.—Apparatus unnecessary.
25.—Tate's Air-pump.
  Bell-jar receiver.
  Two india-rubber balls.
26.—Jar with neck and flange.
  Two pieces of india-rubber for it.
27, 28, 29.—Box with strings.
30.—Magdeburg hemispheres.
  Brass cock for hemispheres.
31.—Tube for barometer.
  Glass mortar for cistern.
  Funnel for filling barometer.
33.—Vibrating wire on support.
37.—Model thermometer.
  Centigrade thermometer.
38.—Bladder two-thirds filled with air.
39.—Further apparatus unnecessary.
40.—Use tin pan of experiment 1.
41.—Use flask of experiment 41.
42.—Flask for boiling water, and cork in duplicate.
  Triangle and wire gauze to support flask.
43, 44.—No special apparatus necessary.
45.—Pan to hold sulphuric acid *in vacuo*, and shallow vessel to hold water.
46.—No apparatus necessary.
47.—Use flask of experiment 42.
48.—Wires to show unequal power of iron and copper to conduct heat.
50.—Use tin pan of experiment 1.
51.—Apparatus to show image of candle.
52.—Apparatus unnecessary.
54.—Electric pendulum.
  Several pieces of elder-pith.
55.—Electroscope.
  Electrical machine, 16-inch plate.
  Box of amalgam.
56.—Rod, half brass, half glass.
  Rod of glass covered with red wax.
  Piece of silk.
  Piece of flannel.
57.—No additional apparatus.
58, 59.—Brass ball, with point, on insulated stand.
60.—No apparatus required.
61.—Leyden jar, pint size.
  Discharger.
62.—Groves battery, 4 cells in frame.
  Yard of fine platinum wire.
63.—Voltameter.
64.—Electro-magnet.
65.—Knitting-needle and thread.
66.—Apparatus for Oested's experiment.
67.—Thirty feet of covered wire.

Mechanics.

Black Rail to suspend the models from the school board.
Lever for suspension.
Lever supported on a fulcrum.
Intermittent Motion, model to explain.
Lock and Key, large working model of.
Compound Wheel and *Axle*.
Tilt Hammer.
Screw, model to explain the action of.
Train of *Wheels*, transmission of motion.
Inclined Plane and *Roller*.
Sets of Pulleys :—
  Four simple pulleys.
  Pair of long three-sheave pulleys.
  Pair of square three-sheave pulleys.
  Pair of White's concentric pulleys.
Centre of Gravity :—
  Irregular board to explain centre of gravity.
  Leaning tower to explain centre of gravity.
  Parallelopipeds, a pair of, to explain centre of gravity.
  Double cone running uphill, to explain centre of gravity.
  Bowl-about, a toy to explain centre of gravity.
  Equilibrist, a toy to explain centre of gravity.
Apparatus to explain *Inertia*.
Plumb-line and *Mercury Mirror*.
Pendulums made of various substances.
Model of a *Vernier*.
Whirling Table of iron, with set of adjuncts, to explain the action of *Centrifugal Force*.
Centrifugal Railway, model of.
Gyroscope for *explaining Inertia*.
Wedge, model, explaining action of.
Percussion Machine for experiments on the *collision of elastic bodies*.
Constructive Mechanics :—
  Tenon and *Mortice*.
  Scarfing.
  Half Lapping.
  Dove-tailing.

Frictional Electricity.

School Electrical Machine with 18-inch glass plate.
Small, Cheap, Cylinder Electrical Machine.
Conical Conductors to show accumulation of electricity near the point.
Cylindrical Conductors to exhibit and explain *induction*.
Set of *three small and simple Conductors, Ball Cylinder* and *Cone*, for Elementary Schools.
Leyden Jar with movable coats to exhibit the separation of the positive and negative electricity. 1. The glass jar. 2. The inner coat. 3. The outer coat.
Unit Leyden Jar to measure strength of a charge.
Set of *Leyden Jars* for obtaining powerful effects.
Biot's Ball and insulated covers to show that the electricity is only on the surface.
Roll of *Tin Foil* to show effect of increasing surface.
Insulated Stool on which to place a boy as a conductor.

Henley's Universal Discharger for passing charges through an insulated body.
Hand Discharger with insulated handle.
Luminous Flash to show discharge in vacuo.
Cheap Electroscope of large size for lectures, a wooden lath supported on a flask.
Set of Bells to be rung by the electrical discharge.
Pith Balls to dance under an excited beaker.
Cheap Conductor, a tea-tray supported by a beaker.
Pith Ball Electroscope.
Gold Leaf Electroscope with condensers.
Gold Leaf Electroscope without condensers.
Vulcanite Electrophorus which gives off sparks when excited by friction.
Glass Rod half roughened to show the two kinds of electricity.
Rod of Sealing Wax for excitation.
Rod of Vulcanite for excitation.
Rod of Shellac for excitation.
Rod of Sulphur for excitation.
Fox's Brush for exciting electricity.
Cats' Skin for exciting electricity.

Pneumatics.

Air-Pump, double barrelled, arranged for rapid action, with an auxiliary *Tate's air-pump* to obtain a very perfect vacuum.
Tate's School Air-Pump; this produces a good vacuum quickly, and is made at a moderate price.
Exhausting and *Condensing Syringe.*
Leslie's Apparatus for *freezing Water* over sulphuric acid in vacuo.
Apparatus to show that a *Guinea* and a *Feather* fall in vacuo with the same velocity.
French Tube Form of *Guinea and Feather Apparatus.*
Apparatus to explain the *Barometer* and the pressure of the atmosphere.
Apparatus to cut an apple in half, by the downward pressure of the atmosphere.
Apparatus to prove that *combustion* cannot take place in *vacuo.*
Heavy Lead Weight which is raised by the expansion of some air contained in a bladder.
Magdeburg Hemispheres for showing that the pressure of the atmosphere is about 15 lbs. on the square inch.
Mercury Cup, in which the pressure of the air forces mercury through the pores of wood.
Marriotte's Apparatus, to show that under the pressure of two atmospheres, air is compressed into half its ordinary bulk.
Marriotte's Apparatus, for experiments under the pressure of half an atmosphere.
Lead Weight supported by the upward pressure of the air.
Fountain in Vacuo, the water being forced up by atmospheric pressure.
Series of Globes to raise water by the alternate expansion and compression of air.
Apparatus showing *expansion* and *compression* of air.
Rubber Balls which expand and contract under varying pressure.
Diving Bell, working model of.
Pair of Water Bottles, the expansion of the air forces the water from one bottle into the other, and the pressure of the atmosphere causes its return.
Tantalus Cup.

Water Hammer, to be exhausted by the air-pump.
Water Hammer; fall of water in a partial vacuum.
Wollaston's Cryophorus, to freeze water by effects of evaporation.

Sound.

Sonometer or *Monochord*, simple form with violin bow.
Sonometer or *Monochord*, superior construction, with violin bow, for experiments on vibrating wires.
Savart's toothed wheels, for use with whirling table, to produce tones.
Paper Disc, perforated, to act as a *Syren*.
Syren, simple, without counter, can be worked by a cheap rubber foot blower.
Syren, with *counter*, can be worked by a cheap rubber foot blower.
Organ Pipe, with *resonance box*.
Resonance Box, with movable piston to explain effect of *length* of *air column*.
Pair of *Lecturer's Tuning Forks*, in *Unison*.
Chladni's Vibrating Plates, a set of three, in brass, with bow, to show figures of vibration.
Chladni's Vibrating Plates, a set of ten in glass, with bow and clamp to hold them.
Hopkins' Forked Tube, to exhibit *interference*.
Speaking Trumpet, model of.
Hydrogen Flame, which *sings* when burning in the glass tube.
Common Gas Jet, which *sings* when burning in the glass tube.
Sound Waves, propagation of, apparatus to explain.

Heat.

Fergusson's Pyrometer, to show *unequal expansion* of various metals.
Compound Bar of iron and copper, which bends when heated, to explain effects of *unequal expansion*.
Set of Bulb Tubes, to exhibit the *unequal expansion* of different liquids.
Bar of Iron and Copper, to explain *unequal conduction* of heat.
Ingenhouz's Apparatus, to show *unequal conducting powers* of various substances.
Cylinders of *various metals*, proving the *unequal conducting powers* of metals, and their capacities for heat.
Apparatus to *break* an *iron bar* by *contraction* of heated bar.
Unequal Absorption, by bright and dull surfaces.
Reflectors, pair of silvered copper, for concentrating rays.
Leslie's Cubes, to explain effects of different surfaces in *radiation*.
Leslie's Differential Thermometer, cheap form.
Count Rumford's Differential Thermometer, improved by Matthieson and Griffin.
Metal Tube, in which water can be *boiled* by the *friction* of wooden boards.
Wollaston's Steam Engine, glass model of, showing *expansion* and *condensation* of steam.
Iron Bottles, very thick, which *burst* when water is *frozen* in them.
Hope's Apparatus, to show when *water* reaches its *maximum density*.
Pulse Glass, showing *expansion* of spirit vapour under *diminished pressure*.
Prince Rupert's Drops, unannealed glass, showing effect of *unequal contraction*.
Candle Bombs, containing spirit, the expansion of which causes explosion.
Dr. Tyndall's Apparatus to show water boiling at a lower temperature under a reduced atmospheric pressure.

Balls, of various metals, which pass through a wax plate in times **varying** as their *capacities* for *heat.*

Faraday's Apparatus, to exhibit *convection* in liquids.

Convection of *Heat* in liquids ; large size, for the lecture table.

Set of Barometer Tubes, to show *elastic force* of various vapours.

Marcet's Steam Boiler, to prove that *temperature* increases with *pressure.*

Ventilation of Rooms, fitted bottle, to explain.

Rain Gauge, simple, to measure fall of rain.

Light.

Electric Lamp, source of light.

Groves Battery, to work *electric lamp.*

Perforated Screen, for supporting objects in the ray of light.

Bottle Prism, for holding *carbon sulphide* and producing long *spectrum* on the screen.

Condensing Lens, to throw enlarged images on the screen or wall.

Ground-Glass Screen, to receive small images.

Oxyhydrogen Gas Burner, as source of light.

Oxyspirit Burner, and lantern, cheap substitute for electric lamp.

Complementary Prisms of *Crown and Flint Glass.*

Oscillating Prism, to produce *white light* by rapid action on the retina.

Bunsen and Kirchoff's Spectroscope, for *Spectrum Analysis.*

Set of Lenses, to explain their shape and action.

Set of Lenses, mounted, with focussing screen to explain their various action.

Glass Mirror, concave, to *concentrate* rays.

Glass Mirror convex, to *disperse* rays.

Curved Mirror, showing *distorted* image.

Plane Mirror, black.

Angular Mirrors, double reflection.

Bunsen's Photometer, to *measure* lighting power of gas, &c.

Ritchie's Photometer, for *comparing* one light power with another.

Colour Disc, to explain *constitution* of *white light.*

Colour Disc, with series of discs, explaining *combinations* of *colour.*

Colour Top, explaining *combinations* of *colour.*

Norremberg's Polariscope, showing colours through plates of *mica, selenite,* &c.

Tourmaline Tongs, showing *polarised* light, in colours, through crystals.

Newton's Coloured Rings, retardation of velocity of rays. *Interference.*

Camera Obscura.

Screens, dull and bright, to explain *absorption* and *reflection.*

Refraction ; a line or stick appears bent when partly immersed in water.

Müller's Apparatus, to show *refraction* in water.

Stereoscope angle of vision.

Revolving Disc, to show rapidity of light rays.

Fluorescence, Uranium-Glass Cube, to produce.

Fluorescence, Permanganate of Potash, to produce.

Fluorescence, Uranium Plate, to produce.

Double Refraction, Iceland Spa, to exhibit.

Water-Bulb Microscope.

4092e. Educational Collection.   *W. and A. K. Johnston.*

12-inch Terrestrial Globe.

Map of New York, Pennsylvania, &c.

Illustrations of Natural Philosophy ; sheet 7—steam engines and book.

Illustrations of Astronomy ; sheet 1 —solar system and book.

Geological Map of the British Isles, and book.

Physiology No. 1, Illustration No. 5, and book.

Properties of Bodies, Illustration No. 1, and book.

Hydrostatics, Illustration No. 3, and handbook.

Physiology No. 2, Illustration No. 6, and book.

Astronomy, sheet 2, Illustration No. 13, and book.

Astronomy, sheet 3, Illustration No. 14, and book.

Mechanical Powers, Illustration No. 2, and book.

Hydraulics, Illustration No. 4, and book.

Case's Map of the United States.

## 4092e. Educational Collection.		*M. Berthelot, Paris.*

Calorimeter for thermo-chemistry, with agitator.

Electric motor for same.

Glass beaker with lip.

Porcelain dish for quicksilver.

Glass apparatus with spiral tube.

Bottle of alcohol.

Two flasks containing fluid.

Large glass apparatus.

Two bent glass electric tubes.

Glass apparatus with spiral tube at top.

Test tube.

Large bent glass tube containing mercury.

Five tubes containing chemicals.

Glass tube or vacuum glass.

Bent glass tube with wire cloth cup at one end.

Small stoppered bottle of formiate of lead.

Thermometer with scale on iron stand with brass base.

Two glass jars.

APPARATUS FOR TEACHING CHEMISTRY.

4093. Working Bench for a **Chemical Laboratory,** for the use of science classes (according to Dr. R. Arendt).

Franz Hugershoff, Leipzig.

4094. Hofmann's Electrolytic Apparatus.

Ch. F. Geissler & Son, Berlin.

4095. Hofmann's Apparatus for liquefying sulphurous anhydride.			*Ch. F. Geissler & Son, Berlin.*

4096. Geissler's Apparatus for determining the greatest density of water.			*Ch. F. Geissler & Son, Berlin.*

4097. Hofmann's Electrolytic Apparatus for decomposing hydrochloric acid.			*Julius Schober, Berlin.*

4098. Hofmann's Apparatus for showing that in the **Synthesis** of **Ammonia Gas** three volumes of hydrogen unite with one volume of nitrogen, to form two volumes of ammonia gas.			*Julius Schober, Berlin.*

4099. Hofmann's Apparatus for showing that in the **Synthesis** of **Water** two volumes of hydrogen unite with one volume of oxygen, to form two volumes of water, gas, or steam.

Julius Schober, Berlin.

4100. Hofmann's Steam Generator, for the lecture-room table. *E. A. Lentz, Berlin.*

4101. Hofmann's Apparatus for decomposing **Phosphuretted Hydrogen.** *Julius Schober, Berlin.*

4102. Distilling Apparatus, for the use of science classes. *F. A. Wolff & Sons, Heilbronn and Vienna.*

This distillation apparatus has been specially constructed at a very cheap rate for the use of science classes, to illustrate distilling operations.

4103. Diagram, showing elevation, section, and plan of a working-bench in the Laboratory of the Strasburg University. *Chemical Institution of Strasburg University.*

4104. Photographs of a portion of the principal room of the **Laboratory** belonging to the Berggewerkschaft Association of Miners at Bochum. *Berggewerkschaftskasse, Bochum, Rhenish Prussia.*

4105. Hofmann's Apparatus for showing the **simultaneous Decomposition** of hydrochloric acid, of water, and of ammonia. *Julius Schober, Berlin.*

4106. Hofmann's Apparatus for showing the **Constancy** of **the Proportions** in the combination of hydrochloric acid gas with water. *Julius Schober, Berlin.*

4107. Hofmann's Apparatus for showing the **Synthesis** of **Water.** *Julius Schober, Berlin.*

4108. Hofmann's Lecture Room Eudiometer. *Julius Schober, Berlin.*

4109. Hofmann's Apparatus for demonstrating that the volumes of oxygen which enter into the composition of carbon and sulphur dioxide are equal to the respective gas volumes of these compound gases. *Julius Schober, Berlin.*

4110. Hofmann's Apparatus for illustrating the **Phenomena** of **Combustion.** *Julius Schober, Berlin.*

4111. Models from Mitscherlich's **Chemico-technical Collection,** representing—
(1.) A puddling furnace.
(2.) A coke oven.
(3.) The lower part of a blast furnace.
Prof. A. Mitscherlich, Münden.

2604. Warmbrunn, Quilitz, and Co.'s Large Glass Gas Holder, described by R. Müncke in Dingler's Polyt. Journ., vol. 218, p. 40. *Warmbrunn, Quilitz, & Co., Berlin.*

2605. Warmbrunn, Quilitz, and Co.'s Large Copper Gas Holder, of bright metal.
Warmbrunn, Quilitz, & Co., Berlin.

2606. Warmbrunn, Quilitz, and Co.'s Large Copper Gas Holder, bronzed, with top water-reservoir.
Warmbrunn, Quilitz, & Co., Berlin.

2607. Warmbrunn, Quilitz, and Co.'s Copper Water Oven, two-walled, with thermostat, thermometer, and side-tube, showing level of water.   *Warmbrunn, Quilitz, & Co., Berlin.*
Described by R. Müncke in Dingler's Polyt. Journ., vol. 219, p. 72.

2608. Water Oven, the same as the above, but made of **Zinc.**   *Warmbrunn, Quilitz, & Co., Berlin.*

2609. Warmbrunn, Quilitz, and Co.'s Copper Drying-Closet, single-walled, provided with thermostat and a thermometer.   *Warmbrunn, Quilitz, & Co., Berlin.*

2610. Bunsen's Universal Clamp and **Stand,** made of iron, lackered, complete.   *Warmbrunn, Quilitz, & Co., Berlin.*

2611. Warmbrunn, Quilitz, and Co.'s Large Universal Clamp and **Stand,** square rods made of brass, square foot-plate of iron and stands, complete.
Warmbrunn, Quilitz, & Co., Berlin.
A description of this apparatus by R. Müncke will be found in Ber. d. Deutsch. chem. Ges. Berlin, 1873, p. 435.

2612. Warmbrunn, Quilitz, and Co.'s Large Universal Stand, with a large double holder, round rods made of brass, and square iron foot-plate, complete (described as above).
Warmbrunn, Quilitz, & Co., Berlin.

2613. Stand of iron, resting on an iron foot-plate, with three two-holed frames or ring-holders, vertical and horizontal ("Doppelmuffen"), rings of three different sizes with straight rods for sliding through the horizontal hole of the frame, binding screws, &c.   *Warmbrunn, Quilitz, & Co., Berlin.*

2614. Stand made of brass, with iron foot-plate, five two-holed frames, three large rings of different sizes, two clamps of varying size, all made of brass.   *Warmbrunn, Quilitz, & Co., Berlin.*

2615. Brass Stand, on an iron tripod, with two simple brass burette holders.   *Warmbrunn, Quilitz, & Co., Berlin.*

2616. Brass Stand, on an iron tripod, with four brass burette holders.   *Warmbrunn, Quilitz, & Co., Berlin.*

2617. Liebig's Condenser, in **Glass,** supported by a brass stand on an iron tripod.   *Warmbrunn, Quilitz, & Co., Berlin.*

2618. Bunsen Burners (two), of simple construction.
Warmbrunn, Quilitz, & Co., Berlin.

2619. Bunsen Burners, with star, chimney, and reduction-cone.
Warmbrunn, Quilitz, & Co., Berlin.

2620. Gas Lamps (two), with arrangement for the simultaneous regulation of gas and air, of simple construction.
Warmbrunn, Quilitz, & Co., Berlin.

2621. Gas Lamps, with star, chimney, and reduction cone.
Warmbrunn, Quilitz, & Co., Berlin.

2622. Two-burner Bunsen Lamp.
Warmbrunn, Quilitz, & Co., Berlin.

2623. Three-burner Bunsen Lamp, with arrangement for the simultaneous adjustment of air and gas.
Warmbrunn, Quilitz, & Co., Berlin.

2624. Five-burner Bunsen Lamp, of similar construction.
Warmbrunn, Quilitz, & Co., Berlin.

2625. Six-burner Bunsen Lamp.
Warmbrunn, Quilitz, & Co., Berlin.

2626. Warmbrunn, Quilitz, and Co.'s Universal Burner, of newest construction, complete.
Warmbrunn, Quilitz, & Co., Berlin.

A description of this burner by R. Müncke, will be found in Ber. d. Deutsch. chem. Ges. 1874, p. 284; in Dingl. Polyt. Journ., vol. 212, p. 141; and in Fres. Zeitsch. f. anal. Chem., XIII., p. 46.

2627. Warmbrunn, Quilitz, and Co.'s Combustion Furnace, with a four-burner gas lamp, complete.
Warmbrunn, Quilitz, & Co., Berlin.

For a description of this furnace by R. Müncke, refer to Dingler's Polyt. Journ., vol. 212, p. 315, and to Fres. Zeitschr. f. anal. Chem., XIII., p. 167.

2628. Warmbrunn, Quilitz, and Co.'s Combustion Furnace, with 12 burners, each provided with simultaneous regulation for gas and air, complete.
Warmbrunn, Quilitz, & Co., Berlin.

2629. Gas-washing Apparatus (comp. R. Müncke, Fres. Zeitsch. f. anal. Chem. XV., Part I.)   Three different sizes.
Warmbrunn, Quilitz, & Co., Berlin.

2630. Hydrofluoric Acid Apparatus, consisting of retort, still-head, and U tube, for condensing the acid.
Warmbrunn, Quilitz, & Co., Berlin.

2631. Hydrofluoric Acid Apparatus, consisting of retort, still-head, and receivers, on a stand.

Warmbrunn, Quilitz, & Co Berlin.

2632. Bardeleben's Gas Evolution Apparatus for laboratories; consisting of a large cylinder with bell-glass, and glass vessel with ground brass-plate and bayonet-joint, gas washing apparatus, &c. Quite new.

Warmbrunn, Quilitz, & Co., Berlin.

2633. Pulse Pump, mounted, with reservoir, manometer, two glass stop-cocks, and unions for connecting with the water supply pipe. *Warmbrunn, Quilitz, & Co., Berlin.*

2634. Kipp's Gas Generator, provided with a Geissler glass stop-cock. *Warmbrunn, Quilitz, & Co., Berlin.*

2635. Set of Beakers, without lip, 16 in the set, up to 680 mm. *Warmbrunn, Quilitz, & Co., Berlin.*

2636. Set of Beakers, not lipped, No. $\dfrac{1-16}{1}$, $\dfrac{1-12}{1}$, $\dfrac{1-8}{1}$, $\dfrac{1-5}{1}$ set. *Warmbrunn, Quilitz, & Co., Berlin.*

2637. Sets of Beakers, lipped, No. $\dfrac{1-12}{1}$, $\dfrac{1-8}{1}$, $\dfrac{1-5}{1}$ set.

Warmbrunn, Quilitz, & Co., Berlin.

2638. Series of Funnels, 10 pieces, from 30 to 130 mm. diameter. *Warmbrunn, Quilitz, & Co., Berlin.*

2639. Series of Lettered (etched) **Reagent Bottles,** 27 pieces. *Warmbrunn, Quilitz, & Co., Berlin.*

2640. Series of Lettered (etched) **Wash Bottles,** for water, alcohol, and ether. *Warmbrunn, Quilitz, & Co., Berlin.*

2641. Graduated Cylinders, of 1,000, 500, 250, and 100 cc.

Warmbrunn, Quilitz, & Co., Berlin.

2642. Graduated Flasks, of $\frac{1}{4}$, $\frac{1}{2}$, and one litre.

Warmbrunn, Quilitz, & Co., Berlin.

2643. Glass-stoppered Burettes, four, of 20 cc. in tenths, 50 cc. in fifths, 50 cc. in tenths, 60 cc. in fifths.

Warmbrunn, Quilitz, & Co., Berlin.

2644. Series of Preparation Glasses for holding liquids, 1-8. *Warmbrunn, Quilitz, & Co., Berlin.*

2645. Series of Preparation Glasses for holding solids, 1-8. *Warmbrunn, Quilitz, & Co., Berlin.*

2646. Series of Flasks, 11 from 30–4,000 cc.

Warmbrunn, Quilitz, & Co., Berlin.

2647. Set of Glass Evaporating Dishes, tipped, 12 pieces, from 55–270 mm.   *Warmbrunn, Quilitz, & Co., Berlin.*

2648. Set of Clock Glasses, for evaporating, from 25–315 mm. diam.   *Warmbrunn, Quilitz, & Co., Berlin.*

2649. Glass Condenser, consisting of three equal-sized bell-glasses connected by ground necks, holding about 2,000 cc.
Warmbrunn, Quilitz, & Co., Berlin.

2650. Drying Apparatus, with glass stop-cock.
Warmbrunn, Quilitz, & Co., Berlin.

2651. Exhaustion Apparatus, in glass, of two litres capacity.   *Warmbrunn, Quilitz, & Co., Berlin.*

2652. Bottle, for keeping mercury over acids.
Warmbrunn, Quilitz, & Co., Berlin.

2653. Levigation Apparatus, with four tubes.
Warmbrunn, Quilitz, & Co., Berlin.

2654. Air-pump Receiver, with tubulures and glass stop-cock, 280 mm. in height and 200 mm. in diam.
Warmbrunn, Quilitz, & Co., Berlin.

2655. Aspirator, consisting of a double tubulated Woulff's bottle and ground in glass stop-cock for the lower tube.
Warmbrunn, Quilitz, & Co., Berlin.

2656. Glass Stop-cock, 330 mm. in length.
Warmbrunn, Quilitz, & Co., Berlin.

2657. Vinegar Stop-cock, 190 mm.
Warmbrunn, Quilitz, & Co., Berlin.

2658. Glass Jar, for mercury, provided with glass stop-cock.
Warmbrunn, Quilitz, & Co., Berlin.

2659. Finkener's Gas Evolution Apparatus, quite new.
Warmbrunn, Quilitz, & Co., Berlin.

2599. Copper Water Bath, nickel-plated, together with a tripod stand and rings of porcelain as well as metal, and provided with an arrangement for keeping the water at a constant level.
Warmbrunn, Quilitz, & Co., Berlin.

2600. Copper Water Bath, together with a tripod stand, and an arrangement for keeping the water at a constant level, a set of copper rings, and also a cover with different sized openings.
Warmbrunn, Quilitz, & Co., Berlin.

2601. Carmichael's Suction Funnel, with improvements.
Chemical Laboratory of the University of Göttingen (Prof. Hübner).

4112. Hofmann's Apparatus (No. 1–16).

Warmbrunn, Quilitz, & Co., Berlin.

(1.) Apparatus for the electrolysis of hydrochloric acid, of water, and ammonia.

(2.) Apparatus for illustrating the synthesis of hydrochloric acid.

(3.) Apparatus for showing that hydrogen and chlorine when united suffer no contraction of volume.

(4.) Apparatus for showing that two volumes of water-gas (steam) are formed when two volumes of hydrogen unite with one volume of oxygen.

(5.) Apparatus for showing that ammonia consists of three volumes of hydrogen and one volume of nitrogen.

(6.) Apparatus for the simultaneous electrolysis of water, hydrochloric acid, and ammonia.

(7.) Apparatus for showing that the composition of hydrochloric acid does not vary.

(8.) Apparatus for showing that hydrogen and oxygen unite with each other only in the proportions in which they are obtained by the electrolysis of water.

(9.) Apparatus for illustrating the deportment of simple and compound gases under the influence of changes of temperature and pressure. Of novel construction.

(10.) Eudiometer for the lecture-room table with stand.

(11.) Apparatus for experimenting with liquid sulphurous acid.

(12.) Apparatus for illustrating the phenomena of combustion.

(13.) Vapour density apparatus complete.

(14.) Apparatus for proving that the volumes of carbonic and sulphurous anhydride equal the volumes of oxygen which enter into them.

(15.) Apparatus for the electrolysis of phosphide of hydrogen.

(16.) Sodium apparatus.

4113. Series of small **Apparatus,** for illustrating in the lecture room projections on a screen by means of Duboscq's lamp. Manufactured by Dr. Geissler of Bonn.

Prof. H. Landolt, Aix-la-Chapelle.

The series consists of:—

(1.) An apparatus for showing the decomposition of water by the galvanic current.

(2.) Apparatus for decomposing water by means of palladium electrodes with absorption of the hydrogen gas.

(3.) Apparatus for the electrolysis of saline solutions.

(4.) Lecture room eudiometer.

(5.) Apparatus for showing the formation of nitric peroxide by passing electric sparks through air.

(6.) Apparatus for showing the action of oxygen upon nitric acid.

(7.) Apparatus for illustrating the absorption of gases by liquids.

(8.) Condenser for bromine and other coloured vapours.

(9.) Apparatus for showing the formation of flowers of sulphur.

(10.) Apparatus for illustrating the manufacture of sulphuric acid, and for showing the decolorising action of sulphurous acid upon nitric peroxide.

4114. Photographs, illustrative of the chemical lecture-room of the Polytechnic School of Aix-la-Chapelle, and of the contrivances for throwing experimental illustrations on a screen by means of a Duboscq's lamp.   *Prof. H. Landolt, Aix-la-Chapelle.*

4115. Photographs of the lecture-room table, showing the provisions made for enabling the audience to see more favourably any experimental demonstration, by employing screens behind the lecture-table.   *Prof. H. Landolt, Aix-la-Chapelle.*

APPARATUS FOR TEACHING MINERALOGY.

4116. " Student's Elementary Collection of Minerals."
J. R. Gregory.

4117. Elementary Collections (four) **of Minerals, Fossils, and Rocks,** systematically arranged, in polished wood cabinets; fitted for the use of the student and science-class teacher, and illustrating the various mineralogical and geological handbooks.
Thomas J. Downing.

4118. Minerals (24) **for Blowpipe Analysis.** In case, for the pocket. Set I.   *Thomas J. Downing.*

4119. Minerals (24) **for Blowpipe Analysis.** In case, for the pocket. Set II.   *Thomas J. Downing.*

4127. Apparatus for demonstrating the physical properties of **Steam** and the **Steam-jet.**
Dr. L. Bleekrode, The Hague, Holland.

This apparatus is designed for the lecture-table, and consists in a copper boiler, which is heated by a common gas-burner with six or seven flames. The quantity of water it contains is sufficient to perform all the experiments during an hour, so that it needs no feeding, and the strength of the material permits the production of steam at a pressure of four to five atmospheres.

The following experiments may successively be taken or illustrated one after another :—

I. The principle of the water-gauge in a boiler.

II. The action of the safety-valve.

III. The relation between the tension of vapour and its temperature.
IV. The boiling of water at a pressure higher than the atmosphere.
V. The latent heat of steam and the warming system with steam.
VI. The form of the steam-jet and its property of extinguishing fire. (Here either the vertical or the horizontal valve may be used to lead the steam on to some burning material.)
VII. The producing of a vacuum by the steam-jet. (A metallic box may easily be filled with the escaping steam, and as the box is afterwards shut, it is depressed by the atmosphere.)
VIII. The Giffard's injector. (A glass model is fixed to the horizontal valve.)
IX. The steam producing sound, illustrating the steam-signal and fog-signals.
X. The steam sand blast. (This interesting experiment is performed in a very simple and yet satisfactory manner, by allowing the sand falling through a vertical funnel to mix with the horizontal escaping steam-jet, at the mouth of the valve.)
XI. The electrical properties of the steam-jet. (Armstrong's hydro-electric machine.)
XII. Heat producing work. (Principle of thermo-dynamics.) (Here a model steam-engine is connected with the horizontal steam valve.)

The chief advantages of this apparatus (already used in several institutions in Holland and Germany) are—
1. It allows, with very little expenditure of time and heat, the illustration of the fundamental principles of heat on the lecture table.
2. It produces steam of high pressure, the properties of which may be shown with very little danger.
3. The relatively small expense (10l.) renders possible the introduction of this apparatus into schools where physics and mechanics form an importan part of the system of education.

MISCELLANEOUS.

4192a. Cosmographic Apparatus for teaching purposes.
Letellier, France.

4193. Stativ (stand), with a fixed and a movable pulley, besides string and counter-weight.

4194. Polyspast, to be appended to the same stand, with four pulleys, string, and counter-weight.

4195. Lever Apparatus, with attached needle (for the balance) and stand.

4196. Twenty-one **Weights** belonging to the foregoing objects.
Royal Prussian Chief Mining Department, Breslau.

4197. Oscillating Machine (centrifugal apparatus), with the following auxiliary apparatus :—
a. Two-fold brass ring with iron axle.
b. Glass globe with axle.
c. Glass cylinder with inserted Sieboglinder (centrifuge).
d. Double pendulum regulator with throttle-valve.
e. Two balls, joined to each other by a chain.
f. Balancing ruler with sliding weights.
Royal Prussian Chief Mining Department, Breslau.

4198. Model of a Suction Pump.
Royal Prussian Chief Mining Department, Breslau.

4199. Model of a Fire Engine (force pump).
Royal Prussian Chief Mining Department, Breslau.

4200. Four Optical Lenses, with stands, and a round **Paper-Screen,** with stand.
Royal Prussian Chief Mining Department, Breslau.

4201. Optical Prism.
Royal Prussian Chief Mining Department, Breslau.

4202. Steel Magnet.
Royal Prussian Chief Mining Department, Breslau.

4203. Two Magnetic Needles, with stands.
Royal Prussian Chief Mining Department, Breslau.

4204. Galvanic Cell, for potassium bichromate.
Royal Prussian Chief Mining Department, Breslau.

4205. Electro-Magnet, with anchor and stand.
Royal Prussian Chief Mining Department, Breslau.

4206. Electrophor, with cover and glass rod.
Royal Prussian Chief Mining Department, Breslau.

4207. Small Induction Apparatus.
Royal Prussian Chief Mining Department, Breslau.

4208. Small Electro-Magnetical Keyboard and Chiming-work, with paper-roll and marking pencil; a primitive telegraph model.
Royal Prussian Chief Mining Department, Breslau.

4209. Air-Pump, with oblique barrel.
Royal Prussian Chief Mining Department, Breslau.

4210. Pair of Magdeburg Hemispheres.
Royal Prussian Chief Mining Department, Breslau.

4211. Fall Tube.
Royal Prussian Chief Mining Department, Breslau.

4212. Two Glass Globes for Air-Pump.
Royal Prussian Chief Mining Department, Breslau.

4213. Mercury Granulation Apparatus, with glass cylinder. *Royal Prussian Chief Mining Department, Breslau.*

4214. Heron's Ball, with glass bowl.
Royal Prussian Chief Mining Department, Breslau.

4215. Freezing Apparatus, brass stand with glass tubes.
Royal Prussian Chief Mining Department, Breslau.

4216. Syphon Apparatus.
Royal Prussian Chief Mining Department, Breslau.

4217. Expansion Apparatus.
Royal Prussian Chief Mining Department, Breslau.

4218. School Microscope, from the Optical Institute of F. W. Schick, Berlin, with two eyepieces and three object-glasses.
Royal Prussian Chief Mining Department, Breslau.

4219. Collection of Forty Microscopical Preparations from the Institute of Rodig, Hamburg, with explanations by Prof. Dr. Ferd. Cohn.
Royal Prussian Chief Mining Department, Breslau.

The collection serves, in connexion with other means of illustration, for the purpose of teaching in the elementary public schools of the Silesian mining districts the rudiments of the physical and mechanical sciences, and thus to prepare the children of miners for their probable future employment as miners, engine attendants, &c. It is worthy of remark that the money required for the acquisition of these collections, as indeed for the founding of the appropriate schools, comes from a special fund, which owes its origin and continued maintenance to some clauses in the older German mining laws, according to which the revenues of certain mines are applied to school and church purposes. Through the rapid growth and development of the Silesian mining industry, this fund has risen to the value of 25,000*l.* per annum.

The above-mentioned schools are not to be mistaken for the mining schools proper, or the schools serving as preparatory institutions to these latter, both of which are intended for grown-up pupils.

4219a. Apparatus employed in **teaching,** for exhibiting the action of **Electric Currents** on **Currents** of **Magnetism.**
W. Gloukhoff, Warden of Russian Standard Measures.

In this apparatus the mode of suspension of the movable conductors is entirely new, and consists of a brass tube A, B, C, D, through which passes an insulated brass wire E, F; the wire and the tube have on their tops the metallic cups E and D, filled with mercury. Into the cup E the steel point of suspension and one end of the movable conductor are plunged, and into the cup D the other end.

4220. Universal Rotation Apparatus with Twenty-six Auxiliary Apparatus, viz.:—

1. Centrifugal pendulum.
2. Spheres of different weight.
3. Globe, 100 grammes in weight, movable horizontally, with spring balance, on which the centrifugal force can be directly measured.
4. Two oblique glass-tubes for different kinds of liquids.
5. Glass-balloon.
6. Apparatus for demonstrating the flat spherical form of the earth.
7. Apparatus for flattening of an oil-globule (Ring of Saturn).
8. Ring, chain, ball, cylinder, suspended on a cord.
9. Ball, suspended on a long string, for Foucault's pendulum experiment.
10. Glass-balloon for demonstrating the rising of fluids.

11. Model of a centrifugal blast.

12. Model of a draining and drying apparatus.

13. Apparatus by Coulomb, for causing water to boil by friction.

14. Syren-disc, playing one accord.

15. Savart's toothed wheels.

16. Disc with mirror surfaces and flame-manometer for acoustic flame-images.

17. Stroboscopic cylinder for demonstrating various oscillations according to Quincke.

18. Colour-discs.

19. Oscillating prism.

20. Glass-balloon for producing Newton's colour-rings in thin metal plates of fluid.

21. Stroboscopical discs on the systems of Dove, Poggendorff, &c., illuminated by Geissler's tubes.

22. Becquerel's phosphoroscope.

23. Three rotatory tubes by Geissler.

24. Apparatus for Arago's rotation magnetism.

25. Apparatus for producing inducted currents in a rotating copper disc, with electro-magnet.

26. Large wire-spirals for the phenomenon of the earth's induction.   *Emil Stöhrer, Leipsic.*

495. Rack and Snail of Clock, to regulate the number of blows struck each hour.   *The Council of King's College, London.*

496. Model of Chronometer Escapement.
    The Council of King's College, London.

497. Model of Lever Escapement.
    The Council of King's College, London.

498. Model of Horizontal Escapement.
    The Council of King's College, London.

499. Model of Locking Plate of Clock, to regulate the number of blows struck each hour.
    The Council of King's College, London.

4220a. Murby's Photo-lithographed Wall Maps (physical).

World.	Africa.
British Isles.	North America.
Europe.	South „
Central Europe.	Palestine.
Asia.	

Thomas Murby.

4220b. Diagrams of Natural History (20).
    Thomas Murby.

FIRST SERIES.—APPARATUS FOR SUPERIOR EDUCATIONAL INSTITUTES.

Kinematics, Statics, and Dynamics.

4221. Model of a Bramah's Hydraulic Press.

4222. Model of a Turbine.

4223. Model of a Centrifugal Pump.

4224. Apparatus for demonstrating the **Archimedean Principle,** consisting of hollow brass floater, glass cylinder with discharge tube, and graduated glass cylinder.

4225. Five different Models of Valves.

4226. Inclined Plane.

4227. Gyroscope, according to Fessel.

4228. Gridiron Pendulum.

4229. Bevelled Gearing.

4230. Windlass, with tooth and movement.

4231. Endless Screw.

4232. Centrifugal Machine.

4233. Apparatus for demonstrating **Foucault's Pendulum Experiments.**

4234. Ball-Regulator.

4235. Apparatus for Oblique Tubes.

4236. Glass Vessel for Liquids of different specific weights.

4237. Syren Disc of brass, sounding major and minor chords.

4238. Apparatus for demonstrating the **oblate form of the Earth.**

4239. Colour Disc.

4240. Apparatus, consisting of two balls of unequal weight moving along a wire.

4241. Apparatus, consisting of a ball sliding on a wire and lifting various weights.   *E. Leybold's Successors, Cologne.*

Molecular Physics.

4242. Stop-cock Air-Pump, with horizontal cylinder and gauge; can be put in motion by a movement.

4243. Hand Air-pump, with oblique cylinder, and gauge.

4244. Pair of Magdeburg Hemispheres, 105 mm. in diameter.

4245. Mercury Granulation Apparatus.

4246. Apparatus for showing the **equal Pressure of the Air** in all directions.

4247. Windmill-sail Apparatus, for showing the resistance of the air.

4248. Apparatus for explaining the **Aneroid Barometer** and **Spring Manometer.**

E. Leybold's Successors, Cologne.

Heat.

4249. Steam Apparatus, of copper, on supports of brass.

4250. Aeolipile, on stand.

4251. Model of a Steam Engine, with oscillating cylinder.

4252. Rotating Steam Ball.

4253. Apparatus for explaining the effects of **Steam.**

4254. Contraction Apparatus.

4255. Pyrometer, according to Muschenbrock.

4256. Mirror Pyrometer, according to Tyndall.

4257. Apparatus for converting work into heat.

4258. Noë's Thermochain, star form.

E. Leybold's Successors, Cologne.

Magnetism.

4259. Graphite Zinc Element, consisting of graphite plate, clay-cell, and zinc cylinder.

Height of the clay-cell   78 millimeter.

„	„	104	„
„	„	130	„
„	„	156	„
„	„	208	„
„	„	260	„

4260. Chromic Acid Element, small, bottle-shaped, with two graphite and one zinc plate.

4261. Chromic Acid Element, larger, with two graphite and one zinc plate.

4262. Chromic Acid Element, with three zinc and one graphite plate.

4263. Coal-light Apparatus, with small reflector (hand regulator).

4264. Water Deposition Apparatus.

4265. Oxy-hydrogen Gas Apparatus, Werner's construction.

4266. Constant Battery, according to Stöhrer.

4267. Twenty pair of Plates.

4268. Tangent-Compass, with movable compass.

4269. Horizontal Galvanometer, box-shaped.

4270. Multiplicator, according to Schweigger, for thermo-electrical currents.

4271. Table Galvanometer.

4272. Electro-Magnet, arranged for a table, and for suspension.

4273. Apparatus for showing the rotation of a magnet.

4274. Apparatus for showing the rotation of two magnets, caused by a current.

4275. Magnet, rotating around its own axis.

4276. Electro-Magnetical Oscillating Apparatus, with supporter for Geissler's tubes.

4277. Electro-Magnet, rotating under the influence of terrestrial magnetism.

4278. Electro-Magnet, rotating in a second, with contrivance for lifting weights.

4279. Machine, according to Page.

4280. Electro-Magnet rotating above a steel magnet.

4281. Barlow's Wheel.

4282. Lightning Wheel, according to Neef.

4283. Apparatus, according to Petrina, showing the attraction of parallel currents.

4284. Ampere's Frame, according to Eisenlohr.

4285. Model of Morse's Writing Telegraph.

4286. Induction Apparatus, according to Dubois-Reymond, arranged for medical purposes.

4287. Pocket Induction Apparatus, according to Werners.

4288. Spark Inductor, according to Ruhmkorff.

4289. Spark Inductor, according to Ruhmkorff.

4290. Induction Apparatus for fundamental experiments.

4291. Steel Magnet, with five thin plates and keeper.

4292. Dipping Needle, with stand.

4293. Magneto-electrical Induction Apparatus.

4294. Blasting Apparatus for **Mining.**
E. Leybold's Successors, Cologne.

Electricity.

4295. Electrical Machine, according to Winter, with different connexions for the condenser.

4296. Holtz's Induction Machine. Diameter of the rotating wheel 520 mm.

4297. Chime with five bells.

4298. Fly-wheel on a circular track.

4299. Lanne's Measure Bottle.

4300. Discharging Rod.

4301. Double Fly-wheel.

4302. Battery of four Leyden Jars.

4303. Pair of fine Gold-leaf Electrometers.

4304. Pair of fine Gold-leaf Electrometers, with **Zamboni's Column,** improved according to Fechner.
E Leybold's Successors, Cologne.

SECOND SERIES.—APPARATUS AND INSTRUMENTS FOR ELEMENTARY SCHOOLS.
Kinematics, Statics, and Dynamics.

4305. Inclined Plane, simple.

4306. Windlass (Göpel).

4307. Rod Windlass.

4308. Wooden Screws, with threads of different section.

4309. Wheel on Axle, with stand.

4310. Reel, with stand.

4311. Six Lever Apparatus, for demonstrating different laws of leverage.

4312. Apparatus for illustrating the action of pulleys.

4313. Glass Model of a Suction and Forcing Pump.

4314. Glass Model of an Hydraulic Press.

4315. Glass Model of a Fire Engine.

4316. Segner's Water-wheel.

4317. Model of a Diving Bell.

4318. Apparatus for illustrating the **Expansion** and **Compression of Air.**

4319. Fly-wheel of glass.

4320. Fly-wheel.

4321. Rotating Suction Siphon.

4322. Apparatus for demonstrating the uniform transmission of pressure in liquids.

4323. Model of a Suction-pump.
E. Leybold's Successors, Cologne.

Molecular Physics.

4324. Hand Air-pump.

4325. Pair of Magdeburg Hemispheres.

4326. Scale Manometer.

4327. Sound Apparatus, for weighing air.

4328. Apparatus for illustrating that a vacuum can be filled by a liquid through atmospheric pressure.

4329. Apparatus for demonstrating that bodies of unequal weight fall at the same rate in vacuo.

4330. Barometer.
E. Leybold's Successors, Cologne.

Light.

4331. Prism.

4332. Astronomical Models, illustrating course of the rays of light.
E. Leybold's Successors, Cologne.

Heat.

4333. Apparatus, glass, for demonstrating the **Effect of Steam.**

4334. Apparatus, globe, and ring, for demonstrating the expansion of metals by heat.

E. Leybold's Successors, Cologne.

Magnetism.

4335. Simple Diving Element.

4336. Double Diving Element.

4337. Galvanoplastical Apparatus.

4338. Pair of Carbon Points, with handles.

4339. Apparatus for making **Wires red hot.**

4340. Water Decomposition Apparatus.

4341. Electro-magnet, with two binding screws.

4342. Electro-magnetical Driving Machine.

4343. Induction Apparatus, for extra current and secondary current.

4344. Bar Magnets.

4345. Horse Shoe Magnet.

E. Leybold's Successors, Cologne.

Electricity.

4346. Simple Electrical Machine.

4347. Electrical Tuft.

4348. Simple Gold-leaf Electrometer.

4349. Fundamental Apparatus, according to Werners, for friction electricity, magnetism, and galvanism.

4350. Electrophorus, 210 mm. in diameter.

4351. Electrical Shower of Balls.

4352. Electrical Pistol.

4353. Electrical Discharging Rod.

4354. Simple Chime of Bells.

4355. Leyden Jar, 156 mm. high.

4356. Isolating Stool.

E. Leybold's Successors, Cologne.

THIRD SERIES.—APPARATUS AND APPLIANCES FOR TEACHING CHEMISTRY.

Chemistry.

4357. Series of **Reagent Jars,** with enamelled labels and inscriptions.

4358. Carbonic Acid Apparatus, according to Bunsen.

4359. Potash Apparatus, according to Liebig.

4360. Potash Apparatus, according to Mohr.

4361. Potash Apparatus, according to Mitscherlich.

4362. Sulphuretted Hydrogen Apparatus.

4363. Drying Apparatus, by means of sulphuric acid.

4364. Combustion Stove, according to Bunsen, with 25 burners.

4365. Combustion Stove, according to Glaser, with 20 burners.

4366. Set of 12 Bohemian Glass Beakers.

4367. Set of 6 Bohemian Glass Beakers.

4368. Bellows, with lamp, for glass-blowing.

4369. Burette, according to Bink's, 50 cc. in $\frac{1}{5}$.

4370. Burette, according to Gay-Lussac, 50 cc. in $\frac{1}{10}$.

4371. Porcelain Burette Stand, with eight burettes (dropping glasses).

4372. Glass Burette, according to Geissler, 50 cc. in $\frac{1}{5}$.

4373. Burette, with porcelain stand.

4374. Burette, with stand of polished wood.

4375. Combination Cylinder, according to Mohr, 1,000 c.

4376. Dephlegmator, according to Staedler.

4377. Azotometer, according to Knop, modified by Wagner.

4378. Filtering Frame, of iron, with leaden base.

4379. Filtering Frame, of brass, with wooden base.

4380. Universal Stand, according to Werners.

4381. Set of 9 Evaporating Dishes, of enamelled iron.

4382. One Litre **Clarifying Bottle,** with glass and stopper.

4383. Series of 8 **Boiling Jars,** according to Erltenmeyer.

4384. Cooling Apparatus, glass.

4385. Set of 7 **Cork Borers,** in brass.

4386. Bunsen Burner, with stand.

4387. Seven Burette-holders, three rings, chimney, retort.

4388.   ,,   ,,   with regulator.

4389.   ,,   ,,   with regulator and stop-cock.

4390.   ,,   ,,   with three pipes.

4391.   ,,   ,,   with seven pipes.

4392.   ,,   ,,   with twelve pipes and tripod.

4393. Gas Stove, with seven burners and stop-cock.

4394. Three Glowing Lamps, according to Masté.

4395. Blow-pipe Case.

4396. Pipette Stand, of polished wood ; porcelain stand.

4397. Polished Revolving Reagent Bottle Frame, to hold 36 bottles.

4398. Unpolished Reagent Bottle Frame, according to Llandolt's, with 18 bottles.

4399. Crucible Tongs, of brass.

4400. Series of Glass Funnels, Bunsen's shape.

4401. Copper Water-bath, 9-inch.

4402. Copper Water-bath, on tripod.

4403. Constant Level.

E. Leybold's Successors, Cologne.

4404. Hestermann's Technological and Physical Science Apparatus :—

4405. Flax and its applications.

4406. The Cotton Plant and its use.

4407. Wool and its application.

4408. The Honey Bee and its industry.

4409. Leather, its preparation and employment.

4410. Silk, its production and application.

4411. Paper, its manufacture and use.

4412. Glass, its manufacture and employment.

4413. Illuminating and Heating Materials, their production and use.

4414. Dyeing and **Cotton Printing.**

4415. Collection of Products, III. Course.

4416. Iron, its production, manufacture, and use.

4417. Collection of Caterpillars, with text, by Director Dr. Bolan.

4418. Cabinet containing Silkworms.

4419. Forest Herbarium, Parts I. and II.

4420. Herbarium of Poisonous Plants.

4421. Herbarium of Grasses.
Chr. Vetter, formerly Ludw. Hestermann, Hamburg.

Apparatus and Instruments for Chemists and Natural Philosophers.

4423. Chemical Scale for 1 ko. weight, indicating 1 milligr., with special balance and scales, and stopping arrangement; gilt; the scales are of platinum, in glass case; the beam works on agate planes.

4424. Set of Weights belonging to the same, gilt, from 1 ko. downwards.

4425. Two Sets of Weights, from 50·0 grm. downwards, nickel plated.

4426. Sets of Weights of Rock Crystal, from 200·0 gr. downwards.

The weights of 200-100 grammes have a cylindrical form. This form has been chiefly chosen in order that, in case of any neglect, injuries might be avoided as much as possible.

4427. Normal 500·0 gr. Weight, of rock crystal.
W. J. Rohrbeck, J. F. Luhme & Co., Berlin (Dr. Herm. Rohrbeck).

Physics.

4428. Single-barrelled Air Pump, with standing glass tube, double acting, on wooden stands.

The arrangement of the air pump is such that the position of the stop-cock of the receiver places it alternately in connexion with the upper and the lower part of the cylinder.

The pump is particularly adapted for exhausting a large number of recipients, as is very frequently requisite at scientific experiments in laboratories.

4429. Attwood's Machine, with pendulum.

4430. Gridiron Seconds Pendulum, on stand.

4431. Mack's Galilei's Fall Kennel.

4432. Centrifugal Machine, with auxiliary apparatus.

The construction of the machine is distinguished by its simplicity. The stand, formed by a cast-iron frame, can be set either horizontally or vertically.

4433. Hydraulic Press.

4434. Air Pump, with two glass tubes, and Babinet's stop-cock.

With apparatus :—*a.* Barometer test ; *b.* Magdeburg hemispheres ; *c.* Mercurial granulation ; *d.* Globe (recipient) ; *e.* Electric egg ; *f.* Cylinder for bursting of bladders ; *g.* Freezing apparatus.

4435. Mach's Wave Apparatus, for illustrating the longitudinal and transversal waves.

4436. Mach's Apparatus for **Sound Reflection.**

4437. Mach's Apparatus for **Refraction and Reflexion.**

4438. Mach's Apparatus for demonstrating the **Law of Refraction** by means of fluorescent liquids.

4439. Mach's Apparatus for demonstrating the **Effects of Lenses.**

4440. Model of a Screw, of wood.

4441. Model of a Screw, of wood.

These two models serve for explaining and illustrating to the student the screw motions. One represents a flat, the other a triangular worm; both serve for demonstrating the longitudinal transverse sections.

4442. Crown Glass Prism.

4443. Flint Glass Prism.

4444. Polyprism.

4445. Hollow Prism.

4446. Universal Kaleidophon, for illustrating Lissajou's figures.

This consists of a thin strip of copper, sliding on a wooden clamp, which may be fastened by a screw to the table, and can be raised to any height desired.

On the upper part of this strip of copper there is attached a second clamp, through which passes a band of steel terminating at the top in a brad. By means of this second clamp the steel band can likewise be fixed at any desired height. If by pulling with the finger the strip of copper, and in a perpendicular level to the same, the steel band is set into oscillation, these two movements will combine into one, which, according as the respective longitudinal relations of the two strips are altered will be different. By the brad attached to the upper part of the apparatus these movements can be clearly observed, and Lissajou's figures demonstrated.

4447. Meyerstein's Heliostat.

4448. Astronomical Telescope, on a wooden stand with scale.

4449. Pocket Spectroscope, according to Professor Emsmann, with only *one* prism, and contrivance for specially placing the separate colours.

4450. Another Specimen of the same kind.

4451. Thermo-electrical Battery, according to Clamond.

This consists of 40 elements, which are arranged in four groups, which can be placed either side by side or one after the other. The heating of the soldered places takes place from the centre by gas, but not *directly*, however, by heated air generated in the interior of the battery. The gas turns from small holes made round a clay cylinder. For an equal supply of gas, particularly in order to avoid at a stronger pressure an excessive heating of the soldered places, an ordinary gas regulator with floats, as is customary with gas meters, has been attached. If the supply of gas increases the valve closes, opening again at the decrease of pressure. As, consequently, the supply of gas regulates itself in this way, the column uninterruptedly acts for weeks and months, while its current is extremely constant at an effective force of 5 cc. oxygen and hydrogen gas a minute, and is therefore particularly adapted for the quantitative separation of metals from solutions.

As regards the durability of the apparatus, it may be remarked that after 18 months daily use, no perceptible alteration has been noticed.

4452. Thermo-electrical Battery, according to Noë.

This differs from the previous one,—

1, by the arrangement of the metals, which in this case are grouped in two horizontal rows, in the centre of which (and this is the second essential difference) a number of burners heat the soldered places *directly*. The cooling of the other soldered places is, as in the previous battery, simply atmospheric, for which purposes, in order to render it as perfect as possible, copper plates are connected with the other soldered places, so that a constant current of air passes through the same.

The number of elements is 88, grouped into 8 elements, which by means of a very handy pachytrope can be intercalated in different manner.

4453. Lane's Measure-bottle, fine adjustment, with scale and vernier.

4454. Separable Leyden Jar.

4455. Electroscope, after Professor Mach, for demonstrating that electricity lies on the surface of excited bodies.

4456. Holtz's Electrical Machine.
Auxiliary apparatus :—*a.* Lightning-tube.   *b.* Lightning-plate.
c. Chime of bells.   *d.* Dancing dolls.   *e.* Contrivance for breaking
glass plates.   *f.* Henley's discharging rod.   *g.* Discharging rod,
with joint.

4457. Pair of Concave Mirrors, brass, 16 inch diameter,
8 inch focal distance, on tripods, with contrivance for adjustment
and equipment for heat experiments.

4458. Two Sectional Models of Steam-engines.

4459. Model of a Screw Steamer.

4460. Lever Pyrometer.
    *W. J. Rohrbeck, J. F. Luhme & Co., Berlin (Dr. H. Rohr-
    beck).*

Chemistry.

4461. Plattner's Assay-balance, for blow-pipe purposes,
in mahogany case, with 1 gr. weight, indicating $\frac{1}{10}$ milligr., with
weights.

4462. Balance, on stand, according to Wackenroder, to
carry 20 gr., and indicating 1 mgr., &c.

4463. Technical Balance, on stand, to carry 500 gr., and
indicating 5 mgr.

4464. Mohr's Tare-balance, in mahogany scale case, with
brass column, bow-scales, and ebonite plates, to carry 1 kilogramme, and indicating 1 centigr.

4465. Scales, for weighing substances affecting metals, to
carry 250 gr., and indicating 5 mgr.

4466. Complete Mohr's Balance, for determining the
specific weight of liquid and solid bodies.

4467. Mohr's Balance, with one arm.

4468. *a.* **Exsiccator,** according to Schrötter, consisting of
tubulated glass jar on a frosted glass plate, framed in wood, with
crucible stand, and chloride of calcium tube.

4469. *b.* **Exsiccator,** according to Fresenius, consisting of
glass cylinder, &c.

4470. *c.* **Exsiccator,** according to Fresenius, with frosted
glass plate.

4471. *d.* **Exsiccator,** according to Luhme, with porcelain
vessel for sulphuric acid, and etagère.

4472. *a.* **Drying Closet,** according to Fresenius, of copper, with two tubes and thermometer.

4473. *b.* **Drying Closet,** according to Fresenius, of copper, with grate, wire net, thermometer, and chloride of calcium tube, for drying pulverised substances.

4474. *c.* **Drying Closet,** according to Rammelsberg, with two tubes and thermometer.

4475. *d.* **Drying Closet,** according to Rose, with double sides, two tubes, and thermometer.

4476. Drying Disc, according to Fresenius, for agricultural chemical purposes, with thermometer.

4477. One Burner to the same.

4478. Burner, Blast, and Lamp.

4479. *a.* **Aeolipile,** of brass plate.

This has a vertical flame for soldering, bending glass, melting, by means of the flame of alcoholic vapour.

They are likewise constructed with horizontal flame of embossed brass or copper with safety-valve, on a stand with lamp.

4480. *b.* **Berzelius' Lamp,** on porcelain plate.

Berzelius lamp on a brass stand on a porcelain plate, consisting of the lamp with spirit reservoir, two rings, and chimney.

4481. *c.* **Lamp,** according to Mitscherlich.

The spirit reservoir of this lamp is separated from the burner by a pipe 230 millimeters in length, and consists of a bottle provided with a conical valve, which is inserted in a reverse way in a brass cylinder, by which means a constant level of the spirit in the lamp has been secured.

4482. *d.* **Simple Burner,** with glass tube.

4483. *e.* **Bunsen's Jet Burner,** uni-radiating, with case for illuminating flame.

4484. *f.* **Bunsen's Burner,** with chimney, and fork-shaped holder.

4485. *g.* **Bunsen's Burner,** with star-ring, chimney, and blow-pipe apparatus.

4486. *h.* **Finkner's Burner,** with gas and air regulator.

4487. *i.* **Finkner's Burner,** with star-ring and chimney.

4488. *k.* **Griffin's Burner.**

4489. *l.* **Burner,** according to the Viennese model.

4490. *m.* **Double-radiating Jet Burner.**

4491. *n.* **Treble-radiating Jet Burner.**

4492. *o.* **Iserlohn Gas-stove Burner,** with one emission ring.

4493. *p.* **Iserlohn Gas-stove Burner,** with double emission ring.

4494. *q.* **Blow-pipe-table Adjustment,** according to Magnus.

4495. *r.* **Blow-pipe-table Adjustment,** with two stop-cocks on stativ, and porcelain plates, according to Rohrbeck.

4496. *s.* **Apel's Stove.**

4497. *t.* **Universal Stand,** of iron, with Finkner's patent burner, four rings, and three retort-holders.

4498. *u.* **Iron Filtering Stand,** with three rings.

4499. *v.* **Schellbach's Retort-holder.**
    W. J. Rohrbeck, J. F. Luhme & Co., Berlin (Dr. Herm. Rohrbeck).

4500. Professor A. W. Hofmann's Apparatus and Instruments.

a. Apparatus for ascertaining the volume proportions of hydrogen and chlorine, consisting of V-shaped tube, with two glass stop-cocks, and iron stand.

b. Apparatus for proving, that at the formation of the HCl, one volume Cl and one volume H combine; consisting of a glass tube, with two stop-cocks, decomposition apparatus, and two pipe supporters, calcium chloride cylinder, and a vessel for the electrolytical analysis of hydrochloric acid.

c. Water-decomposition apparatus for demonstrating that 2 vol. hydrogen and 1 vol. oxygen are the volume proportions of water; consisting of a triangular tube, with two stop-cocks, and platinum electrode (the tube also graduated); iron stand.

d. Steam-tight apparatus, consisting of one graduated barometer-tube with envelope-pipe, iron stand with holder, measuring apparatus, boiling and cooling vessel and serpentine-pipe.

e. Apparatus for ascertaining the volume proportions of nitrogen and hydrogen in ammonia; consisting of tube with stop-cock and stopper, a glass cylinder, movable table, and glass bowl.

f. Apparatus for ascertaining the volume proportions of the elementary constituents of HCl, H_2O, and NH_3, by electrolytical means; consisting of two glass apparatus with coal-electrodes, a water-decomposition apparatus with platinum electrodes, and three iron stands.

f^1. Apparatus for proving that at the combination of the respective elementary gases HCl, H_2O, and NH_3, always two vol. gas are formed; consisting of three V-shaped tubes with platinum electrodes, and three iron stands.

g. Apparatus for demonstrating that the gases H and Cl combine in the constant proportion of 1 : 1 into hydrochloric acid; consisting of a glass tube with stop-cock, and two ground-in stoppers, with stand.

h. Apparatus for the decomposition of nitric acid; consisting of a little platinum piston, a stand, a funnel-tube with stop-cock, a pneumatic trough, and a cylinder.

i. Apparatus for domonstrating the equal volume proportions of simple and compound gases; consisting of steam apparatus, steam-conducting pipe, glass apparatus with five stop-cocks, and an iron stand.

k. Apparatus for making experiments with sulphurous acid.

l. Apparatus for the decomposition of hydrogen phosphide; consisting of V-shaped tube with two electrodes, and a stand.

m. Apparatus for showing the capability of oxidisation and reduction of the gas dame; consisting of a copper crucible, with tripod and gas-burner.

n. Apparatus for the electrolytical decomposition of the HCl, H_2O, and NH_3; consisting of V-shaped tube with platinum electrodes, and iron stands.

o. Reagent étagère for bottles; with octagonal stopper, ribbon-label, etched inscription, and formula on the stoppers.

4501. Apparatus for demonstrating the Valency of Atoms, according to Kekulé.

Consisting of a number of differently coloured balls for demonstrating the importance of the elements and the origin of the combination of the elements, on an iron stand.

4502. *a.* **Cooler of Zinc,** after Liebig, on stand.

4503. *b.* **Cooler of Brass,** after Liebig.

4504. *c.* **Cooler of Glass,** according to Hofmann.

4505. *a* **Two Agate Mortars,** with pestle, 65, 150 mm. diameter.

4506. *b.* **Three Glass Mortars,** with pestle, 60, 100, 120 mm. diameter.

4507. *c.* **Three Powder Mortars,** of porcelain, with pestle.

4508. *d.* **Pill Mortar of Iron,** 155 mm. diam.

4509. *e.* **Diamond Mortar.**

Diamond mortars for facilitating the pulverisation of minerals difficult to triturate, a process absolutely necessary in making quantitative analysis.

4510. e^1. **Diamond Mortar.**

4511. f. **Caustic Potash Form,** of iron, 12 channels.

4512. a. **Filtering Pump,** according to Bunsen.

For filtration under high pressure with barometer scale and regulating cock screwed on sheet iron.

4513. b. **Pump,** according to Liebig, for drying organic substances.

A small hand air pump with screw for fastening the same to the table.

4514. Alkalimetrical Apparatus.

4515. a. **Burette Stand,** with eight movable burettes.

4516. b. **Four graduated Cylinders,** 250·0, 500·0, 1000·0, 3000·0 cc.

4517. c. **Mixing Bottle** (cylinder), 1000 cc.

4518. d. **Measuring Piston,** two litres.

4519. e. **Measuring Piston,** one litre with stopper.

4520. f. **Carbonic Acid Apparatus,** according to Oechel-häuser.

For determining the per-centage of small quantities of carbonic acid contained in coal-gas, by means of the absorption of potash-lye.

4521. g. **Carbonic Acid Apparatus,** according to Rüdorff.

4522. h. **Carbonic Acid Apparatus,** according to Scheibler.

4523. Glass Gasometer, with brass fittings for use in chemical laboratories.

4524. Three different polished Retort-holders.

4525. Four different Blow-pipes.

These are of brass, nickel silver with veneered or inlaid platinum plates, as well as arranged for direct use with gas.

4526. Two crucible Pincers of steel and nickel silver.

4527. One Dozen Funnels.

4528. One Dozen Boiling Jars.

4529. One Dozen Pistons.

4530. Six Sets of Glass Bowls.

4531. One Dozen of Woulff's Jars.

4532. One Dozen Retorts.

4533. Six Glass Stop-Cocks.

4534. Six Separation Funnels.

4535. Six Potash Apparatus.

4536. Sulphuretted Hydrogen Apparatus, according to Kipp, with ground in glass cock.

4537. Displacement Apparatus, according to Robiquet, with funnel provided with ground in glass stopper.

W. J. Rohrbeck, J. F. Luhme & Co., Berlin (Dr. Herm. Rohrbeck).

Glass wool, very convenient for filtering acids and alkalies in place of asbestos, much cleaner, and practical. If washed after use, it can be employed various times.

Technology.

4538. Wooden Model of a Blast Furnace.

4544. Wooden Model for demonstrating the Manufacture of Sulphuric Acid.

4545. Wooden Model of a Glass Furnace (Siemens' regenerative system).

All these wooden models can be conveniently taken asunder, for demonstrating the interior arrangement of the stoves.

4539. Wooden Model of a Shaft Furnace.

4540. Wooden Model of a Puddling Furnace.

4541. Wooden Model of a Rolling and Stamping Mill.

4542. Wooden Model of a Silver-Ore Refining Furnace.

4543. Wooden Model of a Double Roasting-Furnace for Copper.

W. J. Rohrbeck, J. F. Luhme & Co., Berlin (Dr. Herm. Rohrbeck).

4546. Apparatus for the graphical demonstration of the **Law** of **Descent.**

4547. Apparatus for determining the **Linear Expansion** by **Heat,** by means of reading by mirror reflection.

4548. Müller's Wave Disc.
(*a.*) Stativ (stand).
(*b.*) Portfolio with stroboscopical discs.

4549. Some **Drawings** which Müller used, in the years 1840 and 1850, in his lectures explanatory of apparatus.

Physical Institute of the University of Freiburg, Baden (Prof. Dr. A. Claus).

Glass Models of the Eye (3.)

Science Schools, South Kensington Museum.

SECTION 20.—MISCELLANEOUS.

WEST GALLERY, UPPER FLOOR, ROOM Ⓞ

PORTRAITS AND RELICS.

4572. Sir Isaac Newton. Portrait engraved by J. Outrim. From the original drawing in the Pepysian Collection at Cambridge. *Robert James Mann, M.D.*

4572a. Photographic Copy of an early portrait of **Galileo.**
J. Norman Lockyer, F.R.S.

BOOKS.

4573. The Works of Van Marum, and the three volumes already issued of the Archives du Musée Teyler.
Teyler Foundation, Haarlem.

For illustration of the instruments that are shown, and in order to give an idea of the great electrical machine and the chemical apparatus which could not be transported, the works of Van Marum "Machine Electrique" (three volumes), and "Appareils Chimiques" (one volume), containing the engravings of that instrument, are sent.

The three volumes of the "Archives du Musée Teyler" contain the refractive indices and wave-lengths determined by Prof. Van der Willigen; the chemical composition of various specimens of crown and flint-glass by the late Prof. Van Kerckhoff; the description and drawings, by Conservator Dr. Winkler, of fossil tortoises, and other palæontological objects, &c.

4573a. Biography of Don Francisco Salva y Campillo, and an account of his studies and inventions upon electric telegraphs. By Don Antonio Suaver Saavedra.
MS. original. 4to.
Direccion General de Correos y Telegrafos, Madrid.

Don Salva, who lived at Barcelona at the end of the last century, devoted most of his time to making observations and experiments upon electric telegraphs. He published several memoirs on this subject, which are of great interest.

4573a. Studies on Electricity by Don Francisco Salva.
Academia de Cuncias Naturales Yartes de Barcelona, Spain.
"Memoria sobre la electricidad aplicada a la telegrafia."
"Disertacion sobre el galvanismo."
"Memoria sobre el galvanismo aplicado a la telegrafia."

These studies on electricity were written by Don Francisco Salva in 1795 to 1804.
Pamphlet in 4to. Barcelona, 1876.

4573b. Report of the Royal School of Mines upon the apparatus sent to this exhibition to determine the right lineal conduction of a given point, invented by the pupil of the same School of Mines, Sr. Bentibol y Vreta. MS. pamphlet, 4to.
H. Bentabol y Vreta Royal School of Mines, Madrid.

4574. Manuals and Treatises on method of instruction of all branches of Natural History, adopted by the military schools in Russia. Published by Fenoult.
Committee of the Pedagogical Museum, Russia.

331a. Pamphlet. "Procedimientos mecanicos de cubicacion."
4to Madrid, 1876. *Don Eduardo Saavedra, Madrid.*

Sr. Saavedra explains in this pamphlet the method adopted by him in taking out quantities for the projected railroads of the central Spanish Pyrenees.

4574a. List of Publications.
Messrs. Macmillan and Co., London.

Count Rumford's works, with memoir. 5 vols.

Works by Sir G. B. Airy, M.A., LL.D., D.C.L., Astronomer Royal, &c.— Treatise on the algebraical and numerical theory of errors of observation, and the combinations of observations.—Popular astronomy.—On Sound and Atmospheric Vibrations.

Reuleaux's Kinematics of Machinery. Translated by Prof. A. B. Kennedy.

Experimental Mechanics. By R. S. Ball, M.A.

The Beginnings of Life. By H. Charlton Bastian, M.D., F.R.S. 2 vols.

Rudiments of physical geography for the use of Indian schools, and a glossary of the technical terms employed. By Henry F. Blandford, F.G.S.

First principles of chemical philosophy. By Josiah P. Cooke, junior, Ervine Professor of Chemistry and Mineralogy in Harvard College.

Cave Hunting. Researches on the evidence of Caves, respecting the early inhabitants of Europe. By W. Boyd Dawkins, F.R.S., Director of the Museum and Lecturer on Geology in Owen's College, Manchester.

An introduction to the Osteology of the Mammalia. Being the substance of the Course of Lectures delivered at the Royal College of Surgeons of England in 1870. By W. N. Flower, F.R.S., F.R.C.S., Hunterian Professor of Comparative Anatomy and Physiology.

Pharmacographia. A history of the principal drugs of vegetable origin found in commerce in Great Britain and British India. By F. A. Flückiger and D. Hanbury, F.R.S.

The Elements of Embryology. By Michael Foster, M.D., F.R.S. and F. M. Balfour, M.A., Fellow of Trinity College, Cambridge.

Meteorographia, or Methods of Mapping the Weather. By Francis Galton, F.R.S.

Hereditary Genius; an Inquiry into its Laws and Consequences. By Francis Galton, F.R.S.

English Men of Science, their Nature and Nurture. By Francis Galton, F.R.S.

Works by Hugh Godfray, M.A.—An Elementary Treatise on the Lunar Theory, with a brief Sketch of the Problem up to the time of Newton.— A Treatise on Astronomy, for the use of Colleges and Schools.

The Forces of Nature. A popular Introduction to the Study of Physical Phenomena. By Amédée Guillemin. Translated from the French by Mrs.

Norman Lockyer; and edited with Additions and Notes, by J. Norman Lockyer, F.R.S.

The Student's Flora of the British Islands. By J. D. Hooker, C.B., F.R.S., M.D., D.C.L., President of the Royal Society.

A Treatise on Ornamental and Building Stones of Great Britain and Foreign Countries. Arranged according to their Geological Distribution and Mineral Character, with illustrations of their application in Ancient and Modern Structures. By Edward Hull, M.A., F.R.S., Director of the Geological Survey of Ireland, &c.

Lessons in Elementary Physiology. By T. H. Huxley, LL.D., F.R.S., Professor of Natural History in the Royal School of Mines.—Lay Sermons, Addresses, and Reviews. By the same.

A Course of Practical Instruction in Elementary Biology. By T. H. Huxley, LL.D., Sec. R.S., assisted by H. N. Martin, B.A., M.B., D.Sc. Fellow of Christ's College, Cambridge.

Elementary Lessons in Logic : Deductive and Inductive. With copious Questions and Examples, and a Vocabulary of Logical Terms. By W. Stanley Jevons, M.A., F.R.S., Professor of Logic in Owen's College, Manchester.

The Principles of Science. A Treatise on Logic and Scientific Method. 2 vols. By W. Stanley Jevons, M.A., F.R.S., Professor of Logic in Owen's College, Manchester.

The Owen's College Junior Course of Practical Chemistry. By Francis Jones, Chemical Master in the Grammar School, Manchester. With Preface by Professor Roscoe.

Journal of Anatomy and Physiology. Conducted by Professors Humphry and Newton, and Mr. Clark of Cambridge, Professor Turner of Edinburgh, and Dr. Wright of Dublin. Vols. 8 and 9.

Introduction to Quaternions, with Numerous Examples. By P. Kelland, M.A., F.R.S., and P. G. Tait, M.A., Professor in the Department of Mathematics in the University of Edinburgh.

Researches on the Solar Spectrum and the Spectra of the Chemical Elements. By G. Kirchoff, of Heidelberg. Translated by Henry E. Roscoe, F.R.S.

By J. Norman Lockyer, F.R.S. Elementary Lessons in Astronomy.— Contributions to Solar Physics.

The Romance of Astronomy. By R. Kalley Miller, M.A., Fellow and Assistant Tutor of St. Peter's College, Cambridge.

Lessons in Elementary Anatomy. By St. George Mivart, F.R.S.

Nature Series.

The Spectroscope and its Applications. By J. Norman Lockyer, F.R.S.

The Origin and Metamorphoses of Insects. By Sir John Lubbock, M.P., F.R.S.

The Birth of Chemistry. By G. F. Rodwell, F.R.S., F.R.A.S.

The Transit of Venus. By G. Forbes, B.A., Professor of Natural Philosophy in the Andersonian University, Glasgow.

The Common Frog. By St. George Mivart, F.R.S.

Polarisation of Light. By W. Spottiswoode, LL.D., F.R.S.

On British Wild Flowers, considered in relation to Insects. By Sir John Lubbock, M.P., F.R.S.

Newton's Principia. Edited by Professor Sir W. Thomson and Professor Blackburn.

Lessons in Elementary Botany. By Daniel Oliver, F.R.S., F.L.S., Professor of Botany in University College, London.

First Book of Indian Botany. By Daniel Oliver, F.R.S., F.L.S., Professor of Botany in University College, London.

On a Method of Predicting, by graphical construction, Occultations of Stars by the Moon and Solar Eclipses for any given place. Together with More Rigorous Methods for the Accurate Calculation of Longitude. By J. C. Penrose, F.R.A.S.

An Elementary Treatise on Steam. By John Perry, Bachelor of Engineering, Whitworth Scholar, &c., late Lecturer in Physics at Clifton College.

Elements of Physical Manipulation. By Edward C. Pickering, Thayer Professor of Physics in the Massachusetts Institute of Technology.

Primers.—Science.

Chemistry. By H. E. Roscoe, F.R.S., Professor of Chemistry in Owen's College, Manchester.

Physics. By Balfour Stewart, F.R.S., Professor of Natural Philosophy in Owen's College, Manchester.

Physical Geography. By A. Geikie, LL.D., F.R.S., Murchison Professor of Geology and Mineralogy in the University of Edinburgh.

Geology. By Professor Geikie, F.R.S.

Physiology. By Michael Foster, M.D., F.R.S.

Astronomy. By J. N. Lockyer, F.R.S.

Botany. By J. D. Hooker, C.B., P.R.S.

The Theory of the Glaciers of Savoy. By M. Le Chanoine Rendu. Translated by A. Wills, Q.C., late President of the Alpine Club. To which are added the Original Memoirs, and Supplementary Articles by Professors P. G. Tait and J. Ruskin. Edited, with Introductory Remarks, by George Forbes, B.A., Professor of Natural Philosophy in the Andersonian University, Glasgow.

A System of Medicine. Vol. I. Edited by J. Russell Reynolds, M.D., F.R.C.P., London. Part. I. General Diseases, or Affections of the whole System. Part II. Local Diseases, or Affections of particular Systems. Vol. II. Part. II. Diseases of the Nervous System. (a.) General Nervous Diseases. b.) Partial Diseases of the Nervous System. I. Diseases of the Head. II. Diseases of the Spinal Column. III. Diseases of the Nerves. II. Diseases of the Digestive System. (a.) Diseases of the Stomach. Vol. III. § II. Diseases of the Digestive System (continued). (b.) Diseases of the Mouth. (c.) Diseases of the Fauces, Pharynx, and Œsophagus. (d.) Diseases of the Intestines. (e.) Diseases of the Peritoneum. (f.) Diseases of the Liver. (g.) Diseases of the Pancreas. § III. Diseases of the Respiratory Organs. (a.) Diseases of the Larynx. (b.) Diseases of the Thoracic Organs.

Lessons in Elementary Chemistry, Inorganic and Organic. By Henry E. Roscoe, F.R.S., Professor of Chemistry in Owen's College, Manchester.

The Spectrum Analysis. A series of lectures delivered in 1868, with four appendices. Third edition, with the most recent discoveries and additional illustrations. By Henry E. Roscoe, F.R.S., Professor of Chemistry in Owen's College, Manchester.

A Manual of the Chemistry of the Carbon Compounds, or Organic Chemistry. By Schorlemmer, F.R.S.

Lessons in Elementary Physics. By Balfour Stewart, F.R.S., Professor of Natural Philosophy in Owen's College, Manchester.

Recent Advances in Physical Science. By Professor P. G. Tait.

Sound and Music. A Non-mathematical Treatise on the Physical Constitution of Musical Sounds and Harmony, including the chief Acoustical

Discoveries of Professor Helmholtz. By Sedley Taylor, M.A., late Fellow
of Trinity College, Cambridge.

Papers on Electrostatics and Magnetism. By Professor Sir W. Thomson,
F.R.S.

The Depths of the Sea. By C. Wyville Thomson, LL.D., F.R.S., &c.,
Director of the Scientific Staff of the " Challenger ' Exploring Expedition.

A History of the Mathematical Theories of Attraction and the Figure of
the Earth, from the time of Newton to that of Laplace. By Isaac Tod-
hunter, M.A., F.R.S.

Contributions to the Theory of Natural Selection. By Alfred Russell
Wallace.

Prehistoric Annals of Scotland. 2 vols. By Daniel Wilson, LL.D.,
Professor of History and English Literature in University College, Toronto.

Prehistoric Man. 2 vols. By Daniel Wilson, LL.D., Professor of History
and English Literature in University College, Toronto.

Wallace's Geographical Distribution of Animals.

4574b. Collection of Scientific Books.

Messrs. Longmans.

Text-books of science :—
 Strength of Materials.
 Inorganic Chemistry.
 Algebra and Trigonometry.
 Organic Chemistry.
 Principles of Mechanics.
 Elements of Mechanism.
 Metals.
 Elements of Geometry.
 Theory of Heat.
 Workshop Appliances.
 Arithmetic and Mensuration.
 Quantitative Chemical Analysis.
 Telegraphy.
 Railway Appliances.
 Electricity and Magnetism.
 Qualitative Chemical Analysis.

Proctor.—The Moon.
Hartley.—Air and its Relation to Light.
Ganot.—Natural Philosophy.
Proctor.—Universe.
Bourne.—The Steam Engine.
Proctor.—Light Science.
Bourne.—Catechism of the Steam Engine.
Proctor.—Saturn, &c.
Helmholtz.—Scientific Lectures.
Butler.—Atlas.
Bourne.—Improvements in the Steam Engine.
Griffin.—Notes on Algebra.
Irving.—Manual of Heat.
Hunter.—Key to Merrifield's Technical Arithmetic.
Watts' Dictionary of Chemistry. (7 vols.)
Dr. Ure's Dictionary of Arts. (3 vols.)
Weinhold's Experimental Physics.
Ganot's Physics. (Atkinson.)
Helmholtz.—The Sensations of Tone.
Crookes.—Handbook of Dyeing.

Bourne.—Steam Engine. (New edition)
Tyndall.—On Sound.
Tyndall.—Heat a Mode of Motion.
Tyndall.—On Light.
Tyndall.—Contributions to Molecular Physics.
Tyndall.—On Magnetism.
Tyndall.—Notes on Light.
Tyndall.—Notes on Electricity.

4574c. Collection of Books.
Messrs. Cassell, Petter, and Galpin.

Davidson.—Educational Series.—Drawing for Carpenters and Joiners; do. Cabinet Makers; do. Machinists; do. Bricklayers; do. Metal Plate Workers; do. Stone Masons; Linear Drawing; Model Drawing; Gothic Stonework; Practical Perspective; Building Construction; Projection; Linear Drawing and Projection.
Ball.—Applied Mechanics.
Church.—Colour.
Ryan.—Systematic Drawing and Shading.
Galbraith & Haughton.—Mathematical Tables; Euclid, Books I.–III. IV.–VI.; Hydrostatics; Tidal Cards; Manual of Algebra; Plane Trigonometry; Arithmetic; Algebra, Part I.; Steam Engine.
Haughton.—Tides; Mechanics; Optics; Astronomy; Natural Philosophy; The Three Kingdoms of Nature.
Hart.—Elementary Chemistry.
Wallace.—Elements of Algebra.
Cassell's Euclid by Wallace.
Tyndall.—Natural Philosophy.

3417. Magazine of Mathematical Instruments. Part II.
upon Reichenbach's Wiederholung's-circle. Part III. upon Breithaupt's Repeating Circle, &c. Part IV. upon Mine Theodolites. Part V. upon Levelling Instruments.

F. W. Breithaupt and Son, Cassel (G. Breithaupt).

PLANS AND DRAWINGS.

4575. Three Plans of Sir Josiah Mason's College of Science
at Birmingham. Ground plan, Ground floor, and Second floor.
George Gore, F.R.S.

4576. Collection of Plans and Programmes of Foreign Laboratories, &c.
George Gore, F.R.S.

1. Das Neue Chemische Laboratorium, zu Berlin, von Albert Cremer. Mit 12 Kupfertafeln. Berlin, Verlag von Ernst D. Korn, 1868.

2. Die Polytechnische Schule zu Aachen, entworfen von Robert Cremer, ausgeführt und herausgegeben von Ferdinand Esser. Mit 11 Kupfertafeln. Berlin, Verlag von Ernst D. Korn, 1871.

3. Das Chemische Laboratorium der Universität Greifswald, von G. Müller. Mit 6 Kupfertafeln. Berlin, Verlag von Ernst D. Korn, 1864.

4. A. M. Kir, Eggetem Vegytani Intézetének Leirasi (öt Táblával), Than Karoly. Pesten, Eggenberger Ferdinand M. Tud. Akad, Köngvárusnál Hoffmann és Molnár, 1872.

5. Das Chemische Laboratorium der K. Ungarischen Universität in Pest, von Dr. Carl von Than. Mit 5 zinkografischen Tafeln. Wien, 1872, Wilhelm Braumüller.

6. Programm der K. K. Technischen Hochschule in Wien, für das Schuljahr 1873–4. Verlag der K. K. Technischen Hochschule von L. W. Seidel & Sohn, Wien, 1873.

7. Das Neue Chemische Laboratorium der Universität Leipzig, von Hermann Kolbe. Mit einen Situationsplan in Lithographie und 7 Holzschnitten. Leipzig, F. A. Brockhaus, 1868.

8. Programm der Polytechnischen Schule zu Aachen für den Cursus 1871–2. Verl. von J. Stercken, Aachen.

9. Das Landwirthschaftliche Studium an der Universität Göttingen, von Gustav Drech. Tler. Mit 4 lithsafeln, Göttingen, 1872. Deuerlichsche Buchhandlung.

10. Über das Physiologische Privat Laboratorium am der Universität Leipzig, von Johann N. Czermak. Mit 5 Holzschnitten, Leipzig. Verlag von Wilhelm Engelmann, 1873.

4576a. Programme of Polytechnic School, Carlsruhe, 1871 (2 copies). *George Gore, F.R.S.*

4576b. Programme of Polytechnic School, Aix-la-Chapelle, 1872. *George Gore, F.R.S.*

4576c. Higher Polytechnic Instruction in Germany, Switzerland, France, Belgium, and England, 1873. By Paul Köristka. *George Gore, F.R.S.*

4576d. Order of Study, Polytechnic School, Carlsruhe, 1869. *George Gore, F.R.S.*

4576e. Polytechnic School, Hanover, 1856. *George Gore, F.R.S.*

4576f. Ground Plan of the University of Leipzig. *George Gore, F.R.S.*

4576g. Personal Register of the University of Leipzig, 1872. *George Gore, F.R.S.*

4577. Plans of Chemical Laboratories at the Owens College, Manchester. *Alfred Waterhouse.*

As will be seen in the plans, the chief features of the building are two large laboratories, each 70 ft. long, 30 ft. broad, and 29 ft. high. No. 1 (Fig. 1.) is devoted to the first year's or qualitative students. In this there are 60 working places. No. 2 (Fig. 1) is arranged for the advanced or quantitative students, and contains ten blocks of four benches, each for the accommodation of 40 students.

The first essential in a laboratory, that of good light, is provided for by large windows and skylights on both sides. The other essential, that of plenty of air and good ventilation, is secured by lofty rooms having cubic contents of upwards of 50,000 ft., and by means of the powerful draught of a high chimney at one end of the laboratory block; the upward current being maintained by means of a furnace in the basement at the foot of the chimney, which furnace in the winter works the hot water heating apparatus, and in the summer simply serves for ventilating purposes.

The ventilation of the laboratory may be divided into two kinds, (a) the general ventilation, and (b.) the special ventilation. (a) The general ventilation is effected by a wooden perforated ceiling running the whole length of both of the main laboratories, and conveying the vitiated air by a large air-trunk to the shaft, within which rises the smoke flue of the furnace. The supply of fresh air is obtained from a high level by a fresh air shaft, down which the air passes, the hot water pipes being drawn by the aspiration of the chimney into the laboratories through gratings placed in the walls. This ventilation acts so successfully, that although no less than 44 beginners are now working in the qualitative (No. 1) laboratory, and 42 advanced students are working in the quantitative (No. 2) laboratory, still the air in both rooms is clear and pleasant throughout the day. In case, through negligence or accident, a large escape of acid fumes should occur, the windows being hung on pivots, can all be opened, and a thorough renewal of the air be effected in a few minutes. (b) The special ventilation is also worked by the main shaft, and is divided between the evaporating niches in the walls shown in plan, and in detail in Figs. 13 to 21, and the sulphuretted hydrogen closets marked A in the detailed drawing on each working table. Each of the niches is provided with an upright glazed earthenware pipe, 4 in. in diameter, running into a horizontal pipe of the same material, 12 inches in diameter, communicating directly with the main chimney, the draught in which is powerful enough to draw air from each one of the niches without the necessity of any gas flame being used for aspiration, and no escape is noticed in the working of these when any fumes, even of sulphuric acid, are evaporated in any quantity. In order to prevent the condensation of acid liquors, or of water, when these substances are boiled in the niches, a porcelain funnel is introduced into the lower part of the earthenware pipe inside the niche; all vapours generated under this funnel are immediately swept into the chimney, and no condensation whatever takes place inside the glass windows of the niche. One of these porcelain funnels is placed in the exhibition. The amount of air which in actual practice is found to pass up these draughts, is as follows:—

The large niches at the west end of the laboratory, aspirate 100 cubic ft. of air per minute, and each of the smaller ones in the three walls running east and west, 12 cubic ft. of air per minute.

The sulphuretted hydrogen closets in each working bench, shown in section in Fig. 11, and in plan in Figs. 7 and 9, are joined together in groups of two or four, and placed in connexion with a 7 in. or 4½ in. glazed earthenware pipe communicating with a horizontal flue shown in section in Fig. 5, running between the fireproof arching under the floor, and passing into the chimney at the point (S), Fig. 1. The down draught in each of these closets is continuous and powerful, each closet in the laboratory No. 1 aspirating at the rate of 5 cubic ft. per minute, whilst those in laboratory No. 2 aspirate on an average 20 cubic ft. per minute, those at the furthest end of the laboratory differing but slightly from those situated nearest to the chimney. The details of the arrangements of the working benches are seen in Figs. 6 to 12. The small block of buildings shown in section in Fig. 5, and on plan in Figs. 1, 2, and 3, consists of two floors, each containing seven rooms divided into, on the ground floor, 1, organic analysis room; 2 and 3, balance rooms; 4, gas analysis room; 5, library; 6, Prof. Schorlemmer's lecture theatre for organic chemistry; 7, Prof. Schorlemmer's private room.

The basement floor contains—1, laboratory for medical students, with accommodation for 50 workers, also ventilated by the main shaft; 2, metal lurgical laboratory containing furnaces, also in connexion with the large chimney; 3, store room; 4, class room; 5, lavatory and cloak room; 6, spectroscopic room; 7, photographic room; 8, dark room for photometry; 9, boiler house; and 10, preparation rooms.

The first floor, Fig. 2, contains Professor Roscoe's private room, private balance room, and private laboratory with window opening into laboratories Nos. 1 and 2.

It will be seen from the plans that the block of working laboratories communicates by a corridor with the large chemical lecture theatre (capable of seating an audience of 380), which has a small laboratory for the lecturer's immediate use behind his platform.

4577a. Drawings of the Lecture Table in the Theatre of the Royal College of Science, Dublin, erected under the direction of Professor Barrett. *Prof. W. F. Barrett.*

No. 1 represents the front elevation.

The shafting S, S, is divided into two parts, which can be thrown into one by the clutch and handle H, H^1. A belt passes round the large wheel E to a three-cylinder water engine situated under the theatre seats; another belt passes round T to a turbine placed beside the water engine; both engines are surrounded by felt and packing to deaden their sound. Water is admitted to the engines by means of the screw handles E^1 and T^1, thus one or both halves of the shaft can be simultaneously driven. The spindles of E^1 and T^1 pass through the floor to the pipes below; a hinged lid provides access to the cocks below for oiling or repair. This lid, &c. also serves as a sloping foot-board for the lecturer.

Gas is laid on to every part of the table, and by means of the connections and cocks G, G, flexible tubes can be attached. The peculiar construction of these cocks enables them to be turned aside below the table when out of use, or to be turned over the table when in use, thus preventing the kinking of the flexible tubing. A large view of the cock is shown at G, Fig. 4.

Oxygen is laid on to different parts of the table for the oxyhydrogen light, or other purposes. One cock is seen at O near the pneumatic trough, other oxygen cocks are seen at O, O^1, in Fig. 2. The supply is obtained from gas-holders in the yard.

Compressed coal gas or hydrogen in like manner is laid on from a gas-holder and comes to the table through the pipes and cocks H, H, Fig. 2.

Battery wires are laid on to the switch C, Fig. 1, a large view of which is shown on the small diagram. By turning the handle of the switch from 10 to 50 cells or 1 to 5, can be thrown into circuit; the binding screws for wire attachments are seen at B, B, B.

Water at high and low pressure is laid to the cocks at W′, W″, W‴, &c.

W is a large cock for screwing on hose, &c.

W′ is a cock for filling pails, and by turning it outwards over the sunk sink 2 the water flows, and is shut off by turning it back.

W″ is a small cock for filling small vessels.

W‴ is a high pressure screw tap admitting water to a pillar screwed on to W‴ in Fig. 2.

W^4 is a cock that turns water on to the pneumatic trough, and W^5 is a screw cock that gives a supply of water to the other end of the table.

A *speaking* tube P is laid on from the table to the battery and gasholders in the yard.

A, A^1, are *cupboards* below the table, A for the purpose of keeping appa-ratus dry and warm, A^1 under the pneumatic trough for a battery closet. In

Fig. 5 this cupboard is shown open; X is a *draught* pipe leading to a flue at the back of the lecture theatre.

In Fig. 5 is also shown the manner of opening the lid of the pneumatic trough; the slide V rests on the solid brass pillar Y, which is hinged at its upper end to the table.

No. 2 shows the surface of the table.

The pneumatic trough is seen open. X^1 is a down draught pipe to carry off noxious gases.

X^{11} is another down draught at the right hand of the lecturer. These draught holes are closed by round lids, over which is a square well-fitting block. When all the openings are closed the top ef the table presents a solid unbroken surface.

R is a slide that can be removed entirely to allow of free access to the lamp which stands on a swivel table, as shown in Fig. 3.

U, U^1, are brass collars wherein fit, by means of a bayonet joint, sliding brass arms, one of which is shown in Fig. 3.

I is an aperture (that can be closed by the lid I^1) to allow a belt to pass from the shafting on to the table.

Fig. 3 gives a perspective view of the table, with some apparatus placed on it to show the use of some of the fittings.

4578. Plan of the Laboratory of Physics of the Imperial University of St. Petersburg, with an index of its sections.   *Imperial University, St. Petersburg.*

In the plan of the room devoted to the experiments of students, there is indicated the arrangement of the several groups of apparatus required for demonstration. A complementary description written separately contains a short sketch of the experimental problems that are successfully laid before the students during their first year's experimental studies.

PHOTOGRAPHS.

4556. Photographs of Scientific Instruments in the Institute of Physical Science, Royal University of Rome.

Director, Prof. Blaserna.

1. Balance of precision made by the mechanician Scateni, of Urbino; the suspension very simple and excellent. With a weight of kilogram 2 per scale. Indicates to $\frac{1}{2}$ milligramme.

2. Universal wheel, made in the workshop of Galileo, in Florence. It allows rotary motion to be produced in cylinders and discs in horizontal and vertical directions with the most different velocities.

3. Scott's Phonautograph, made by the machinist De Palma, of Naples. The writing lever has a special arrangement devised by Mr. Campbell, lecturer at the Physical Science Institute, by which it rests only upon the centre of the vibrating membrane, while the other support transforms the vibrations into movements parallel to the axis of the rotating cylinder.

4. Instruments of precision, made by Starke and Kaumerer, of Vienna:—
(A.) Complete theodolite; horizontal and vertical circle, which permits, by the aid of microscopes with a movable ocular thread, to read up to $1''$.
(B.) Apparatus for the measurement of the indices of refraction. For the reading are added two microscopes with movable ocular threads, allowing a direct reading of $1''$. Repeating movement, with excellent arrangement of the

plate bearing the prism. (C.) Cathetometer, with movable ocular thread, allowing a reading of $\frac{1}{100}$ millimetres.

5. Magnetic instruments, according to Lamont, made by Dr. Carl, of Munich, Bavaria : —(A.) Theodolite, for the absolute measurement of declination and of the horizontal component. The two microscopes allow the reading of the second. (B.) Instruments to measure the variations of the horizontal component and of the inclination.

6. Differential interruptor, made by Prof. Blaserna, for the study of currents induced and of the extra currents. Full description given in the "Giornale di Scienze Naturali ed Economiche," Palermo, 1870. Sensitive from $\frac{1}{100 \cdot 000}$ to $\frac{1}{100 \cdot 000}''$.

7. Electric and magnetic apparatus, for measurement and by projection: — (A.) Quadrant galvanometer, by Dr. Carl, for instruction purposes, with two threads, one thick and short, the other fine and long. (B.) Sine and tangent compasses. (C.) Apparatus of projection, by Prof. Blaserna, for the demonstration of magnetic phenomena, provided with two arresters of the oscillations, one by water and the other of thick brass, for experiments in deviation and oscillation. (D.) Electrometer (Thomson), according to Kirchhoff, for projection. (E.) Wiedemann's compass, with three pairs of spirals by Kipp.

8. Electric apparatus, for measurement and by projection :—(A.) Bifilar electrometer of Palmieri, made by De Palma, of Naples. (B.) Edelmann's compass. (C.) Thomson's galvanometer. (D.) Wheatstone's rheostat.

9. Electric apparatus, for measurement and by projection :—Wheatstone's bridge, made by Stöhrer, of Leipsic, with its commutator, Kipp's rheostat. (B.) Astatic galvanometer, with a very short thread, to show the thermo-electric currents, and currents of terrestrial induction by projection, made by Campbell. (C.) Astatic galvanometer, with a long thread and reflector, made by Ruhmkorff. (D.) Electric multiplier, by Belli.

10. (A.) Perfected Kipp's chronoscope, with apparatus for the fall of weights. (B.) Kipp's chronograph.

4562. Photographs representing various Scientific Instruments in the Cabinet of Geodesy and Hydrometry, Royal University of Padua.   *Prof. Legnazzi, Director, Padua.*

GROUP I.—TELESCOPES, SQUARES, AND PRISMS.

1. Ramsden's Dynamometer, made in the workshops of the Royal Observatory of Padua (1866).
2. Dioptric with pointers by the Giuseppe Stefani, of Padua (1827).
3. Squares with triangular prism, by Ertel, of Monaco (1872).
4. Square of Wollaston, by Merlo, of Milan (1869).
5. Model of telescope for public instruction, designed by Professor Legnazzi, and constructed by Francesco Pasini (1875).
6. Achromatic telescope by the celebrated Geo. Dollond, of London (1740).
7. Telescope with distance-measurer, by Rochon (1828).
8. Land surveying squares, simple, from the workshop of the Royal Observatory of Padua.
9. Heliotrope, by Ertel, of Monaco (1873).
10. Camera Lucida, by the distinguished optician Amici, of Florence (1836).
11. Reflection square from the Tecnomanasio of Milan (1868).
12. Reflection square and allineator, by Goldschmid, of Zurich (1872).
13. Seconds counter, by Ertel, of Monaco (1874).
14. Telescope with distance-measurer, by Rochon (1869).
15. Terrestrial telescope, by the optician Plössl, of Vienna (1845).
16. Pantometer, with compass from the Tecnomanasio of Milan (1864).

17. Ordinary telescope—Cornet.
18. Achromatic terrestrial telescope, by Frescura, of Padua, 1872.
19. Hypsometer from the Tecnomanasio of Milan.
20. Pantometer from the Tecnomanasio of Milan.
21. Studies of the human eye considered as a telescope, made by Professo Legnazzi in Padua (1873).

GROUP II.—GRAPHOMETERS, MULTIPLYING CIRCLES, AND THEODOLITES.

1. Antique graphometer; more than three centuries old; instead of verniers, it is furnished with the diagonal scale designed by Don Pedro Nunnez, and gilt with sequin-gold. The tripod is worthy of observation.
2. English sextant (1798).
3. Ebony sextant with ivory edge, by Shuttleworth, of London (1840).
4. Repeating circle, by Lerebours, of Paris (1832).
5. Reflecting circle, by Baumann, which gives five seconds with telescope (1845).
6. Theodolite of the celebrated mechanician Reichenbach, of Monaco, which gives 10 seconds (1828).
7. Repeating theodolite, by Carlo Starke, constructed in the Imperial and Royal Polytechnic Institute of Vienna; the horizontal circle gives 4″, the vertical 10″. The bubble has a sensibility of 4″ (1852).
8. Multiplying circle, by Reichenbach, of Monaco; a remarkable machine the circle gives 4″ (1832).
9. Multiplying circle, by Gambey, of Paris, which gives 3″ (1822).
10. Repeating theodolite of Troughton and Sims, which gives 10″: a well designed and well constructed machine.
11. Eccentric theodolite, by Ertel, of Monaco, gives 10″; an excellent and most convenient instrument, constructed with great care and elegance (1843).
12. Multiplying circle, by Lenoir, of Paris; is an embryo instrument, but well designed (1836).
13. Theodolite by Nairne and Blunt, of London; besides degrees, the horizontal circle furnishes the tangents of the various angles. With this theodolite the engineer Valle executed the topographical survey of the city of Padua (1828).

GROUP III.—PRÆTORIAN TABLETS.

1. Steel band, length 20 meters, with divisions, constructed in the Tecnomanasio of Milan (1870).
2. Prætorian tablet on the Italian system, constructed by Giuseppe Stefani, of the Royal Observatory of Padua (1833). This is completely mounted to follow the configuration of the ground.
3. Gnomon, for the rotation of the dioptric.
4. Dioptric, with pointers and telescope, by Giuseppe Stefani.
5. Small spherical level from the Tecnomanasio of Milan (1869).
6. Dioptric, with telescope, by Alemanno, of Turin (1870).
7. Small level, with bubble of invariable length, designed by Professor Legnazzi, and constructed by his pupil Francesco Pasini (1875).
8. Small common level of the Italian tablet.
9. Prætorian tablet of the Italian system by Rochetti, of Padua, without the mirror and the accessory instruments in order to show its construction.
10. Small metrical chain of precision, of 10 meters, which was employed in the operations of the measurement of the arc of meridian of France (1802–1817).

11. Topographical compass, by Giuseppe Rodella, of the Royal Observatory of Padua (1795).
12. Prætorian tablet of the German system by Carl Starke, of Vienna, without the mirror, in order to show its construction (1856).
13. Prætorian tablet of German system, by Carl Starke, of Vienna, complete (1842).
14. Small level, by Ertel, of Monaco (1874).
15. Dioptric, with telescope and distance-measurer, by Carl Starke, of Vienna (1852) ; an excellent instrument.
16. Small level of the Starke tablet.
17. Gnomon and compass of the Starke tablet.
18. Dioptric and telescope of the Starke tablet.
19. Metrical chain of 20 mètres.
20. Odometer, by Adams, of London (1791).

Group. IV.—Levels.

1. Aneroid, by J. Goldschmid, of Zurich ; where can be read an eight-thousandth part of a millimètre of movement in the sides of the box (1874).
2. Naudet's olosteric barometer, by Feigestock, of Vienna, which gives the one-tenth of a millimètre of atmospherical pressure (1874).
3. Metallic thermometer of Bourdon.
4. Naudet's olosteric barometer, by Hirsch, of Florence; gives half a millimètre of pressure (1872).
5. Naudet's olosteric barometer, by Frescura, of Padua (1873).
6. Naudet's olosteric barometer, by Feigestock, of Vienna ; like No. 2.
7. Aneroid barometer, by Bourdon (1859).
8. Aneroid barometer, by Naudet; like No. 2.
9. Metallic barometer, Bourdon, from the Tecnomanasio of Milan.
10. Olosterical barometer, Naudet, by Feigestock ; like No. 2.
11. Level with telescope, by Rocchetti, of Padua, with thread micrometer (1853).
12. Level with distance measurer, by Carl Starke, of the Imperial and Royal Polytechnic Institute of Vienna (1855) ; an excellent instrument.
13. Pendulum level with two telescopes ; an historical piece ; must be very old ; the telescopes are not achromatic.
14. Level with telescope, by Rodella, of Padua (1820).
15. Level with mirror, rather old.
16. Level with distance measurer, by Starke and Kammerer, of Vienna (1876) ; an exquisite instrument.
17. Pocket level, by Carl Starke, of Vienna (1850).
18. Level with communicating tubes ; an old model (1810).
19. Level, system Egault, with telescope (1861).
20. Bubble, fitted by Rocchetti, of Padua (1844).
21. Barometer, Gay Lussac (1830).
22. Barometer, Fortin, by Rocchetti, of Padua (1850).
23. Water level with telescope ; an historical instrument ; must be at east two centuries and a half old.
24. Ecclimeter, by Rocchetti, which also serves as a level (1847).
25. German Biffa, system Stampfer (1856).
26. Italian Biffa, by Rocchetti, of Padua (1849).
27. Speaking Biffa (1862).
28. Egyptian Biffa (1862).
29. Old Italian Biffa (1812).
30. German Biffa, system Stampfer (1856).
31. Biffa, Milesi (1874).

GROUP V.—PLANIMETERS, COMPASSES, AND INSTRUMENTS OF PRECISION.

1. Octagonal Planimeter, of the Florentine Tito Gonella, by Carl Starke, of Vienna (1852); an exquisite instrument.
2. Three polar planimeters (1) from the Tecnomanasio of Milan; (2) by Amsler-Laffon, of Schafhausen; (3) by Ertel, of Monaco (1869).
3. Planimeter, by Amsler-Laffon, of Schafhausen; a very delicate instrument of recent invention (1873).
4. Tychonic scale, designed by Giuseppe Pavan, of Padua (1874).
5. Box compasses, by Montan, of Padua (1827).
6. Crystal grating, by Merlo, of Milan (1870).
7. Compass fitted by Montan, of Padua (1827).
8. Calculating circle, constructed by Professor Legnazzi (1856).
9. Calculating rule, Gravet-Lenoir (1860).
10. Crystal grating, at the workshop of the Observatory of Padua (1849).
11. Apparatus to graduate the small levels designed by Professor Legnazzi (1866).
12. Box compass complete, by Toni, of Milan (1853).
13. Box compass, by Bordogna, of Milan, a celebrated maker (1780).
14. Ellipsograph, designed by Prof. Legnazzi, and constructed at the Royal Observatory of Padua in 1870.
15. Compass of precision to take thicknesses.
16. Compass of Bauernfeind, constructed by Mayer, of Dünkelsbühl, which serves for the prætorian tablet (1871).
17. Rule to draw a parallel at a given distance, by Rodella, of Padua (1819).
18. Wide compass for drawings of precision (1872).
19. Steel callipers with vernier (1870).
20. Models of spikes with measuring wedges for the measurement of a base; invented by Bessel, constructed by Ertel (1874).
21. Models of rectilinear verniers for public instruction (1848).
22. Ebony pantograph from the workshop of the Royal Observatory of Padua (1830).
23. Rod Compass, by the distinguished mechanician Montan, of Padua (1824).

GROUP VI.—HYDRAULICS.

1. Model of a careening basin, constructed 70 years ago.
2. Model of a shell fitted in the Canale di Navigazione of Pavia; designed by Leonardo da Vinci (1485).
3. Hydrometrical pendulum of Venturoli, constructed by the engineer Albanesi, of Venice, in 1790.
4. Model of the Cremona regulator for dispensing water.
5. Model of the Piedmontese regulator for dispensing water.
6. Model of the regulator of the Milanese Bocca Magistrale for dispensing water.
7. Ball plummet, 30 mètres long, with small brass chain for centimètres and millimètres (1860).
8. Pilot tube, modified by Mallet, length 14 mètres, with float, constructed by the Royal Observatory of Padua (1852).
9. Tachymeter, of Brünnings, constructed in the workshop of the Royal Observatory of Padua, in 1483.
10 Portable hydrometer, in several pieces, length 8 mètres, of recent construction, to take the surface of the water of rivers with perfect accuracy, whence to deduce the declivity of the superficies, and to

introduce it into the hydraulic formulæ ; constructed in the workshop of the Royal Observatory in 1874. The cabinet has three of them, which admirably answer the purpose.

11. Retrometrical rod, of Teodoro Bonati, of wood, composed of 10 pieces, constructed at the Royal Observatory of Padua in 1848.

12. Small mill by Woltmann, constructed in the workshop of the Royal Observatory of Padua in 1858.

13. Float, simple and compound.

14. Retrometrical Asta, of Teodoro Bonati, composed of tin tubes, 16 mètres long, divisible into 38 pieces of various lengths, constructed in 1859.

15. Pitot Tube, modified by Darcy, constructed by Sarran of Paris, in 1871.

16. Small mill of Woltmann, constructed in 1870, by J. Kern of Aaran, with two flyers and two wheels unequal in size.

17. Tachymeter of Brünnings, modified by Professor Turazza, constructed in the workshop of the Royal Observatory of Padua, in 1859.

4563. Photographs of Scientific Instruments in the Cabinet of Physical Science, Royal University of Padua.

Director, Prof. Rossetti.

(1.) INSTRUMENTS OF GALILEO.

a. Apparatus to demonstrate the theorem that a body falls in equal time from a chord and from the vertical diameter of a circle.

b. Telescope.

c. Atmospheric and water thermometer.

d. Natural loadstone, armed, weighs $2\frac{1}{2}$ oz., supports 100 oz. The scale is about mill. 100 to a mètre.

(2.) INSTRUMENT for dividing by Musschenbroeck, improved by Poleni, 1740–1760.

(3.) INSTRUMENT by s'Gravesande for central force; improved by Poleni, 1740–1760.

(4.) CONDENSING ELECTROMETERS, by Prof. Salvatore Dal Negro, 1804.

These instruments are interesting, historically ; upon the results obtained by them, Dal Negro disputed the theory of Volta, and established from the first the foundations of the chemical theory of the pile. (*See* Description of a new electrometer and some experiments in charging the voltaic column, by Ab. Dal Negro, T. X. I., Italian Society of Science, Modena, 1804.)

(5.) 1 and 2. OLIGOCHRONOMETER, by Prof. Salvatore Dal Negro, applied to researches on the laws of weights falling freely. By cutting the thread which holds the weights suspended, the pendulum is set at liberty, which, being stopped by the weight arriving at the end of its course, cuts another thread.

3. Oligochronometer. (*See* New method for measuring the smallest fractions of time, invented by the Ab. Dal Negro, Padova, 1816. New Oligochronometer, Padova, 1809.)

(6.) MODEL OF OLIGOCRONOMETER, by Dal Negro, applied to the measurement of the velocity of projectiles. 1. Cannon, the ball

of which on coming out of the mouth breaks a thread which sets
the pendulum of the oligochronometer in motion (2). 3. Movable
hinged target, which on being hit breaks a thread which stops the
pendulum. 4. Details of the oligochronometer. (*See* New
method for measuring the velocity of projectiles, 1824, by Sig. Ab.
Dal Negro, Padova, 1824.)

(7.) 1. ELECTRO-MAGNETIC MOTOR, by Prof. Salvatore Dal
Negro. Two permanent magnets suspended like a pendulum
can oscillate in the proximity of the poles of a temporary magnet,
in which the current circulates alternately in contrary directions, by means of a swinging commutator with small cups of
mercury moved by the pendulum itself. The attractions and
repulsions which result render the movement of the pendulum
continuous, and this, by means of levers, is transformed into
continuous rotatory motion. (*See* Description of electro-magnetic
rams (arieti), both simple and compound, by Prof. Ab. Dal Negro.
Padova, 1838. Also Annali delle Scienze del Regno Lombardo-
Veneto, tomo III. bimestre I., 1838.)

2. Electro-magnetic Pendulum, by Prof. Dal Negro. The rod
of the pendulum is formed by a bundle of permanent magnets,
one extremity of which can oscillate between the poles of an
electro-magnet. The inversion of the current is effected as in
the above-described motor. This pendulum is interesting, historically, as recording the first attempts at using electricity as a
motive power.

(8.) ELECTRO-MAGNETIC MOTOR, by Prof. Dal Negro. The
apparatus consists of a lever of the first order, to one extremity
of which is attached the anchor of an electro-magnet, and to the
other a hammer which by alternate elevations and descents strikes
a rod, and produces a movement like that of a shuttlecock, and
afterwards the lifting of a weight. This movement of the lever
produces the inversion of the current in the electro-magnet, and
by means of a swing commutator with small cups of mercury.
(*See* Description by Prof. Dal Negro, Padova, 1838, and in
Annali delle Scienze del Regno Lombardo-Veneto, tomo VIII.
bimestre 1°, 1838.)

(9.) 1. MACHINE for DIVIDING. (Salleron, Paris, 1870.)

2. Cathetometer of the Tecnomanasio of Milan, improved in the
workshop of the Astronomical Observatory at Padua; added to it
is an internal counterpoise.

3. Balance of precision, by Deleuil, of Paris.

4. Quadrant Areometer, by Poleastro (1803). It is a hydrostatic balance with quadrant. It is used to ascertain the density of
fluids, and is much more correct than ordinary areometers.

5. Spherometer, by Brunner (1870).

6. Goniometer, by Babinet.

(10.) 1. MODEL of a PUMP by suction and pressure.
2. Barometer, by Poleni (1740–1760).*
3. Barometer, by Amontons.
4. Barometer, by Cassini.
5. Barometer (quadrant).
6. Barometer of Fortin.
7. Barometer of Fortin, constructed by Bellani.
8. Syphon barometer, constructed by Bellani.

(11.) 1. AIR-PUMP by suction and pressure, by Deleuil. Gives rarefaction to 2 mill., and compression to 10 atmospheres. The effect is double, and the piston does not touch the walls.
2. Differential manometer of Regnault.
3. Metallic manometer.

(12.) 1. DOUBLE SIREN of Helmholtz (Kœnig, Paris).
2. Apparatus by Kœnig for the combination of vibrating forces. May also be used for the experiments of Lissajous.
3. Apparatus by Kœnig for the interference of sounds.
4. Organ pipes with manometrical capsules, by Kœnig.
5. Apparatus for musical flames.
6. Analyser of sounds (Kœnig).
7. Acoustic interrupter.

(13.) 1. HELIOSTAT, made at Padua in 1828 by the younger Tessarolo, resembling, with modifications, that by Prandi. The reflected ray may be thrown in any direction. (*See* Opere Scient. di Bologna, 1824, p. 244, and Santini, Strumenti Ottici, vol. 1, Padova, 1828.)
2. Heliostat of Silbermann; made at Paris.
3. Lantern for electric light, with Foucault's regulator, constructed by Duboscq.†
4. Lamp for oxyhydrogen magnetic light, by Carlevaris.
5. Lamp for the Drummond light.
6. Regulator for electric light, by Archereau, with reflectors, invented by Prof. Rossetti so as to throw the light in all directions.

(14.) 1. COMPLETE OPTICAL BANK.
2. Microscope, by Hartnack.
3. Microscope, by Amici.
4. Microscope, by Plössl.
5. Dialytic telescope, by Plössl.
6. " Lentiprisma."
7. Prism for projections.
8. Spectroscope, for direct vision.
9. Bunsen's Spectroscope.

---

* A constant level in the vessel is obtained by the immersion of a screw.
† To this is attached a projection microscope.

(15.) 1. Large Apparatus for Polarisation, made by Soleil.
2. Apparatus by Nörremberg.
3. Apparatus for polarisation, Biot.
4. Apparatus for polarisation, Arago.
5. Cianopolarimeter, Arago.
6. Apparatus for polarisation, by Malus.
7. Saccharometer, by Soleil.
8. Phosphoroscope, by Becquerel.

(16.) 1. Atmospheric Thermometer, by Jolly, made at Monaco.
2. Apparatus by Clement and Desormes.
3. Mercury calorimeter, by Favre and Silbermann.
4. Complete apparatus of Melloni, made by Ruhmkorff.

(17.) Working model of stationary Steam Engine.

(18.) 1. Compass, by Prazmowsky, for measuring the intensity of the horizontal components of terrestrial magnetism.
2. Compass for declination.
3. Magnetic theodolite, of Lamont, made at Monaco. This instrument is portable, and serves to determine all the magnetic constants.
4. Apparatus, Faraday's, for diamagnetism.

(19.) 1. Large Electric Machine, by Winter, with two discs. Made at Vienna.
2. Machine, Holtz's, of the second kind, with horizontal discs. Made at the Tecnomanasi of Milan.
3. Double machine, Poggendorff's. Made at Berlin.

(20.) Holtz's Machine, of the first kind, worked by a wheel.
It was used in this way by Prof. Rossetti to measure the amount of work employed to produce a current of given intensity, and for other experiments described in his memoir "Nuovi Studii sulle Correnti delle Macchine Elettriche" (Atti dell' Istituto Veneto, Vol. III., Ser. IV., 1874 ; Nuovo Cimento, Vol. XII., 1874. Ann. de Chim. et de Phys. T. IV. 1875., Journal de Physique, 1875).

(21) 1. Rheotonu of Bréguet and Masson.
2. Sine compass.
3. Tangent compass.
4. Thermometer, Riess's.
5. Reometric Balance, Bernardi's.
A magnetic needle, working upon horizontal pivots, is surrounded by a frame, on which is wound a wire, and is united in action with the yoke of a sensitive balance. The intensity of the current traversing the wire is measured by the weight that must be placed in one of the scales of the balance, in order that the beam, and consequently the needle, may rest horizontally.
6. Galvanometer (vertical) for instruction.
7. Galvanometer of Du Bois Reymond, made by Sauerwald.
8. Torsion galvanometer.

40075.   3 Z

(22.) 1. FOUCAULT'S APPARATUS for transforming heat into work by means of induced currents. Electricity.

2. Electro-magnetic motor, applied to Geissler's tubes.

3. Electro-magnetic motor, by Page.

4. Apparatus of De la Rive for the rotation of the voltaic arch.

5. Matteucci's induction coil.

6. Magneto-electric machine, by Stöhrer.

7. Dynamo-magneto-electric machine, by Ladd.

(23.) Large RUHMKORFF'S COIL, with Foucault's interrupter, and accessories; made at Paris. Gives sparks 40 centim. long.

(24.) 1. HYGROMETER, Regnault's.

2. Hygrometer, Daniel's.

3. Psychrometer.

4. Condensation hygrometer, Belli's. (*See* Belli, Elementary Course of Experimental Physics, Vol. II., Milan, 1838.)

5. Psychrometer, Belli's. (*See* Belli, Elementary Course of Experimental Physics, Vol. II., Milan, 1838.)

6. Parchment hygrometer, Bellani's.

7. Maximum and minimum thermometer, Bellani's.

8. Maximum and minimum thermometer, Walferdin's.

9. Atmidometer of Landriani, perfected by Bellani. (*See* Belli, Elementary Course of Experimental Physics, Vol. II., Milan, 1838.)

(25.) ROSSETTI'S APPARATUS for the display of luminous electric figures obtained by the discharge of positive or negative electricity upon the naked surface of a glass, the other surface of which is furnished with an armature. (*See* Di una curiosa esperienza elettrica :—Atti delle Società Veneto-Trentina di Scienze Naturali, Vol. I., 1872. Nuovo Cimento, Vols. V., VI., 1872. Carl's Repertorium der Physik, 1873. Journal de Physique, Paris, 1873, Novembre.)

4564. Photographs of Scientific Instruments in the Physical Cabinet of Science, University of Pavia.

Director, Prof. Giovanni Cantoni.

1. Small apparatus for the study of reciprocal electrical influence, with two electrometers. Volta.

2. Two electrophori which recharge themselves and work as unfailing sources of electricity (first example of an electric machine by induction). Volta.

3. Measurer of powerful electric tension, by Volta.

4. Apparatus for the study of the electric spark. Volta.

5. Apparatus for the dilatation of gaseous fluids. Volta.

6. Electric duplicator by Giuseppe Belli. (Attempt at an electric multiplier which served to introduce the next machine.)

7. Electric machine by induction, G. Belli. (First example of an electrometer by induction, somewhat like the first machine of Holtz).

8. Hygrometer, with durable indications (by the same).

9. Psychrometer with bellows (by the same).

10. Collector of atmospheric heat, with continuous indications, by Con. Bellani.

11. Psychrometer with continuous indications, by Con. Bellani.

12. Balance barometer, modified by Prof. Giovanni Cantoni.

13. Normal barometer for Meteorological Stations, by Prof. Giovanni Cantoni.

14. Psychrometer with ventilator, by Prof. Giovanni Cantoni.

15. Electric balance, by Prof. Giovanni Cantoni.

16. Electric scale, for the measurement of electrical adherence.

17. Maximum and minimum thermometer, by Prof. Giovanni Cantoni.

18. Anemometrograph, by Prof. Parnisetti and Ferd. Brusotti.

19. Modification of the machine of Attwood, by Ferdinand Brusotti.

20. Hygrometer worked with sulphuric acid, improved by Brusotti.

4565. Photographs of Scientific Instruments in the Royal Observatory of Palermo. *Director, Prof. G. Cacciatore.*

1. Great circle by Ramsden, 1790. The principal parts are, first, a vertical axis of conics, parallelopiped form, destined to support two graduated circles, one for altitude, the other for azimuth. 2nd. The upper support of this axis, formed of four columns with four arches, and a collar in the middle. 3rd. The lower support, composed of three circles placed horizontally one on the other. 4th. A balustrade. 5th. An achromatic telescope. 6th. Three microscopic micrometers, with other smaller pieces. (*See* annexed description by Piazzi)

2. Catalogue of the stars made by circle No. 1.

3. Equatorial, by Merz. (Aperture of object glass 9·665 inches.) Set up in 1865.

4. Spectroscopic observations by the above equatorial.

5. Seismograph of Niccolo Cacciatore, 1826. A circular receiver, perforated laterally in eight parts, indicating the four cardinal points, and the intermediate points. Placed on a perfectly horizontal plane, and filled with mercury which touches the holes, and which therefore can spill at the least oscillation, and in the direction of the oscillation. A corresponding number of small cups placed below the holes to receive the spilt mercury, which may afterwards be replaced. The mercury received in the cups indicates the undulatory shocks of earthquake.

6. Anemometer by Niccolo Cacciatore, 1832. Attached is a description of the instrument by the inventor.

4567. Photograph of Registering Evaporimeter, according to the last plan adopted by the inventor, Prof. Ragona. *Observatory of the Royal University of Modena ; Director, Prof. Domenico Ragona.*

4568. Photographs of Scientific Instruments in the Cabinet of Physical Science of the University of Naples. *Director, Prof. Giuliano Giordano.*

1. Fresnel lens used by Melloni in his researches upon radiant heat, in particular with reference to the heat of the rays of the moon.

2. Floating plummet, by Paolo Anania de Luca.

3. A most accurate balance, by Bandini, machinist of the University, resembling that of Steinheil, but much less costly.

4569. Photographs and Drawings of Scientific Apparatus in the Vesuvian and Meteorological Observatory at Naples.

Director, Prof. Luigi Palmieri.

1. Portable seismograph by Palmieri, with a printed description.
2. Stationary seismograph, made by order of the Hydrographic Department of the English Admiralty, by Prof. Palmieri.
3. Stationary seismograph, made for the Vesuvian Observatory. (*See* explanatory note.)
4. Anemograph with hydrograph annexed, by Prof. Palmieri, with an illustrated description.
5. Bifilar electrometer and apparatus, with movable conductor, by the same, with two printed descriptions.

4570. Photographs of Scientific Apparatus.
University of Pisa; Director, Prof. Riccardo Felici.

Very ancient thermometer. It is impossible to say whether it works by water or alcohol. For graduation it has a string of glass beads. In the upper part are enamelled the characters, D. 20, A. 25.

II., III., IV. Galvanic chronographic interrupter, described in the Nuovo Cimento, t. xii., p. 115, and t. xiii., p. 266.* It opens and shuts successively electric circuits, measuring the intervals of time by the diapason (*d*), which by vibration marks by a stile the oscillations on the smoked cylinder (*c*). Perpendicular to the axis of the cylinder are eight rays, which rotate with it, and are furnished with small stops of steel isolated on ebonite. These, striking on the levers (*l*), detach them from the pieces (*p*) opening and closing the circuits, since a mechanical method of closing reduces itself to one of opening. For the ordinary way of closing, the piece (*g*) is used, which is to be found on fol. iii. (see description quoted). The surfaces of contact at the points of interruption are of silver, and to make them fit better the cylinders (*p*) are movable on their own axis. The diapason is movable around a horizontal axis (fol. iv.), and can slide along it. The plate which carries it is movable round a vertical axis; the correcting screw *h* can incline it slightly. The screw *z* regulates the contact of the stile with the cylinder. The piece *a*, which turns with the cylinder, at one point of the experiment meets the piece *b* lifting the diapason. The small bit of wood *s* serves to set the diapason in vibration. The base of the apparatus is of cast zinc, length 0·64, breadth 0·19 meter. The remainder is of brass, except the isolating pieces, which are of ebonite.

---

* In the apparatus described in the Nuovo Cimento there is a Froment's Siren instead of the diapason, the piece *a* is wanting, and there are only two radii *r*.

LONDON:

Printed by George E. Eyre and William Spottiswoode,
Printers to the Queen's most Excellent Majesty.
For Her Majesty's Stationery Office.
[7183.—2000.—8/77.]

NOTES ON THE MODEL

OF

NEWCOMEN'S STEAM-ENGINE (1705).

EXHIBITED IN THE

"SPECIAL LOAN COLLECTION OF SCIENTIFIC APPARATUS,"

SOUTH KENSINGTON MUSEUM,

1876.

BY

THOMAS LIDSTONE, Dartmouth,

ARCHITECT, DIOCESAN SURVEYOR, ETC.

LONDON:

E. & F. N. SPON, 48, CHARING CROSS.

NEW YORK:

446, BROOME STREET.

1876.

Price Sixpence.

Note.—"The house in which Mr. Newcomen lived when he invented the steam-engine, was situate in Lower Street, in Dartmouth, South Devonshire. It was sold and taken down, by order of the Local Sanitary Authority, when Mr. Thomas Lidstone, of that place, purchased the ancient carved and moulded woodwork of its street frontage, &c., which he rebuilt (in 1868) on the Ridgeway, in the same town. . . . . The house is named 'Newcomen Cottage.'"—*Extract from Tracts II. and III. on Newcomen, the Inventor of the Steam-engine.*

May 15, 1876.

AMIDST the many attractions of the Special Loan Collection of Scientific Apparatus this day opened to the public at the South Kensington Museum, there is one small model, situate:—

SOUTH COURT. SECTION XII. [Room B.]

Classed, *Applied Mechanics.* Dated "about 1705."

Numbered 1943; in Official Catalogue, *page* 258 (*Edition "under revision"*).

to which the present writer may be pardoned for calling attention.

It is distinguished as NEWCOMEN'S ORIGINAL MODEL OF HIS STEAM-ENGINE: a model, in fact, of the first machine by means of which steam-power could be safely applied for mechanical purposes: and it is so called because it was produced by MR. THOMAS NEWCOMEN, of Dartmouth, South Devonshire; an ironmonger there; a gentleman of good family and resources, who applied himself scientifically to the accomplishment of this purpose, and with complete success.

The model exhibited is from the Museum of KING'S COLLEGE, the Exhibitors being the "*Council of King's College, London*"; and it is said to have been presented to KING GEORGE III. (reigned 1760–1820). It is probable, therefore, that THOMAS NEWCOMEN, the inventor of the steam-engine, had nothing to do with this particular model, for he died in 1729. But his second son ELIAS, who followed him in directing the erection of steam-machinery, may have been engaged about a model prepared as a present to so august a personage. ELIAS NEWCOMEN

lived until 1765. This model bears evidence in some parts
of later work.

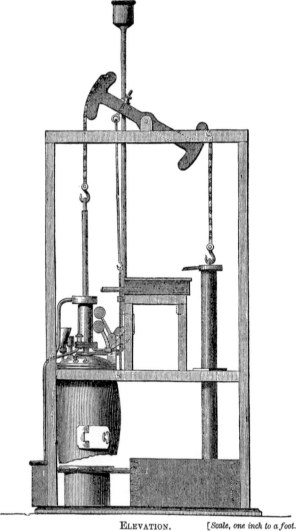

ELEVATION.   [*Scale, one inch to a foot.*

5

An older model—and described as a " rude " one—
namely, " the model of Newcomen's engine which WATT
was repairing when he invented his improvements " (as the
late Professor RANKINE expressed himself, in a letter of
January 4, 1870, in answer to an inquiry from the present
writer), is in the Museum of the Glasgow University. A
drawing of it is on the preceding page.

Other names have been associated with NEWCOMEN'S
in the invention of the steam-engine. SAVERY'S is one of
these ; he was also a Devonshire man : and West country-
men observe with some satisfaction that the united counties
of Devon and Cornwall (forming the diocese of Exeter)
produced the pioneers of that elaborate system of steam-
engineering and of locomotion by which the material
wealth of this kingdom has been enormously developed,
and civilization so much advanced, in the persons of :—

Newcomen ; Dartmouth, Devon; stationary engine.
Trevithick; Illogan, Cornwall; locomotive.
Gay; Exeter, Devon ; railroads.
[Savery ; Modbury, Devon ; steam navigation.]

NEWCOMEN was well acquainted with the progress of
science in his day ; but it has not been shown that he was
associated in any patent, nor that he obtained one himself.
CALLEY, a Dartmouth brazier, and How, another Dartmouth man, assisted NEWCOMEN ; and it would seem in a
somewhat higher capacity than that of mechanics. He
appears, however, at the later stage of his experiments to
have taken matters entirely into his own hands. In a
former Tract (No. III.), the present writer produced
a letter by which it was shown that NEWCOMEN made his
experiments alone, " in secret, on the leads of his house."
Since that letter was printed, this fact has been singularly

confirmed by another and older letter, for copy of which, and for many interesting facts concerning this Great Man's personal history, the writer is indebted to some of his descendants now residing in Devonshire, and these may become sufficiently important, it is believed, to warrant publication. The following is an extract from the last-named letter :—

> " When [Newcomen] was engaged on his great work, which took him three years from its commencement until it was completed, and was kept a profound secret, some of his friends would press Mrs. Newcomen to find out what her husband was engaged about, and ' for their part, they would not be satisfied to be kept in ignorance.' Mrs. Newcomen replied, ' I am perfectly easy. Mr. Newcomen cannot be employed about anything wrong; and I am fully persuaded, when he thinks proper, he will himself, unasked, inform me.' "

Remains of his experimental Models were in Dartmouth, down to a recent date. A correspondent, still living, assures me that they formed a portion of his playthings, or toys, in his youth.

The importance of this invention has never been publicly recognized in connection with NEWCOMEN's memory; and in no place is there greater apathy on the subject than in the town where he perfected his wonder-working Machine.

Milton Keynes UK
Ingram Content Group UK Ltd.
UKHW032320161024
449665UK00001B/25

9 781108 042420